AF290911

Be STARS

INTERNATIONAL ASTRONOMICAL UNION
UNION ASTRONOMIQUE INTERNATIONALE

SYMPOSIUM No. 98

HELD IN MÜNICH, FEDERAL REPUBLIC OF GERMANY,
APRIL 6–10, 1981

Be STARS

EDITED BY

MERCEDES JASCHEK

Strasbourg Observatory, France

and

HANS-GÜNTER GROTH

Münich Observatory, Federal Republic of Germany

D. REIDEL PUBLISHING COMPANY

DORDRECHT : HOLLAND / BOSTON : U.S.A. / LONDON : ENGLAND

Library of Congress Cataloging in Publication Data

Main entry under title:

Be stars.

Includes index.
1. B stars—Congresses. 2. Shell stars—Congresses.
I. Jaschek, Mercedes, 1926–   . II. Groth, Hans-Günter.
III. International Astronomical Union.
QB843.B12B43   523.8′44   81–19936
ISBN 90-277-1366-9   AACR2
ISBN 90-277-1367-7 (pbk.)

*Published on behalf of
the International Astronomical Union
by
D. Reidel Publishing Company, P. O. Box 17, 3300 AA Dordrecht, Holland*

*Sold and distributed in the U.S.A. and Canada
by Kluwer Boston Inc.,
190 Old Derby Street, Hingham, MA 02043, U.S.A.*

*In all other countries, sold and distributed
by Kluwer Academic Publishers Group,
P. O. Box 322, 3300 AH Dordrecht, Holland*

D. Reidel Publishing Company is a member of the Kluwer Group.

Printed in The Netherlands

TABLE OF CONTENTS

PREFACE

The idea of the symposium came during the XVIIth General Assembly
of the IAU at Montreal. The Working Group on Be stars adopted both the
proposal of holding a meeting, and of having it at the Universitäts-
sternwarte München.

The meeting was organized under the auspices of IAU Comm. 29 (Stellar Spectra) and the sponsorship of Comm. 45 (Stellar Classification).
The Scientific Organizing Committee was composed of Mercedes Jaschek
(chairman), W. Bonsack, C. de Loore, A. Feinstein, H. G. Groth,
P. Harmanec, L. Houziaux, A. M. Hubert-Delplace, L. S. Luud, A. Slettebak
and A. Underhill.

The members of this committee are to be thanked for their devotion
to the organization of what turned out to be a very successful meeting.
The program was organized on an observational approach, comprising
sessions on photometry, polarization, spectroscopy, infrared observations,
rotation and binarity, X-ray observations, UV observations and mass loss,
and atmospheric models. Each session started with an invited summary
paper, followed by a number of contributions. The different sessions
were chaired by A. Feinstein, R. Stalio, C. de Loore, Ch. Fehrenbach,
J. P. Swings, C. Jaschek, A. Sapar, G. T. Traving, M. de Groot and
H. G. Groth.

Upon request of the Working Group, a special session was devoted
to bibliographic problems and observing campaigns.

The Dean of the Faculty for Physics of the Ludwig-Maximilians-
Universität welcomed the participants at the beginning of the Symposium.
The meeting was closed by a summary talk, delivered by T. P. Snow.
The local organization was handled very efficiently by the Local Organizing Committee composed of H. G. Groth (chairman), M. Jaschek, K. Metz
and P. Wellmann. The work of the committee was greatly helped by the
collaboration of a number of volunteers from the Institut für Astronomie
und Astrophysik of the University of München.

The presence of participants from many countries was supported by
financial aid from the International Astronomical Union and the Deutsche
Forschungsgemeinschaft. The meeting took place, free of charge, at the
Mathematisches Institut of the University of München. We thank all
these institutions for their support.

September 1981   M. JASCHEK

                                                  H.-G. GROTH

xi

SCIENTIFIC ORGANIZING COMMITTEE

    M. Jaschek (chairperson), W. Bonsack, C. de Loore, A. Feinstein,
H.-G. Groth, P. Harmanec, L. Houziaux, A.M. Hubert-Delplace, L.S. Luud,
A. Slettebak, A. Underhill.

LOCAL ORGANIZING COMMITTEE

    H.-G. Groth (chairperson), M. Jaschek, K. Metz, P. Wellmann.

LIST OF PARTICIPANTS

ANDRILLAT, Y., Observatoire de Haute-Provence, France
BAADE, D., ESO, München, F.R.G.
BAHYL, V., Czechoslovakia
BALLEREAU, D., Observatoire de Meudon, France
BARKER, P., University of Western Ontario, Canada
BAUSER, S., Institut für Astronomie und Astrophysik, München, F.R.G.
BOLTON, C.T., David Dunlap Observatory, Canada
BOSSI, M., Osservatorio Astronomico di Brera, Italy
BRIOT, D., Observatoire de Paris, France
CARDON, M.P. (S.J.), Specola Vaticana, Vatican
COYNE, G.V. (S.J.), Specola Vaticana, Vatican
CHKHIKVADZE, J.N., Abastumani Astrophysical Observatory, USSR
DACHS, J., Astronomisches Institut, Bochum, F.R.G.
DANEZIS, E., Laboratory of Astrophysics Athens, Greece
DIVAN, L., Institut d'Astrophysique de Paris, France
DOAZAN, V., Observatoire de Paris, France
DUCATI, J.R., Observatoire de Strasbourg, France
EGRET, D., Observatoire de Strasbourg, France
ENDAL, A.S., Louisiana State University, USA
FEINSTEIN, A., Observatorio Astronomico, La Plata, Argentina
FEHRENBACH, Ch., Observatoire de Haute-Provence, France
FINKENZELLER, U., Landessternwarte Heidelberg Königstuhl, F.R.G.
FRANCO, M.L., Osservatorio Astronomico di Trieste, Italy
FRACASSINI, M., Osservatorio Astronomico di Brera, Italy
de FREITAS PACHECO, J.A., Observatorio Nacional, Brazil
GALKINA, T.S., Crimean Astrophysical Observatory, USSR
GIOVANELLI, F., Instituto de Astrofisica Spaziale, Italy
GRANES, P., CERGA, France
de GROOT, M., Armagh Observatory, Northern Ireland
GROTH, H.G., Institut für Astronomie und Astrophysik, München, F.R.G.
GUARNIERI, A., Osservatorio Astronomico, Bologna, Italy
HABETS, C.M.H.J., Sterrenkundig Instituut, Amsterdam, Netherlands
HARMANEC, P., Astronomical Institute Ondrejov, Czechoslovakia
HENRICHS, H.S., Sterrenkundig Instituut, Amsterdam, Netherlands
HIRATA, R., Observatoire de Maudon, France
HOUZIAUX, L., Institut d'Astrophysique de Liège, Belgium
HUBERT, H., Observatoire de Meudon, France
HUBERT-DELPLACE, A.M., Observatoire de Meudon, France
ISOBE, S., Tokyo Astronomical Observatory, Japan
JASCHEK, C., Observatoire de Strasbourg, France
JASCHEK, M., Observatoire de Strasbourg, France
JENKNER, H., Institut für Astronomie, Wien, Austria
JERZYKIEWICZ, M., Astronomical Observatory, Warsawa, Poland
KAISER, D., Astronomisches Institut, Bochum, F.R.G.

KOEPPEN, J., Lehrstuhl für Theoretische Astrophysik, Heidelberg, F.R.G.
KOUBSKY, P., Astronomical Institute Ondrejov, Czechoslovakia
KOWATSCHEW, B., National Observatory, Bulgaria
KOZOK, J.R., Astronomisches Institut, Bochum, F.R.G.
LACOARRET, C., Observatoire de Nice, France
LOPEZ ARROYO, M., Observatorio Astronomico, Madrid, Spain
de LOORE, C., Astrophysical Institute, Bruxelles, Belgium
LOOSE, A., Astronomisches Institut, Bochum, F.R.G.
MACCONNELL, D.J., Centro Astronomico, Merida, Venezuela
MANTEGAZZA, L., Osservatorio Astronomico di Brera, Italy
MARLBOROUGH, J.M., Astronomy Center, Falmer-Brighton, U.K.
MENDOZA, E., Instituto de Astronomia, Mexico City, Mexico
METZ, K., Institut für Astronomie und Astrophysik, München, F.R.G.
MERMILLIOD, J.C., Observatoire de Genève, Switzerland
MOLARO, P., Osservatorio Astronomico di Trieste, Italy
MOTCH, C., ESO, München, F.R.G.
MUMINOVIC, N., Univerzitetsko Astronomisko Drnstvo, Yugoslavia
PAKULL, M., ESO, München, F.R.G.
PASINETTI, L.E., Osservatorio Astronomico di Brera, Italy
PASTORI, L., Osservatorio Astronomico di Brera, Italy
PATERSON-BEECKMANS, F., ESTEC, Noordwijk, Netherlands
PAVLOVSKI, K. Observatory Hvar, Yugoslavia
PERSI, P., Instituto de Astrofisica Spaziale, Italy
PETERS, G.J., University of Southern California, Los Angeles, USA
POECKERT, R., Dominion Astrophysical Observatory, Victoria, Canada
POLLOCK, H., Astronomisches Institut Münster, F.R.G.
RAPPAPORT, S., Joint Institute of Laboratory Astrophysics, Boulder, USA
RUUZALEPP, M., Tartu Observatory, USSR
SAPAR, A., Tartu Observatory, USSR
SAREYAN, J.P., Observatoire de Nice, France
SCHEID, P.L., Institut für Astronomie und Astrophysik, München, F.R.G.
SCHUMANN, J.D., Observatoire Hoher List, F.R.G.
SELVELLI, P.L., Osservatorio Astronomico di Trieste, Italy
SLETTEBAK, A., Ohio State University, Columbus, USA
SNOW, T.P., LASP University of Colorado, Boulder, USA
SONNEBORN, G., Ohio State University, Columbus, USA
STALIO, R., Osservatorio Astronomico di Trieste, Italy
SOUFFRIN, P., Max Planck Institut für Astrophysik, München, F.R.G.
STERKEN, C., Astrofysich Institut, Bruxelles, Belgium
SWINGS, J.P., Institut d'Astrophysique de Liège, Belgium
THOMAS, R.N., Institut d'Astrophysique de Paris, France
TRAVING, G., Lehrstuhl für Theoretische Astrophysik, Heidelberg, F.R.G.
VIOTTI, R., Laboratorio di Astrofisica, Frascati, Italy
WOLF, B., Landessternwarte Heidelberg Königstuhl, F.R.G.
ZEUGE, W., Hamburger Sternwarte, F.R.G.
ZOREC, J., Institut d'Astrophysique de Paris, France

SOME ASPECTS OF ASTRONOMY VERSUS SPORT

A. Sapar
Tartu Observatory
USSR

Before going to the reports I would like to make some introductory remarks. As all of us know, Munich is an olympic town. As a testimony of that fact can serve the bamboo-cane used by the speakers as index for slides.

Both sport and astronomy have ancient traditions and much common features.

Olympic principles: Citius, altius, fortius! mean efforts to gain further progress in extensive (R), intensive ($\dot{R}$) and accelerative aspects. Thus, olympic principles can be formulated as an extremum for R, $\dot{R}$, $\ddot{R}$.

Astronomical principles: Computers, Instruments, Astronomers guarantee the very same progress by giant telescopes and high sensitivity devices with spectral range from γ-rays to radio waves (R), by high speed computers, hardware and software included ($\dot{R}$) and by highly intelligent brains ($\ddot{R}$) as a very soft ware.

As you see very tightly linked to olympic principlers are also the speakers's principles; slowly, distinctly, loudly!
This means that all speakers are asked :
1) to try to maximize the signal to noise ratio whereby only Oxford English is considered as a signal,
2) to accomodate their gigabytes per second output facilities to a decabyte per second input needed for translation to different software languages and for filtering,
3) to speak forte fortissimo. Thus, speakers principles can be quantitatively formulated as

$$\text{Decabyte/sec,} \quad \frac{S_{\text{Oxford English}}}{N} \rightarrow \infty \;,\; ff \precsim fff.$$

(These introductory remarks were given by Dr. Sapar as chairman of the session on Thursday morning (April 9))

I. PHOTOMETRY

SOME PHOTOMETRIC CHARACTERISTICS OF Be STARS
(Review Paper)

Eugenio E. Mendoza V.
Instituto de Astronomía
Universidad Nacional Autónoma de México

Abstract. Homogeneous observational data on four photometric systems
which cover the spectral range from 0.14 to 4 microns through narrow,
intermediate and broad passband filters are used to review and derive
some photometric characteristics of Be stars. We find that 60% of the
stars under discussion have ultraviolet and infrared excesses.

1. INTRODUCTION

The subject "Photometry of Be Stars" is too extense to give a full
account of it at this symposium. Therefore, I have to restrict myself
to a selection, which naturally is biased by my own knowledge and
interests. However, some facts derived from photometric measurements
will be discussed.

2. FOUR PHOTOMETRIC SYSTEMS

We are including in our discussion observational data taken in four
photometric systems, namely,

1) Absolutes fluxes in four passbands centered at $\lambda\lambda 1565$, 1956, 2365
and 2740A observed by the Ultraviolet Sky Survey Telescope. Its main
photometric characteristics are given in the Catalogue of stellar
Ultraviolet Fluxes compiled by Thompson et al. (1978).
2) The UBVRIJHKL-system (Johnson et al., 1966).
3) The 13-color photometric system (Johnson and Mitchell, 1975).
4) The $\alpha(16)$, $\Lambda(9)$ photometric system (Mendoza, 1977, 1979). These
photometric systems cover the spectral range from 0.14 to 4 microns,
approximately, through broad intermediate and narrow passbands filters.

3. THE OBSERVATIONAL DATA

We have reduced the observational data very much to be discussed herein.
The criterium was to include 09.5-B8 type stars which have $\alpha(16)$ and
$\Lambda(9)$ measurements made by ourselves, and which are also suspected or

3

M. Jaschek and H.-G. Groth (eds.), Be Stars, 3–17.

confirmed to have Balmer lines contaminated by emission. We have
collected 56 stars altogether. They are listed in Table 1. The star's
designations are HD numbers; spectral types are from Mendoza (1958),
and Jaschek et al. (1971). The $\alpha(16)$ and $\Lambda(9)$-indices are new, except
for the four Be Pleiades stars (Mendoza, 1979). The UBVRIJHKL
photometry has been taken from Mendoza (1967, 1969), and Johnson et al.
(1966). H-magnitudes from Iriarte (1969), and Morel and Magnenat (1978).
The 13-color photometry from Johnson and Mitchell (1975). The
ultraviolet fluxes were transformed to the visual magnitude scale using
the absolute calibration of Hayes and Lathman (1975):

$$m(\lambda) = -2.5\log F(\lambda) - 21.175$$

the derived magnitudes are referred in Table 1 and herein as 16, 20, 24
and 27-magnitudes.

The photometry given in Table 1 is characterized by its homogeneity
and high quality. We illustrate in Figure 1, as an example, a comparison
of the $\alpha(16)$-indices and the strengths of Hα given by Andrews (1968).
This comparison includes 100 normal early type stars, plotted as a solid
line. The actual scatter is very small around it, less than 0.03
magnitudes, no a single normal star departs from the line more than this
amount. The open circles in Figure 1 represent the Be stars listed in
Table 1. Its scatter is not unexpected since the strength of the
emission in Hα varies as function of time. The time difference between
Andrew's data and ours in about 15 years. We notice from figure 1 that
around one-third of the stars listed in Table 1 have changed their Hα

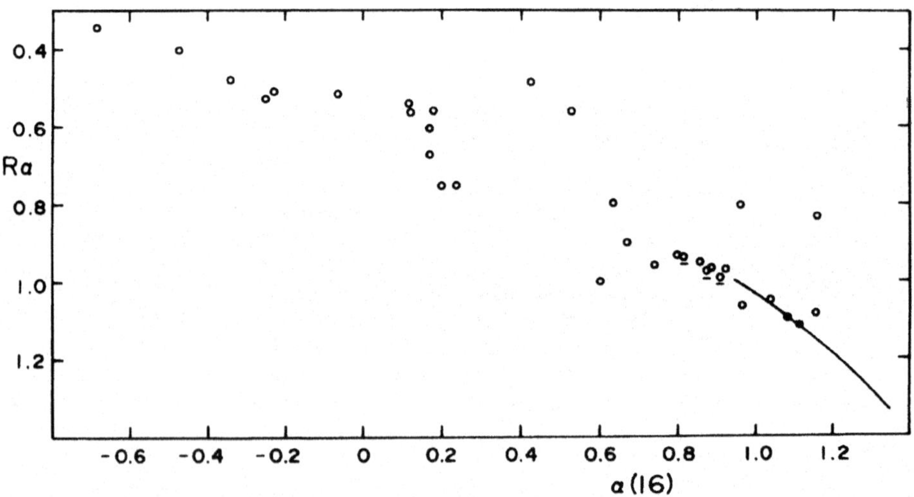

Fig. 1.– A comparison of two Hα-photometric systems

TABLA 1.- *OBSERVATIONAL DATA*

HD / MX	α(16) / Λ(9)	16 / 33	20 / 35	24 / 37	27 / 40	U / 45	B / 52	V / 58	R / 63	I / 72	J / 80	H / 86	K / 99	L / 110
2905	0.794	2.60	3.15	3.18	3.11	3.50	4.30	4.16	4.02	3.96	3.99		3.92	3.31
B1 IA	0.288	3.177	3.234	3.638	4.288	4.362	4.220	4.128	4.056	4.001	3.949	3.959	4.002	3.934
5394	0.199	-1.01	-0.56	-0.17	-0.42	1.18	2.29	2.39	2.32	2.40	2.47	2.20	2.20	2.00
B0.5 IV E	0.244	0.695	0.759	1.365	2.105	2.292	2.270	2.262	2.184	2.141	2.079	2.181	2.301	2.260
10144	0.994	-2.13	-1.64	-1.28	-0.89	-0.35	0.32	0.47	0.50	0.57	0.79	0.86	0.88	0.95
B5 V E	0.351	-0.508	-0.482	-0.215	0.248	0.459	0.471	0.491	0.562	0.593	0.644	0.646	0.723	0.699
10516	-0.473	0.98	1.68	1.91	2.44	3.10	4.02	4.06	3.90	3.88	3.83	3.57	3.35	2.83
B1 III,IV PE	0.217	2.738	2.792	3.319	3.984	4.136	4.095	4.086	3.917	3.924	3.831	3.904	3.999	3.889
13854	0.867					6.10	6.76	6.48	6.14	5.95				
B1 IAB	0.291													
18552	0.233	4.43	4.78	5.15	5.38	5.55	5.94	6:						
B8 V E	0.264													
20336	0.856	1.88	2.45	2.84	3.30	3.93	4.70	4.85	4.88	5.01		5.17	5.25	
B2 V E	0.294	3.645	3.732	4.105	4.631	4.805	4.841	4.867	4.899	4.904	4.937	4.978	5.073	5.063
22192	-0.225	2.17	2.67	3.03	3.29	3.61	4.17	4.23	4.13	4.14				
B5 E	0.245	3.355	3.392	3.649	4.109	4.237	4.234	4.243	4.168	4.198	4.154	4.178	4.232	4.164
23180	1.082	1.59	2.36	2.65	2.74	3.13	3.88	3.83	3.71	3.71			3.76	3.89
B1 III	0.299	2.805	2.868	3.280	3.837	3.945	3.857	3.828	3.774	3.754	3.712	3.728	3.767	3.767
23302	1.115	1.75	2.16	2.59	2.92	3.18	3.58	3.70	3.71	3.81	3.88	3.92	3.94	4.01
B6 III	0.314	3.044	3.100	3.211	3.549	3.689	3.709	3.722	3.739	3.751	3.780	3.755	3.842	3.800
23480	0.869	2.37	2.83	3.19	3.46	3.70	4.12	4.18	4.11	4.15	4.16	4.21	4.06	3.97
B6 IV NN	0.302	3.523	3.569	3.711	4.104	4.184	4.184	4.172	4.131	4.151	4.120	4.139	4.160	4.149
23630	0.670					2.43	2.78	2.87	2.84	2.88	2.97		2.96	2.90
B7 III	0.284	2.313	2.354	2.376	2.724	2.864	2.875	2.887	2.869	2.882	2.929	2.894	2.931	2.933
23862	0.529				4.83	4.73	5.01	5.09	5.02	5.08	5.11	5.15	5.06	4.95
B8 PE	0.511	4.311	4.359	4.545	4.884	5.003	5.007	5.014	4.992	5.027	5.020	5.039	5.093	5.049
24398	1.034	1.06	1.82	1.96	1.94	2.19	2.97	2.85	2.99	3.08	3.05		3.03	3.04
B1 IB	0.295	1.924	1.963	2.358	2.972	3.054	2.927	2.844	2.765	2.707	2.647	2.648	2.703	2.646

TABLA 1.- CONTINUED

| HD | α(16) | 16 | 20 | 24 | 27 | U | B | V | R | I | J | H | K | L |
MK	Λ(9)	33	35	37	40	45	52	58	63	72	80	86	99	110
24534	0.421	4.36	5.10	5.57	5.49	6.03	6.87	6.76	6.51	6.40		5.43	5.28	5.14
B0: PE	0.259													
25940	0.167	1.91	2.53	2.97	3.12	3.45	4.00	4.03	3.91	3.93				3.70
B3 V PE	0.270	3.240	3.285	3.564	4.005	4.120	4.084	4.064	3.996	4.033	3.982	3.979	4.047	3.992
28497	-0.106	2.22	2.75	3.06	3.62	4.36	5.30	5.44	5.41	5.53				
B1 E	0.226													
29866	1.154	4.82	5.46	5.95	5.81	5.89	6.18	6.08						
B7: E	0.325													
30614	0.814	2.06	2.46	2.90	2.92	3.44	4.32	4.29	4.18	4.18				
09.5 IA	0.278	3.043	3.125	3.604	4.280	4.416	4.343	4.271	4.224	4.185	4.132	4.144	4.226	4.192
32343	0.170	2.64	3.15	3.53	3.93	4.39	5.12	5.20				5.29	5.15	
B2 V: PE	0.251													
32991	-0.066	4.52	5.26	5.54	5.33	5.46	6.02	5.82				5.27	5.08	4.70
B2 V P	0.252													
34085	0.917	-1.04	-0.93	-0.79	-0.62	-0.55	0.10	0.13	0.12	0.14	0.22	0.11	0.20	0.17
B8 IA	0.414	-0.773	-0.731	-0.519	0.062	0.218	0.172	0.151	0.125	0.092	0.099	0.072	0.168	0.172
34959	0.643	4.40	4.87	5.24	5.55	6.11	6.57	6.65	6.61	6.65				
B5 P	0.403													
35165	0.860	3.03	3.63	4.00	4.50		5.92	6.11				6.30	6.40	
B5 NE	0.425													
35411	1.083	-0.10	0.64	1.02	1.55	2.25	3.18	3.35	3.40	3.65	3.75		3.87	3.92
B1 V	0.287	1.814	1.918	2.391	3.050	3.276	3.324	3.368	3.395	3.432	3.466	3.507	3.652	3.700
35439	0.965					3.83	4.76	4.96	5.04	5.26		5.44	5.47	
B1 V: PE	0.298	3.456	3.573	4.019	4.641	4.871	4.934	4.996	5.044	5.090	5.113	5.174	5.323	5.352
36576	0.118	3.28	3.92	4.25	4.36			5.50						
B1 E	0.251													
37128	0.905	-1.50	-1.16	-0.85	-0.39	0.48	1.51	1.69	1.76	1.93	2.06		2.17	2.15
B0 IA	0.276	0.100	0.209	0.722	1.437	1.666	1.694	1.740	1.763	1.807	1.849	1.908	2.036	2.037

TABLA 1.- CONTINUED

HD / MK	α(16) / Λ(9)	16 / 33	20 / 35	24 / 37	27 / 40	U / 45	B / 52	V / 58	R / 63	I / 72	J / 80	H / 86	Y. / 99	L / 110
37202	0.177	0.26	1.01	1.27	1.75	2.22	2.84	3.03	3.06	3.15	3.20		2.94	
B3 IV E	0.409	1.741	1.824	2.106	2.688	2.916	2.955	3.013	2.961	3.124	3.158	3.104	3.207	3.123
37490	0.603	1.89	2.41	2.65	3.06	3.73	4.49	4.59	4.57	4.67	4.76		4.82	4.80
B2 III E	0.278	3.379	3.467	3.831	4.397	4.575	4.568	4.587	4.564	4.579	4.544	4.614	4.684	4.653
37742	0.908	-1.74	-1.34	-0.92	-0.43	0.50	1.56	1.77	1.85	2.05	2.21	2.27	2.33	2.31
09.5 IB	0.282	0.082	0.196	0.733	1.450	1.694	1.744	1.795	1.824	1.871	1.913	2.007	2.110	2.145
37795	0.544	0.64	1.02	1.38	1.80	2.05	2.52	2.64	2.66	2.76	2.81		2.84	
B8 NE	0.280	1.940	1.949	2.125	2.426	2.625	2.635	2.642	2.660	2.691	2.709	2.696	2.770	2.734
37967	-0.340	3.72	4.31	4.69	4.98	5.29	5.94	6:						
B3 PE	0.218													
41117	0.870	3.97	4.71	4.54	4.08	4.24	4.91	4.63	4.32	4.10	4.04		3.86	3.71
B2 IA	0.310	3.902	3.907	4.294	4.912	4.919	4.680	4.542	4.417	4.329	4.214	4.160	4.162	4.035
43285	0.960	3.64	4.11	4.53	4.96	5.40	5.93	6.05						
B5 E	0.305													
44458	-0.120	3.23	3.76	3.97	4.25	4.72	5.55	5.63	5.37	5.32				
B2 V E	0.227													
45546	0.813					4.10	4.87	5.05	5.15	5.32				
B2 V	0.288	3.775	3.881	4.281	4.788	4.988	5.032	5.094	5.129	5.215	5.237	5.273	5.402	5.328
45725-7	0.105					2.87	3.61	3.76	3.77	3.93		3.86	3.76	3.66
B3 E	0.287	2.496	2.578	2.961	3.491	3.664	3.720	3.776	3.723	3.841	3.833	3.888	3.999	3.999
45995	0.112					5.20	6.06	6.14						
B2 V E	0.237													
47129	0.807					4.75	5.22	6.1:	6.06					
B0 P	0.270													
48917	0.042	2.28	2.77	3.03	3.52	4.27	5.19	5.35						
B3 NE	0.246													
50013	0.272	0.36	0.91	1.28	1.91	2.79	3.72	3.95	4.05	4.25		3.79	3.57	3.07
B2 NE	0.259	2.135	2.201	2.869	3.518	3.730	4.228	3.765	3.712	3.702	3.665	3.776	3.981	3.892

TABLA 1.- CONTINUED

HD	α(16)	16	20	24	27	U	B	V	R	I	J	H	K	L
MK	Λ(9)	33	35	37	40	45	52	58	63	72	80	86	99	110
56139	0.342	1.17	1.63	1.99	2.50	2.90	3.64	3.82	3.80	3.90	4.23		4.28	
B3 E	0.270	2.657	2.719	3.172	3.697	3.864	3.894	3.937	3.882	3.887	3.861	3.963	4.086	3.940
58715	0.739	1.30	1.62	1.97	2.31	2.52	2.80	2.89	2.90	2.96	3.06	3.04	3.06	
B8 NE	0.270	2.383	2.455	2.524	2.731	2.833	2.863	2.912	2.915	2.949	2.951	2.945	2.990	2.970
60606	-0.266	3.05	3.54	3.77	4.21	4.67	5.37	5.44	5.33	5.32				
B5 E	0.235													
60855	1.132	2.95	3.66	4.11	4.32	4.82	5.55	5.66						
B2 IV E	0.322													
63462	0.422	1.11	1.66	2.06	2.67	3.45	4.47	4.52	4.37	4.39		4.16	3.98	3.73
B3 E	0.255													
68980	-0.406	1.64	2.15	2.47	3.01	3.68	4.66	4.77	4.68	4.77	4.71	4.73	4.37	
B3 NE	0.181													
83953	0.451	2.36	2.80	3.14	3.57	4.08	4.66	4.78	4.76	4.85		4.87	4.82	4.77
B5 E	0.267													
105435	-0.260				0.91	1.64	2.56	2.65	2.61	2.73				
B2 V PE	0.185	1.096	1.211	1.705	2.315	2.489	2.509	2.568	2.516	2.514	2.537	2.547	2.701	2.630
120324	1.085	0.10	0.65	1.09	1.73	1.88	2.78	2.94	2.91	3.00				
B2 V PE	0.301	2.008	2.102	2.604	3.186	3.391	3.429	3.478	3.457	3.579	3.618	3.685	3.808	3.871
127972,3	0.823	-0.93	-0.40	-0.04	0.67	1.30	2.12	2.31	2.41	2.59		3.08	3.04	2.99
B2 V E	0.359	0.979	1.080	1.509	2.053	2.278	2.339	2.399	2.429	2.500	2.558	2.611	2.756	2.812
142983	0.215	2.35	3.07	3.54	3.80	4.61	4.77	4.87	4.91	4.96		4.67	4.54	4.15
A PE A	0.478	4.552	4.657	4.450	4.827	4.922	4.946	4.978	4.960	4.999	4.990	4.949	4.984	4.999
193237	-0.685	4.87	5.60	5.47	4.72	4.66	5.24	4.82	4.28	4.02	3.78	3.49	3.31	2.87
B1 EQ	0.270	4.339	4.326	4.720	5.301	5.239	4.956	4.659	4.426	4.390	4.272	4.213	4.097	3.849
198478	0.894	5.23	5.88	5.72	4.99	4.83	5.28	4.86	4.41	4.10				
B3 IA	0.356	4.534	4.516	4.742	5.272	5.202	4.933	4.706	4.546	4.393	4.255	4.185	4.116	4.015
217675	1.154	1.51	1.96	2.31	2.63	3.00	3.53	3.62	3.61	3.69		3.80	3.86	3.86
B6.5 E	0.330	2.756	2.831	2.996	3.458	3.615	3.637	3.645	3.663	3.651	3.665	3.661	3.722	3.730

strength in this lapse of time, a few of them as much as 0.4 mag. other
Be stars show only a faint emission, if any. It should also be pointed
out that several of the Be stars listed in Table 1 have changed their Hα
strength from one year to the next. For instance X Persei brightened by
0.12 mag. from 1978 to 1979, and ζ Tauri by 0.06 mag. from 1977 to 1978.
The probable error of a single observation of the α(16)-index is around
0.004 mag. (Mendoza 1979). Thus these changes are 30 and 15 times this
observational error. It is interesting to mention that two stars
listed in Table 1 with α(16)-indices contaminated by emission do not
appear in the Be catalogue (Jaschek et al. 1971), namely, BS 1761 and 10
Monocerotis. The first one is in Andrew's (1968) catalogue with an Rα
value which indicates its Be nature.

4. SOME PHOTOMETRIC PROBLEMS

Normal stars are often obscured by interstellar extinction, thus, their
colors depart from the intrinsic colors, which in most cases are not
accurately known. A study of 800 normal Main Sequence stars, taken
mostly from the Bright Star Catalogue (Mendoza, 1975), which contains
many unreddened stars, indicates that mean colors deviate from Johnson's
(1966) as much as 0.3 magnitudes. The exact amount depends on wavelength
and spectral type. For instance, at B1 V mean colors are redder than
intrinsic colors from 0.03 (visual) to 0.17 (infrared) mag. This can be
interpreted either as there are not unreddened early type stars or
Johnson's intrinsic colors are too blue or the "cosmic scatter" is too
large. More work is needed in this direction.

Additional problems are encountered in the Be stars, per example,
1) Variability. Hα emission variations were stated above. In addition,
Banhg (1976), among others, have reported rapid variations of Hα
Variability in the UBVRI has also been reported by many authors, see
for instance Feinstein (1968). A systematic study of a number of Be
stars by Alvarez and Schuster (1981) in the 13 color system shows that a
number of Be stars are undoubtedly variable stars. Variations in the 52
and 58 magnitudes are the most common form of variability. Variations
in the Balmer continuum and near infrared are the most common color
changes.
2) Emission. Several filters of the Photometric systems under
consideration allow the pass of spectral features contaminated by
emission, such as the R-filter which transmits at the Hα-wavelength.
Thus, the colors containing the R-magnitude may be different in Be
stars than in normal B-type stars.
3) Color excesses. Ultraviolet and infrared color excesses have been
reported by several authors (Mendoza 1958, 1969; Feinstein 1968) which
are unexplainable solely by interstellar extinction.
4) Binaries. Several Be stars are binaries with a probable red
companion (Polidan and Peters, 1976). If this is the case their colors
may be distorted.

5. TWO COLOR DIAGRAMS

The analysis of the 28-color photometry contained in Table 1 could
include many comparison of the several color indices. We begin by
plotting $\alpha(16)$ versus $\Lambda(9)$ in Figure 2. The symbols in Figure 2
represent: the solid line, the standard relationship for normal O9.5-B8
Main Sequence early type stars. In addition, extreme shell Be stars lie
in Figure 2 to the right, even further than the early type supergiants.
This probably means that these Be stars have larger but similar
envelopes than early type supergiant stars. A number of Be stars have

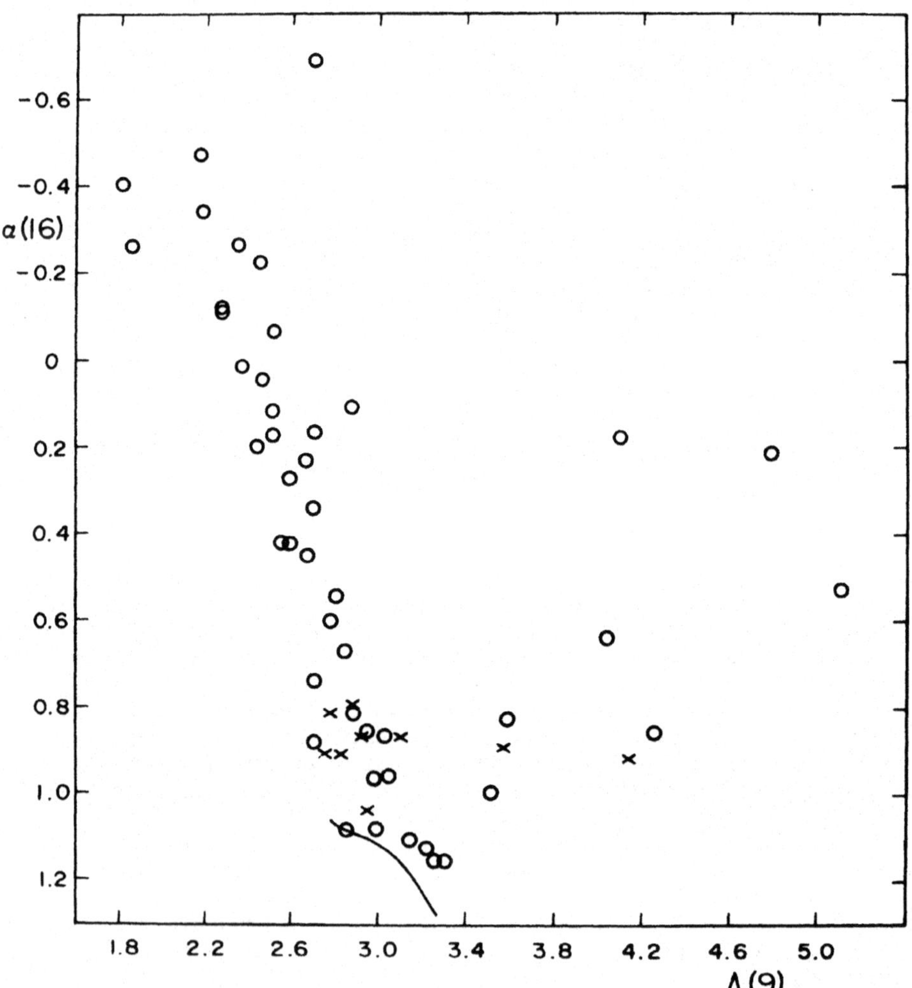

Fig. 2.- The $\alpha(16),\Lambda(9)$-plane

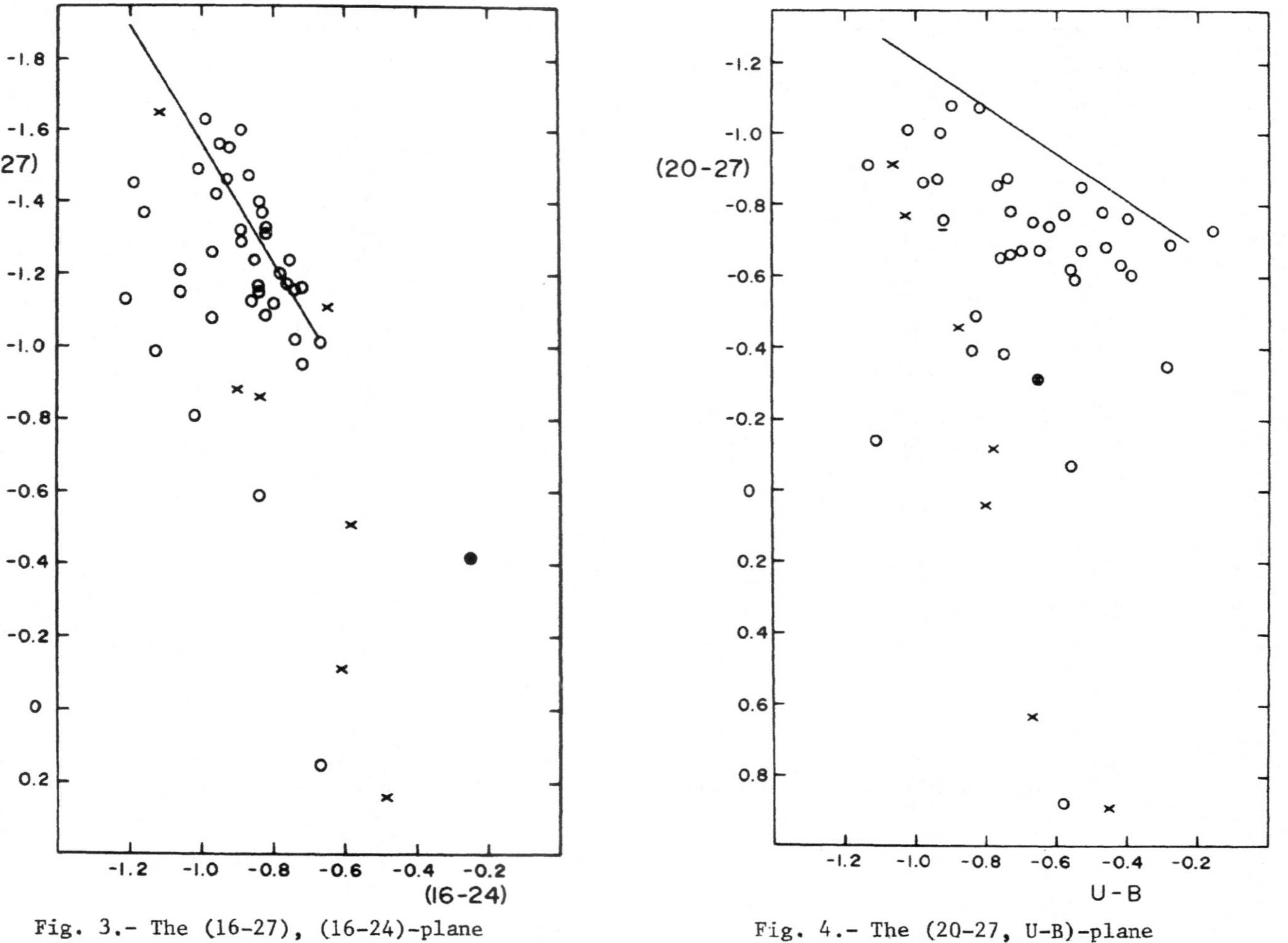

Fig. 3.- The (16-27), (16-24)-plane

Fig. 4.- The (20-27, U-B)-plane

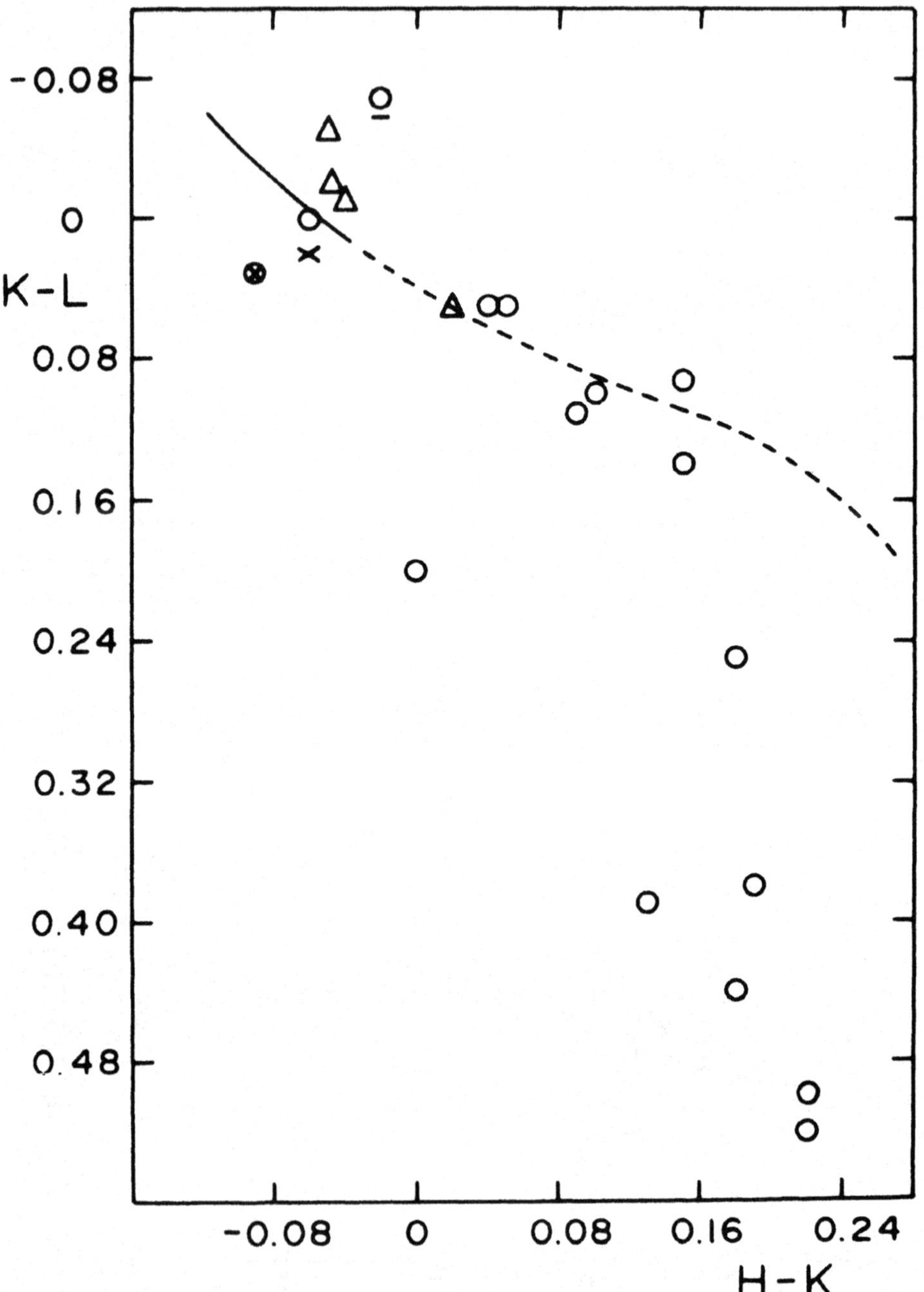

Fig. 5.- The (K-L, H-K)-plane

an $\Lambda(9)$-index contaminated by emission, such as those shown in the
extreme left in figure 2. However, the reverse is not true. For
instance P Cygni has an OI (λ7774A) both in emission and in absorption
(Mendoza and Johnson, 1979) which yields an $\Lambda(9)$-index similar to that
of a Main Sequence B-type star.

The photometry from the Ultraviolet Sky Survey Telescope indicates
that roughly half of the Be stars listed in Table 1 have UV-Main
Sequence colors like. The other half sets aside from the unreddened
Main Sequence B-type stars. We show, as an example, the relationship
between the (16-27) and (16-24) colors in Figure 3. The symbols in
this Figure represent: solid line, an approximate Main Sequence
relationship for unreddened B-type stars; open circles, Be stars;
crosses, supergiant stars (the crossed cirle represents β Orionis). In
the (20-27, U-B)-array Be stars are better separated than in the (16-27,
16-24)-diagram. This is illustrated graphically in Figure 4 (the
symbols in Fig. 4 have the same meaning as in Fig. 3). Infrared colors

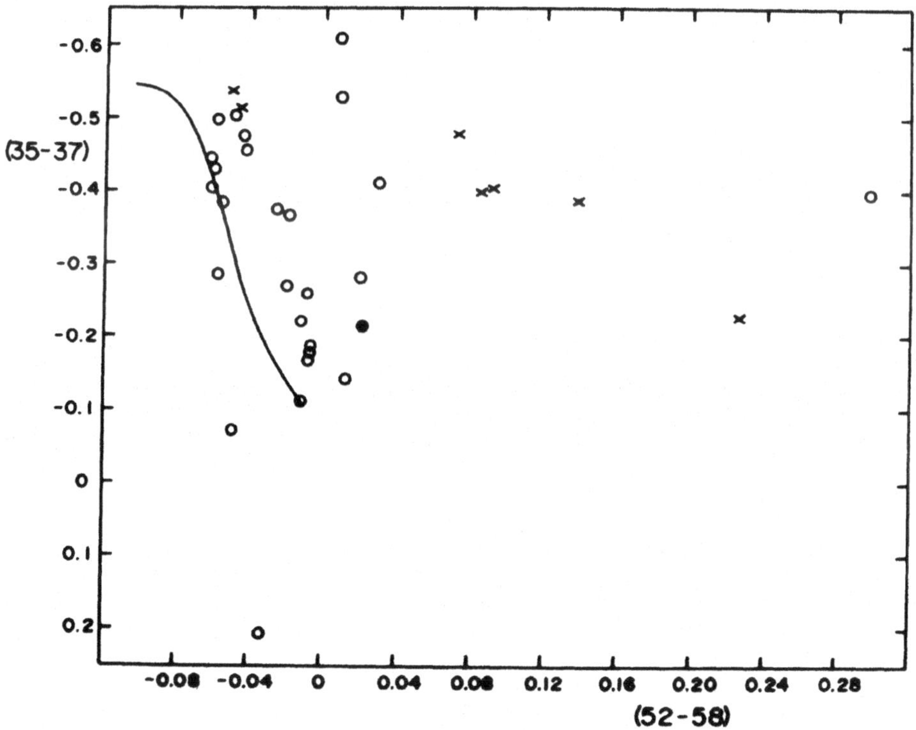

Fig. 6.- The (35-37), (52-58)-plane

can also be used to divide color diagrams in two regions, namely, Be
and normal B-type stars. In figure 5 the stars listed in Table 1 are
plotted in the (K-L, H-K) array. Other stars which are not contained
in Table 1 also show the same characteristics. Early type supergiant
stars perhaps do not lie far from the unreddened Main Sequence in the
(K-L, H-K)-plane. The triangles in Figure 5 represent early type
supergiant stars not listed in Table 1. The broken line represents the
Main Sequence for spectral type later than B8. Another interesting two-
color diagram is (35-37) vs (52-58) shown in Figure 6, where again a
number of Be stars are separated from normal B-type stars.

6. COLOR-EXCESSES

Two-color diagrams which involve either ultraviolet or infrared colors
(see for instance Figs. 3-6) give a qualitative account of color
excesses for a number of Be stars. These excesses can be investigated
further with the aid of intrinsic colors. Since the infrared excesses
seem to be too large to be explained by solely interstellar obscuration,
then it is also necessary to know the extinction law valid in the
direction of each Be star.

We are going to assume a single normal extinction law, such as
the one of Cygnus (Mendoza 1968) and to use Johnson's (1966) intrinsic
colors for the UBVRIJHKL-photometry, and provisional values, derived
from hopefully unreddened Main Sequence early type stars for the far
ultraviolet and 13-color photometries. We are also going to assume that
the colors least disturbed by the Be-phenomenon are (V-R) and (52-58).
Let us once more recall that variability, less accurate types for Be
stars, and other items mentioned above may affect the end product.

The results indicate that the Be stars listed in Table 1, around
60% of them have ultraviolet and infrared excesses, the 13-color
photometry shows that in the 0.3-1.1 micron region some of the Be stars
with color excesses have secondary peaks in both the Balmer and Paschen
continua (filters 37 and 80), probably contaminated by emission. The
broad passband photometry does not show these features; however, it
indicates that larger excesses exist in the neighborhood of $\lambda 2365$ in
the UV, and 4 microns and at longer wavelengths in the infrared. γ Cas
and ϕ Per have these properties. Their color excesses versus $1/\lambda$ are
shown in Figure 7.

7. SUMMARY AND CONCLUSIONS

The four photometric systems which yield 28 color indices in the wavelength interval from 0.14 to 4 microns, approximately show that Be
stars have as a group:
1) Light variations in their continua and in some spectral lines.
2) Ultraviolet and infrared color excesses (see Fig. 7) for about 60%
of the Be stars under discussion.
3) Some extreme shell Be stars have the OI $\lambda 7774$A enhanced absorption
stronger than in supergiant stars of the same spectral type. This

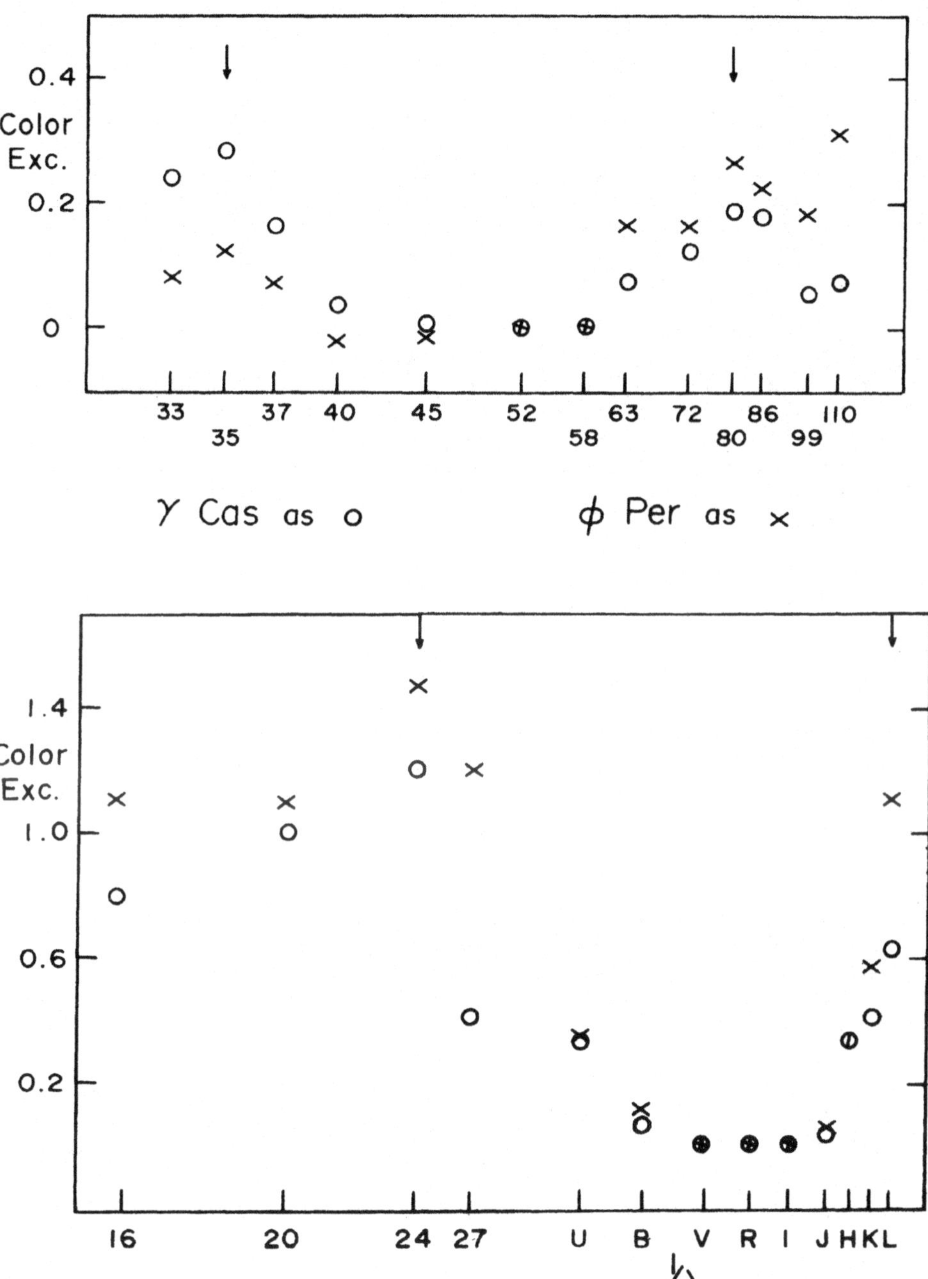

Fig. 7.- Color excesses for two Be stars in two photometric systems

information may be obtained from the $\Lambda(9)$-index.

It should be pointed out that a few specific Be stars have quite
different photometric characteristics than the average. P Cygni is one
of them. Its (V-R) and (52-58) seem to have an additional reddening,
other than interstellar, which perhaps will make that the derived color
excesses from these indices may be wrong by a considerable amount (too
large in the far UV and too small in the near IR). P Cygny in most
two-color diagrams lies close to the 55 Cygni locus (see Figs. 3, 4, 6)
the supergiant star listed in Table 1 with the strongest Hα-emission
(Mendoza and Johnson, 1979).

Current work on Be stars at the University of Mexico is:
1) Light variability on the 13-color (Alvarez and Schuster)
2) Michelson Fourier Spectra.
3) H, He and OI-line strength measurements (Mendoza and Ortega).

More new observations and explanations of the photometric
characteristics presented here, most likely, will be treated in this
symposium.

ACKNOWLDEGEMENT

It is a pleasure to thank T. Gómez for inking the Figures, to M. Alvarez
and W. Schuster for a pre-print of their work read at the Second Latin
American Regional Meeting; to M. Moreno for the UBVRI observations of X
Persei, to A. Quintero for his valuable help with the $\alpha(16)$ and $\Lambda(9)$
observations and to Miss P. Guerrero for typing the manuscript.

References

Alvarez, M., and Schuster, W.: 1981 in Second Latin American Regional
  Meeting. Rev. Mex. Astron. Astrof. Vol. 6, in press.
Andrews, P. H.: 1978, Mem. R. astr. Soc., Vol. 72, pp. 35-99.
Bahng, J. D. R.: 1976, in A. Slettebak (ed), Be and Shell Stars, D.
  Reidel Publ. Co. Dordrecht-Holland, pp. 41-49.
Feinstein, A.: 1968, Z. Astrophysics, Vol. 68, pp. 29-47.
Hayes, D. S., and Latham, D. W.: 1976, Ap. J., Vol. 197, pp. 593-601.
Iriarte, B.: 1969, Bol. Tonantzintla y Tacubaya, Vol. 5, pp. 89-100.
Jaschek, C., Ferrer, L., and Jaschek, M.: 1971, Catalogue and Biblio-
  graphy of B type emission line stars. Obs. Astr. La Plata.
Johnson, H. L.: 1966, Ann. Rev. Astr. and Ap. Vol. 4, pp. 193-206, Ed.
  L. Goldberg (Palo Alto, Calif.: Annual Reviews).
Johnson, H. L., and Mitchell, R. I.: 1975, Rev. Mex. Astron. Astrof.
  Vol. 1, pp. 229-324.
Johnson, H. L., Mitchell, R. I., Iriarte, B., and Wiśniewski, W. Z.:
  1966, Comm. Lunar and Planetary Lab., Vol. 4, pp. 99-110.
Mendoza, E. E.: 1958, Ap. J., Vol. 128, pp. 207-218.
Mendoza, E. E.: 1967, Bol. Tonantzintla y Tacubaya, Vol. 4, pp. 149-196.
Mendoza, E. E.: 1969, Publ. Astr. Dept. U. Chile Nr 7, pp. 106-126.
Mendoza, E. E.: 1975, Publ. Astr. Soc. Pacific, Vol. 87, pp. 495-497.

Mendoza, E. E.: 1977, Rev. Mex. Astron, Astrof., Vol. 2, pp. 259-261.
Mendoza, E. E.: 1979, Astron. Astrophys., Vol. 71, pp. 147-150.
Morel, M., and Magnenat, P.: 1978, Astron. Astrophys. Supp., Vol. 34,
    pp. 477-478.
Polidan, R. S., and Peters, G. J.: 1976, in A. Slettebak (ed), Be and
    Shell Stars, D. Reidel Publ, Co. Dordrecht-Holland, pp. 59-65.
Thompson, G. I., Nandy, K., Jamar, C., Monfils, A., Houziaux, L.,
    Carnochan, D. J., Wilson, R.: 1978, "Catalogue of Stellar Ultraviolet
    Fluxes". The Science Research Council.

DISCUSSION
Jaschek: Which is the least affected passband by colour excess, both
ultraviolett and infrared?

Mendoza: Perhaps the (52-58) colour index of the 13 colour system.
For the broad passband photometric system I have assumed that V-R
is the least affected by the colour excess.

Kozok: How does a star behave in your colour system, if it loses its
emission?

Mendoza: A Be star without emission is not seperated from the normal
early type star relationship, when an emission index is used. However
it may still keep a colour excess, thus, for instance it behaves like
a Be star in the K-L vs. H-K colour diagram.

Traving: Did you include in your program Herbig-Be stars?

Mendoza: No, only Be stars from the Bright Star Catalogue.

Harmanec: Referring to the probable variability of most Be stars on
different time scales, it would be important to publish exact J D of
every observation even in studies of statistical character. Otherwise
an important piece of infromation is lost.

Mendoza: The J D is included in the broad band pass photometry and 13
colour photometry. The J D for the $\alpha(16)$, $\Lambda(9)$ indices will be given
on request.

Feinstein: Do the same stars have infrared and ultraviolet excesses?

Mendoza: Be stars have sometimes only UV excess other times only IR
excess, and also sometimes both UV and IR excess.

Viotti: In your sample of Be stars you included some objects that
cannot be considered properly as Be stars, like P Cygni which has no
signs of "photospheric" lines in its optical spectrum, and even β
Lyrae. But it could be useful to find out some statistical properties
of these stars, in particular to have a basis for modelling, looking
at their time variations.

A STUDY OF Be STAR VARIABILITY

Joachim Dachs
Astronomisches Institut, Ruhr-Universität
D-4630 Bochum, West Germany

ABSTRACT. Variations of Balmer emission line strength for Be stars are
compared to simultaneously measured photometric light-curves in the visual
and infrared regions of the spectrum. Three different types of correlations between photometric and spectrometric variations are described. Variable intensity of continuous recombination radiation from ionized hydrogen in the circumstellar envelope is shown to be a major source of
light variations for Be stars.

Variability is a well established fact for many if not all Be stars
(see, e.g., Feinstein and Marraco, 1979). Yet systematic investigations
of light-curves and associated changes of emission-line strength are
lacking, and the few known studies on correlations between photometric
and spectrometric variations for Be stars (Sharov and Lyuty, 1976; Nordh
and Olofson, 1977) lead to contradictory results.

Therefore new measurements were obtained in 1978 and 1979 for 35
bright southern and equatorial early-type Be stars, by means of the 61
cm and 1 m reflectors at the European Southern Observatory in Chile. For
these stars, photoelectric UBV magnitudes (unpublished) were measured
nearly simultaneously with Balmer emission line profiles (Dachs et al.,
1981) and infrared JHKLM magnitudes (Dachs and Wamsteker, 1981). As far
as possible, these data are compared to published observations of the
same stars at other epochs, in particular from the lists of Allen (1973),
Feinstein (1974, 1975) and Gehrz et al. (1974). For the study of spectrometric variations, the photometric α indices of Hα emission line strength
determined by Feinstein (1974) are converted into equivalent widths of Hα
emission by using the empirical transformation formula

$$W_e(H\alpha) = 173 \ \overset{\circ}{A} \ (\alpha_{Feinstein} - 1.455) \tag{1}.$$

Evaluation of these data permits the following conclusions:
1. Amplitudes of light variations in the V photometric band exceed $0\overset{m}{.}5$
for many well observed Be stars (μ Cen, κ CMa, χ Oph, 31 Peg).
2. If V brightness increases, Be stars generally become redder. The ratio

19

M. Jaschek and H.-G. Groth (eds.), Be Stars, 19–22.
Copyright © 1982 by the IAU.

of (B-V) colour index to V magnitude variations, Δ(B-V)/ΔV, ranges between
about -O.1 and -O.3 for individual Be stars.
3. Photometric V band variations are accompanied by changes of Balmer
emission line strengths and of infrared excess radiation from the circumstellar envelope. Infrared and V magnitude variations are always positi-
vely correlated.
4. Infrared excess radiation of early-type Be stars as measured, e.g., by
the (J-M) colour index is strongly correlated to the equivalent width of
the Hα emission line.
5. Photometric and spectrometric variations of large amplitude (exceeding
$|\Delta V| = O^m.1$ or $|\Delta W_e(H\alpha)/W_e(H\alpha)| = O.2$) proceed rather slowly on time scales
of more than about one month. For the study of large-amplitude variations
of Be stars, measurements obtained within less than one month can therefore usually be considered to refer to the same epoch.

According to the sign of the correlation between photometric variations and changes of Balmer emission line strength three different types
of Be star variations can be distinguished:

Type I - Positive correlation: Simultaneous increase or decrease in
V brightness, near infrared brightness, and equivalent width of Balmer
line emission. Type I variations are observed for Be stars showing weak
to moderate emission-line strength, with equivalent widths of their Hα
line emission of no more than about -3O $\AA$. As an example, the variations
for μ Cen from 1968 until 1979 are shown in Fig. 1; this star completely
lost its Balmer line emission in 1978 when also its V and infrared bright-
nesses reached minimum values. For type I variations, the ratio of amplitudes in the L band (at wavelength 3.6 microns) to those in the V spectral range, $\Delta L/\Delta V$, is of the order of 1O.

Type II - Hα emission saturation: The intensity of Balmer line emission from the circumstellar envelope remains constant while large amplitude changes occur in visual and infrared magnitudes. Variability of this
type is observed for the extreme B2Ve star χ Oph in 1972 - 1979 (Fig. 2).
The ratio of amplitudes for V and L infrared variations is of order unity.

Type III - Anticorrelation: Maximum strength of metallic shell absorption lines coincides with a minimum of V brightness as observed, e.g.,
for Pleione in late 1973 by Sharov and Lyuty (1976).

Variations of type I indicate that in this case broad-band magnitude
variations are caused by variable amounts of continuous hydrogen recombination radiation emitted from a stellar envelope of variable density and
dimension, as first suggested by Nordh and Olofson (1977) in order to explain the variations observed for π Aqr. Analysis of the slope for the
correlation between photometric and Balmer emission line variations of
type I shows that Be star envelopes are optically thick for Hα line radiation.

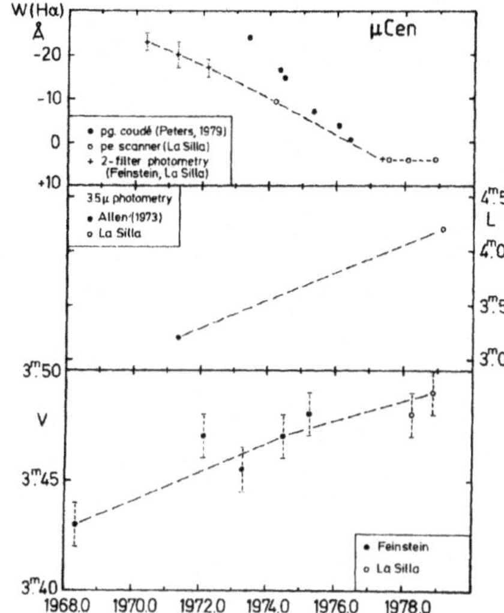

Figure 1. Variations of type I
for μ Cen (B2IV-Ve).

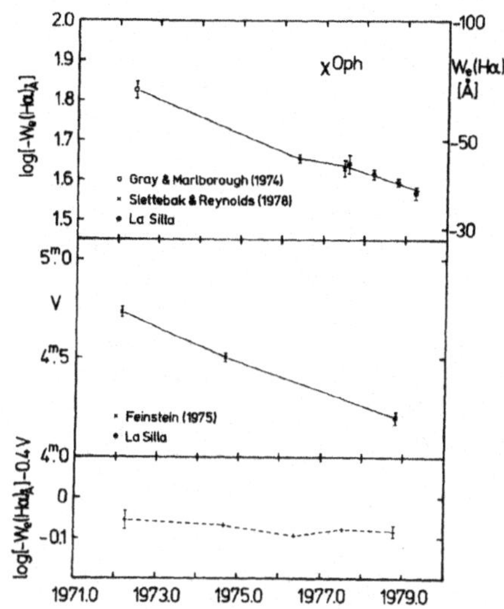

Figure 2. Variations of type II
for χ Oph (B2Ve).

Variations of type II point to variable amounts of continuous light
scattered by electrons in the circumstellar envelope at constant level of
saturated Balmer line emission. Type III variations were attributed by
Sharov and Lyuty (1976) to absorption of visual starlight in the ejected
shell material.

Support by the European Southern Observatory and by the Deutsche
Forschungsgemeinschaft under grant Da 75/5 is gratefully acknowledged.

REFERENCES

Allen, D.A.: 1973, Monthly Notes Royal Astr. Soc. 161, 145
Dachs, J., Eichendorf, W., Schleicher, H., Schmidt-Kaler, T., Stift, M.,
    Tüg, H.: 1981, Astron. & Astrophys. Suppl. 43, 427 (1981)
Dachs, J. Wamsteker, W.: 1981, to be submitted to Astron. & Astrophys.
Feinstein, A.: 1974, Monthly Notes Royal Astr. Soc. 169, 171
Feinstein, A.: 1975, Publ. Astr. Soc. Pacific 87, 603
Feinstein, A., Marraco, H.G.: 1979, Astron. J. 84, 1713
Gehrz, R.D., Hackwell, J.A., Jones, T.W.: 1974, Astrophys.J. 191, 675
Gray, D.F., Marlborough, J.M.: 1974, Astrophys.J.Suppl. 27, 121
Nordh, H.L., Olofson, S.G.: 1977, Astron. & Astrophys. 56, 117
Peters, G.J.: 1979, Astrophys. J. Suppl. 39, 175
Sharov, A.S., Lyuty, V.M.: 1976, IAU Symposium 70 (ed. A. Slettebak),105
Slettebak, A., Reynolds, R.C.: 1978, Astrophys.J.Suppl. 38, 205

DISCUSSION

A.M.Hubert: I think it is difficult to determine with accuracy the
equivalent widths of the emission lines in χ Oph, at some time,
because of an effect of "veiling" temporary observed.

Dachs: For photometrically variable Be stars, one should try to
determine the intensity of Hα emission by simultaneously measuring
the Hα equivalent width and the actual brightness of the star in
Johnson's R band.

Peters: Are you continuing to observe μ Cen? It is currently, once
again, developing an emission line envelope.
The B-V colour for μ Cen appears to be anticorrelated with U-B just
as you showed for χ Oph.

Dachs: I am continuing my observations.

Viotti: χ Per has a long term behaviour similar to that of your stars,
being redder when brighter, with smaller variations in U-B, but
larger IR variations. These results are a clue to a model of the
atmospheric envelopes, but it is difficult for the moment to say if
the variations are due to changes in the mass loss rates or in the
envelope structure or both.

Snow: Have you looked for variations on shorter time scales than you
have discussed in this presentation?

Dachs: I looked for night-to-night variations and found a few cases
of photometric and spectroscopic variations within 24 hours.

COMPOSITE COLOUR-MAGNITUDE AND COLOUR-COLOUR DIAGRAMS FOR Be STARS
IN OPEN CLUSTERS

J.-C. Mermilliod
Institut d'Astronomie de l'Université de Lausanne et
Observatoire de Genève

By the end of 1980, the total number of Be stars discovered in the
field of open clusters amounted to 180 stars distributed in 60 clusters.
Among these, 110 Be stars belong to 32 clusters included in the sample
I studied, which contains 75 open clusters younger than the Hyades.
But only 88 stars with complete *UBV* photoelectric photometry have been
taken into consideration here.   The concept of age groups, defined
elsewhere (Mermilliod 1981a), is used throughout the present analysis,
as well as the new estimate of the colour excesses and distance moduli
I obtained for these 32 clusters.   Under the assumption of uniform
reddening across the clusters, absolute magnitudes and dereddened colour
indices have been calculated for the 88 Be stars.

Fig. 1 shows a composite M_V vs $(U-B)_o$ diagram.   The lower sequence
is the ZAMS (Mermilliod 1981b) and the upper dashed one is the theoretical TAMS from models with overshooting in the core (Maeder and
Mermilliod 1981).   The four curves represent the empirical time-lines
I determined for the so-called NGC 884, 457, 3766 and Pleiades age
groups.   Close examination of my individual and composite diagrams
shows that the Be stars do not occur at random on the main sequence.
Several classes and sub-classes have been distinguished according first-
ly to the ages of the parent clusters and secondly to the location in
the colour-magnitude diagram.   They are represented by various symbols
in fig. 1:

a) B8.5-A2 stars: mostly shell stars ("+")

b) B3-B8e stars: they occur at three different places:

   1) on the main sequence, at M_V=-0.5 $\pm$ 0.5, and spectral type
      B8 ("o"),
   2) near the termination of the main sequence ("x"),
   3) in the position of blue stragglers.

c) B1-B2e stars: three sub-classes are distinguished:

M. Jaschek and H.-G. Groth (eds.), Be Stars, 23–26.

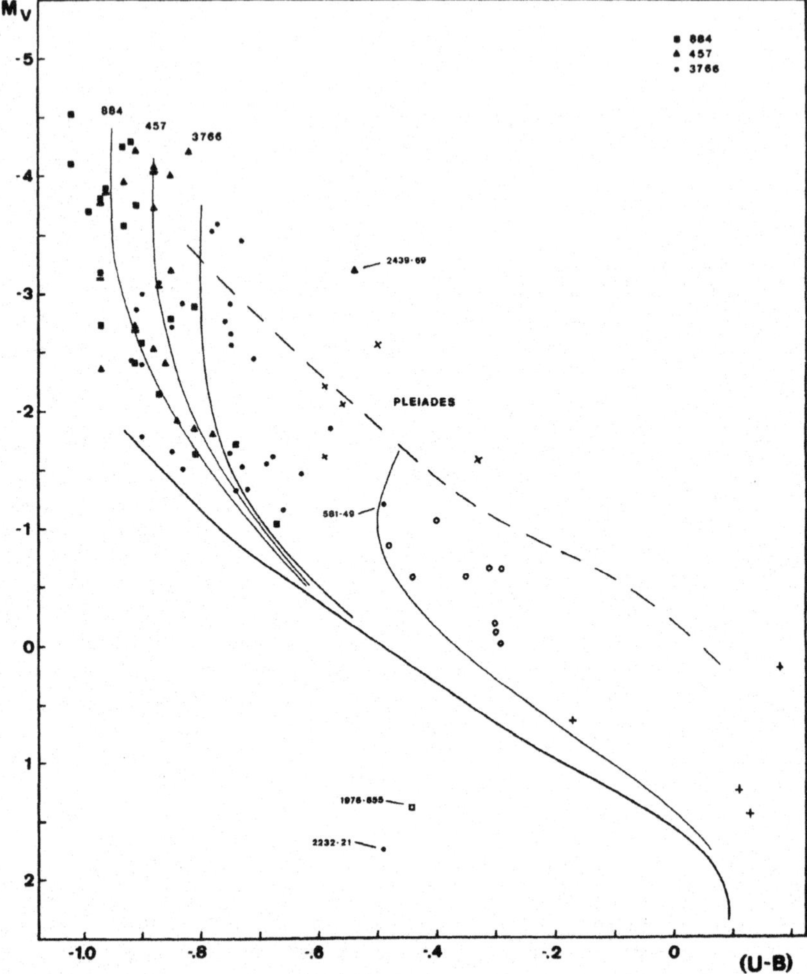

Fig. 1   Composite colour-magnitude diagram for Be stars in open clus-
         ters.   Curves and symbols are explained in the text.

   1) the nearly unevolved stars: $M_V > -2$,
   2) the semi-evolved stars: $-3.25 < M_V < -2.25$,
   3) the stars near the TAMS: $M_V < -3.5$.

The B1-B2e stars present a strong frequency maximum, which reaches 34%
in NGC 663 (Sanduleak 1979).   The earlier O-B0e stars are much less
frequent.

   These classes are closely related to the spectroscopic groups

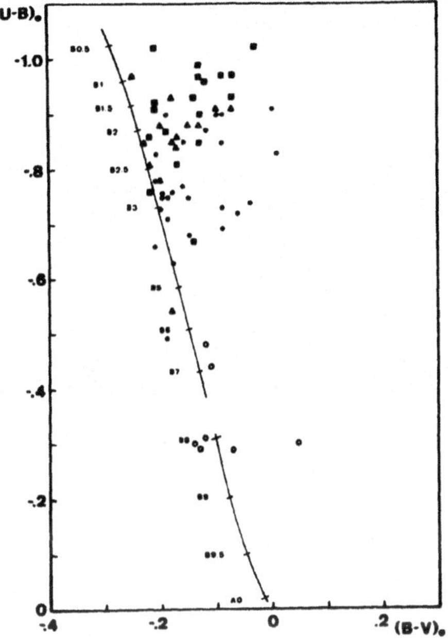

Fig. 2 Colour-colour diagram for Be stars in open clusters. The interstellar reddening has been subtracted.

formed by Jaschek et al. (1980). This results in establishing a connection between the location of the Be stars in the HR diagram and their spectroscopic properties.

Fig. 2 shows the distribution of the Be stars "intrinsic" colours (interstellar reddening subtracted) in the UBV plane. It is similar to the figure obtained by Feinstein and Marraco (1979) for bright field Be stars. The analysis of this diagram confirms that the colours of the B1-B2e stars are different from those of the normal stars of similar spectral type: $B-V$ is redder while $U-B$ is bluer. Although the hypothesis that the pseudo-reddening depends on both the stars' orientation and the envelope density appears as a tempting one, an application of the models of Poeckert and Marlborough (1978) is not straightforward.

Much more data on $V\sin i$, emission-line strength and photometry for stars in open clusters are badly needed to refine and extend the present picture.

REFERENCES

Feinstein A., Marraco H.G. 1979, Astron. J. 84, 1713
Jaschek M., Hubert-Delplace A.M., Hubert H., Jaschek C. 1980, Astron. Astrophys. Suppl. 42, 103
Maeder A., Mermilliod J.-C. 1981, Astron. Astrophys. 93, 136

Mermilliod J.-C. 1981a, Astron. Astrophys. Suppl. (in press)
Mermilliod J.-C. 1981b, Astron. Astrophys. (in press)
Poeckert R., Marlborough J.M. 1978, Astrophys. J. Suppl. 38, 229
Sanduleak N. 1979, Astron. J. 84, 1319

DISCUSSION

Houziaux: How do you determine intrinsic colours of the Be stars?

Mermilliod: Intrinsic colours have been determined by using the B-V
colour excesses estimated for the parent clusters. The assumption
of uniform reddening seems to be justified in most cases.

STATISTICAL ANALYSIS OF THE DATA AVAILABLE FOR Be STARS

Daniel Egret
Observatoire de Strasbourg

1. INTRODUCTION

A comprehensive catalogue of more than 1100 Be stars with known MK
classification has been prepared at the Strasbourg Stellar Data Center,
and is described in an other session of this Symposium (Jaschek and
Egret, 1982). The compilation of photometric and spectroscopic data
available for these stars has made possible a general statistical approach of the Be star spectral group.

2. DISTRIBUTION IN MAGNITUDES AND SPECTRAL TYPES

The sample studied here is not exhaustive to any given limit in
magnitude or distance, and is affected by several observational biases.
In the following we will often use the subsample of stars in the region
south of $\delta=-40°$ covered by the Michigan Spectral Survey for the HD stars
(Houk and Cowley, 1975; Houk, 1978): for this subsample statistical
comparisons with normal B stars should be more significant.

The distribution in visual magnitudes is illustrated by Figure 1.
There are 160 stars brighter than V=6.5 and 750 stars brighter than 9.5.
In the region south of -40° the number of detected Be stars is the following: 45 Be stars brighter than 6.5 (=9% of the B III to V stars) and
only 420 HD stars (=4%). These numbers do not take into account the
variability of the Be feature.

The distribution within the spectral types B0-B9 and the luminosity
classes III-V is presented in Figure 2. The main features are the maxima
at spectral types B2 and B8 and the increasing proportion of giants over
dwarfs towards the later types.

3. THE UBV PHOTOMETRY OF Be STARS

Careful studies of the photometric behavior of the brightest Be
stars have already been published (Feinstein and Maracco, 1979). We have
extended the study to some 600 stars with UBV measures from the compilation by Nicolet (1978) and compared the positions of these stars in a
(U-B) vs (B-V) diagram with those of main sequence B stars as given by

M. Jaschek and H.-G. Groth (eds.), Be Stars, 27–31.
Copyright © 1982 by the IAU.

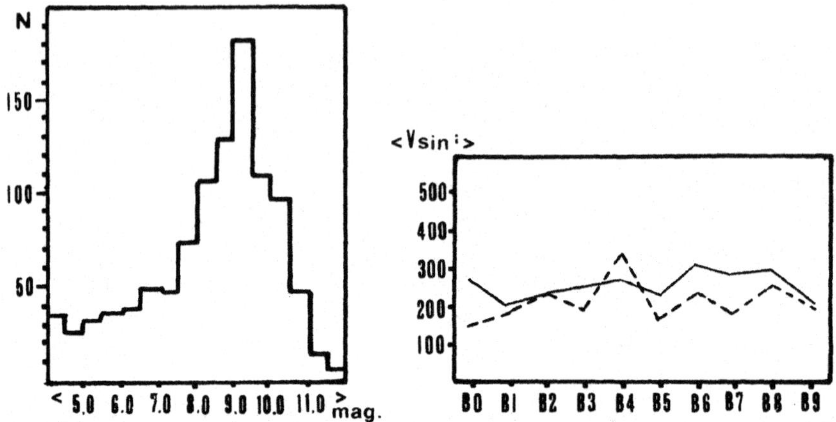

Fig. 1. Histogram of the
visual magnitudes of 1000
Be stars.

Fig. 3. Mean rotational velocity
for each spectral group (luminosity classes III and V).

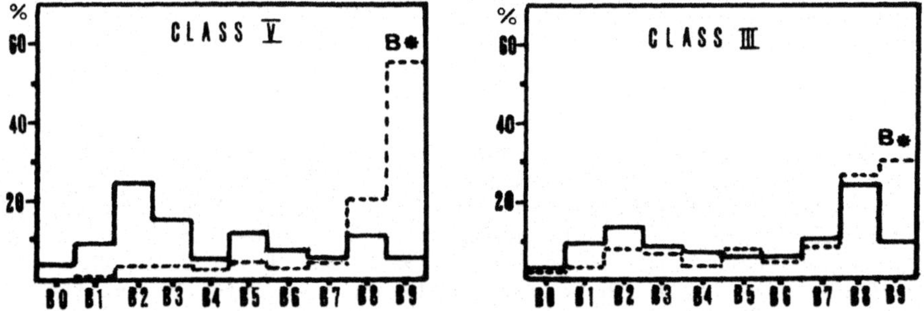

Fig. 2a and 2b. Distributions of the spectral types B0 to B9 in
the region south of -40° (2a: class V; 2b: class III). The dashed
line is the distribution of normal B stars.

FitzGerald (1970). For a number of stars (and especially in the range B2-B4) the photometric classification is apparently earlier than the MK classification. This should be generally interpreted (provided that both oservations refer to the same state of Be features) by an ultraviolet colour excess of these stars. The (B-V) excess is often large and the distinction between circumstellar and interstellar reddening for field stars is quite delicate; however in the hypothesis of a normal interstellar reddening Av=3E(B-V), we can easily detect those stars for which the total absorption as a function of the magnitude exceeds the current values for B stars. An analysis of these data is in preparation (Egret, 1981).

4. THE ROTATIONAL VELOCITY OF Be STARS

The recent edition of the catalogue of stellar rotations (Uesugi and Fukuda (1981) contains values of $V \sin i$ for about 300 Be stars. The distribution of $\langle V \sin i \rangle$ vs spectral type is shown in Figure 3: the mean rotational velocity is not sensibly different for all the spectral types between B0 and B9, but the giants are apparently slightly slower rotators than the dwarfs.

5. THE SPATIAL DISTRIBUTION OF Be STARS

The major concentrations in the galactic plane are found in the direction of Perseus (ℓ=130°) and Carina (ℓ=290°, included in the Michigan Survey). This is illustrated by Figure 4.

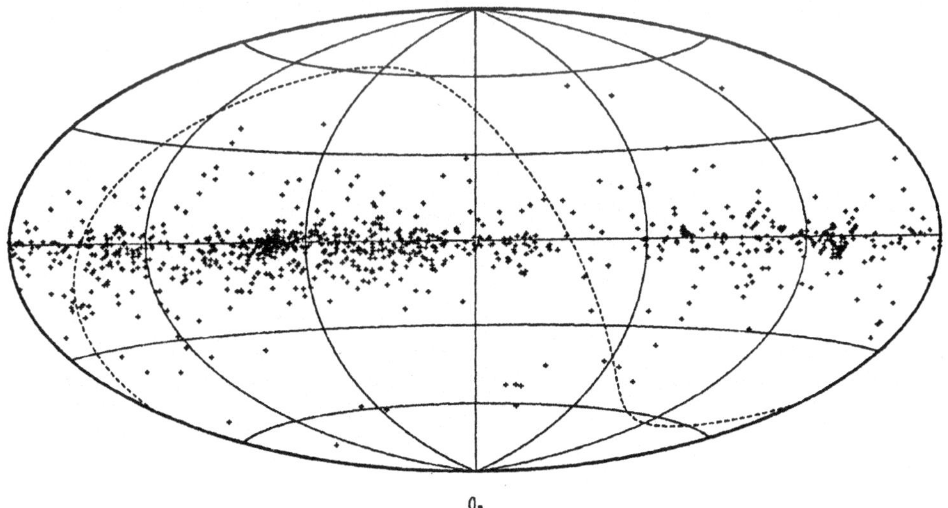

Figure 4.   Distribution in galactic coordinates of Be stars.
(The dashed line is the curve δ=0°).

The galactic concentration of Be stars in the region south of $\delta = -40°$ appears similar to the one of normal B stars. For the whole catalogue the galactic concentration parameter is $\beta = 80pc$ (60pc for B stars) but there is a selection effect in favour of the stars outside the plane. 30 detected Be stars have a galactic latitude higher than 25°. A study of their kinematics is in preparation.

6. FINAL REMARK

Our statistical approach does not take into account the variability of the Be characteristics and does not pretend to supersede the detailed studies of some objects frequently observed. Anyway we hope that this will encourage future observing programs of more and more stars, the only way to improve our understanding of the common properties of Be stars.

REFERENCES

Egret,D.: 1981, thesis (in preparation).
Feinstein,A. and Maracco,H.,J.: 1979, Astron.J. 84, 1713.
FitzGerald,P.: 1970, Astron.Astrophys. 4, 234.
Houk,N. and Cowley,A.,P.: 1975, *Michigan Catalogue of Two-dimensional Spectral Types for the HD Stars*, vol. I, Michigan Univ.
Houk,N.: 1978, *Michigan Catalogue of Two-dimensional Spectral Types for the HD Stars*, vol. II, Michigan Univ.
Jaschek,M. and Egret,D.: 1982, IAU Symp. 98 "*Be stars*", Jaschek,M. and Groth,H.G. eds., p. 261
Nicolet,B.: 1978, Astron.Astrophys.Suppl.Ser. 34, 1.
Uesugi,A. and Fukuda,I.: 1981, Proc. of 7th International CODATA Conference (in press).

DISCUSSION

Snow: You said there are 30 Be stars in your catalogue that have galactic latitudes greater than 25°. Are these bright foreground stars, or are some of them sufficiently distant to really be out of the galactic plane?

Egret: Part of these stars are really distant stars (half are fainter than 6^m, and half a dozen fainter than 8^m). A detailed analysis of these stars is in preparation.

Metz: Are the rotational velocities presented in your catalogue derived both from shell emission lines and photospheric emission lines as well?

Egret: We have used the v sin i values by Bernacca and Perinotto (1973) and by Uesugi (1973) generally derived from photospheric absorption lines.

MacConnell: Did you include B stars in your analysis which Nancy Houk has indicated as having Hβ filled in? My observations on H_α objective-prism plates show that the majority of these stars have H_α in emission and are therefore Be stars.

Egret: Yes. In fact we used simultaneously MK classification catalogues (such as the one of N. Houk which does not cover the H_α regions)and catalogues or lists of H_α emission stars.

Doazan: You have considered only Be stars which have MK spectral types to define the population of Be stars relative to normal B's in the Jaschek's et.al. catalogue. These statistics bias your results severely and lowers considerably the percentage of Be/B stars. Can you say what results you would obtain by using Wackerling's catalogue for example, and all Be stars detected whatever the system of classification used?

Egret: I have not yet done such statistics concerning all the H_α emission line stars. The only thing we can say for the moment is that Wackerling's catalogue, for instance, obtains about 3 times more B stars (including supergiants) than our catalogue of "classical" Be stars, for comparable limits in magnitude.

ABSOLUTE MAGNITUDES AND INTRINSIC COLOURS OF NON-SUPERGIANT Be STARS[*]

J.R. Kozok
Astronomisches Institut, Ruhr-Universität
463o Bochum 1, FRG

ABSTRACT. 1o1 normal Be stars, probable members of 56 galactic clusters and OB-associations, and more than 2o extreme Be stars in the Large Magellanic Cloud were used to derive intrinsic colours of O9-B9(III-V)e stars. Furthermore, the correlation between the intrinsic colour $(U-B)_o$ and the absolute magnitude of non-supergiant Be stars was confirmed to be

$$M_V = 4.55 \cdot (U-B)_o + 0\overset{m}{.}81 \qquad \text{for} \quad < (U-B)_o < -0\overset{m}{.}53$$

$$\text{and} \qquad M_V = 1.18 \cdot (U-B)_o - 0\overset{m}{.}98 \qquad \text{for} \quad (U-B)_o > -0\overset{m}{.}53 \ .$$

The aim of the present investigation is to enlarge the basis for the determination of intrinsic colours and absolute magnitudes by providing a large sample from the southern sky.

The program stars were taken out of the catalogues of Wackerling (197o) and Jaschek et. al. (1971). Probable cluster membership was found by cross-checking with the Alter and Ruprecht catalogue of star clusters and associations (197o). Stars lying within two cluster radii of the center were regarded as possible candidates for membership. Additionally we have used all Be stars, which were regarded by other authors as members of clusters and associations. Some of these have been measured once more during this research. The objects of the northern hemisphere, which have been investigated in previous papers e.g. by Schmidt-Kaler (1964a,b,c), Schild (1966), Schild and Romanishin (1976) have also been included.

This research also included Be-supergiants of luminosity classes I-II, but these are not presented here. To obtain a sufficient sample I have observed Hα -emission-line stars in the Large Magellanic Cloud (LMC) compiled in a catalogue by Bohannan and Epps (1974). Besides the supergiants this catalogue may contain a bigger sample of dwarf B(III-V)e (dBe) stars earlier than B1, if we consider its limiting magnitude. The colour-colour diagram (figure 1) for the measured stars fainter than V = 13^m makes visible a group of stars having an (U-B)-excess. Using the normal dereddening procedure we would obtain intrinsic colours above the values for O5 stars. These stars are according to their photometric appearance equivalent to the extreme Be stars as defined spectroscopically

M. Jaschek and H.-G. Groth (eds.), Be Stars, 33–36.

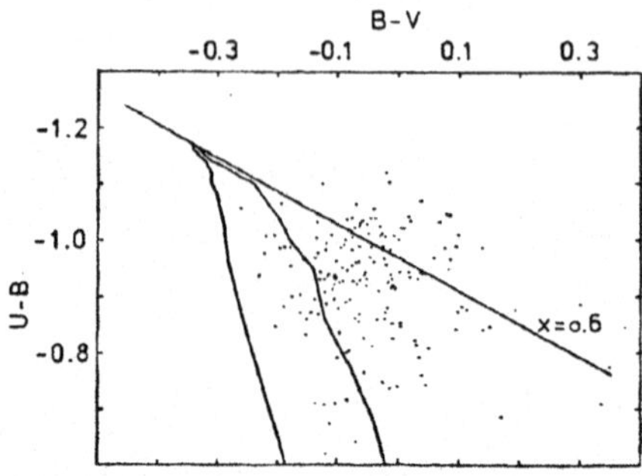

Figure 1. Two colour diagram for LMC emission-line stars with visual magnitudes $V = 13^m - 15^m$. The line of reddening for $E_{U-B}/E_{B-V} = 0.60$ is plotted.

by Schild (1966). Taking into account the excesses given by Schild and Romanishin (1976) it is possible to deredden these stars which yields reasonable E_{B-V} values for the LMC. Moreover the assumption that these stars are extreme Be is consistent with the absolute magnitudes we derive after correcting the basic visual magnitudes for the distance modulus of the LMC $m_O - M_V = 18\overset{m}{.}6$ (Crampton, 1979). Looking at the brighter LMC stars which were measured, the percentage of these objects substantially reduces and disappears for stars with $M_V < - 6^m$ completely. This is another hint that we deal partly with extreme dBe stars and that these objects are very well suited to investigate the intrinsic colours of the early Be stars.

The program stars in the galactic clusters and OB-associations have been measured during three observing runs in 1979 and 1980 in the UBV system. For the analysis of the data I have taken mean reddenings $\bar{E}_{B-V}$ and distance moduli as given by Becker and Fenkart (1971) and Fenkart and Binggeli (1979). The values for the clusters not listed there were taken from the latest papers. Dereddening the stars with the corresponding E_{B-V} gives the colour-colour relation (figure 2) for the intrinsic colours of the dBe stars. It can easily be seen that most of the stars are lying above the main sequence. After calculating floating means we obtain relation 1 between $(B-V)_O$ and $(U-B)_O$ shown in this diagram. Relation 2 is derived from stars for which cluster membership seemed quite sure. Both graphs are very much alike and for further analysis we will use line 2. Both diagrams show the same behaviour for stars earlier than Bo as the main sequence, but this is defined mostly by stars in associations for which the mean reddening is not that reliable.

To get better accuracy for this part of the sequence we reduce the LMC stars, which have been characterized as extreme Be, with a mean $\bar{E}_{B-V} = 0\overset{m}{.}07$ given by Isserstedt (1975). For the bluest stars we obtain positions that lie somewhat above the turn-off point. After applying the excess values for extreme Be stars, they all lie on the main sequence with reasonable E_{B-V}'s. The adopted intrinsic colours for dBe stars in this part of the diagram are therefore to be modified for the extreme Be stars, which results in line 3. The higher dispersion for the data in this region is caused by a mixture of extreme and normal Be stars. But the upper end is completely dominated by extreme Be stars.

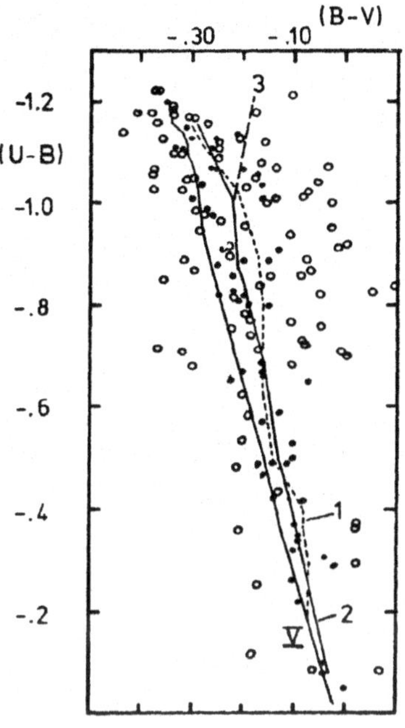

Figure 2. Two colour diagram for
dereddened non-supergiant Be stars,
probable members of galactic
clusters and OB-associations.
(Dots: the cluster membership for
the Be star is quite sure;
open circle: the cluster membership is uncertain.)

The last part of the paper concerns the correlation
between the absolute magnitude M_V and the intrinsic colour $(U-B)_0$. For
this purpose we take first the values dereddened by $\bar{E}_{B-V}$ and correct
for the distance modules, and second we use the relation 2 to deredden
each star individually and then correct for its distance (figure 3)
and then compare both values of absolute magnitude, which do not differ
substantially.

$$M_V = 4.55 \cdot (U-B)_0 + 0\overset{m}{.}81 \qquad \text{for} \qquad (U-B)_0 < - 0\overset{m}{.}53$$

$$\text{and} \quad M_V = 1.18 \cdot (U-B)_0 - 0\overset{m}{.}98 \qquad \text{for} \qquad (U-B)_0 > - 0\overset{m}{.}53 \ .$$

Figure 3. The absolute magnitude M_V is plotted as a
function of the intrinsic
colour $(U-B)_0$ for non-supergiant normal Be stars,
probable members of galactic
clusters and OB-associations.
(Dots: own measurements;
open circles: data given by
other authors.)

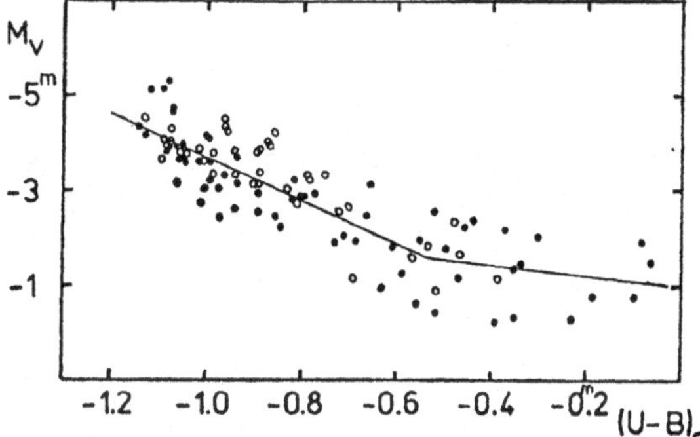

For early type stars the mean deviation is about $\pm$ $0\overset{m}{.}80$. The later type
Be's show worse correlation, but we should not forget that this is also
because we have less data for this part of the relation. Uncertain
distances of clusters contribute essentially to the presented dispersion.

The analysis for the Be stars of luminosity class I-II in the LMC
is under way. Additional measurements in both Magellanic Clouds and
a comparison to the few galactic Be supergiants, which have been
discovered in clusters,is intended. As Iam limited by time, it is not
possible to give a full presentation of all data here.

***** Based on observations obtained partly at the European Southern
Observatory, La Silla.

REFERENCES

Alter, G., Balázs, B., and Ruprecht, J.: 197o, "Catalogue of star clusters
      and associations", Akadémiai Kiadó, Budapest.
Becker, W. and Fenkart, R.: 1971, Astron.Astrophys.Suppl. 4, pp.241-252.
Bohannan, B. and Epps, H.W.: 1974, Astron.Astrophys.Suppl. 18, pp.47-79.
Crampton, D.: 1979, Astrophys. J. 23o, pp.717-723.
Fenkart, R.P. and Binggeli, B.: 1979, Astron.Astrophys.Suppl. 35,pp.271-275
Isserstedt, J.: 1975, Astron.Astrophys. 41, pp.175-182.
Jaschek, C., et.al.: 1971, Obs. Astron. Univ. Nacional La Plata,
      Ser. Astron. (Argentina) Vol. 37, pp.1-69.
Schild, R.: 1966, Astrophys. J. 146, pp.142-151.
Schild, R. and Romanishin, W.: 1976, Astrophys. J. 2o4, pp.493-5o1.
Schmidt-Kaler, Th.: 1964a, Zeitschr.Astrophys. 58, pp.217-24o.
Schmidt-Kaler, Th.: 1964b, Zeitschr.Astrophys. 58, pp.241-247.
Schmidt-Kaler, Th.: 1964c, Veröff. Bonn Nr. 7o, pp.1-43.
Wackerling, L.R.: 197o, Mem.Roy.Astr.Soc. 73, pp.153-319.

LUMINOSITY CLASSIFICATION OF Be STARS BY BALMER LINE NARROW BAND
PHOTOMETRY*

Wolfgang Zeuge
Hamburg Observatory, Gojenbergsweg 112, D-2050 Hamburg 80,
FRG

Summary: The absolute luminosity of most Be stars can be
determined by using Balmer line narrow band photometry with an
accuracy of about 0.4 mag. The few cases in which this method fails
can be detected.

It is well known that the Be stars can be easily discrimi-
nated from other stars with the help of at least two hydrogen line
indices (Abt and Golson 1966).

In the course of a research program about the photoelectric
determination of the absolute luminosities of OB stars, the H_α, H_β,
H_γ lines of 65 Be stars were observed (Zeuge 1981). The filter system
is somewhat different from e.g. the H_β standard system (Crawford and
Mander 1966). The narrow filters are wider than usual (40, 35 and
25 Å for H_α, H_β and H_γ respectively). Therefore this system is more
stable against effects like the rotational broadening of the lines.

In order to determine the emission of the Be stars, I
applied techniques similar to those of Feinstein (1976). The non
emission stars are located on a narrow straight band in the H_α - H_β
H_α - H_γ diagrams:

$$H_\alpha = 0.741 + 0.558 \cdot H_\beta,$$
$$\text{(instrumental system)} \qquad (1)$$
$$H_\alpha = 0.757 + 0.537 \cdot H_\gamma.$$

Next I determined average indices for the spectral classes
of those stars which do not show any emission and for which a complete MK classification is available. Now we can calculate the difference (E_α, E_β, E_γ) between the observed indices and the expected

---

* Based on observations obtained at the European Southern Observatory,
La Silla

37

M. Jaschek and H.-G. Groth (eds.), Be Stars, 37–40.
Copyright © 1982 by the IAU.

indices in the case of no emission for those Be stars, which are
included in the MK classification system.

The dependance of the emission E_β in H_β from E_α and E_γ
from E_α respectively can be described very well by a quadratic
expression where the linear term depends on H_β or H_γ:

$$E_\beta = 0.286 \cdot E_\alpha^2 + (1.501 - 0.80 \cdot H_\beta) \cdot E_\alpha \;,$$

$$E_\gamma = 0.294 \cdot E_\alpha^2 + (1.204 - 0.63 \cdot H_\gamma) \cdot E_\alpha \;,$$

$$(2)$$

with s.d. 0.022 and 0.020 respectively (one star).

We now have two different possibilities to determine the
emission in H_α by solving the equations (1) and (2) simultaneously,
one from $H_\alpha - H_\beta$ and one from $H_\alpha - H_\gamma$. In general these two values
coincide quite well (see fig. 1). Stars with differences greater than
0.015 mag are marked by an "X".

If the emission in H_α is known, the emission-free values
H_{β_0}, H_{γ_0} can be calculated. In the following I shall use only H_{γ_0}
because then the corrections are smaller.

22 of the Be stars are members of open clusters. For those
I could calculate quite good absolute luminosities from the distances
of the clusters. If one now applies the absolute luminosity calibration of no emission stars (Zeuge 1981) to Be stars using H_{γ_0}, then

one sees that the luminosities of stars with strong emission are
brighter than expected. The correction term is " $+ 1.10 \cdot E_\alpha$ "
(E_α is negative). In this way we get the M_v calibration for Be stars:

$$M_v (\alpha\beta\gamma) = M_v (H_{\gamma_0}) + 1.10 \cdot E_\alpha \;. \qquad (3)$$

This relation is valid if the two different values of E_α have nearly
the same value. Fig. 2 shows the residues between this calibration
and the absolute luminosities from the distances of the open clusters
against E_α. " + " shows stars where the calibration is valid. The s.d.
is 0.4 mag. " X " shows stars for which the calibration is not
applicable because the results from the two calculations of E_α are
too different.

In addition, some remarks about the percentage of the Be
stars for which this method can be applied:
14 of 22 stars in open clusters are included in the calibration.
For all of the 43 field stars the method seems to be applicable,
for all have small differences between the two ways of calculating E_α.

So, one might draw the conclusion:

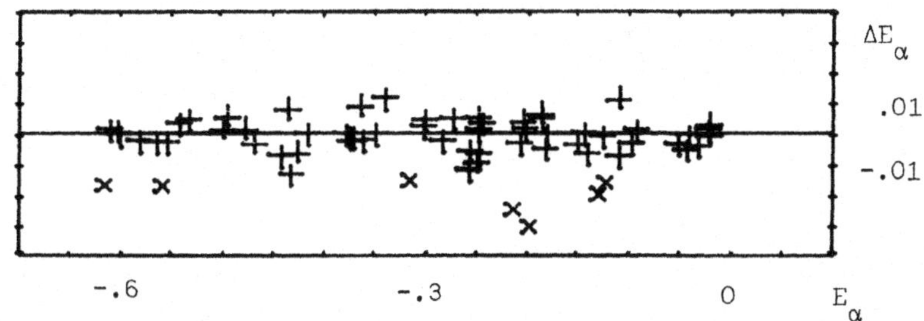

Fig. 1:   The differences between the two ways of calculating E_α
           against E_α. " X " shows stars with $\left| \Delta E_\alpha \right| > 0.015$ mag.

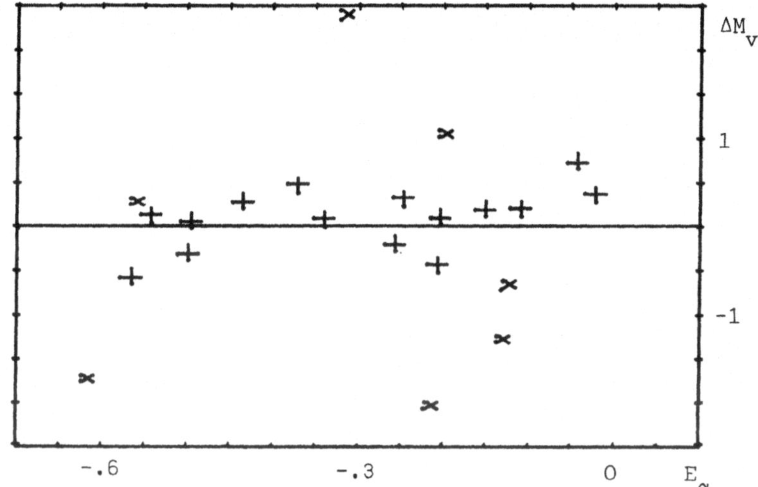

Fig. 2:   The residues of the luminosity classification against
           E_α (see text)

The absolute luminosities of most of the Be field stars and of the
majority of the Be cluster stars can be determined with an error of
about 0.4 mag and the Be stars for which this method is not applicable can be recognized. It also seems that the photometric behavior
of some of the cluster Be stars is significantly different from that
of the field stars.

Literature:
Abt, H.A., Golson, J.C.: 1966, Astrophys. J. 143, p.306
Crowford, D.L., Mander, J.: 1966, Astron. J. 71, p.114
Feinstein, A.: 1976, in IAU Symposium 70, Be and shell stars,
    ed. A.Slettebak (D. Reidel, Dordrecht, Holland), p.249
Feinstein, A.: 1979, Astron. J. 84, p.1713
Zeuge, W.: 1981, thesis to be published

DISCUSSION following Kozok

Jerzykiewicz: 1. How does your $(U-B)_0$ correlate with the effective
temperature?
2. For normal B stars your $M_V-(U-B)_0$ diagram would be a form of the
HR diagram. Is it also the case for your last slide?

Kozok: 1. I do not know the effective temperature of these stars.
2. No, it is just an observational correlation of the photometrically
derived absolute magnitudes and intrinsic colours $(U-B)_0$.

Slettebak: I still have a problem with the concept of "extreme Be
stars", which have been defined in various ways. I hope that these
objects will be discussed during this symposium so that we can decide
whether such a physical group actually exists, or whether we are
simply discussing various aspects of normal Be stars. My question is,
how do you define "extreme Be stars"?

Kozok: I regard those stars as "extreme Be stars" which show a clear
photometric excess. That means for fig. 1 that the stars which cannot
be dereddened to the main sequence but are lying above are named by
me as "extreme Be stars".

Harmanec: A note of warning: We obtained more or less systematic
UBV photometry of CX Dra (HD174237) covering the period 1964-1980.
If you plot the B-V and the U-B values in the U-B versus B-V diagram,
they represent perfectly the cloud defined by your 150 stars.

Kozok: The stars have been measured on a time scale of one year and
I have used for my analysis only stars which did not show any variation of more than $\overset{m}{.}01$. Furthermore, the dispersion of the stars taken
for line 2 is about $\pm \overset{m}{.}05$.

DISCUSSION following Zeuge

Harmanec: Is your relation fulfilled also by supergiant stars?

Zeuge: No, but these stars are seperated from the other Be stars on
the $H_\alpha - H_\gamma$ diagram.

LONG-TERM VARIATION OF Be STARS ON THE COLOR-MAGNITUDE DIAGRAM

Ryuko Hirata
DEPEG, Observatoire de Paris, Meudon,
on leave from Department of Astronomy, University of Kyoto,
Japan.

Abstract The ratios $\alpha=\Delta V/\Delta(B-V)$ and $\beta=\Delta(U-B)/\Delta(B-V)$ for the long-term variation of Be stars have been derived. The inclination effect on α is found. The dependence of α on the spectral type is also inferred. The signs of α and β are essentially the same.

The long-term variations of Be stars on the color-magnitude and color-color diagrams have been re-examined, by adding newly published data (e.g.,Harmanec et al. 1980) after the work of Nordh and Olofsson(1977). Taking into consideration of the influence of the short-term variation and the photometric accuracy, we selected the Be stars which fulfil the following three conditions, i) total span of observations$\geqslant$10 years, ii) number of years in which the observations were made$\geqslant$5, iii) total range of variation, $\Delta V \geq 0\overset{m}{.}15$ or $\Delta(B-V)\geq0\overset{m}{.}05$. 37 Be stars were thus selected. The values of V, B-V, and U-B were plotted against time to check whether they reflect the long-term variation or not. When many data were available in one observational season, the average was taken. Then, the regression curve for $\alpha=\Delta V/\Delta(B-V)$ was obtained for 32 stars. We could determined the regression curves of 25 stars for both α and $\beta=\Delta(U-B)/\Delta(B-V)$. In figure 1a,we plot α against Vsini. The rotational velocities, Vsini, are taken from Uesugi(1978). The broken line corresponds to α=-3. The dependence of α on Vsini is clearly seen, i.e., α<-3 for the stars with Vsini<270 km/s except one star(HD184279), and α>-3 for the stars with Vsini>270 km/s with one exception(CX Dra). The mean Vsini values are shown in table 1, together with the numbers of stars in parentheses. The Vsini-dependence of α is seen in each spectral type. It is also noticed

Table 1. Mean rotational velocities (*HD184279)

	B0-B1.5	B2-B2.5	B3	B4-B5	B6-B8
α<-3	153(2)	193(6)+?(1)	170(3)+?(1)	270(3)	–
-3<α<0	323(2)	350(2)	285(2)+?(2)	?(1)	–
α>0	228(1)*	–	295(2)	369(1)	323(3)

M. Jaschek and H.-G. Groth (eds.), Be Stars, 41–44.

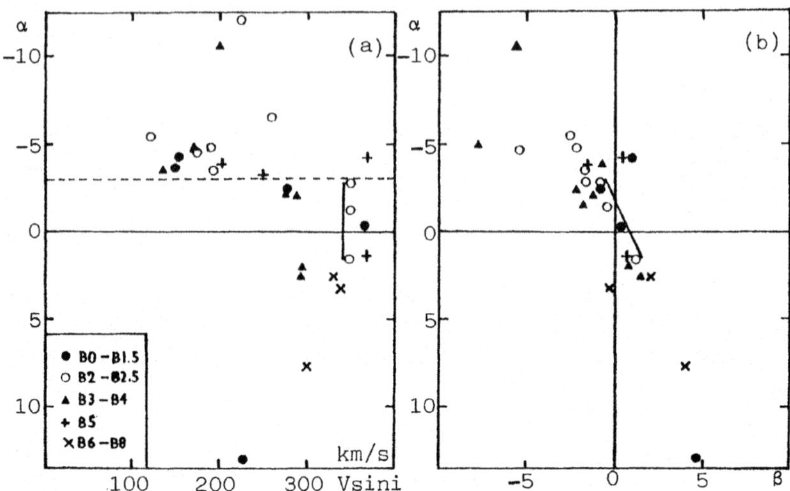

Fig.1 a) α=ΔV/Δ(B-V) vs. Vsini, b) α vs. β=Δ(U-B)/Δ(B-V).

that α<0 for BO-B2 stars except HD184279, α>0 for B6-B8 stars, and B3-B5
stars take all signs for α. This is a tentative suggestion, because our
sample stars contain only several early Be stars with large Vsini values,
and no late Be stars with small Vsini values. Figure 1b shows the correlation between α and β. The signs of α and β are essentially the same.

Let us discuss briefly some possible interpretations of the present
results. Variations in the B and V bands have their origin in the inner
dense part of the envelope(Poekert and Marlborough 1978), or even in the
photosphere. The U-mag. is affected also by the outer part of the envelope because of the larger opacity and emissivity in the Balmer continuum.
The existence of the stars with α>0 indicates the opaqueness of the continuum formation region.
1) Variation of density and temperature in the envelope.   Models of
Poekert and Marlborough(1978) give α,β<0 for the density change(Teff=
25000K), which is in accordance with the observational results in early
Be stars. It is to be noted that the brightness increases when i<45°,
and decreases when i>60°, while the emission-line intensity always increases as the density increases in their model. It is interesting to
calculate the analogous models for the late Be stars.
2) Variation in the photospheric level.   The weakening of the Balmer-
line wings and the appearance of the very broad component of the CaII K
line, accompaning the brightness decrease, in the shell phase of Pleione
imply such a possibility. It is interesting to point out that the rapidly
rotating model atmospheres of Collins and Sonneborn(1977) can account the
observed tendencies on both Vsini- and spectral type- dependencies of α.
Figure 2 shows the variations on the color-magnitude diagram for the
change of ω/ω_c (ω_c:critical angular velocity) from 0.9 to 1.0(i=0°,30°,
45°,60°, and 90°). The pole-on case is designated by open circle. In this

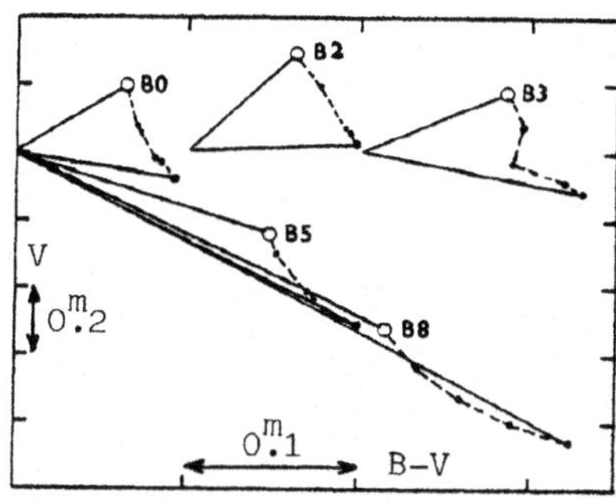

Fig.2 Photometric behaviors of the rapidly rotating stars for the change of ω/ω_c:0.9→1.0(Collins and Sonneborn 1977).

picture, the photometric variation is caused by the gravity darkening due to the angular momentum transport.

3) Precession of the elongated disk.   When mostly extended region faces to the observer, one can expect the considerable diminution in the continuum, especially in the Balmer continuum. Shell stars accompaning cyclic V/R variation are the most probable candidates. In fact, such an effect was detected in the U-mag. of 48 Lib(Delplace and Chambon 1976, Feinstein 1968,1970,1975), and in the far-ultraviolet region of ζ Tau(Beeckmans 1978).

It is obvious that the comparative study of the spectroscopic and photometric materials is absolutely necessary to clarify the mechanism of the long-term photometric variations of the Be stars.   In fact, it now becomes clear that, in some well-observed stars like EW Lac and 88 Her, their photometric behaviors depend on their spectral phases.

REFERENCES

Beeckmans,F.: 1978,Memoire de Thése,Liège,Belgique.
Collins II,G.W., and Sonneborn,G.H.: 1977,Astrophys.J.Suppl.34,pp.41-94.
Delplace,A.M., and Chambon,M.Th.: 1976,IAU Symp.No.70,pp.79-85.
Feinstein,A.: 1968,Zs.Astrophys.68,pp.29-47.
Feinstein,A.: 1970,Publ.Astron.Soc.Pacific,82,pp.132-135.
Feinstein,A.: 1975,Publ.Astron.Soc.Pacific,87,pp.603-605.
Harmanec,P.,Horn,J.,Koubský,P.,Ždárský,F.,Kříž,S.,and Pavlovski,K.: 1980,
    Bull.Astron.Inst.Czech.31,pp.144-159.
Nordh,H.L.,and Olofsson,S.G.: 1977,Astron.Astrophys.56,pp.117-121.
Poekert,R.,and Marlborough,J.M.: 1978,Astrophys.J.Suppl.38,pp.229-252.
Uesugi,A.: 1978,Revised Catalogue of Stellar Rotational Velocities-
    Preliminary Edition-.

NOTE ADDED IN PROOF: During the Symposium, Dr.A.Feinstein provided me additional photometric data of a number of southern Be stars. Preliminary analysis, after adding these data, indicates,a) the number of sample stars, which fulfil the conditions (i) and (ii), increases to about 70 stars, b) almost all stars observed by Dr.A.Feinstein fulfil the third condition, c) present conclusions on Vsini- and spectral type- dependencies of α are not altered for total 50 stars.

DISCUSSION

Snow: I have a comment and a question. The comment is that it is
clear from this morning's talks and, I am sure, from what we will
hear from the rest of the week, that time variability is a very
important feature of Be stars, and must be included in our efforts
to understand these objects. In your study you have deliberately
overlooked short-term variations, and it is worth keeping in mind
that they do occur, and should not be ignored. My question is whether
you looked at any stars that did not show variability, and if so,
what percentage of your sample fell into that category?

Hirata: We must make an effort to reveal the nature of the short-time
variations, as well as that of the long-term variations. There are
many Be stars which do not show large variations in V or B-V. But
it is uncertain whether it reflects no real variation or the lack of
the data.

Endal: Is there any observational evidence for Be stars changing
their rotational velocities (as measured by absorption line width,
for example)?

Hirata: Not obvious. It is difficult to detect the change of absorption widths by the conventional method. But I think that the
appearance of a very broad component of the CaIIK line in Pleione,
prior to the shell appearance, gives such a clue.

Hubert-Delplace: It seems to me that in the case of Be stars in a
shell phase, the relation $\Delta V / \Delta (B-V)$ which separate early Be stars
from late Be stars, breaks down.

Hirata: Sample stars could be added in future. I also showed phase
dependency in the case of 88 Her, and future data can clarify such
details.

THE VATICAN EMISSION STAR SURVEY: REVIEW AND COMMENTS

Pierre Cardon de Lichtbuer
Vatican Observatory

Eight years ago a program was started at the Vatican Observatory to
search for faint emission-line stars close to the galactic plane. Indeed
at that time few early type emission stars were known fainter than V=12.
This survey consists in a double program:the search for emission line
stars with the Vatican Schmidt using the objective prism and an appropriate combination of plate and filter in order to have a narrow passband
around Hα(McCarthy Treanor 1970);and UBVRI photometry at the Mt. Lem-
mon facilities of the Univ. of Arizona observatories. Five papers have
already been published(Coyne et al 1974, 1975, 1978;Wisniewski et al 1976,
1977). Two more will be published later this year. The Vatican Emission
Star Survey(VES) resulted in the discovery of about 1000 new emission-
line stars; for 700 of those the photometry is finished.

We want to present the results of a part of the VES with l^{11} between
55° and 115° and b^{11} from -5° to +5°. The Wackerling catalogue(1971)lists
for the same region about 675 emission-line stars; the VES added to these
about 550 new stars. Table 1 gives the distribution of the Wackerling
catalogue and of the VES stars according to galactic longitude.

Table 1. The distribution of emission-line stars.

l^{11}	60	65	70	75	80	85	90	95	100	105	110
WAC	40	38	58	93	45	60	40	47	84	98	
VES	52	43	75	63	39	86	45	27	30	42	

It is too early to decide whether the survey succeeded in covering the
area in an homogeneous way.

The UBVRI photometry has been finished for 516 stars. Table 2 gives the distribution according to visual magnitude.

Table 2. The distribution of the visual magnitude.

Vis.Mag	8	9	10	11	12	13	14	15	16
VES	3	15	74	189	161	57	12	5	

M. Jaschek and H.-G. Groth (eds.), Be Stars, 45–48.
Copyright © 1982 by the IAU.

Figure 1 shows a two color diagram for 500 stars of the VES. The main
sequence and an average reddening-line are drawn. The distribution of the
stars in this diagram has been divided in four somewhat arbitrary groups.
The most numerous group(I) contains all the early type stars on which
the regular calibration methods(Heintze 1973) can be applied. About 380
stars belong to this group. Stars of group (II) show very large absorption
when calibration and reddining correction are straightforwardly applied;
many of these stars are late type stars, like dMe. The other two groups
contain stars of which the colors are too red(III) or too blue(IV) to fit the
calibration. The Vatican Observatory is now planning an extensive study
of the stars belonging to the last three groups. With the Johnson(1966)
calibration of the multicolor photometry one can produce an approximate
spectral classification of the VES-stars. Table 3 gives the distribution
for 380 stars of group (I).

Table 3. The distribution according to spectral type.

Spectral type	O8	O9.5	B0	B1	B2	B3	B5	B7	B8	B9	A0	A1
VES stars	2	2	19	26	29	39	33	52	60	42	31	38

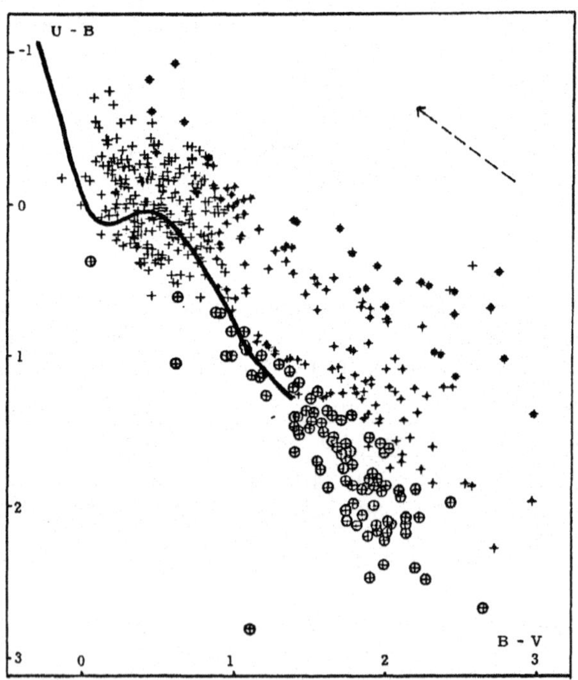

Fig. 1. Two colour diagram of VES stars. Group I :+;
Group II:+ ; Group III:⊕; Group IV :+ .

Using the Schmidt-Kaler(1964) absolute magnitudes for emission stars, and using a constant R=3.1 value, one can produce a distance estimate for these stars. The results for O to B3 stars are shown on figure 2.

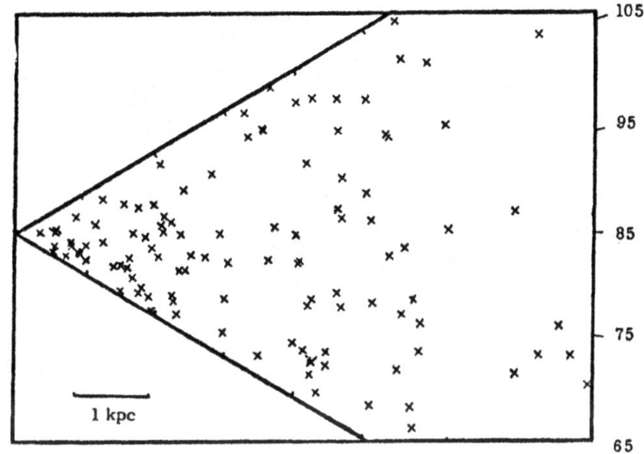

Fig. 2.   Spatial distribution of O to B3 stars.

Comparing these results with earlier studies of the galactic spiral structure(Schmidt-Kaler 1964, Humphreys 1979) one can reach two conclusions. The VES is definitely going deeper; 153 stars are farther than 2kpc, of which 74 belong to the O-B3 group. But one should not forget the imprecision inherent in the method and calibration. Accepting these distance estimates, one finds the same spiral structure as previously known of course for these longitudes one is looking along the arms of the galaxy. Later portions of the VES are looking across the arms of the galaxy; this will give more information about the methodology of the VES.

REFERENCES

G. V. Coyne, T. A. Lee, E. De Graeve, 1974, Vat. Obs. Publ. Vol 1, 5.
G. V. Coyne, W. Wisniewski, C. Corbally, 1975, Vat. Obs. Publ. Vol1, 6.
G. V. Coyne, W. Wisniewski, L. B. Otten, 1978, Vat. Obs. Publ. Vol 1, 12.
H. L. Johnson, 1966, in Annual Review of Astronomy and Astrophysics, vol 4, p 193.
J. R. W. Heintze, 1973, in IAU Symposium 54, Reidel Publ. , p 231.
R. M. Humphreys, 1979, in IAU Symposium 84, Reidel Publ., p 93.
M. F. McCarthy, P. J. Treanor, 1970, The Observatory 90, p108.
L. R. Wackerling, 1970, Mem. R. Astr. Soc. 73, p 153.
W. Wisniewski, G. V. Coyne, 1976, Vat. Obs. Publ. Vol 1, 8.
W. Wisniewski, G. V. Coyne, 1977, Vat. Obs. Publ. Vol 1, 11.

DISCUSSION

<u>Bolton</u>: Magnitude limited surveys tend to have supergiants over-represented relative to main-sequence stars. Have you attempted to estimate the contamination of your Be star sample due to this effect?

<u>Cardon</u>: No estimate has been done yet. We hope to use the RI colours to separate the classes.

LIGHT VARIATIONS IN SEVERAL BROAD-LINED B STARS

M. Jerzykiewicz
Wrocław University Observatory, Wrocław, Poland
C. Sterken
Vrije Universiteit Brussel, Brussels, Belgium

In our photoelectric searches for the β Cephei variables (JS77, JS79, and J82) several early B stars with nebulous lines were included. A summary of their photometric behaviour is presented in Table 1, columns four to six. Two of the stars, HR 2963 and HR 3476, are

Table 1. The photometric behaviour of a sample of Bn stars.

HD	HR or Name	MK	Range	Time Scale	Type	Ref.
24190	–	B2 Vn	$0^{m}\!.015$ $\underline{b}$	days?	var?	J82
28446	1 Cam	BO IIIn	0.010 $\underline{y}$	hours?	var?	J82
36646	1863	B4 Vn	0.015 $\underline{y}$	unknown	var	J82
45871	2364	B4 Vnp	0.055 $\underline{b}$	~10 hours	var	JS77
61878	2963	B5 Vn	0.010 $\underline{b}$	–	cst	JS77
67341	3179	B3 Von	~0.02 $\underline{u}$	unknown	var	JS79
67536	3186	B2.5 Vn	0.040 $\underline{b}$	~10 hours	var	this paper
68324	3213	B1 IVn	0.015 u	$P=0^{d}\!.108$	β Cep	JS79
72014	–	B3 Vnnek	0.050 $\underline{b}$	days	var	JS77
74455	3462	B1.5 Vn	0.045 $\underline{b}$	$P=0^{d}\!.563$	Ell	JS77
74753	3476	BO IIIn	0.010 $\underline{b}$	–	cst	JS77
85860	–	B9	0.080 $\underline{b}$	~10 hours	var	JS77
201819	8105	BO.5 IVn	0.020 $\underline{y}$	hours?	var	J82
219634	8854	BO Vn	~0.10 $\underline{b}$	$P=2^{d}\!.3913$	EA	J82

apparently constant in light, HR 3213 is a short-period β Cephei star, HR 3462 shows ellipsoidal variations with a period of $0^{d}\!.563$, which implies an orbital period twice that long (cf. also Shobbrook 1981), and HR 8854 is an eclipsing binary with $P = 2^{d}\!.3913$.

Of the remaining nine cases, in which the type of variability is somewhat less clear, four seem to be related to the topic of this symposium. These are HR 2364, HR 3186, HD 72014, and HD 85860. All were discovered by JS77 from observations carried out in November and

M. Jaschek and H.-G. Groth (eds.), Be Stars, 49–52.
Copyright © 1982 by the IAU.

December 1975 at the European Southern Observatory with the four-channel uvby spectrograph-photometer, attached to the Danish National 50-cm reflecting telescope. About 20 observations on four or five nights were obtained of each star. HD 72014, the only star on our list with the emission line MK classification, showed a night-to-night light variation typical of Be stars. For HR 2364, HR 3186, and HD 85860 we found somewhat irregular variations with the time scale of the order of ten hours. In fact, HD 85860 is included in Table 1 not on the basis of its MK class, which is not available, but because it varied in a manner similar to HR 2364 and HR 3186.

The above-mentioned 1975 observations indicated that in the light variation of HR 3186 a strong component with the period of either $0^{d}.264$ or $0^{d}.359$ may be present. This made the star a candidate for membership in the β Cephei group, and therefore it was re-observed by JS79 along with a few other β Cephei suspects. The observations were carried out on seven nights in November and December 1977 at the same observatory and with the same equipment as in 1975. Neither of the two periods was confirmed. In fact, no periodicity at all was present in the light variation. This is clearly seen from Fig.1, where the b

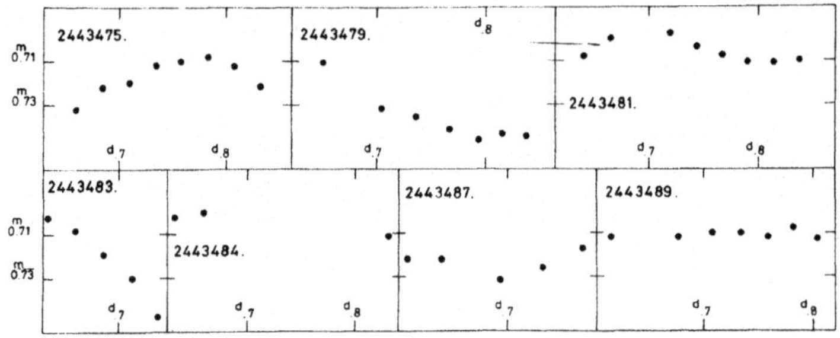

Fig.1. The b light-curves of HR 3186 on seven nights in 1977.

magnitudes of HR 3186, referred to the same comparison stars as in 1975, are plotted as a function of heliocentric Julian date.

The uvby light-curves of HR 3186 on JD 2443479 and JD 2443483, the two nights in 1977 on which the star showed the largest variation, are displayed in Fig.2. On both nights the light range was independent of wavelength to within a few $0^{m}.001$.

On JD 2443475, concurrently with the photometry, a series of eight coudé spectrograms was taken of HR 3186 with the ESO 152-cm telescope at the dispersion of 12.3 Å/mm. As have already been mentioned by JS79, on these spectrograms the Hβ line appeared with a central absorption feature flanked by blue and red emission components. The other lines, λλ 3819, 4026, 4143, and 4471 of He I, as well as the higher members of the Balmer series, were broad and shallow. The

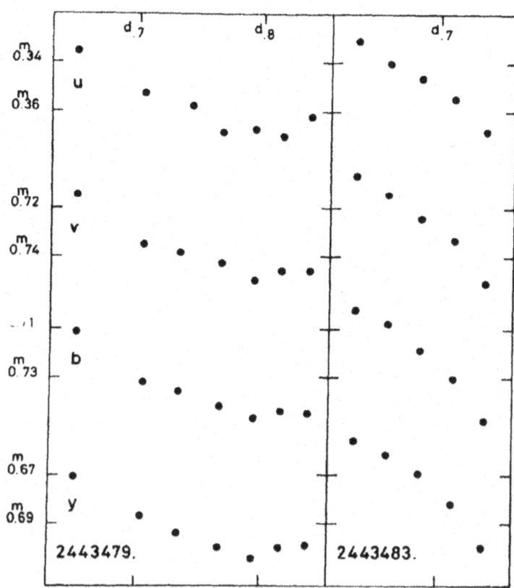

Fig.2.   The uvby light variation of HR 3186 on two nights in
1977.

radial velocity showed considerable amount of scatter around a mean
value of 35.4 + 3.1 km/s, due mainly to errors of measurement. The
interstellar Ca II K line had a mean velocity equal to 15.4 + 1.0 km/s.

These results show that variability of HR 3186 may be due to photospheric and/or shell activity. Fast rotation and large-scale surface
motions may also play a role. It is possible that HR 2364 and HD 85860
are variables of the same kind. In fact, such stars may be quite common.
Other examples are HD 34626 (Percy 1974), HR 9070 (Percy 1979), HR 3593
(Burki et al. 1980), and λ Crucis (Shobbrook 1981). Whether some of
these objects are related to the β Cephei stars, as suggested by Percy
(1979) and by Burki et al. (1980), is not known.

REFERENCES

Burki, G., Burnet, M., and Rufener, F. : 1980, Comm. 27 IAU Inf.Bull.
    var. Stars, No.1820.
Jerzykiewicz, M. : 1982, in preparation (J82).
Jerzykiewicz, M., and Sterken, C. : 1977, Acta astr., 27, 365 (JS77).
Jerzykiewicz, M., and Sterken, C. : 1979, Proc. IAU Colloquium No.46,
    eds. F.M. Bateson, J. Smak, and I.H. Urch, Univ. of Waikato, Hamilton,
    New Zealand, p. 474 (JS79).
Percy, J.R. : 1974, Comm. 27 IAU Inf. Bull. var. Stars, No. 894.
Percy, J.R. : 1979, Comm. 27 IAU Inf. Bull. var. Stars, No. 1530.
Shobbrook, R.R. : 1981, Mon. Not. R. astr. Soc., in press.

DISCUSSION

<u>Bolton</u>: John Percy has been observing a number of northern hemisphere
Be stars. He has observed these in a single colour (Strömgren b)
several times per night for up to 10 nights at a time. He has
characterized the variations to me as follows. γ Cas and 59 Cyg:
variations $< 0^{\rm m}\!.01$ on time scales of hours up to 10 days measured on two
different runs. HR2142: short period variation with amplitude $> 0^{\rm m}\!.03$
almost certain. 28 Cygni: the amplitude of the variation has been
steadily increasing, $P \approx 0^{\rm d}\!.7$ and the amplitude is now $0^{\rm m}\!.06$. EW Lac:
noticable variations with range $0^{\rm m}\!.05$ and $P \approx 0^{\rm d}\!.7$. HR9070: very variable
with well-defined $P \approx 0^{\rm d}\!.3$, probably a β Cephei star.

<u>Jerzykiewicz</u>: $P = 0^{\rm d}\!.7$ is certainly too large for a β Cephei star and $0^{\rm d}\!.3$
would be the longest period known for this type of variability. Of
course, in order to prove that HR9070 is a β Cephei star one would
have to find out whether its radial velocity also varies with the $0^{\rm d}\!.3$
period.

<u>Sareyan</u>: Apart from its spectral type, which is quite different, HR3186
seems to show a behaviour similar to V986 Oph. Especially the impossi-
bility to define a stable period. Can you comment this.

<u>Jerzykiewicz</u>: The MK type of V986 Oph is B0 II-IIIn and the period is
probably $0^{\rm d}\!.2907$. However, it is not certain, that the star is a β Cephei
variable, as has been pointed out by Pike and Lloyd (Comm. 27 IAU Inf.
Bull. Var. stars, No. 1716). Still, the U amplitude in the case of
V986 Oph is considerably greater than the B and V amplitudes, while
HR3186 shows no wavelength dependence of amplitude.

<u>Slettebak</u>: Are rotational velocities of your stars known, so that
a check could be made as to whether the time scales of the variations
you find are consistent with rotational modulation of a bright or dark
area on the star?

<u>Jerzykiewicz</u>: Some are. The problem is that the periods are not well
defined, so that it is difficult to answer your question. However,
rotational modulation alone, won't account for the light variation
of these stars.

CORRELATIONS BETWEEN BCD PARAMETERS OF THE CONTINUOUS SPECTRUM AND THE BALMER DECREMENT OF Be STARS.

L. DIVAN[1] , J. ZOREC[2] , D. BRIOT[3] .
[1]Institut d'Astrophysique, Paris , France ; [2]Laboratoire d'Astrophysique, Collège de France ; [3]Observatoire de Paris France.

I . INTRODUCTION

One of the greatest difficulties in interpreting the continuous spectrum of Be stars is to separate the effects of interstellar reddening from the effects due to the presence of the envelope. This difficulty has been avoided in the two types of correlations considered here. In the first one, parameters not affected by interstellar reddening are used (the Balmer jump and the Balmer decrement). In the second one, the parameters used can be affected by the interstellar extinction but comparisons are made only between values which correspond to the same (but variable) Be star, at different epochs, with different amounts of emission.

II CORRELATION BETWEEN THE BALMER JUMP AND THE BALMER DECREMENT.

It has been shown (Divan, 1979) that Be/shell stars sometimes show two Balmer discontinuities : a longwavelength one, D_o, as in ordinary B stars of luminosity class V, IV or III, and a shortwavelength one, d, occuring very near the theoretical Balmer limit, as in supergiant stars. This second discontinuity can be either in emission or in absorption. For a given star, it can vary and even completely disappear, but these changes never affect the longwavelength discontinuity which remains constant. The longwavelength discontinuity D_o is due to the underlying star ; the shortwavelength discontinuity, d, corresponds to a low density plasma and is due to the variable envelope. The resulting discontinuity D is equal to D_o + d, and varies like d. Figure 1 shows the correlation between d (in dex) and the Balmer decrement for 12 different stars. Difficulties due to the variability of the stars are avoided here because d and the Balmer decrement (taken from Briot, 1971 and 1981) have been measured almost simultaneously. A small wavelength overlap between the two discontinuities D_o and d renders the measurement of d somewhat difficult, except when D_o is known accurately from observations done at a phase without, or almost without emission. This was the case for all the stars in Fig. 1 except υ Cyg, φ Per and HD 194335. For these three stars, d is a lower limit and can be in error by a few times 0.01dex. For 9 of the 10 stars having the second Balmer jump in emission (d$\langle$0), the correlation seems quite good, the smallest Balmer decrements corresponding to the

M. Jaschek and H.-G. Groth (eds.), Be Stars, 53–56.

d[-D-D₀]
-0.10
-0.05
0.00
0.05
0.10
59 Cyg
X Per
γ Cas
11 Cam
υ Cyg
φ Per
194335
48 Per
ψ Per
1H Cam
ζ Tau
217050
Be stars
Shell stars
I(Hα)/I(Hβ)
1.0   2.0   3.0   4.0   5.0   6.0

Fig. 1.

X Per
V
6.00
6.40
V
1   DEC. 52
2   NOV. 57
3   OCT. 65
4   FEB. 77
5   NOV. 77
Φrb
1.80
1.60
Φrb
Φuv
1.40
1.20
1.00
Φuv
D₀
-0.100   0   +0.100   D

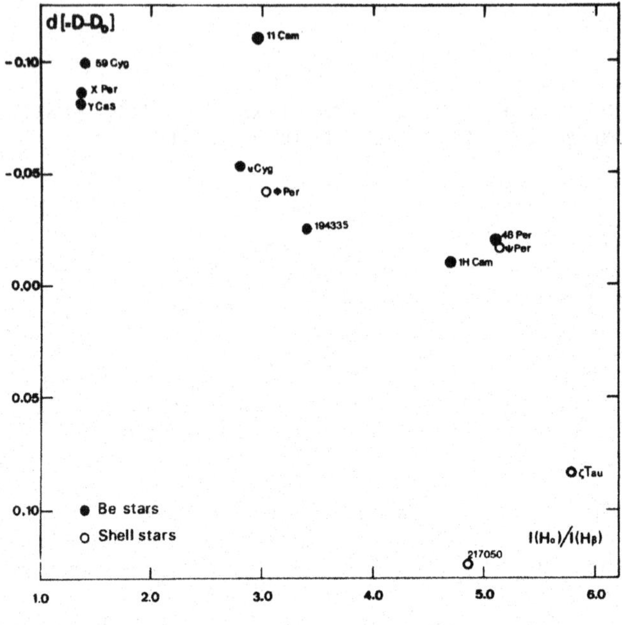

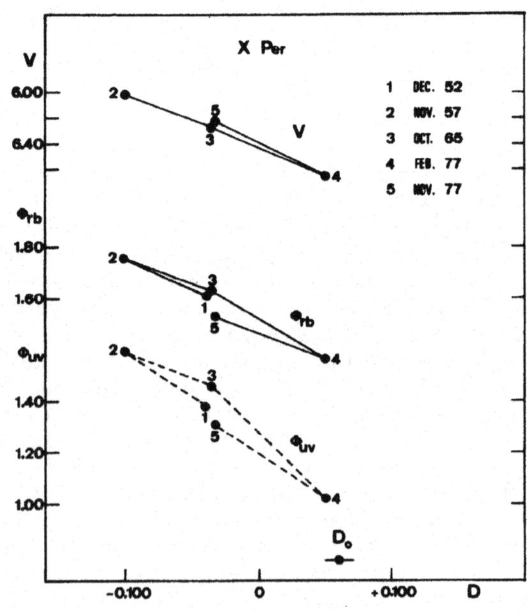

Fig. 2.

strongest emission in the Balmer continuum ; even stars showing shell phases like φ Per or ψ Per do not deviate. The exception is 11Cam which also behaves differently in the other correlations (see below). This correlation between d and the Balmer decrement is definitely better than the correlations between d and the equivalent width of H_α. Only two stars (ζ Tau and HD217050) have the second Balmer discontinuity in absorption and nothing can be said in this case. However it seems improbable that a correlation should exist between this absorption and the Balmer decrement corresponding to emission lines.

III CORRELATIONS BETWEEN BCD PARAMETERS FOR A VARIABLE Be STAR.

Figure 2 shows the variation of
1- the V magnitude (measured at λ 5500Å on the BCD spectra).
2- the red-blue gradient $\emptyset_{rb}$ corresponding to the wavelength range 6200-4000Å ($\emptyset$ increases when the color temperature decreases).
3- the ultraviolet gradient $\emptyset_{uv}$ corresponding to the wavelength range 3700-3150Å
as a function of D (equal to d + the constant value D_0), for five different states of the spectrum of X Per, observed in Dec. 1952, Nov. 1957, Oct. 1965, Feb.1977 and Nov. 1977. The value of D_0 is indicated, with the error bar due to the possible existence of a small emission in the Balmer continuum, not detected in the almost "B normal" phase of Feb. 1977.

The same correlations, with the same slopes, have been obtained for 59 Cyg (6 different phases) and 1HCam (3 different phases). The star Y Cas also seems to behave in the same manner. But for 11 Cam (3 different phases) the slopes are quite different. The sample of stars is too small to allow any conclusion. However the behaviour of X Per seems to be the rule for many of the Be stars for which emission in the Balmer continuum can be present ; they become brighter and redder when the emission increases. This intrinsic reddening explains the B-V colors of Be stars which often seem too red for their MK spectral types and their suspected interstellar reddening. Inversely, the U-B colors of the same stars are too blue ; this is due to the influence of the second Balmer jump in emission on the U magnitude, and Fig.2 shows clearly that the color of the star continuum is redder on both sides of the Balmer jump.

Finally it must be noted that many Be stars have no second Balmer jump, neither in emission nor in absorption, and in this case their colors are normal and correspond to the MK spectral type.

REFERENCES :
Briot,D.: 1971, Astron. Astrophys. 11,57
Briot,D.: 1981, Astron. Astrophys. in press
Divan,L.: 1979, IAU Colloquium 47 : Spectral classification of the
    Future ; Mc Carthy, Philip, Coyne(eds), Vatican Observatory,
        p.247

DISCUSSION

<u>Viotti</u>: It is important to note that if the Balmer discontinuity and
the continuum gradients of X Per were close to that of a reddened
B0 star also there was no excess.

<u>Divan</u>: I am very glad to learn that. It shows that when X Per shows
only one Balmer discontinuity (in absorption) as in Feb. 1977, its
energy distribution is that of a normal B star, from the IR to 3100 A
at least.

<u>Harmanec</u>: Is it not possible, that the different dependence of the V
magnitude on the Balmer discontinuity for 11 Cam in comparison to
X Per and 59 Cyg is a consequence of an aspect effect -- the observed
v sin i is small for 11 Cam and large (and similar) for X Per and
59 Cyg.

<u>Divan</u>: Yes, it is quite possible.

<u>Andrillat</u>: Parmis les étoiles O, il existe une classe spectrale
d'étoiles (Oe) qui s'apparentent aux étoiles Be. Avez-vous étudié
la discontinuité de Balmer dans les étoiles et l'avez-vous comparée
à celles des étoiles Be?

<u>Divan</u>: Very few O stars are really Oe stars in the same sense that
B stars are Be stars. I observed only one of them, HD60848. It has a
short wavelength Balmer discontinuity in emission. But as we observed
it only once we do not know whether it varies.

<u>INTRINSIC REDDENING OF Be STARS AND ITS RELATION WITH H_α</u>
<u>EMISSION INTENSITIES</u>

L. Divan[1], V. Doazan[2], J. Zorec[3].
Institut d'Astrophysique[1]; Observatoire de Paris[2]; Laboratoire
d'Astrophysique Théorique, Collège de France[3].

ABSTRACT :
     The intrinsic reddening and its relation with H_α emission has been de-
termined for 32 Be/Shell stars observed simultaneously at high disper-
sion at H_α and between $\lambda\lambda 3000$-and $6200\,\text{Å}$ with the Chalonge spectrograph.
A trend toward higher reddenings when $H\alpha$ emission is stronger is observed,
together with a large scatter which indicates that for a given emission
at $H\alpha$, intrinsic reddening can be quite different for different stars.

I . <u>INTRODUCTION</u>
     The continuous spectrum of Be stars have long been known to show
a different energy distribution, in the visual, relative to normal B
stars of same spectral type. The intrinsic nature of the observed reddening has been demonstrated by the large variations in colour temperature of γ Cas in 1936 (Chalonge and Safir, 1936). The statistical properties of the energy distribution of Be stars (intrinsic reddening in
the visible and near UV, and peculiar Balmer jump), clearly established
by Barbier and Chalonge (1941) have been largely confirmed by subsequent
studies (see for ex., Mendoza 1958, Feinstein 1968, Schild 1978, Divan
1979). Individual stars have shown during the course of their variations,
that an increase in Balmer line emission is accompanied by an increase
in reddening and in luminosity in the visual. Such a behavior can be
accounted for by b-f and f-f emission produced in the outer atmosphere
of the star. We have raised the question to see whether there exists, for
a sample of Be/shell stars, some relation between the intensity of $H\alpha$
emission and the intrinsic reddening of the Paschen continuum i.e between the regions where continuous and line spectrum are formed in the
outer atmosphere.

II. <u>THE OBSERVATIONS</u>
     We have carried out observations of 32 Be/shell stars at the Haute-
Provence Observatory simultaneously in the region of $H\alpha$ (dispersion 12.2
and 20 Å/mm) and between $\lambda\lambda 3000$ and $6200\,\text{Å}$ with the Chalonge spectrograph at a dispersion of 220 Å/mm at $H\gamma$. The $H\alpha$ equivalent widths were
converted to intensities using the fluxes of Kurucz's models (1979) at
the effective temperatures of the stars, normalized to the type B2V.

57

M. Jaschek and H.-G. Groth (eds.), Be Stars, 57–59.
Copyright © 1982 by the IAU.

We have used the calibration of spectral type vs. T_{eff} given by Underhill et al.(1979). To deredden the fluxes, the values of interstellar reddening were taken from Beeckmans (1978) using the interstellar bump at $\lambda 2200$ Å. The minimum error estimated for the derived E(B-V) is $\simeq 0.04$. We have determined, after dereddening of the flux, the (intrinsic) gradient of the star Φ°rb, in the range $\lambda\lambda 4000 - 6200$ Å and its BCD spectral type. The intrinsic reddening of the Be star is then $\Delta\phi^{\circ}_{rb} = \phi^{\circ}_{rb} - \phi^{ST}_{rb}$, where ϕ^{ST}_{rb} is the gradient of a normal star having the same spectral type.

Figure 1. shows the behavior of $\Delta\phi^{\circ}_{rb}$ as a function of Hα emission intensity for the sample of Be/Shell stars considered. A large scatter is observed in the diagram. Part of the scatter can be attributed to the uncertainty of the interstellar extinction correction, indicated by the error bar. The uncertainty in the equivalent widths of Hα comes principally from saturation effects of the photographic plate when emission is strong, ie. for values of $WH_{\alpha} \gtrsim 50$ Å. This uncertainty can be quite large, up to rougthly 20 % for the strongest lines ; for $WH_{\alpha} < 50$ Å the uncertainty is less than 10 %.

There exists a trend toward large reddening for strong Hα emission, but the fact that a given value of the Hα emission can correspond to quite different intrinsic reddenings seems to be well established. No systematic effect with vsini could be found.

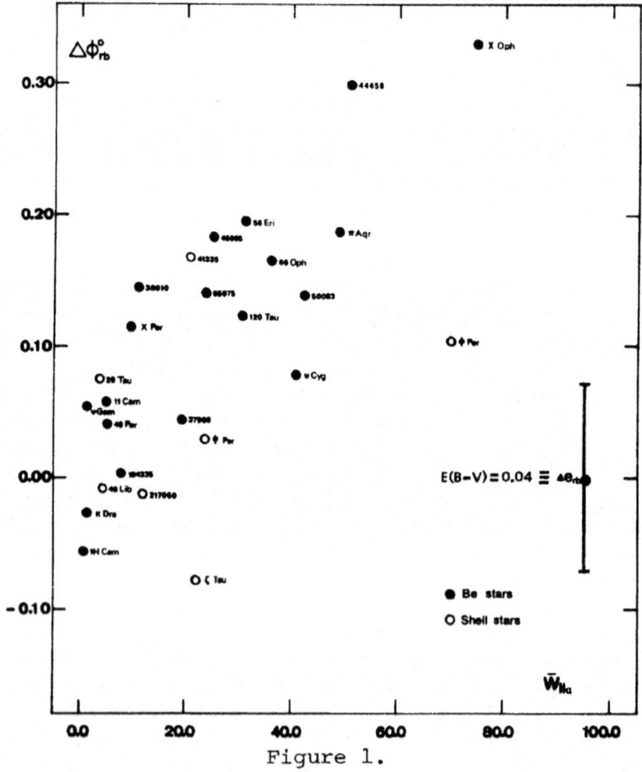

Figure 1.

Shell spectra deserve particular attention : they show a different degree of reddening or even a bluening, and a different behavior of their Balmer jump characteristics (for details see Divan, 1979). Moreover, for ζ Tau, Φ Per and HD 41335, shell characteristics are not seen in the Hα line, but only in the Hβ line (for HD41335) and higher members (ζ Tau and Φ Per). This will be discussed in a separate paper.

REFERENCES
Barbier, D., Chalonge, D.: 1941, Ann. Astrophys. 4,30
Beeckmans, F.: 1978, Thesis, University of Liège, Belgium
Chalonge, D., Safir, H.: 1936, C.R. Acad. Sc. Paris, 203, p1329
Divan, L. : 1979, IAU Colloq. N°47, Ed. G.V. Coyne, M.F. McCarty, A.G.D.
     Philip, p247
Feinstein, A.: 1968, Z. Astrophys. 68, 29
Kurucz, R.L.: 1979, Astrophys.J.Suppl. 40,1
Mendoza, V.E.E. : 1958, Astrophys.J. 128, 207
Schild, R.E.: 1978, Astrophys.J.Suppl. 37, 77
Underhill, A.B., Divan, L., Prévot-Burnichon, M.L., Doazan, V.: 1979;
     M.N.R.A.S. 189, 601

DISCUSSION

Peters: The "shell" lines seen in HR2142 (HD41335) and ϕ Per are "gas stream" features which are quite variable with phase. Strong visible shell features in HR2142 persists for less than 5% of the binary period of 80.86 days.

Zorec: With the term "shell" we distinguish these objects that had some "shell absorption features" at the time we observed them. I do not know whether these stars may or may not be considered like other objects (binaries or not) that have "shell" characteristics during longer time scales.

BCD SPECTROPHOTOMETRY OF THE Be-SHELL STAR 88 Her

L. Divan[1], J. Zorec[2]
Institut d'Astrophysique de Paris, France[1]; Laboratoire d'Astrophysique Théorique, Collège de France.[2]

I. INTRODUCTION.

The spectrophotometric behavior of the Be/shell stars seems to be different when their second Balmer discontinuity is in emission from that when it is in absorption. We have analysed here a particular case in which the second Balmer jump is in absorption. In this note we compare the results of a BCD analysis concerning the variations of the Be-Shell star 88 Her (HD162732), to the results of the UBV photometry.

This star is a spectroscopic binary, which has displayed a long-term light variation, that did not correspond to an eclipsing pattern (Harmanec et al., 1978, 1980 ; Nakagiri and Hirata, 1979 ;Magalashvili and Kumsishvili, 1980). The light variations were as large as O.15 in V and B band and about O.30 in the U magnitude, simulating a change in the spectral type (photometrically determined) from B6 to B8 and back (Harmanec et al., 1978). These variations are shown in Fig.1.

It is interesting to note the apparent change in the B-V colors, the star being redder when its photometric spectral type is later.

The light curves (Harmance et al., 1978 ;Magalashvili and Kumsishvili, 1980) show that the photometric spectral types are different for two states of the star corresponding to the same V magnitude (before and after the light minimum). This apparently paradoxal result is discussed below.

II . RESULTS AND DISCUSSION

The Be-shell star 88 Her has been observed in the BCD system in July 1977 and July 1980 (cycles 275 and 288 ; before and after the light minimum) at two dates corresponding approximately to the same V magnitude, but showing some important differences in the energy distribution.

The BCD parameters (Divan, 1979 ; Divan et al., 1982) obtained at both observing epochs are listed in table 1. We first note that the spectral type of the underlying star, given by D_0 and λ_1 (Chalonge and Divan, 1973), is exactly the same in the two cases, B6 IV. However, between the two observations of 88 Her, a great change occurs in the Balmer region. In july 1977 only one Balmer discontinuity is visible, that due to the underlying star. In july 1980 (see Fig.2) a second Balmer discontinuity

M. Jaschek and H.-G. Groth (eds.), Be Stars, 61–63.
Copyright © 1982 by the IAU.

Table 1

	D	d	D_0	$\lambda_1 - 3700$ Å	$\emptyset_{rb}$	$\emptyset_{uv}$
July 1977	0.29	0.00	0.29	48	0.90	1.03
July 1980	0.40	0.09	0.31	50	0.90:	0.93 :

d, due to the enveloppe has appeared.

The second discontinuity ($d \sim 0.1$ dex) is situated in the domain of the U filter (as can be seen in Fig.2) and should produce an increase of 2.5d = 0.25 in the U magnitude of 88 Her. For july 1977, UBV photometric gives U =−0.45. In July 1980, U should then be equal to −0.45 + 0.25 = −0.20, a value which corresponds to an apparent photometric spectral type around B9 (if the small possible interstellar reddening is ignored), later than the real spectral type B6 by several subclasses. Thus, for 88 Her and for all Be/shell stars which have the second Balmer discontinuity in absorption, the apparent photometric spectral type is too late. Inversely, for Be/Shell stars which have the second Balmer discontinuity in emission, the apparent photometric spectral type is too early, when the parameters λ_1 and D_0 define unambiguously the non-variable spectral type of the photospheric part of these objects.

The B-V color variation of 88 Her is not very well established (see Fig.1, in which error bars are indicated). A reddening in B-V between cycles 275 and 280 seems to be indicated, but this would be in contradiction with the constancy of the gradient $\emptyset_{rb}$ (Table 1.).

In Figure 1. we have plotted the reddening vector corresponding to the color excess deduced from the absorption bump at λ 2200 Å (Beeckmans, 1978). If 88 Her was dereddened with this vector in the U-B, B-V plane, the points corresponding to the different states of 88 Her would be on the left side of the main sequence line. The reddening may have been overestimated, and more observations are necessary to know the real intrinsic positions of 88 Her in the (U-B,B-V) plane.

In the BCD system, 88 Her shows approximately constant values of $\emptyset_{rb}$ and $\emptyset_{uv}$, with a slight indication of a bluer color when the second discontinuity exists. The actual values of $\emptyset_{rb}$ and $\emptyset_{uv}$ for July 1980 are provisionnal and shall be improved, but quite similar results are obtained for another star, Pleione, which has an increasingly strong second Balmer discontinuity in absorption, the colors becoming only very slightly bluer, in contrast with the important reddening which accompanies the apparition of a second Balmer jump in emission. In modeling objects like 88 Her, it will be necessary (and probably difficult) to imagine emission-absorption compensating mechanisms that conserve an almost constant slope of the energy distribution around the Balmer discontinuity when its magnitude is strongly affected.

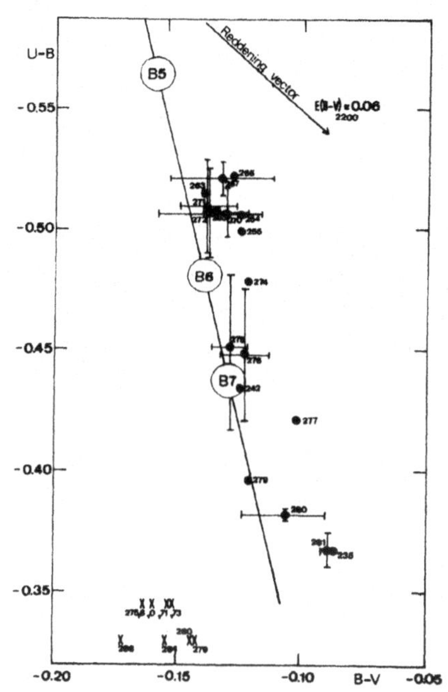

Fig. 1.

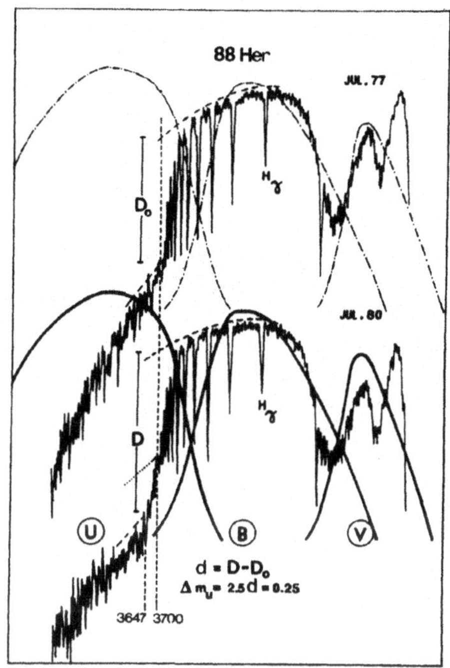

Fig. 2.

REFERENCES

Beeckmans, F.: 1978, Thesis, University of Liège, Belgium

Chalonge, D., Divan, L.:1973, Astron. Astrophys. 23, 69

Divan, L.: 1979 IAU Colloq. 47, Ed. G.V. Coyne, M.F. McCarthy, A.G.D.
     Philip

Divan, L., Zorec, J., Briot, D.: 1982, IAU Symposium N°98, Be Stars, 53.

Harmanec, P. Horn, J., Koubshy, P., Kriz, S., Zdarsky, F., Papousek, J.,
     Doazan, V., Bourdonneau, B., Baldinelli, L., Ghedini, S., Pavlovski,
     K.:1978, Bull. Astron. Inst. Czech. Vol 29,278

Harmanec, P. Horn, J., Koubsky, F., Zdarsky, F., Kriz, S., Pavlovsky, K.:
     1980, Bull. Astron. Inst. Czech. Vol 31,144

Magalashvili, N.L., Kumsishvili, J.I.: 1980, IAU IBVS N° 1868

Nakagiri, M., Hirata, R.: 1979, IAU IBVS N° 1565

CORRELATION BETWEEN SPECTRUM CHARACTERISTICS AND
PHOTOMETRIC BEHAVIOUR OF Be STARS

G.A. Ponomareva
Sternberg Astronomical Institute - Moscow

Based upon a sample of 140 BO-AO emission-line stars contained in "A
Photographic Atlas of Be stars" statistical relationships between variable stars and Be stars are examined. It is found that 44 % of all
the stars of the sample show some kind of photometric variations. The
stars with narrow metallic absorption lines display relatively more
intense photometric activity. Proceeding from this conclusion, the photometric variability of two stars HD 50138 and HD 193182 can be predicted.

It is known that photometric changes and rapid variations of the
structure of the emission lines indicate instabilities in the envelopes
of Be stars. Nevertheless, not all of them display the same activity.

The purpose of the present note is to determine what kind of spectrum
characteristics are more closely associated with the photometric variability of these stars. To answer this question we have examined the
photometric behaviour of the sample of 140 stars contained in "A Photographic Atlas of Be stars" (A.M. Hubert-Delplace, H. Hubert, 1979)
using all information on variability published up to 1981.

Based upon a discussion of spectrum characteristics of the stars of
this sample M. Jaschek et al (1980) proposed for the first time to sub-
divide the Be star group into five physically significant subgroups.
The distribution of the stars over the subgroups is given in Table 1.
Under the group number are summarized the main spectral features. The
number of variable stars among the stars in each subgroup and the
fraction of variable stars expressed in percentages are also given in
the second and third lines of table 1. The total fraction of variable
stars is 44 %. This quantity is close to the 53 % obtained by L. Ferrer
and C. Jaschek (1971) from a comparison of individual brightness estimates presented for 130 bright Be stars in the catalogue of Blanco et
al (1968).

M. Jaschek and H.-G. Groth (eds.), Be Stars, 65–67.

Table 1

Subgroups	I	II	III	IV	V	Total
Stars	α,β...+Fe IIem	α,β+Sh H	α,β+Sh H+met	α,β..	B→Be	
Be	44	15	7	37	37	140
Be var.	24	6	6	8	18	62
Be var/Be	54 %	40 %	86 %	21 %	51 %	44 %
Be des.var	17	4	3	4	12	40
Be des.var/Bevar	67 %	66 %	50 %	50 %	66 %	64 %

According to Table 1, the maximum photometric activity is displayed by
the stars of the subgroup III, (86 %) the least one by the stars of the
subgroup IV (19 %). The sample of our information about variability is
not homogeneous ; however, the following considerations allow us to
conclude on the absence of bias, i.e. any preferential study of the
stars of a certain group. In fact, an estimate of the fraction of designated variables, i.e. of the stars whose variability has been confirmed by special observations, in each subgroup is pratically the same
(fifth line of table 1).

The stars of subgroup V changed from B to Be or vice versa during the
period of study, and some of them even underwent two changes (Be→B→Be
or B→Be→B). During their Be phase they can belong to subgroups II,
III or IV. In Table 2 the distribution over subgroups of these objects
is presented, as well as a similar distribution of the variable stars
of the same group.

Table 2

Subgroups					
Stars	II	III	IV	Abs.	Total
Be	23	2	5	7	37
Be var.	13	2	2	1	18
Be var/Be	56 %	100 %	40 %	14 %	51 %

It is obvious that the photometric activity of the stars having variable
emission is higher than that of the permanent emission members of same
subgroups. Also it is interesting that among the stars that had no emission during 20 years the relative occurrence of variable stars is 14 %,
being close to that of the subgroup IV (stars with emission).

The examination of the two tables permits to estimate the probability
of a manifestation of the light variability of Be stars which possess
certain spectral features. For instance, from Table 1 one can see that
the variability is most probable for the members of the subgroup III
having narrow metallic absorption lines in their spectrum. Although the
statistical significance of the quoted estimate of the variable stars
fraction for this group is not very reliable owing to the small number
of objects in the group, additional data for the stars from the subgroup I and IV indirectly confirm our conclusion. Namely, in Table 2,
according to the spectrum features in the Be phase, two of the 37 stars
are attributed to the subgroup III, and both of these stars are variable, although the fraction of these in the group as a whole is 51 %.
Among 44 members of the subgroup I, four also have shown metallic shell
lines, three of them are variable, the occurrence of variables in the
group being 54 %. Proceeding from this, one can expect manifestations
of light variability for the two stars with narrow metallic absorption
lines found in the group of non-variables, HD 50138 (subgroup I) and
HD 193182 (subgroup III). At present, for the star HD 50138 there are
spectral observations pointing to an instability of its envelope.

REFERENCES.

Blanco, V.M., Demers, S.; Douglass, G.G., Fitzgerald, M.P. 1968, Publ.
     U.S. Naval Obs., 2-d series, 21
Ferrer, L., Jaschek, C. 1971, Publ. Astron. Soc. Pacific 83,346
Hubert-Delplace, A.M., Hubert, H., 1979 "Un Atlas des Etoiles Be"
     Observatoire de Paris-Meudon and Observatoire de Haute-Provence
Jaschek, M., Hubert-Delplace, A.M., Hubert, H., Jaschek, C. 1980,
     Astron. and Astrophys. Suppl. Ser. 42,103

OPTICAL VARIATIONS OF THE Be STAR HDE 245770/A 0535+26

A. Guarnieri, C. Bartolini, A. Piccioni
Osservatorio Astronomico dell'Università di Bologno

A. Giangrande, F. Giovannelli
Istituto di Astrofisica Spaziale, CNR, Frascati

ABSTRACT

Results of photoelectric observations of HDE 245770 spreaded
over six years are presented. No variations greater than some hundreths
of magnitudes are evident in contrast with the ancient light history,
which is more complex. A sensible variation of the colour indices is
present on a time scale of tens of years. The temporal analysis of these
data shows some evidence of orbital periodicity.

INTRODUCTION

The 9th magnitude early type star HDE 245770 was proposed as the
optical counterpart of the recurrent transient X-ray pulsar A 0535+26
by several authors (e.g. : Liller, 1975 ; Murdin, 1975 ; Soderblom,
1975 ; Giangrande et al. 1976). The identification became virtually
certain after the observation, made by Bartolini et al. (1978) and
independently by Rössiger (1978a, 1978b), of an optical enhancement during
the December 1977 X-ray flare-up (Chartres and Li, 1977 ; Rakhamimov
et al., 1980). The X-ray source, discovered by the ARIEL V satellite,
was found to be modulated at 104 s (Rosemberg et al., 1975). The observed assimmetry of the UV resonance lines (Giovannelli et al., 1980) and
the IR excess of HDE 245770 (Persi et al., 1979) confirm the presence
of an expanding envelope surrounding the star. The main characteristics
of HDE 245770 are reported in the papers by Giangrande et al. (1980)
and Giovannelli et al. (1980) and in the references therein.

Under the commently admitted hypothesis of the binarity of this
system, it is crucial to know its orbital elements for a better understanding of the underlying physical processes.

In this paper we present the results of photoelectric measurements
of HDE 245770 spread over six years after the discovery of the X-ray
source A 0535+26, in order to complete the long term history of this
system and to find evidence of orbital motion.

M. Jaschek and H.-G. Groth (eds.), Be Stars, 69–74.

OBSERVATIONS AND DISCUSSIONS

Fifty eight nights of photoelectric observations were carried out
at the Loiano 60 cm telescope of Bologna Observatory during the years
1976-1981. The blue annual mean values are reported in Fig. 1 along
with the photoelectric observations of Rössiger (1976), Lenouvel and
Flogere (1957) and Hiltner (1956). For the sake of completness, the
historical photographic light curve by Stier and Liller (1976) is also
reported. In Table 1 the photoelectric results of these last years are
summarized.

Table 1. Summary of photoelectric data

Year	V	B-V	U-B	Authors
1953	9.39	+0.45	-0.54	Hiltner (1956)
1956	9.38	+0.46	-0.53	Lenouvel and Flogère (1957)
1970	8.86	+0.53	-0.66	Rössiger and Wenzel (1974)
1973	8.93	+0.57	-0.55	Rössiger (1976)
1974	8.94	+0.57	-0.61	" "
1975	8.92	+0.54	-0.57	" "
1976	8.91	+0.57		This paper
1977	8.91	+0.56		" "
1978	8.91	+0.54		" "
1979	8.94	+0.53		" "
1980	8.89	+0.56		" "
1981	8.91	+0.54		" "

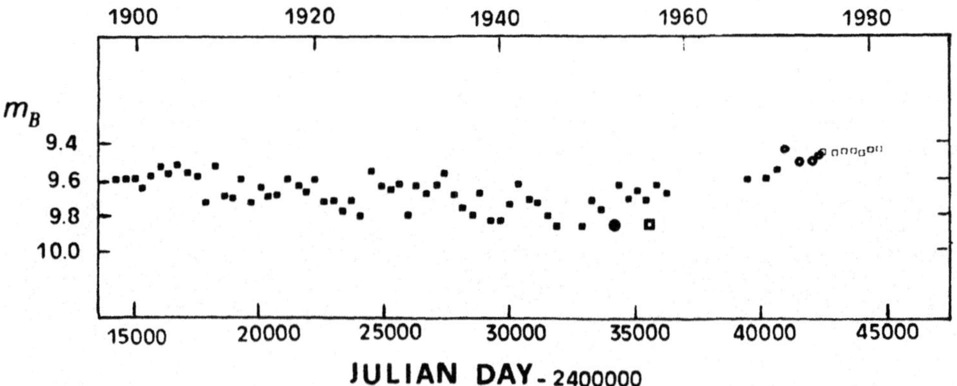

Black squares : Stier and Liller - Black circles : Hiltner - Empty
squares : Lenouvel and Flogere - Double circles : Rössiger - Small
squares : This work.

Fig. 1. Long term photometric history of HDE 245770.

From Fig. 1 some features can be outlined :

1. the star is at present at its highest luminosity level ;
2. no long term variations greater than some hundreths of magnitude are evident in the last years, in contrast with the ancient light history, which appears more complex ;
3. a light-minimum can be located around the beginning of the fifties. After that time the source brightened (regularly ?) by about 0.4 mag in twenty years ; a less marked secular fading is observed from 1900 to 1950. This behaviour on a time scale of decades is not atypical for a Be star and can be probably connected with circumstellar matter or shell ;
4. from the photoelectric data quoted in the literature and ours, a diverging trend of colour indices is also evident : the star had B-V = +0.45 mag and U-B= -0.54 mag in the fifties (Hiltner, 1956 ; Lenouvel and Flogère, 1957) and now it has mean values B-V=+0.56 mag and U-B≈0.60 mag (Rössiger, 1976 ;this paper).

Fig. 2 shows our V and B-V photoelectric light-curves (dayly mean values) and Rössiger's (1976, 1978). From all our data we deduce the mean magnitudes : V=8.903 $\pm$ 0.002 and B = 9.461 $\pm$ 0.003 (mean errors of the mean). The standard deviation of the dayly individual means reported in Fig. 2 is 0.018 mag and 0.025 mag in V and B colours, respectively. Since dayly individual means are much better defined, these figures are representative of an actual scatter in the luminosity of the star. Therefore, we emphasize the substantially quiet trend of the optical light-curve in this time interval, which is characterized by eight observed X-ray flares, marked with arrows in the plot. The optical enhancement observed simultaneously with the December 77 X-ray flare-up can be considered from this point of view only as an episode without consequences on a long time scale light-curve.

The binary nature of this system is generally admitted (see e.g. Rappaport et al., 1978) and many authors (e.g. Rappaport et al., 1976 ; Hutchings et al., 1978 ; Li et al., 1979) have made serious efforts to search for a direct evidence of it. From period variations of the X-ray 104 s pulse, Rappaport et al. (1976) suggested a permitted range 17 d P 77 d for the orbital period and the more probable range 39 and 77 d. From spectral lines anlysis Hutchings et al. (1978) suggested some possible periods, the more convincing ones, after their thorough discussion, being 28, 48 and 94 days.

We have made such a search on our photometric data in the range 17-200 d. Indications for about 32, 63 and 77 days periodicities were found, with some better evidence for 32 d. Clearly this deserves more investigations and more time-extened observations because the lenght of such periods, because of the shallowness of the variations (0.03 mag)

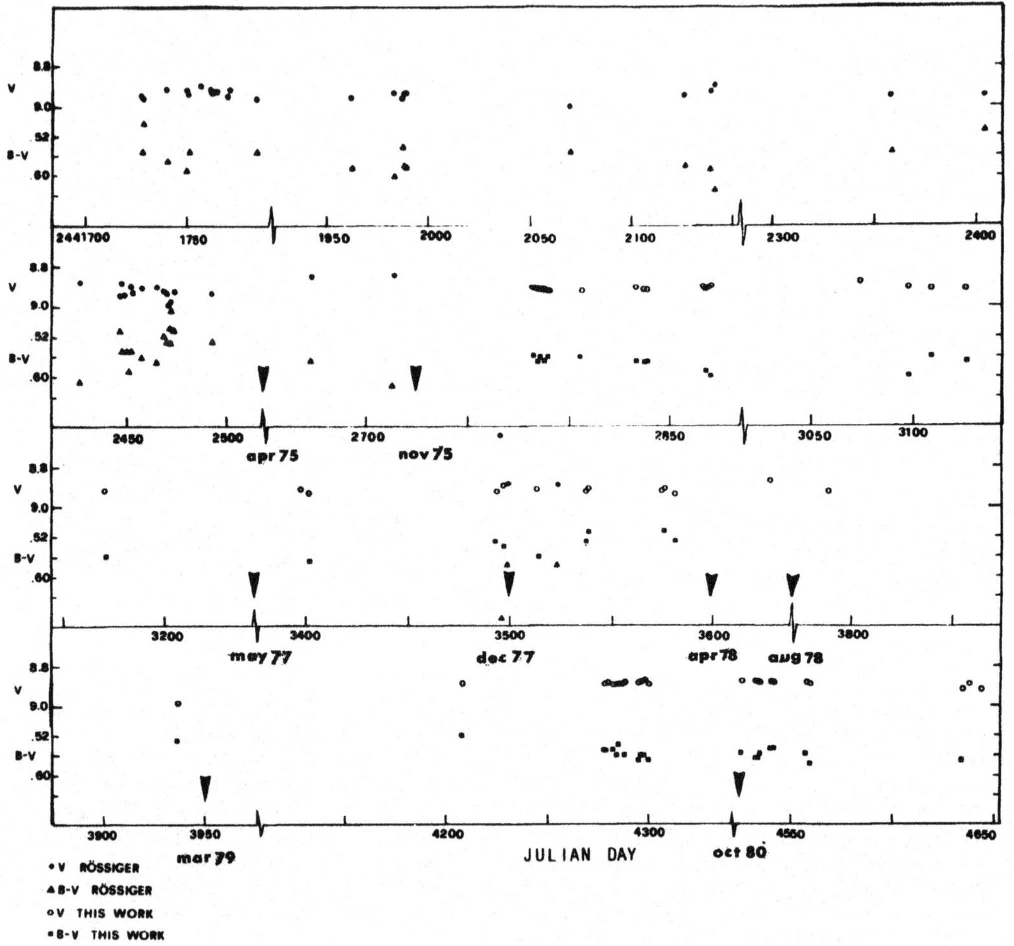

Fig. 2.   Short term photoelectric history of HDE 245770.

and because of possible complications connected with Be-type like
phenomena.

Ellipsoidal type light variations of the primary component of the
binary system (Avni and Bachall, 1975 ; Hutchings, 1978) could be invoked to explain such a periodic variation, if confirmed.

REFERENCES

Avni Y. and Bachall J.N. : 1975, Ap. J. 197, 675
Bartolini C., Guarnieri A., Piccioni A., Giangrande A. and Giovannelli
    F. : 1978, IAU Circ. N° 3167
Chartres M. and Li F. : 1977, IAU Circ. N° 3154
Giangrande A., Giovannelli F., Bartolini C., Guarnieri A. and Piccioni
    A. : 1977, IAU Circ. N° 3129
Giangrande A., Giovannelli F., Bartolini C., Guarnieri A. and Piccioni
    A. : 1980, Astron. Astrophys. Suppl. 40; 289
Giovannelli F., Ferrari Toniolo, M., Giangrande A., Persi P., Bartolini
    C., Guarnieri A., Piccioni A. and Rucinski S.M. : 1980, ESA SP-157,
    159
Hiltner W.A. : 1956, Ap. J. Suppl. 2,389
Hutchings J.B., Bernard J.E., Crampton D., and Cowley A.P. : 1978, Ap.
    J. 223,530
Hutchings J.B. : 1978, Ap. J. 226, 264
Lenouvel F. and Flogère C. : 1957, Jour. des Obs. 40, 37
Li F., Rappaport S., Clark G.W. and Jernigan J.G. : 1979, Ap. J. 228,
    893
Liller W. : 1975, IAU Circ. N° 2780
Persi P., Ferrai Toniolo M. and Spada G. : 1979, Mass Loss and Evolution
    of O-Type Stars, P.S. Conti and C.W.H. de Loore eds. 139
Murdin P. : 1975, IAU Circ. N° 2784
Rakhamimov S.Y., Estulin V., Vedrenne G. and Niel M. : 1980, Soviet
    Astron. Letters, 6,10
Rappaport S., Joss P.C., Bradt H., Clark G.W. and Jernigan J.G. : 1976
    Ap. J. Letters 208, 109
Rappaport S., Clark G.W., Cominsky L., Joss P.C. and Li F. : 1978, Ap.
    J. Letters 224, 1
Rosemberg F.D., Eyles C.J., Skinner G.K. and Willmore A.P. : 1975,
    Nature, 256, 628
Rössiger S. : 1976, Mitteil. Veränd. Sterne, 7, 105
    "   : 1978, IAU Circ. N° 3184
    "   : 1978, IBVS N° 1393
Rössiger S., and Wenzel W. : 1974, Astr. Nach. 295, 47
Soderblöm D.R. : 1976, IAU Circ. N° 2971
Stier M. and Liller W. : 1976, Ap. J. 206, 257

DISCUSSION

<u>Pakull</u>: How large was the "optical flare" you observed coincident
with the x-ray outburst in 1977?

<u>Guarnieri</u>: Combining our data with Rössiger's the amplitude of the
"optical flare" was about .15 magnitudes. Our own observations were
made during the decay phase of the flare and we observed a fading of
about $^m04$ in four hours. Considering the recently published results
of the satellite Prognos 6, the source reached the maximum about on
January first 1978, so the optical flare of HDE245770 entered the
maximum brightness about 10 days earlier than the x-ray source. The
FWHM of the optical flare was of the order of 2-3 days.

<u>Thomas</u>: Your mass-loss figure is <u>very</u> interesting, lying around the
highest quoted for Be stars. Could you say how you got it?

<u>Guarnieri</u>: The value $M \leqslant 5 \times 10^{-6}$ $M_\odot$/year is a upper limit derived by
IR measurements at 3.6 μm by Persi and colleagues.

<u>Persi</u>: (comment): We recently derived the mass-loss rate of HDE245770
from our observations at 10 μm using a spherically envelope model.
Our value of 3×10^{-6} $M_\odot$ /year is in agreement with the upper limit
reported by Guarnieri.

II. POLARIZATION

POLARIMETRY AND PHYSICS OF Be STAR ENVELOPES
(Review paper)

George V. Coyne, S.J.
University of Arizona and Vatican Observatory

Ian S. McLean
Royal Observatory Edinburgh

A review of the most recent developments in polarization studies
of Be stars is presented. New polarization techniques for high-resolution spectropolarimetry and for near infrared polarimetry are described
and a wide range of new observations are discussed. These include broad-
band, intermediate-band and multichannel observations of the continuum
polarization of Be stars in the wavelenght interval 0.3 - 2.2 microns,
high resolution (0.5 Å) line profile polarimetry of a few stars and
surveys of many stars for the purposes of statistical analyses. The
physical significance of the observational material is discussed in the
light of recent theoretical models. Emphasis is placed on the physical
and geometrical parameters of Be star envelopes which polarimetry helps
to determine.

1. INTRODUCTION

The mere fact that the optical radiation from Be stars is observed
to be intrinsically linearly polarized provides perhaps our most con-
clusive clue that the circumstellar shells of these stars are disk-like
in nature. It is not surprising then that Be stars are relatively well-
studied for polarization although, as we will demonstrate, this is a
tool which still remains under-exploited.

At the time of IAU Symposium 70 on Be stars and Shell stars
(Slettebak, 1976) where the last review of polarization in Be stars was
presented (Coyne, 1976), polarimetry at high spectral resolutions was
just in its infancy. Wide-band polarimetry at optical wavelenghts had
barely provided sufficient observational material for the construction
of preliminary models of the shell structure about Be stars (Capps,
Coyne and Dick, 1973 ; Cassinelli and Haisch, 1974). Since 1975 a number
of new polarimeters have been placed in operation so that not only has
there been a dramatic increase in the data base but spectropolarimetry
has also been extended into the infrared while, across the H-beta
spectral feature, resolutions as good as 0.5 Å have been achieved.

M. Jaschek and H.-G. Groth (eds.), Be Stars, 77–93.

Simultaneous with these developments in polarimetric observations, there
have occurred a number of new approaches to the modelling of shells
about Be stars which attempt to predict and account for intrinsic polarization. These have resulted, for example, in stellar wind models for
Be stars which, in contrast to earlier models, take account of a star
of finite size, of line and continuum emission and absorption in the
circumstellar shell, of the variation of density and temperature with
distance from the star, of scattering of both stellar and circumstellar
radiation and, significantly, of the kinematic properties (rotation and
expansion) of the shell.

Since there is a specific review paper on atmospheric models of Be
Stars to be given by Poeckert elsewhere in this symposium, we will not
deal critically or at lenght with models. Rather, while reviewing the
observational results, we shall emphasize those physically significant
parameters of Be stars which polarimetry helps to determine or even, at
times, exclusively determines.

First we present a brief summary of the instrumental developments
that have affected substantially the accumulation of observational data
on the polarization of Be stars. Then, after a review of observational
material acquired since 1975, we discuss the physical significance of
the new results in terms of general physical principles and some currently proposed models for Be star shells.

2. INSTRUMENTAL DEVELOPMENTS

Since 1975 the principal instrumental developments which have directly contributed to Be star studies have been the diversification and
progress in techniques of high resolution spectropolarimetry at optical
wavelenghts and improvements of IR polarimeters. McLean (1980) has recently reviewed modern polarimetric techniques. Here we present only
the highlights of those techniques which have been specifically applied
to measurements of Be stars.

Already in 1973 at the 5m Hale telescope, Landstreet and Angel had
used a multichannel spectrophotometer (the Oke Scanner) with a Pockels
cell and Glan-air prism before the entrance aperture (Angel and
Landstreet, 1974) to obtain polarization measurements of Gamma Cas at
resolutions of 20-40 A across the entire visible spectrum. Those
results were not published until later (Poeckert and Marlborough, 1978a)
As of 1975 several polarimeters, employing the technique of scanning
across atomic lines by tilting a narrow-band interference filter, had
been used successfully to measure polarization sequentially at discrete
wavelength intervals across the hydrogen emission lines (H-alpha and
H-gamma) ; resolutions of 2 to 12 A were achieved. Such polarimeters
and their earlier results were discussed in the review by Coyne (1976).
Tilt-scanning polarimeters have continued to be extremely useful since
1975 in the study of many emission-line stars and binaries, e.g. Beta

Lyrae (McLean, 1977), a detailed study of H-alpha in Gamma Cas (Poeckert and Marlborough, 1977), and a survey of southern hemisphere Be stars (McLean and Clarke, 1979). Poeckert, Bastien and Landstreet (1979) have employed a set of narrow-band filters to survey a large number of Be stars, whereas Hayes (1978) has used a broad-band technique to search for intrinsic polarization in O-type emission-line stars. All of the above polarimeters use a photomultiplier tube as the detector and, with the exception of the Oke Scanner, all of the measurements are made sequentially.

More recently, attention has focussed on the application to spectro-polarimetry of new multi-element photoelectric detectors which, coupled to spectrographs, permit simultaneous measurements at many wavelengths with high spectral resolution.

The highest spectral resolution polarimetry to date has been obtained by McLean et al. (1979) with a Digicon echelle-spectropolari-meter. Resolutions of 0.5 A across the H-beta line of several Be and shell stars have been obtained. In that instrument the polarization modulator consisted of a superachromatic halfwave plate continuously rotating at 3.5 Hz and followed by a Glan prism. An echelle spectrograph in an off-plane configuration forms a spectrum on the photocathode of a Digicon image tube where photoelectrons emitted by the photocathode are then focussed to form an electron image on a linear array of 106 diodes. Another Digicon (Beaver et al. 1972) has been used with a Pockels cell and a Cassegrain spectrograph at the University of Arizona (Stockman et al. 1979) for the measurement of circular polarization due to the Zeeman effect. Unfortunately, these Digicons had a fairly limited life span and are no longer commercially available. Some, however, are being manufactured under contract to the Space Telescope. Other multichannel devices which have proved successful with polarimeters but which have not yet been applied to Be stars include Reticon self-scanning linear photodiode arrays and image intensifier dissector scanners (IDS). Two IDS systems, both employing Pockels cells, are in use, one at University of Western Ontario (Tomaszewski, Landstreet and Symonds, 1980) and one at Lick Observatory (Miller, Schmidt and Robinson 1980). More recently there has been a move towards intensified Vidicons and image photon counting devices as well as towards solid state sensors such as Charge Injection and Charge Coupled Devices.

New developments have also occurred in infrared polarimetry. For example, beat frequency photo-elastic modulators (Kemp et at. 1977) have been introduced for the 1-10 micron range, although they have so far been little used for Be star studies. Jones (1979) introduced improvements into an existing IR polarimeter by placing the rotating polarization analyzer in front of all the beam splitting and beam directing mirrors and by using a quartz Lyot depolarizer to reduce any spurious effects introduced by the mirror. This instrument has been used to obtain the polarization of Be stars at wavelengths of 1.25 and 2.2. microns.

3. OBSERVATIONAL RESULTS

Over the past six years the principal contributions to the polarimetric data on Be stars have been : (1) further and higher spectral
resolution measurements of polarization across the hydrogen emission
lines ; (2) additions to the meagre data prviously available in the infrared ; (3) more information, for more stars, on polarization variability both in wide bands and at high spectral resolution ; (4) statistical analyses of polarization in comparison with other parameters.

Before reviewing the new observations, it is useful to know that in
almost all cases it is possible, by a careful combination of techniques,
to remove the contaminating effects due to interstellar polarization.
Only then can the intrinsic polarization be modelled in detail. In ge -
neral these techniques consist in : (1) comparison of the magnitude and
directions of the polarizations for other stars at about the same distance and in the same direction as the program star ; (2) variations
in the observed polarization and/or position angle of the plane of polarization with time and wavelength ; (3) variation of the observed polarization across spectral lines when the effect of the line is known. Recent applications of these techniques have been presented by McLean
and Clarke (1979) who surveyed the fifteen brightest southern Be stars,
separated the intrinsic and interstellar components and discussed the
possible difficulties and misuse of the methods of analysis. A numerical method based on technique (2) above has also been applied to 70 Be
stars by Poeckert, Bastien and Landstreet (1979).

The reduced polarization observed in the lower Balmer emission
lines of a few bright Be stars was found to be approximately inversely
related to the emission-line strengths, hence implying that the emission
line flux was essentially unpolarized (cf. Coyne, 1979). Using a two-
channel Pockels cell polarimeter, Poeckert and Marlborough (1976)
obtained additional polarization measures for a number of Be stars in
the H-alpha line and neighbouring continuum and, employing the inverse
relationship mentioned above, they direved intrinsic polarizations
and correlated these with Vsini. Already, however, improvements in the
observations were beginning to reveal evidence that a simple unpolarized
emission model was inadequate. "Filling-in" of the line polarization
was reported for Phi Per by Coyne and Mc Lean (1975) and variations
or oscillations in position angle across the emission lines of Beta
Lyrae were observed by McLean (1977). Poeckert and Marlborough (1977)
made polarimetric scans across the H-alpha line in Gamma Cas using an
interference filter with 1.4 Å full width at half-maximum (FWHM) tuned
to an accuracy of 0.5 Å. Those observations clearly indicated that the
decrease of polarization at the line centre could not be attributed
solely to unpolarized emission. Small variations in the position angle
of the linear polarization across the line were observed which,
because of their antisymmetry with respect to the line centre, could
not be the product of interstellar polarization but must in fact
result from the rotation of a disk-like envelope.

While monitoring the continuum polarization in a band near H-alpha variations in the polarization amounting in some cases to 0.1 percent in a day were observed. Furthermore, changes with time in the position angle of the line polarization were detected and the largest such changes seem to occur when V/R of the emission-line profile is not close to unity.

Using the same techniques, Poeckert and Marlborough (1978a) measured simultaneously the polarization in two intermediate continuum bands located on either side of the Balmer discontinuity. These date were then combined with their previous H-alpha data and with polarimetric scans with Oke Scanner in the discussion of an envelope model. The combined observational data are shown in Fig. 1 where for the position angle (top) and polarization (bottom) the individual symbols represent measures made on different nights from January 1976 to January 1977. The light solid line represents the spectropolarimetry of Landstreet and Angel and the heavy solid line represents a model fit. The polarization changes abruptly across the series limits and there is also a slight position angle change. Actually, at higher spectral resolution the polarization discontinuities are not so abrupt (McLean 1981). Shortwards of the Balmer absorption edge the polarization remains nearly constant with time while that in all other bands changed. There is a sharp increase with time in the linear polarization in all measured bands but especially in the continuum just longwards of the Balmer limit where, in about five months, the polarization increased by 0.3 percent. The authors find that this is correlated with an increase in the equivalent width of the H-alpha line in emission and with an increase in the V/R asymmetry of the line profile.

Employing a two-channel, rotating-waveplate polarimeter with tilt-scanning of narrow-band interference filters, McLean and Clarke (1979) measured the flux and polarization across H-alpha (8.5 Å FWHM) and H-beta (2.3 Å FWHM) and at four continuum wavelengths in most of the Southern hemisphere Be and shell stars brighter than about fifth magnitude. For stars with extremely strong emission such as Alpha Ara and Delta Cen, the line center polarization was not as weak as might be expected from unpolarized emission and the shell star Eta Cen, which was undergoing an outburst, actually exhibited an increase of polarization at H-beta as well as a change of continuum polarization of 0.2 percent in two weeks. These results argue that the emission line flux is itself partly polarized and/or that there is further shell absorption in the line of direct unpolarized starlight. For several stars they found that the intrinsic polarization position angles are variable, indicating a non-axisymmetric distribution of scattering material which may be due to a binary character for these stars.

Recently McLean (1981) has discussed polarization observations of Zeta Tau made with 20-50 Å resolution across the Balmer jump using the single channel spectropolarimeter of the Royal Observatory Edinburgh. These observations, like the data for Gamma Cas, reveal that the

intrinsic polarization is remarkably stable just shortward of the
Balmer jump but that it varies markedly with time in the Paschen continuum. Also, the expected abrupt discontinuity in polarization at the
Balmer limit is considerably "rounded". This is due undoubtedly to the
confluence of the Balmer lines as the series limit is approached.

The highest spectral resolution polarimetry to date was obtained
across the H-beta line for four Be stars by Mc Lean, Coyne, Frecker
and Serkowski (1979) with the Digicon Echelle spectropolarimeter described earlier. A spectral resolution of about 0.5 Å was obtained by
averaging over 1.5 diodes (3 channels) of the 106 element (212 channel)
array. The results are displayed in Fig. 2 where for each star there is
presented from top to bottom, the position angle of the polarization
plane, the percent linear polarization and the normalized intensity
profile across the H-beta line. While there are significant differences
from one star to the next the following common features are present :
(1) the linear polarization decreases as one approaches the line center
from the wings and the decrease is not usually symmetrical for the two
wings ; (2) the linear polarization increases at the line center and
this generally associated with a central absorption core in the line
profile ; (3) secondary minima in the linear polarization occur in the
line wings ; (4) there are complex variations in the position angle
across the lines which in general are neither symmetric nor antisym-
metric with respect to the line center, although there is a tendency
toward antisymmetry associated with certain localised turning points
in the intensity profiles. The physical significance of these these
new observations is discussed in the next section.

As of 1975 linear polarization measurements at wavelengths greater than one micron existed for only two stars, Phi Per (Coyne and
Mc Lean 1975) and Zeta Tau (Capps, Coyne and Dick, 1973). Jones (1979)
has published polarization observations at 1.25 and 2.2 microns for
eight Be stars and has used his data to improve upon previously existing
models of the continuum flux and polarization.

There have three noteworthy observational surveys and statistical
analyses of the polarization in Be stars. Poeckert and Marlborough
(1976) determined the intrinsic linear polarization for 48 Be stars
from H-alpha line-to-continuum measurements and compared these with
Vsini in the context of a specific, numerically calculated model. McLean
and Brown (1978), using the methods discussed by Mc Lean and Clarke
(1979), analysed the existing data on 67 Be stars and derived intrinsic
polarizations. On the basis of general physical considerations they
derived an analytic expression relating intrinsic polarization and the
inclination of the stellar rotation axis which is valid for all optically thin axisymmetric Be star envelopes (cf. Brown and Mc Lean, 1977).

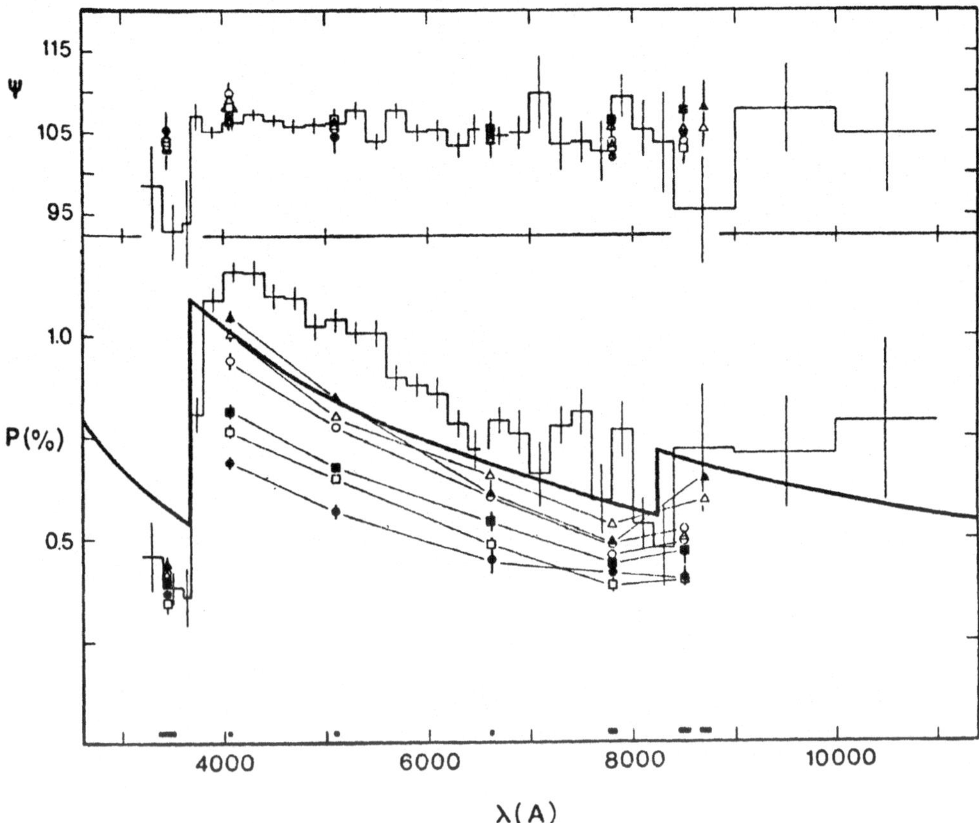

Fig. 1. - Position angle of polarization (top) and percent polarization (bottom) versus wavelength for Gamma Cas. The solid line represents
a model fit by Poeckert and Marlborough (1978a) and the thin line the
observations they present from scans made by Landstreet and Angel with
the Oke scanner. The other individual symbols represent observations
made on different nights between January 1976 and January 1977 by
Poeckert and Marlborough (1978a).

The results of both studies may be summarised as follow : (1) the
intrinsic polarization is never greater than about 2 percent indicating
that extremely oblate envelope do not occur ánd/or direct unpolarized
starlight always makes a strong contribution ; (2) low values of the
intrinsic polarization dominate, indicating that there are more objects
in which the circumstellar envelopes are more nearly spherical shells

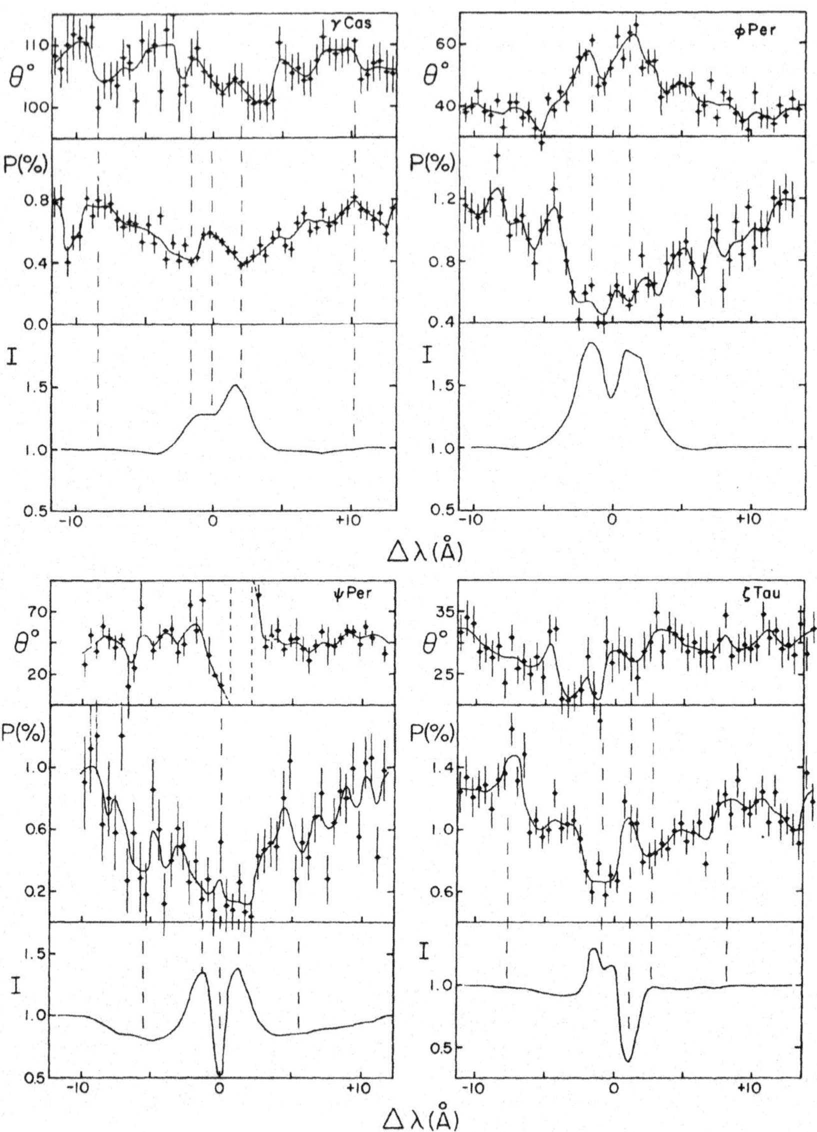

Fig. 2. For each of the four designated stars the position angle of
the polarization, the percent linear polarization and the normalized
intensity profile (top to bottom) across H-beta are presented. The
observations at a spectral resolution of about 0.5 Å were made with a
Digicon Echelle spectropolarimeter (McLean et al., 1979).

(as opposed to highly flattened disks) and/or the electron density is
usually quite low ; (3) there is a definite upper and lower boundary to
the plot of intrinsic polarization versus rotational velocity. There are
no Be stars with low rotational velocities and high polarization, which
probably indicates that a certain minimal rotational velocity is required to produce a flattened shell.

Poeckert, Bastien and Lanstreet (1979) surveyed the polarization
for 70 bright northern hemisphere Be stars using intermediate-band
filters at the Balmer and Paschen edges and in the Paschen continuum.
This survey demonstrated quite clearly that the size of the polarization
discontinuities and the slope of the Paschen continuum polarization
varies considerably from star to star, indicating differences of electron
density and temperature among the various stellar envelopes. A number
of the stars observed showed time variable polarization and, more
importantly, some of them showedan apparent time lag between polarimetric and spectroscopic variations. Again, those stars with large intrinsic polarizations have large rotational velocities and strong emission lines.

Another possible source of correlation is between polarization and
the strength or equivalent width of the Balmer emission lines which, in
a tenuous hydrogen envelope, is proportional to N_e^2. In the denser Be
star envelopes the situation is more complex. Mc Lean (1979), from a
plot of intrinsic polarization versus equivalent width of H-alpha, finds
a tendency for the most highly polarized stars to have large equivalent
widths. Variations of equivalent width and changes in polarization with
time have also been noticed. For instance, in Pi Aqr (North and Olofsson, 1977, Mc Lean, 1979) both have increased, while in Gamma Cas
(Poeckert and Marlborough, 1977 ; Mc Lean, 1979) both have decreased
with time.

Before terminating this observational section it should be remarked
that, although significant advances have been made in the past five
years in the polarimetric observations of Be stars, there remains a
good deal to do. This can be seen by studying the polarimetric data
for 67 stars assembled by Mc Lean and Brown (1978). In that list of
stars only onehalf had been measured in more than one pass-band and
only one-third had repeated measurements.

4. PHYSICAL SIGNIFICANCE OF NEW OBSERVATIONS

It is obvious that any self-consistent model of the shells about
Be stars must account for all observed parameters : the continuum flux
at all wavenlengths and its variability, line profiles and their variability, continuum and line polarization and its variability, etc...
Our limited purpose here, however, is to highlight those observations
discussed above which provide physical insight into the nature of the
shells about Be stars.

Let us start with the now-accepted notion that the polarization
in Be stars is due to scattering of the stellar radiation by free
electrons in an oblate, axisymmetric envelope of ionized hydrogen about
the star. This model, first proposed by Coyne and Kruszewski (1969), has
been elaborated upon by Capps, Coyne and Dyck (1973) and further discussed in an analytic form, by Mc Lean (1979). Most recently it has received a more sophisticated treatment by Jones (1979). All of these models
are ad hoc in the sense that physical and geometrical parameters of the
shell are arbitrarily varied to try to fit the observations. On the
other hand, Brown and Mc Lean (1977) provide a very useful generalized
analytical treatment for the polarization resulting from single electron
scattering in any axisymmetric, optically thin, envelope distribution
about the star considered as a point source. By neglecting absorption
and emission in the shell, they show that the resultant polarization
depends on the shape of the density distribution function, the mean
scattering optical depth and the inclination of the rotation axis to
the line of sight. This same approach was extended by Mc Lean (1979)
to include the physical effects of absorptive opacity and emission. He
pointed out that, although the range of applicability of such analytic
methods is limited, they are useful for gaining an insight into the
physical characteristics of the envelopes. For example, the analysis of
Mc Lean (1979) reveals the same dependence on the inclination of the
rotation axis as that revealed by the stellar wind models of Poeckert
and Marlborough (1976). Also, Daniel (1980) has recently performed
Monte Carlo calculations for multiple electron scattering in Be star
envelopes which, in the limit of single scattering with a small central
source, approach the analytic results. Again, approximate expressions
for the wavenlength dependence of polarization in an optically thin
envelope containing hydrogen atoms undergoing ionizations and recombi-
nations (Mc Lean, 1979) reveal that it is the average electron temperature, electron density and size of the envelope that control the
form of the polarization curve. Suc analytic results also tend to reveal
more explicitly the number of free parameters and hence the limits on
uniqueness of model fits.

Turning to numerical model calculations, consider the model proposed by Jones (1979) which has the following elements : (1) the
source star has a finite size and is spherical ; (2) the star is surrounded by an axisymmetric flat disk of uniform thickness except that
it tapers near the star ; (3) the envelope density varies with distance
from the star ; (4) light from the star is allowed to scatter twice
before exiting from the disk ; (5) absorption and emission processes
take place in the disk and the disk emission is allowed to scatter once
before leaving. Again this is an ad hoc model but it is important because it introduces further sophistication into the processes producing
polarization. Jones (1979) has provided a useful discussion of the dependence of various observed parameters on the physical and geometric
properties of this particular disk model but much of the discussion
will relate to any axisymmetric model.

Jones (1979) discusses the application of his disk model to 7 Be stars with the following general conclusions : (1) the same electrons that polarize by scattering also provide the IR excess by free-free emission ; (2) the electron density must increase near the star in order to explain the drop in polarization beyond 1 micron ; (3) the temperature of the scattering regions of the disk can be determined from the Balmer and Paschen continua polarizations. According to Jones, details of the spectroscopic observations may refer to other parts of the disk.

Several comments on particular stars are of interest. For Gamma Cas the inclination is 45°. This value allows a simultaneous fit to the overall level of polarization and the wavelength dependence. This is in good agreement with the model fit to the H-Alpha line with stellar wind calculations by Poeckert and Marlborough (1977) and is consistent with the position angle structure in the H-Beta line observed at 0.5 Å resolution. Another interesting case is 48 Librae. According to Jones (1979) the disk may actually be a detached equatorial ring whose inner edge begins at 2 stellar radii. This deduction is made on the basis that a cool enough disk temperature to explain the optical polarization does not explain the infrared polarization and excess. If the disk is a ring then the IR polarization is due mainly to scattering of radiation emitted from within the ring. The proposed geometry is similar to the ring models of Hazlehurst (1967) and Huang (1972).

Totally independent constraints from those discussed above can be placed on shell models of Be stars by information derived from the photometric and polarimetric profiles of the Balmer emission lines in these stars. Using a development of the Brown, Mc Lean and Emslie (1978) non- axisymmetric electron scattering models of close binaries with gas streams, Mc Lean (1979) was able to demonstrate that complex patterns of variation in line profile polarizations are to be expected in all envelopes undergoing large-scale mass motions irrespective of the embedded star. This follows because the severe Doppler broadening of envelope lines effectively affords a means of spatially resolving the circumstellar structure. For instance, Mc Lean (1979) has shown analytically that for a rotating star having an arbitrary axial inclination to the line of sight, since one sees deeper into the receding part of the envelope at wavelengths in the blue wing of a spectral line and deeper into the approaching part at wavenlengths in the red wing, one obtains a different spatial integration of the scattered light at different wavelengths over the spectral line and the Stokes parameters of the polarized light will vary accordingly.

In a series of papers Poeckert and Marlborough (1977, 1978a, 1978b, 1979) have developed radiative transfer models involving a stellar wind-electron scattering axisymmetric envelope, including a prescription for the envelope kinematics, and have applied the model to observations principally of Gamma Cas but also Phi Per. In order to appreciate the contraints placed on such a model by the high resolution Digicon/

echelle H-beta observations presented in Fig. 2, it is worthwhile to
review all the ways in which polarization structure may be generated
in a line by processes in the envelope : (1) more or less of the direct
starlight may be obscured causing respectively an increase or decrease
in the polarization ; (2) polarized flux may be absorbed, hence redu-
cing the polarization ; (3) the emission line radiation may be unpolarized or it may be scattered and polarized which will respectively
decrease or increase the overall polarization at that wavelength ;
(4) in a line profile dominated by Doppler broadening from mass motions in the envelope we see different parts of the polarizing envelope at different wavelengths as discussed above. Hence we see different
polarization planes and different contributions of effects (1) to (3)
depending on wavelength. More importantly, note that in an envelope
containing differential rotation about an axis plus radial expansion
plus other distortions, the structure in the position angle of polarization across the line will be complex. With these features in mind, it
is possible to unterstand much of the detailed structure in the H-beta
polarizations shown in Fig. 2 by following the discussion by Poeckert
and Marlborough (1978a) who demonstrate the gradual "build-up", according to the stellar wind models, of the intensity profile and associated polarization parameters as the extent of the envelope is increased.
The discussion centres upon Fig. 3 and Fig. 4. Various profiles of
H-alpha in Fig. 4 have been derived by cutting off the transfer solution at the radial point in the envelope indicated by the number on
each profile which corresponds to the radial distance from the star
centre (in units of the stellar radius) shown in Fig. 3. The extreme
wings of the line in Fig. 4 are formed within $3R_*$ (where R_* is the
stellar radius), whereas the principal line centre assymetries occur
mainly between $6R_*$ and $9R_*$. This can be understood by referring to the
radial-velocity contour map of the equatorial plane shown in Fig. 3.
Referring to the polarization "profiles" of Fig. 4 we see that 70 percent of polarized continuum flux occures within $3R_*$. Minima in the
polarization curve first appear in the wings and then move towards the
line centre with increasing envelope size. These changes can be explained as follows : (1) absorption by those parts of the envelope along
the observer's line of sight to the star will tend to increase the
polarization while absorption at the edges of the envelope will tend
to decrease the polarization. Since the envelope is wedge shaped the
front part of the envelope to the observer absorbs more than the back
part. Thus the polarization minima in the line wings are due to absorption at the edges of the envelope. Since the rotational velocity of
the envelope is decreasing with increasing radius the absorption at the
edges will have a smaller and smaller Doppler shift and the polarization
minima will therefore move towards the line centre. Obviously, other
velocity patterns will produce slightly different effects and the
absorption process is a function of electron temperature and density.
Indeed as pointed out by McLean et al. (1979), even this very detailed
model does not predict all of the structure observed in the H-beta
lines of Gamma Cas and Zeta Tau. This is partly interpreted as indicating that the kinematical effects have been oversimplified. Note

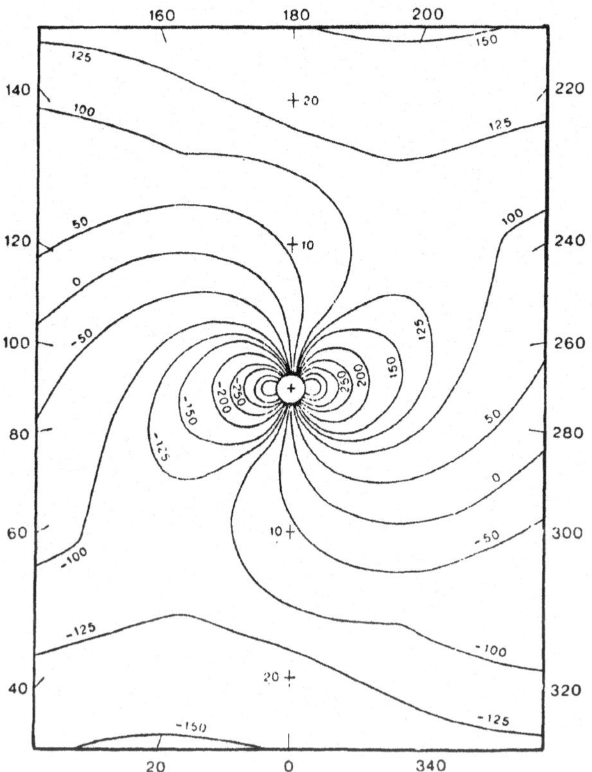

Fig. 3 - Radial-velocity contours in the equatorial plane of a star with an expanding envelope and an equatorial rotational velocity of 569 km/sec (Poeckert and Marlborough, 1978a). The observer is at azimuthal angle zero where the angles are labelled around the outside of the figure. Contours are labelled in km/sec.

from Fig. 4 that the central polarization maximum is still present at $6R_*$ even though there is considerable line emission. It is apparent, therefore, that line absorption is more important than line emission (recall the observations of Eta Cen) in determining the line polarization. In retrospect this is not surprising since the continuum polarization, especially at the Balmer jump, is itself dominated by absorption effects. This means that the higher Balmer lines should also show decreases in polarization, even though very little emission is present. This is indeed the case (Mc Lean, 1981) and is the cause of the "rounding" of the abrupt polarization jump as the Balmer limit is approached.

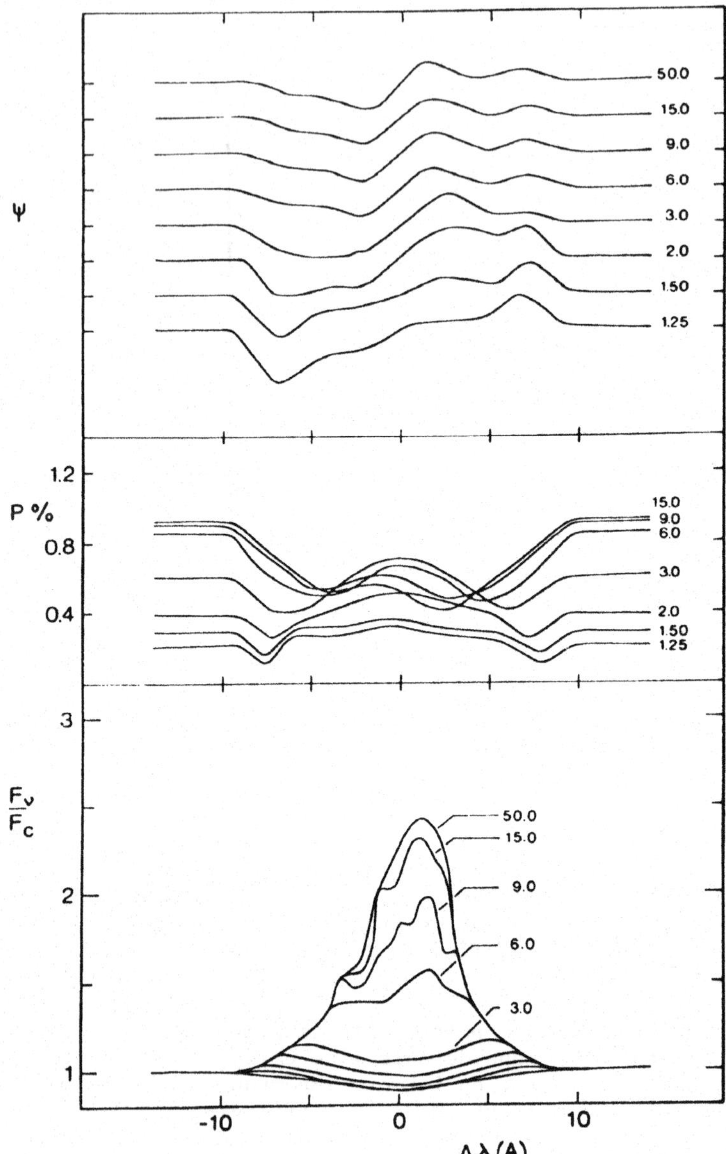

Fig. 4 – Position angle of polarization, percent polarization and line profile (top to bottom) derived by cutting off the transfer solution at the radial point in the envelope indicated by the number on each profile corresponding in Fig. 3 to the distance in stellar radii from the star's center (Poeckert and Marlborough, 1978a).

5. CONCLUSIONS

Polarimetry has proven to be particularly valuable in our attempt to understand the nature of the extended envelopes about Be stars, since these envelopes are flattened due to the high rotational velocity of these stars. In particular the wavelength dependence of the polarization depends upon the electron density, the electron temperature and the size and shape of the envelope. Furthermore, detailed variations of polarization across atomic lines serve as a probe of the dynamics, particularly the rotational and expansion velocities, of the gas in these envelopes. Although significant advances in observations have been made, especially in the area of high-resolution spectropolarimetry, a great deal remain to be done observationally. Especially needed are systematic and detailed observations of more stars, since only four stars, Gamma Cas, Zeta Tau, 48 Lib and Phi Per, have been observed in more than a routine manner. While there may have been significant progress in developing analytical models and impressive advances with stellar wind models, incorporating rotational and expansion velocities of the envelope gas, an explanation of the detailed spectral polarimetry observed across atomic lines will require more sophisticated models.

REFERENCES

Angel, J.P.R. and Landstreet, J.D. : 1974, Ap.J. 191, 457
Beaver, E.A., McIlwain, C.E., Choisser, J.P., Wysoczanski, W. : 1972
    Adv. Electronics and Electron Physics, 33B, 863
Brown, J.C. and McLean, Ian S. : 1977, Astron. Ap. 57, 141
Brown, J.C. and McLean, Ian, S. and Emslie, A.G. : 1978, Astron. Ap.
    68, 415
Capps, R.W., Coyne, G.V. and Dyck, H.M. : 1973, Ap.J., 184, 173
Cassinelli, J.P. and Haisch, B.M. : 1974, Ap.J., 188, 101
Coyne, G.V. : 1976, in IAU Symposium 70, Be and Shell Stars, ed.
    Slettebak, A. (Dordrecht : Reidel) p. 233
Coyne, G.V. and Kruszewski, A. : 1969, Astron. J., 74, 528
Coyne, G.V. and McLean, Ian S. : 1975, Astron. J., 80, 702
Daniel, J.Y. : 1980, Astron. Ap., 86, 198
Hayes, D.P. : 1978, Ap.J. 219, 952
Hazlehurst, J. : 1967, Zs. Ap., 65, 311
Huang, S.S. : 1972, Ap. J., 171, 549
Jones, T.L. : 1979, Ap.J. 228, 787
Kemp, J.C., Rieke, G.H., Lebofsky, M.J. and Coyne, G.V. : 1977, Ap. J.
    215, L107
McLean, Ian S. : 1977, Astron. Ap., 55, 347
McLean, Ian S. : 1979, M.N.R.A.S., 186, 265
McLean, Ian S. : 1980, in New Techniques in Stellar Photometry and
    Polarimetry, Ricercher Astronomiche, Vol. 10, ed. McCarthy, M.F.
    (Specola Vaticana : Città del Vaticano), p. 71

McLean, Ian S. : 1981, M.N.R.A.S. in press
McLean, Ian S. and Brown, J.C. : 1978, Astron. Ap., 69,291
McLean, Ian S. and Clarke, D. : 1976, IAU Symposium 70, Be and Shell
        Stars, ed. Slettebak, A. (Dordrecht : Reidel), p. 261
McLean, Ian S. and Clarke, D. : 1979, M.N.R.A.S., 186, 245
McLean, Ian S., Coyne, G.V., Frecker, Jack E. and Serkowski, K. : 1979,
        Ap. J. 228, 802
Miller, J.S., Robinson, L.B. and Schmidt, G.D. : 1980, P.S.A.P.,
        92, 702
Nordh, H.L. and Olofsson, S.G. : 1977, Astron. Ap., 56, 117
Poeckert, R. : 1975, Ap. J., 196, 777
Poeckert, R., Bastien, P. and Lanstreet, J.D. : 1979, Astron. J., 84,
        812
Poeckert, R. and Marlborough, J.M. : 1976, Ap. J., 206, 182
Poeckert, R. and Marlborough, J.M. : 1977, Ap. J., 218, 220
Poeckert, R. and Marlborough, J.M. : 1978a, Ap. J., 220, 940
Poeckert, R? and Marlborough, J.M. : 1978b, Ap. J. Suppl., 38, 229
Poeckert, R. and Marlborough, J.M. : 1979, Ap. J., 233, 259
Slettebak, A. : 1976, IAU Symposium 70, Be and Shell Stars (Dordrecht:
        Reidel)
Stockman, H.S., Angel, J.R.P. and Hier, R. : 1979, B.A.A.S., 10, 689
Tomaszewski, L., Landstreet, J.D. and Symonds, D. : 1980, in press.

DISCUSSION

<u>Stalio</u>: Did you compute your radiative transfer problem in rotating and expanding atmospheres with Sobolev's approximation? If not, could you give a very brief description of the method?

<u>Poeckert</u>: No, we did not use the Sobolev approximation. We assume that the source function and opacity is a linear function of geometric distance between sample points along a line of sight. The sampling frequency depends on the density and velocity gradients along the line of sight. The phase is evaluated at 2000 frequencies within the line.

<u>Peters</u>: Is it not true that as a class Be-shell stars show the highest degree of polarization?

<u>Coyne</u>: As best I know it is not true that the largest polarizations are associated with the shell phases of Be stars or that only those stars which have shell phases (shell stars) have the largest polarizations.

<u>Hirata</u>: The elliptic ring-(or disk) model for the V/R variables could be checked by the polarimetric observations. Is there any indication which supports or denies such model in the polarimetric observations?

<u>Coyne</u>: I know of no polarimetric observations that would exclude an elliptical ring model. Infact the elliptical ring model can be considered a specification of the more general scattering flattend envelope models being applied to explain the polarization of Be stars.

<u>Sonneborn</u>: I wish to comment on the net linear polarization of the flux from a rotating star. Model stellar atmospheres for rapidly rotating stars show the intrinsic polarization of the photospheric radiation is at most .05% in the visual, but may be as large as 2.5% in the far ultraviolet. I will discuss this further on my talk on Friday.

<u>de Freitas Pacheco</u>: Could you comment on the observational fact you have mentioned that the polarization decreases on the line core and the result of model calculations.

<u>Coyne</u>: The polarization generally decreases to the line center due to the addition of unpolarized emission flux which dilutes the polarization. One cannot, however, exclude, infact in some cases we observe this, that the emission flux is partially polarized. A further effect is that in the presence of a central absorption core imposed on the emission line we detect an increase in polarization due, apparently, to the further absorption of the direct flux.

SIMULTANEOUS SPECTROSCOPIC AND POLARIMETRIC
OBSERVATIONS OF π Aqr*

Klaus Metz
Institut für Astronomie und Astrophysik,
Universität München
8000 München 80, Scheinerstr. 1, FRG

Simultaneous spectroscopic and polarimetric observations
of about 40 bright southern Be stars have been carried out
by Dr. Pöllitsch and myself with the ESO 1.5m coudé spectrograph and with the ESO two channel polarimeter.

From our measurements we can conclude that the continuum
polarization seems to be much more stable than the emission
spectrum: Whereas several stars changed from an emission to
an absorption spectrum and vice versa within a relatively
short time interval, no dramatic variations of the polarization could be observed. however, if we measure by means of
interference filters the polarization shortly before and
after the Balmer jump and determine the difference of these
two measurements (the socalled Balmer jump of polarization),
then this jump is remarkably different from zero only for
those stars which exhibit a pronounced emission. Since the
difference is independent of interstellar polarization, it
can be used to determine the intrinsic polarization of the
star. This has been pointed out already by Poeckert et al.
(1979). The most important result of our work is that the
observed emission line profiles of $H\alpha$ and $H\beta$ of all our
programme stars can be reproduced best by a shell which is
not a disc but rather a sphere extending 5 to 6 photospheric
radii. The shell rotates differentially and probably with
conservation of angular momentum and continuous mass flow.
The density is proportional to $1/r^2$.

Though this model is extremely simple, it reproduces the
observed line profiles better than other more complicated
models. However, it conflicts of course with the usual
interpretation of the observed intrinsic polarization of
Be stars. This contradiction becomes evident, if we consider the results of our measurements of π Aqr in 1977.
This star exhibits the highest polarization of all stars
measured during our campagne.

M. Jaschek and H.-G. Groth (eds.), Be Stars, 95–99.

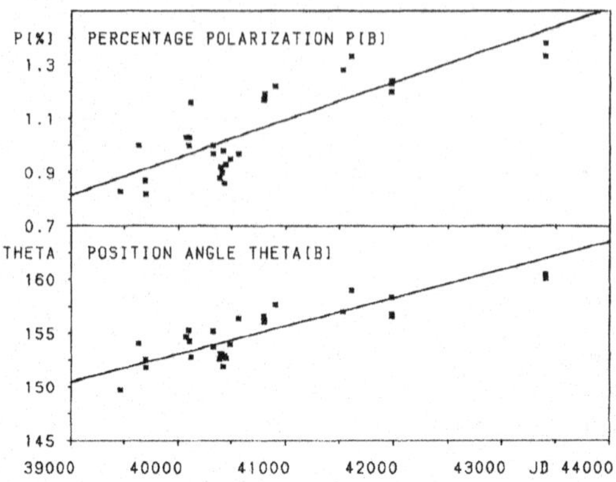

Fig. 1: The increasing polarization of π Aqr in the blue colour observed over a period of 12 years by different authors.

Fig. 1 shows its polarization compiled from measurements of several authors during 12 years. After an elimination of the interstellar component the resulting intrinsic polarization was 1.3% in 1977.

To explain such a strong polarization in the usual way, a disc with a radius at least 3 to 5 times its thickness would be necessary. On the other hand we have the contrary spectroscopic results (see fig. 2).

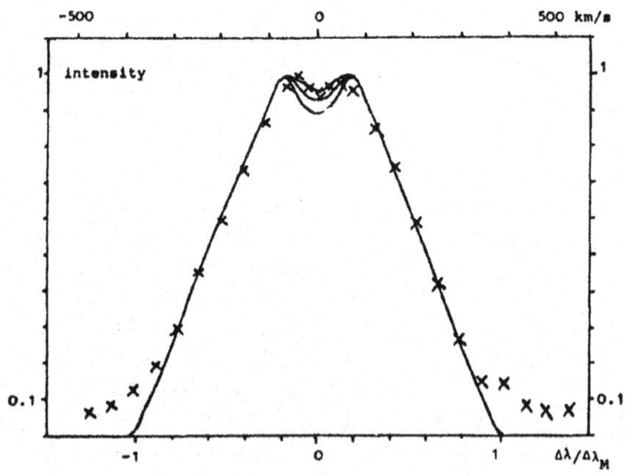

Fig. 2: Measured Hα profile of π Aqr normalized to intensity 1 for maximum emission (1977, Sept. 19, ESO 1.5m coudé, 3.3 A/mm). Solid lines derived from model calculations.

The crosses represent the measured Hα profile. Solid lines are derived from model calculations of shells with a ratio of polar to equatorial diameter between 1 (line above) and 0.6 (line below). The ratio 1 fits the observed profile better than the ratio 0.6. From the line width of Hα we get a projected radial velocity of about 430 km/sec and this is very precisely the same value, which we can derive from the photospheric absorption line He Iλ4472. The consequence is

that the emitting region begins very close to the stellar
photosphere. Further the measured velocity is not too far
from the critical break up velocity of the star and this
indicates that we observe π Aqr nearly edge on.
            A star rotating near the break up velocity
and viewed edge on provides exactly the asymmetry which
is necessary to produce a net polarization even for a
complete spherical shell: Dependent on the angular velocity,
the star will be rotationally distorted. In consequence of
the resulting geometry the outgoing flux is stronger in
polar directions and produces therefore an excess polarization with an angle of vibration perpendicular to the axis
of rotation. This effect will be very essentially enhanced
according to the Von Zeipel theorem which predicts a flux
directly proportional to the local gravity.

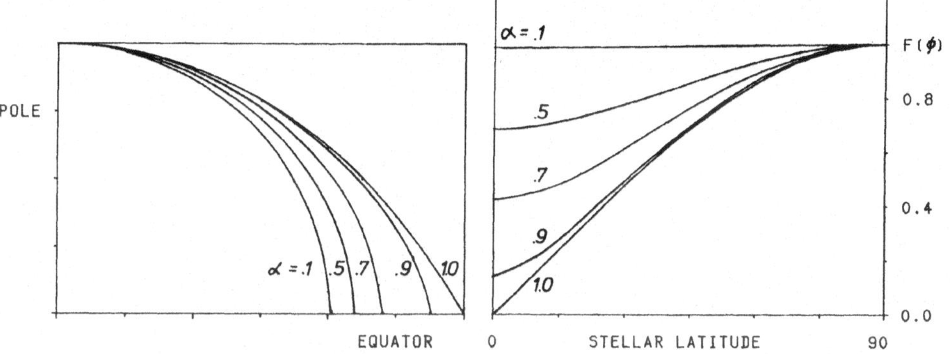

Fig. 3: a) The shape of a star rotating with $\omega/\omega_{cr}= 0.1$,
           0.5, 0.7, 0.9, 1.0.
        b) Theoretical surface flux $F(\phi)$ as a function of
           stellar latitude ϕ for $\omega/\omega_{cr}= 0.1$, 0.5, 0.7, 0.9,
           1.0 (according to Von Zeipel).

        In fig. 3 the shape and local fluxes of stars,
rotating with different velocities, are plotted. As can be
seen the difference of fluxes in polar and equatorial
directions is a steep function of $\alpha = \omega/\omega_{crit}$.

For a quantitative determination of the resulting polarization of the system, the corresponding scattering angle for
each point of the surface of the star and the envelope as
well has to be calculated. Further the different local flux
according to Von Zeipel and the dilution factor have to be
determined. Finally it has to be checked which points of
the shell are hidden by the star. Altogether this procedure
needs several hours computer time even at a fast computer
and therefore double scattering or the wavelength dependent

absorption within the shell has not been included. For the
numerical integration a stellar radius of 5 solar radii,
an electron density of 10^{12} at the inner limit of the shell
and an outer radius of the shell in the order of 5 stellar
radii has been taken.

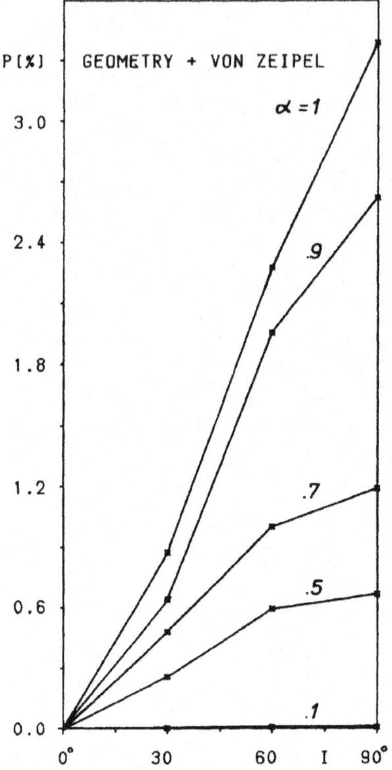

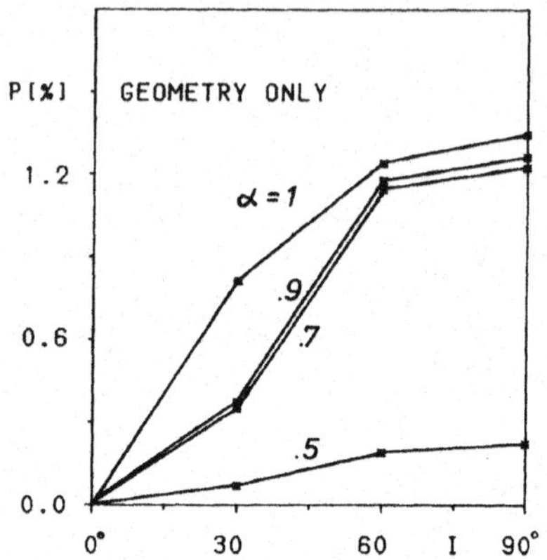

Fig. 4: Polarization by scattering
of light in a spherical shell for
inclination angles i=0°, 30°, 60°,
90° and different rotational velocities. Asymmetry caused a) by
rotation and gravity darkening as
well b) by rotation only.

The result is that a high polarization as observed in π Aqr
can be produced by our model even for angles of inclination
relevantly smaller than 90° and also for rotational velocities quite different from the break up velocity. This
holds especially, if we apply the Von Zeipel theorem.

  * Based on observations obtained at the European
Southern Observatory, La Silla.

References

Poeckert, R., Bastien, P., and Landstreet, J. D. (1979),
     Astron. J. 84, 812
Metz, K., Pöllitsch, G. (1979), ESO Messenger 18, 19.

DISCUSSION

<u>Fehrenbach</u>: Le spectre de π Aqr est variable. En décembre 1980 le contour de H_α comporte 3 absorptions et non pas une centrale.

<u>Metz</u>: In our model the lack of a central absorption only indicates that the dimensions of the shell are greater than 5 stellar radii.

<u>Snow</u>: This object is one of the Be stars whose mass–loss rate I have attempted to deduce from the analysis of the ultraviolet line profiles, to be reported later in the symposium. In this regard it is also unusual, with a high mass–loss rate compared to most of the rest of my sample stars.

<u>Poeckert</u>: What is the resolution of your data for H_α? The $H\alpha$ line in π Aqr is highly variable and one should perhaps be careful in aplying your model to all Be stars.

<u>Metz</u>: The dispersion was 3.3 A/mm. The time resolution was about 90 min. With our model we could reproduce the observed H_α profiles of stars which exhibit fairly symmetric lines.

THE STRONGLY POLARIZED P CYGNI STAR WITH INFRARED EXCESS CPD -52° 9243

J.P. Swings
European Southern Observatory and
Institut d'Astrophysique, Université de Liège

ABSTRACT

The visible spectrum of the point-like source with infrared excess CPD⁻52°9243 is identified; emission and absorption line strengths are given together with various ion velocities of P Cygni profiles; a strong polarization is detected : $P \simeq 5$ %, $\theta \simeq 36°$. From stellar and interstellar features, a spectral type around B8Ia and a distance of about 3.1 kpcs are derived.

A detailed publication is presently in press in Astronomy and Astrophysics.

M. Jaschek and H.-G. Groth (eds.), Be Stars, 101–102.

DISCUSSION

<u>Viotti</u>: The star CPD-52°9243 has a very positive (B-V) colour and I could conclude a very large interstellar extinction. If you correct your IR magnitudes for this extinction, the "dust" excess is depressed and the possibility for free-free emission (or for no IR excess at all) is still open. I just derived $E_{B-V} \approx 1.9$ and $A_k \approx .7$ which is quite large.

<u>Swings</u>: It is much more the colour that matters than the magnitudes themselves.

<u>Coyne</u>: Can the IR excess in CPD-52°9243 be due to free-free emission in hydrogen, rather than dust? Can the disc polarization be due to dust?

<u>Swings</u>: The location of CPD-52°9243 in the H-K/K-L diagram indicates that the majority of the IR excess must be due to dust. The effect of dust is certainly important to produce the high polarization that is measured. The IR colours of CPD-52°9243, however, are not extreme, so that geometry must play a big role as well, I would guess.

<u>Hubert-Delplace</u>: You mentioned that the star CPD-52°9243 could be a B supergiant star with dust and forbidden emission lines -- B8Ia[e] , are there other B supergiants with such properties: dust and forbidden emission lines?

<u>Swings</u>: RX Puppis could have been considered as a similar case, but it seems to be evolving back to a symbiotic star. There are a few supergiants exhibiting similar properties in the Magellanic Clouds. For the galactic objects, the problem is that quite often their spectrum does not reveal the phospheric lines enabling one to know the spectral type and luminosity.

POLARIZATION IN PECULIAR EMISSION-LINE OBJECTS[*]

R. Barbier, European Southern Observatory;

J.P. Swings, European Southern Observatory, and
Institut d'Astrophysique, Université de Liège

ABSTRACT

We report on visual polarimetric measurements of B[e]
and related stars, and on an unsuccessful search for
correlating the observational data with characteristics of
the objects or of their surrounding dust shells.

INTRODUCTION

Be stars in general are known to show intrinsic linear
polarization (see e.g. Coyne, 1976, 1982) which arises from
Thomson scattering in a flattened plasma disk around those
stars ($P_{max} \simeq 1$ % at about 5000 A). One case of a B[e]star,
HD 45677, was studied very carefully by Coyne and Vrba
(1976) : its polarization, which varies on a time scale of
months, and which is always larger at wavelengths longer
than 5000 A, is attributed to scattering from patchy clouds
of dust located in a ring around the star. In the present
study we observe a series of B[e]stars in order to confirm
the presence of the dust surrounding them that was detected
via near infrared photometry (see e.g. Allen and Swings,
1976).

OBSERVATIONS

The observations were made with two different
polarimeters of the European Southern Observatory :

1) a four channel photometer-polarimeter attached to the

____

[*]Based on data collected at the European Southern
Observatory (La Silla, Chile)

103

M. Jaschek and H.-G. Groth (eds.), Be Stars, 103–106.
Copyright © 1982 by the IAU.

TABLE 1 : "u, b, y" POLARIMETRY (ESO 3.6 M; DEC. 1979)[x]

Object	p(u) %	θ(u) °	p(b) %	θ(b) °	p(y) %	θ(y) °	No.of meas.
HD 44179	1.5± .4	148±2	1.45±.25	148±1	1.64±.4	152±1	2
HD 45677	.75 .15	169 1	1.1 .2	173 1	.85 .25	177 2	2
RX Puppis	1.65 .5	113 1	1.9 .5	118 1	1.8 .5	123 2	2
HD 87643	.8 .4	134 1	.9 .4	138 1	1.0 .4	139 1	2
GG Car	1.0 .3	93 1	2.0 .4	98 1	1.9 .3	98 1	2

[x]values from a quick look reduction program written by R.B. giving the values of p and θ with their absolute errors.

TABLE 2

UBV POLARIMETRY (ESO 1 M, MARCH 1980)[x]

Object	p(U) %	θ(U) °	p(B) %	θ(B) °	p(V) %	θ(V) °	No. of meas.
HD 37806	.7	116	1.0	120	.9	113	1
17 Lep	.5	37	.7	35	.7	29	1
HD 44179	3.1	?	1.8	36	1.7	26	1
HD 45677	1.2±.1	172±2	1.5 ±.1	173±1	1.2 ±.1	170±3	8
HD 50138	.9	162	1.2	164	1.0	159	1
Z CMa	1.2	147	1.7	?	.8	140	1
3 Pup	1.3	91	1.6	94	1.7	90	1
RX Pup	2.1±.1	120±2	2.3 ±.1	124±3	2.1 ±.1	121±2	6
HD 87643	.7±.1	132±6	.75±.1	133±7	.6 ±.1	134±5	6
HR Car	2.2±.2	127±1	2.6 ±.1	129	2.8 ±.5	124	2
GG Car	.8±.1	99 ±2	1.6 ±.1	99±1	1.65±.4	95±2	5
AG Car	.5	135	.8	132	.8	130	1
X Oph	.7	138	.9	136	.9	130	1
CPD −52°9243	5.2±.3	39±3	5.6 ±.1	38±1	5.8 ±.1	34	2
HD 326823	3.0	30	3.3	33	2.2	?	1
HD 163296	.7	143	.8	152	.5	156	1

[x] - using the calibration standards HD 80558 and HD 111613;

- after taking into account an interstellar contribution;

- errors indicated : rms on the average value when multiple measurements; individual measurements < .5 % on p; from < 1° to 20° on θ.

3.6 m telescope (with only three channels in operation
at the time of the observations, so that u, b and y were
chosen);

2) a two channel polarimeter attached to the 1 m telescope,
where the broad bands U, B, V were used.

Both instruments are described in the E.S.O. Users'
Manual.

In the first case, five objects were measured, whose
values are listed in Table 1; in five other objects, HK Ori
LkHα208, MWC 137, MWC 819 and CD-24°5721 polarization
was detected, but additional measurements are required.

In the second case, sixteen objects were measured,
four on several occasions, as well as a series of nearby
objects in order to be able to subtract the "environmental"
polarization (sometimes quite large, but always with
different angles). The reduction was performed as for
CPD-52°9243 (Swings, 1981), and the results are listed in
Table 2.

REMARKS

We searched for correlations between the polarization
values and several quantities related to the extended
atmosphere of the objects, i.e. the emission class
(excitation, emission lines, see Allen and Swings, 1976) or
the infrared excess, i.e. the H(1.6 μ) - K(2.2 μ) and V-K
indices : only negative results were obtained.

We then searched for systematic differences between
the polarizations in B[e] and in Be stars (of our sample
or on a more general basis). From our data on 12 B[e]'s and
4 Be's, we find that the difference between the peculiar
and the classical Be stars is essentially and(surprisingly)
independent of wavelength : 1.0 % in U, .9 % in B and 1.1 %
in V.

Data at longer wavelengths are required since the
effect of dust would manifest itself more for λ > 5000 A.

BIBLIOGRAPHY

Allen, D.A. and Swings, J.P. : 1976, Astron. Astrophys.,
   <u>47</u>, p. 293.

Coyne, G.V. : 1976, in Be and Shell Stars, I.A.U. Symp. 70,
    ed. A. Slettebak, p. 233.
Coyne, G.V. and McLean, I.S. : 1982, this volume, p. 77.
Coyne, G.V. and Vrba, F.J. : 1976, Astrophys. J., 207,
    p. 790.
Swings, J.P. : 1981, Astron. Astrophys., in press.

DISCUSSION

<u>Metz</u>: For a study of the difference between peculiar Be stars and "normal" Be stars, it is necessary to select stars with nearly equal angles of inclination. Did you pay attention to that ?

<u>Swings</u>: The problem is that for most of the B[e] stars one knows very little about how one sees the object. What I did, was to take the average polarization for 12 B[e]'s and compare it to an average of 4 B|e|'s chosen at random, so to speak.

<u>Poeckert</u>: Could the B[e] stars be a hot extension of the T Tau stars ?

<u>Swings</u>: My feeling is that B[e] are much older, and are evolving towards proto-planetary nebulae. They may bare similarities to T Tauries (IR excess, [SII] emissions, polarization, etc.), but they are not the same type of objects.

III. SPECTROSCOPY

SPECTROSCOPIC OBSERVATIONS OF Be STARS IN THE PHOTOGRAPHIC AND VISUAL
REGIONS
(Review Paper)

Arne Slettebak
Perkins Observatory
Ohio State and Ohio Wesleyan Universities
Delaware, Ohio 43015, U.S.A.

1. INTRODUCTION

The term "Be star" has been used at times to describe classes of
objects which are physically rather different from one another. While
it could include early-type supergiant stars with $H\alpha$ emission, early-
type pre-main sequence nebular variables, or quasi-planetary nebulae
like MWC 349, I will limit this review paper to a discussion of the
"classical" Be stars. These are defined as stars of luminosity classes
III to V, usually rapid rotators, which show normal B-type spectra with
superposed Balmer (and sometimes Fe II) emission. Included also,
however, will be the Oe stars and the A-type shell stars, which seem to
represent extensions of the classical Be phenomenon to higher and lower temperatures, respectively.

The proceedings of IAU Symposium No. 70 on Be and Shell Stars (D.
Reidel Pub. Co., Dordrecht, Holland, 1976) contain papers and references to Be-star research done up to 1975. A later review paper on Be
stars (Slettebak, 1979) includes more recent references plus some historical comments. Also, Hubert-Delplace (1979) has written an excellent
review paper on spectroscopic studies in the visible. I should like to
limit my discussion in this paper to work done since the 1975 IAU Symposium and omit also much of the historical discussion in my aforementioned review paper.

Finally, I would like to make the usual disclaimer with regard to
completeness which all reviewers must feel. Research in Be-star spectroscopy has been very active in recent years and I realize that my own
particular interests could well have introduced a bias causing me to
overlook significant work done in this field. We can all agree that Be
star researchers represent an unusually fine group of human beings --
since I do not want to lose any of you as friends, I hope that you will
regard any omissions in my paper as entirely unintentional.

M. Jaschek and H.-G. Groth (eds.), Be Stars, 109–124.
Copyright © 1982 by the IAU.

2. SURVEYS AND ATLASES

Large numbers of new Be stars continue to be reported each year, mostly as a result of surveys for Hα emission-line objects. Recent work includes that of Arkhipova et al. (1976), Coyne et al. (1978), Doazan et al. (1977), Dolidze et al. (1977), Gomez and Mendoza (1976), Henize (1976), Irvine and Irvine (1979), Kucewicz (1980), MacConnell (1981), Martinez et al. (1980), Sanduleak and Bidelman (1980), Stephenson and Sanduleak (1977), and Vega et al. (1980). It seems likely that new Be stars, including relatively bright objects, will continue to be discovered in future years, for at least two reasons: (1) Many Be stars show spectrum variability over long time periods, reverting to the appearance of normal B-type stars between shell episodes. An Hα survey at any given time would therefore miss a certain fraction of potential Be candidates. (2) Hα emission is often weak and narrow, requiring relatively high-dispersion spectra for detection. Objective-prism spectroscopy would be likely to miss such objects.

A beautiful and useful atlas of Be-star spectra has been published by Hubert-Delplace and Hubert (1979). The atlas includes 51 plates of photographic spectra of 35 Be stars, showing spectrum changes over 10-20 years for most of the stars. A description of spectrum changes for 148 stars since about 1953 is also included.

3. THE UNDERLYING STARS

Spectral classification of the stars underlying the Be shells is very difficult because of the great line broadening typically shown by these objects, as well as the disturbing effects of line emission in their spectra. Yet it is important to know the distribution of spectral types and, especially, luminosities, of the Be stars if we are to understand their relationship to normal stars and how they evolve. Programs of spectral classification of Be stars are underway at several observatories.

Divan (1979) has developed a quantitative system of spectral classification of Be stars from low-dispersion spectra, involving measurements around the Balmer discontinuity. Of special interest is the fact that two Balmer jumps are observed in some Be stars, one from the underlying star and one from the shell.

Jaschek et al. (1980) have developed a classification scheme for Be stars based upon the visual inspection of several thousand spectrograms collected at the Meudon Observatory between 1953 and 1976 of 140 northern Be stars brighter than about magnitude 7 (the same plate material that was used for the aforementioned atlas of Be-star spectra). They derive MK spectral types and then establish five groups of Be stars, using spectrum characteristics and the time scale of variations as parameters, and suggest that predictions of the future behavior of a given star can be made once it is assigned to a particular group.

Spectral classification of southern Be stars has been done or is in process at several observatories. Garrison et al. (1977) classified 148 Be stars among 1113 OB stars brighter than the 10th magnitude observed at the Cerro Tololo Inter-American Observatory. Jaschek and Jaschek (1979) are observing and classifying southern Be stars brighter than magnitude 8.0 with dispersion 20 Å/mm at the European Southern Observatory.

I have recently completed the spectral classification (Slettebak, 1981) of all the known northern and southern Be stars brighter than magnitude 6.0. Greatly-widened, fine-grain (IIIa-J emulsion) spectrograms of dispersion 40 Å/mm were obtained at the Lowell Observatory in Flagstaff, Arizona and at the Cerro Tololo Inter-American Observatory in Chile, in order to permit accurate spectral classification plus visual estimates of line broadening for rotational velocity determinations. Preliminary results indicate that of the 168 stars with spectral types in the range B0 to A0, there are about twice as many B0-5 as B6-A0 emission-line stars. The number of stars estimated to be more luminous than class V was greater than the number of main-sequence stars for both the B0-5 and B6-A0 groups. This confirms earlier work on Be stars in galactic clusters and binary systems which shows these objects to be somewhat above the main sequence, on the average.

Axial rotation of Be stars will be discussed at greater length in the review paper by Dr. Harmanec, but its importance in the Be phenomenom is certainly one thing we can all agree upon. As a class, Be stars are the most rapidly rotating stars known (excepting degenerate stars, of course), and the presence of shells around them must be related in some way to their rapid rotation. My spectrograms of the brighter Be stars (Slettebak, 1981) confirm their rapid rotation but show very little difference in mean rotational velocities between the early-type and late-type Be stars, on the one hand, and between the main-sequence and subgiant-giant Be stars in both groups, on the other.

While we puzzle at this Symposium as to what Be stars are and how they throw off their shells, we should remember that the Be phenomenon apparently extends to both earlier and later spectral types than B. Our theories and ideas should encompass also the Oe and the A-type shell stars. Both are also characterized by rapid rotation, have luminosities in the main-sequence to giant band, and show shell features (though not necessarily emission lines) in their spectra. Frost and Conti (1976) estimate that 14 percent of O-type stars have exhibited the Oe phenomenon (including Balmer emission) at some time. This is somewhat less than the approximately 20 percent estimated for the early B-type stars (Slettebak, 1979). The fraction falls off toward the B stars of later type, and is quite uncertain for the A-type stars. Although only a few (e.g., 17 Lep, 14 Com) were known in Struve's time, their number has increased significantly since then, thanks to the work of Abt and Moyd (1973) and others. Recent papers, which include references to the earlier work, include Andersen and Nordström (1977), Clayton and Marlborough (1980), Dominy and Smith (1977), and Slettebak (1979).

4. THE SHELLS

I should like to make some general remarks about the shells surrounding Be stars at this point; recent work on individual objects will
be discussed in Section 6 of this paper.

Struve's (1931) rotational model for Be stars suggested a rapidly-
rotating underlying star surrounded by an equatorial ring or shell which
gives rise to the emission lines. In addition to the Balmer line (and
sometimes Fe II) emission from the shell as a whole, that portion of the
shell which is seen projected across the photosphere of the underlying
star gives rise to absorption lines which are relatively sharp as compared with the broad absorption lines from the rapidly-rotating underlying star.

Be stars which show such narrow absorption lines in their spectra
have been called "shell stars". This is somewhat misleading since all
Be stars can properly be called shell stars insofar as their emission
arises from some sort of surrounding shell. Be stars with the most
pronounced absorption-line shell spectra always show considerable broadening of the absorption lines from the underlying star, however, implying that we are viewing them nearly equatorially. It is these objects
(e.g., Pleione, 48 Lib, etc.) which historically have been called "shell
stars".

The absorption-line spectra from Be shells may look very different
from star to star. The hottest Be shells show lines arising from metastable energy states of He I (e.g. λ 3889 and 3965) in addition to the
sharp central absorption cores in the Balmer series. Somewhat cooler
shells (generally surrounding underlying stars in the spectral-type
range B2-9) also show sharp Balmer core absorption plus a metallic-line
spectrum (Fe II, Ti II, Ni II, Cr II, etc.) which resembles somewhat the
spectrum of the A2 supergiant α Cyg. Typical metallic lines which are
usually strong in such shell spectra include Fe II 4179, 4233, 4549, and
4584, all of which arise from metastable levels 2.6 to 2.8 eV above the
ground state. The coolest shells are found surrounding rapidly-rotating
stars of spectral-type A. These generally show neither Balmer emission
nor Balmer absorption cores (an exception is 17 Lep which, however, is
atypical in having an M-type companion plus an expanding shell). Shells
surrounding A-type stars may show sharp absorption cores in the Ca II
H and K lines (which arise from the ground state) and sharp lines of
Ti II at 3685, 3759, and 3761 Å. The latter have lower levels which are
metastable and only about 0.6 eV above the ground state. Clearly these
shells are now so cool that very little hydrogen nor Fe II is to be
found in excited energy levels.

Physical parameters for Be shells as determined by various techniques were summarized in my review article (Slettebak, 1979) as of the
literature of 1977-78. Much important work has been done since and we
will hear about it during this Symposium. Ultraviolet observations have
been especially important in emphasizing the role of mass loss,

particularly for the Be stars of early type. There are still problems
associated with the mechanism by which Be shells are produced, however.
The discontinuous nature of shell ejection and shell dissipation (even
though slow) suggests that something in addition to stellar winds must
be operating. Even more serious problems arise in trying to explain
shells around the later B-type stars (e.g., Pleione) and, especially,
around A-type stars, since these do not seem to have sufficiently strong
stellar winds to move matter out into the shells. Is it possible that
magnetic fields play some role here (as well as in the hotter Be stars),
as was suggested by Limber (1974, 1976) and by Saito (1974)? Recent
work by Clayton and Marlborough (1980) on polarization in A-type shell
stars places an upper limit of about 300 gauss on any longitudinal fields
which may exist undetected. Could smaller fields be effective in Be and
A-shell production?

5. VARIABILITY

     Variability is an inherent part of the Be phenomenon. I would like
to make a few general remarks and then discuss individual stars in the
next section of this paper. This discussion is restricted to variability
in Be spectra in the photographic and visual regions. A great deal of
work has been done on photometric and polarimetric variability as well
as variability in the X-ray, ultraviolet, and infrared spectral regions
of Be spectra, which will be discussed in other papers during this
Symposium.

     Be spectra appear to vary on several time scales. The variations
are sometimes periodic or quasi-periodic but more often irregular. Be
shells have been observed to come and go in intervals of a decade or two
(e.g., γ Cas and Pleione), leaving a fairly normal-looking, rapidly-
rotating B-type star between shell episodes. A decade is also the
approximate time interval between quasi-periodic spectrum and radial
velocity changes for the well-studied Be stars 48 Lib and ζ Tau. Quasi-
periodic changes in V/R (the ratio of the violet to the red emission
components in the Balmer lines of Be spectra) have been observed in many
Be stars, again on time scales of the order of a decade or so. Many examples of Be star variability may be found in the aforementioned Hubert-
Delplace and Hubert (1979) atlas. Additional references to earlier work
are in my review paper (Slettebak, 1979) -- more recent work will be
discussed in the next section.

     Spectroscopic observations of Be stars also suggest variability
over shorter time scales. Changes in time periods of months seem well
established; references to the earlier work may be found in Slettebak
and Reynolds (1978), whereas recent papers describing such variations
include those by Doazan et al. (1980), Elias et al. (1978), Hirata and
Kogure (1979), Metz and Pöllitsch (1979), and Reynolds and Slettebak
(1980). Additional papers will be discussed in the next section.

     Variations in Be Spectra on time scales of days, hours, and even

minutes have also been reported, but here there is rather conflicting
evidence. Such variability is generally irregular, which makes it
difficult to separate real spectrum changes from random variations.
Clarke and Wyllie (1977), and Lacy (1977), in particular, have cautioned
against interpreting all observed changes as real. Again, references to
the earlier work may be found in Slettebak and Reynolds (1978). Papers
written since IAU Symposium No. 70 in support of rapid spectroscopic
changes include those by Aydin and Faraggiana (1978), Baliunas and
Guinan (1976), Bijaoui and Doazan (1979), Cowley and Houk (1976), Dachs
et al. (1977), Fraquelli (1979), Gulliver and Bolton (1978), Gulliver
et al. (1980), Harmanec et al. (1977), Mamatkazina (1978), Schoembs and
Spannagl (1976), Slettebak and Snow (1978), and Vojkhanskaya (1976). On
the other hand, Kitchin (1976), Luud (1978), and Reynolds and Slettebak
(1980) have searched for rapid changes in Be spectra, without success.
Such irregular changes, if confirmed, would give us information about
turbulence and physical conditions in the Be shells.

Not all observed rapid changes in Be spectra are irregular, of
course. Baade (1979), for example, reports changes in the spectrum of
ω CMa with period 1.36 days, which he attributes to non-radial pulsations in that star. Short period, interacting-binary Be stars or rotation of an active area on the surface of a single Be star could also
produce changes in their spectra on time scales of hours or days.

One method of guarding against the instrumental or terrestrial
atmospheric effects which could be confused with real changes in Be
spectra is to observe these objects simultaneously with independent
telescopes and recording systems. This has been done by Haefner et al.
(1975), Metz and Pöllitsch (1979) Reynolds and Slettebak (1980), and
Slettebak and Snow (1978), for example, but more observations would be
desirable.

6. RECENT WORK ON SOME INDIVIDUAL Be STARS

Pleione. One of the most studied of the Be stars, Pleione continues to be entertaining. The star entered a new shell phase in 1972
and the shell has developed during the 1970's. In a series of papers,
Hirata and Kogure (1976, 1977, 1978) and Higurashi and Hirata (1978)
have studied Pleione's latest shell phase. They report a gradual development of the shell from 1972 to 1976, the envelope concentrated toward
the equatorial plane and continuing its slow expansion in that plane as
well as in the vertical direction, with a nearly constant mass-loss rate
of 4×10^{-11} $M_\odot$ yr.$^{-1}$. They suggest that the shell consisted of a relatively compact, dense and cool region concentrated toward the equa-
torial plane and a hotter extended region, with mean excitation temperature 9000-10,000 °K in 1973-1976. Gulliver (1977) studied the spectrum
variations of Pleione from 1938 to 1975 and reported that his observational material rules out the elliptical ring and binary models for
Pleione, whereas good agreement was found between the stellar wind model
and the observed variations, with the exception of the emission intensity

and the Balmer progression. Other recent research on Pleione includes
the publication of an atlas of the shell spectrum between 3167 Å and
4924 Å by Ballereau (1980) and a spectrophotometric study by Sapargalieva
(1978).

o And. This very interesting object has a long history of spectrum
variations. It has changed from a normal B-type star to a shell star
and back again a number of times, and also shows more rapid variations.
Various models have been proposed to explain the spectroscopic and photometric behavior of o And, but no general agreement exists. Recent radial
velocity and equivalent width measurements by Fracassini, Pasinetti,
and Pastori (1977, 1979) suggest a 1.6 day period(supporting an earlier
contact-binary system hypothesis) as well as a shell period of about 30
years. The latter has been questioned by Harmanec et al. (1977) and by
Gulliver and Bolton (1978), however, whereas the binary hypothesis was
also challenged in earlier photometric work and by Gulliver et al.
(1980). Other hypotheses which have been proposed (and attacked) to explain the photometric and spectroscopic variations of o And include
rotation of a photospheric spot, pulsation, and variable shell absorption.
Obviously, more observations of this bright and interesting star would
be very desirable.

γ Cas. Discovered by Secchi to show Hβ emission in 1866, this
first Be star continues to be a fascinating object. γ Cas was announced
to be an X-ray source by Jernigan (1976). Cowley, Rogers, and Hutchings
(1976) then looked for radial velocity variations on spectrograms taken
over an 18-year period and concluded that γ Cas is probably a single
star, although they could not rule out the possible existence of a low-
mass companion. Marlborough et al. (1978) suggested a coronal model for
the emission from γ Cas, based on the variability of ultraviolet data
from Copernicus and Hα scans. They reviewed and rejected as implausible
for γ Cas the detached-ring hypothesis and the late-type companion mass-
exchange model for explaining the observed V/R variations. Instead,
they proposed a modified stellar wind model in which turbulence produced
by differential rotation near the equatorial plane leads to the production of a coronal layer above and below the equatorial plane. The V/R
variations are explained in terms of a 1 $M_\odot$ neutron star companion with
orbital period of about 4 years, which also produces the observed
transient X-ray flux.

ϕ Per. This is another controversial object. At IAU Symposium No.
70, one investigator (Hendry, 1976) suggested on the basis of radial
velocity measurements that the system is made up of a B1 primary and a
B3 secondary, both emission-line stars. But Peters (1976) maintained
that the UCLA series of spectrograms showed no evidence of duplicity.
More recently, Suzuki (1980) has reinterpreted Hynek's data on the assumption that the circumstellar envelope of the star is a gas ring in a
stable periodic orbit of the restricted three-body problem. He deduces
the masses of the primary and secondary stars to be 20 $M_\odot$ and 4 $M_\odot$,
respectively, with the gas ring revolving around the primary star in the
orbital plane of the system. Poeckert (1979, 1981), using new

spectrographic material, finds masses of 21 $M_\odot$ and 3.4 $M_\odot$, very similar
to Suzuki's values, but suggests that the secondary is peculiar in that
He II 4686 emission arises from within its vicinity. He believes that
the secondary may be the helium core of a once more massive star, its
mass having been transferred to the primary.

48 Lib. This is another classical shell star, which has been
studied since the time of Struve. Recent work by Aydin and Faraggiana
(1978) suggests that although the radial velocity variations are repeated cyclically, the period is not constant, and therefore the simple binary model must be rejected. They consider the multiple-system hypothesis to be attractive but do not have enough data to elaborate a
physically reliable model. Variations in the shell density are considered to be responsible for many of the observed spectrum peculiarities.
Garcia-Alegre and Lopez Arroyo (1980) have also studied the shell spectrum of 48 Lib during the period 1962-1967, and find an indication of
weaker variations in radial velocity superimposed on the 9.6 year major
variation.

Other Stars. In addition to the aforementioned objects, the
following Be stars were included among recent spectroscopic investigations in the photographic and visual regions: θ CrB (Poeckert, 1979;
Poeckert and Duric, 1980), 59 Cyg (Hubert-Delplace, 1981), 27 CMa (Danks
and Houziaux, 1978), 88 Her (Hirata, 1978), ζ Oph (Ebbets, 1980 -- note
that the star is erroneously listed as θ Oph), HD 51480 (Welin, 1979),
HD 58050 (Ballereau and Hubert, 1981), HD 118246 (Turner et al., 1978),
HD 183656 (Aab and Vojkhanskaya, 1977), HD 200775 (Altamore et al.,
1979), and HDE 245770 (Giangrande, 1980). Pöllitsch (1979) also reported
spectroscopic observations of six bright Be stars.

7. CLUSTERS AND EVOLUTION

The evolutionary status of the classical Be stars is still not
clear. There is considerable observational evidence (c.f. Slettebak,
1979) that, on the average, they are located one-half to one magnitude
above the main sequence. Is this an evolutionary effect? Earlier
attempts to explain Be stars as partially evolved objects, in the secondary contraction phase following hydrogen exhaustion in the core,
have been criticized on both theoretical (mass loss may also occur in
models of rotating stars in earlier stages of evolution) and statistical
(the observed number of Be stars relative to B stars is too large to be
consistent with the relatively short-duration secondary contraction
stage) grounds. It is also significant that several investigators have
shown that Be stars in clusters do not occur only above the main sequence but are also found near the zero-age main sequence.

In a study of Be stars in clusters, Schild and Romanishin (1976)
suggest that rotating stars can become Be stars in their early hydrogen-
burning evolution away from the main sequence, but are more likely to
do so after the onset of gravitational core contraction. Lloyd Evans

(1980) investigated Be stars in two open clusters and also concluded that Be stars are most likely to appear in the core contraction stage of evolution.

Abt and his co-workers have found a number of Be stars in open clusters; his paper on the occurrence of abnormal stars in open clusters (Abt, 1979) summarizes his findings and includes references to his earlier papers. For 13 Be stars in five clusters, Abt finds the mean frequency of Be stars to be about 9 percent (Schild and Romanishin had found about 7 percent), which he considers to be comparable to the frequency for field Be stars. There is a considerable variation from cluster to cluster, however; Abt obtains values between 5 and 16 percent, and Sanduleak (1979) finds at least 34 percent of the early B-type stars in NGC 663 to have shown Be characteristics at one time or another. Sanduleak and Bidelman (1979) also estimate that the open cluster χ Per has about a 25 percent representation of Be stars. It seems clear that the Be phenomenon, whatever it might be, is not an isolated event that occurs to an occasional star.

Abt (1979) also finds that the frequency of Be and shell stars in open clusters shows no strong dependence upon age. Such stars occur among the youngest ($10^{5.7}$ yr.) and oldest ($10^{8.8}$ yr.) clusters in his sample. These investigations suggest that Be stars form early in the cluster histories, with a possible formation spurt at the onset of gravitational core contraction.

Finally, it should be mentioned that rotationally-induced gravity darkening may also play a role in understanding the average position of Be stars above the main sequence. A recent paper by Collins and Sonneborn (1977), based on extensive model calculations, suggests that the position of the Be stars above the main sequence can be interpreted as the result of rotation alone, without invoking evolutionary arguments. Slettebak et al. (1980), in a study of the effects of stellar rotation on spectral classification, provide support for this idea from line strengths: the combination of predicted Balmer line weakening plus the predicted behavior of He I/Si II and He I/Mg II line ratios suggests that rapidly rotating B-type stars viewed equatorially will be assigned luminosity classes which place them somewhat above the main sequence. These rotational effects could also combine with evolutionary effects to explain the position of the Be stars on the H-R diagram -- unfortunately, we cannot distinguish between them at this time.

8. CONCLUDING REMARKS

Even though many exciting results and ideas regarding classical Be stars have come from the opening up of new wavelength regions, it seems clear that spectroscopic observations in the photographic and visual regions are still of the greatest importance. Further study of the Oe stars and the A-type shell stars, which appear to represent extensions of the Be phenomenon, seems important to me, particularly with respect

to the problem of shell formation. More observations to define the
nature of Be-star variability on various time scales also seem very
desirable. We now have much better coverage of changes in the spectra
of the brighter Be stars than was available a few decades ago, which
makes it possible to attempt modeling some of them. Others have so far
defied any reasonable explanation of their spectroscopic behavior, and
continued observations would be desirable. More work on Be stars in
clusters might throw additional light on their evolutionary status.

Although Be-star spectroscopists have been very active since our
last Symposium in 1975, there is obviously still much to be done. It
would be fascinating to hear and difficult to guess the content of the
papers presented at IAU Symposium No. 523 in the year 2066 (on Be stars,
naturally, in honor of the 200th anniversary of the discovery of the
first Be star), but we can be certain that most of them would call for
more observations.

ACKNOWLEDGEMENTS

It is a pleasure to acknowledge very useful discussions with
A. D. Code and with G. W. Collins, II in preparing this paper.

REFERENCES

Aab, O. E. and Vojkhanskaya, N. F.: 1977, Astrofiz. Issled. Izv. Spets.
    Astrofiz. Obs. 9, p. 22.
Abt, H. A.: 1979, Astrophys. J. 230, p. 485.
Abt, H. A. and Moyd, K. I.: 1973, Astrophys. J. 182, p. 809.
Altamore, A., Baratta, G. B., Cassatella, A., Grasdalen, G., Persi, P.,
    and Viotti, R.: 1979, Mem. Soc. Astron. Italiana 50, p. 223.
Andersen, J. and Nordström, B.: 1977, Astron. Astrophys. Suppl. 29, 309.
Arkhipova, V. P., Dokuchaeva, O. D., and Saveleva, M. V.: 1976, Pisma
    Astron. Zh. 2, p. 122.
Aydin, C. and Faraggiana, R.: 1978, Astron. Astrophys. Suppl. 34, p. 51.
Baade, D.: 1979, ESO Messenger, Dec. 1979.
Baliunas, S. L. and Guinan, E. F.: 1976, Publ. Astron. Soc. Pacific 88,
    p. 10.
Ballereau, D.: 1980, Astron. Astrophys. Suppl. 41, p. 305.
Ballereau, D. and Hubert, A. M.: 1981, IAU Circ. No. 3565.
Bijaoui, A. and Doazan, V.: 1979, Astron. Astrophys. 70, p. 285.
Clarke, D. and Wyllie, T. H. A.: 1977, Observatory 97, p. 21.
Clayton, G. C. and Marlborough, J. M.: 1980, Astrophys. J. 242, p. 165.
Collins, G. W., II and Sonneborn, G. H.: 1977, Astrophys. J. Suppl. 34,
    p. 41.
Cowley, A. P. and Houk, N.: 1976, Publ. Astron. Soc. Pacific 88, p. 37.
Cowley, A. P., Rogers, L. and Hutchings, J. B.: 1976, Publ. Astron. Soc.
    Pacific 88, p. 911.
Coyne, G. V., Wisniewski, W., and Otten, L. B.: 1978, Vatican Obs. Publ.
    1, p. 257.

Dachs, J., Maitzen, H.-M, Moffat, A. F. J., Sherwood, W. A., and Stift,
    M.: 1977, Astron. Astrophys. 56, p. 417.
Danks, A. C. and Houziaux, L.: 1978, Publ. Astron. Soc. Pacific 90,
    p. 453.
Divan, L.: 1979, in M. F. McCarthy, G. V. Coyne, and A. G. D. Philip
    (eds.), "Spectral Classification of the Future", IAU Colloq. 47,
    Specola Vaticana Ricerche Astronomiche 9, p. 247.
Doazan, V., Bourdonneau, B., and Letourneur, N.: 1977, Astron. Astrophys.
    56, p. 481.
Doazan, V., Kuhi, L. V., and Thomas, R. N.: 1980, Astrophys. J. 235, L17.
Dolidze, M. V., Dokuchaeva, O. D., and Kimeridze, G. N.: 1977, Astron.
    Cirk. Bjuro Astron. Soobshch Akad. Nauk USSR No. 943.
Dominy, J. F. and Smith, M. A.: 1977, Astrophys. J. 217, p. 494.
Ebbets, D.: 1980, IAU Circ. No. 3469.
Elias, J., Lanning, H., and Neugebauer, G.: 1978, Publ. Astron. Soc.
    Pacific 90, p. 697.
Fracassini, M., Pasinetti, L. E., and Pastori, L.: 1977, Astrophys.
    Space Sci. 49, p. 145.
Fracassini, M., Pasinetti, L. E., and Pastori, L.: 1979, Publ. Obs.
    Astron. Strasbourg 6, p. 62.
Fraquelli, D. A.: 1979, Publ. Astron. Soc. Pacific 91, p. 502.
Frost, S. A. and Conti, P. S.: 1976, in A. Slettebak (ed.), "Be and Shell
    Stars", IAU Symp. 70, p. 139.
Garcia-Alegre, M. C. and Lopez Arroyo, M.: 1980, Astron. Astrophys. 83,
    p. 163.
Garrison, R. F., Hiltner, W. A., and Schild, R. E.: 1977, Astrophys. J.
    Suppl. 35, p. 111.
Giangrande, A., Giovannelli, F., Bartolini, C., Guarnieri, A., and
    Piccioni, A.: 1980, Astron. Astrophys. Suppl. 40, p. 289.
Gomez, T. and Mendoza, V, E.E.: 1976, Rev. Mex. 1, p. 381.
Gulliver, A. F.: 1977, Astrophys. J. Suppl. 35, p. 441.
Gulliver, A. F. and Bolton, C. T.: 1978, Publ. Astron. Soc. Pacific 90,
    p. 732.
Gulliver, A. F., Bolton, C. T., and Poeckert, R.: 1980, Publ. Astron.
    Soc. Pacific 92, p. 774.
Haefner, R., Metz, K., and Schoembs, R.: 1975, Astron. Astrophys. 38,
    p. 203.
Harmanec, P., Koubsky, P., Krpata, J., Zdarsky, F., and Dolenska, A.:
    1977, Inf. Bull. Var. Stars No. 1296.
Hendry, E. M.: 1976, in A. Slettebak (ed.), "Be and Shell Stars", IAU
    Symp. 70, p. 429.
Henize, K. G.: 1976, Astrophys. J. Suppl. 30, p. 491.
Higurashi, T., and Hirata, R.: 1978, Publ. Astron. Soc. Japan 30, p. 615.
Hirata, R.: 1978, Info. Bull. Var. Stars No. 1496.
Hirata, R. and Kogure, T.: 1976, Publ. Astron. Soc. Japan 28, p. 509.
Hirata, R. and Kogure, T.: 1977, Publ. Astron. Soc. Japan 29, p. 477.
Hirata, R. and Kogure, T.: 1978, Publ. Astron. Soc. Japan 30, p. 601.
Hirata, R. and Kogure, T.: 1979, Inf. Bull. Var. Stars No. 1575.
Hubert-Delplace, A. M.: 1979, Publ. Obs. Astron. Strasbourg 6, p. 18.
Hubert-Delplace, A. M.: 1981, Astron. Astrophys. (in press).

Hubert-Delplace, A.-M. and Hubert, H.: 1979, Paris-Meudon Obs., Meudon,
    France.
Irvine, N. J. and Irvine, C. E.: 1979, Publ. Astron. Soc. Pacific 91,
    p. 105.
Jaschek, C. and Jaschek, M.: 1979, Publ. Obs. Astron. Strasbourg 6,
    p. 41.
Jaschek, M., Hubert-Delplace, A.-M., Hubert, H., and Jaschek, C.: 1980,
    Astron. Astrophys. Suppl. 42, p. 103.
Jernigan, J. G.: 1976, IAU Circ. No. 2900.
Kitchin, C. R.: 1976, Astrophys. Space Sci. 45, p. 119.
Kucewicz, B.: 1980, Bull. Argentine Astron. Assn. 18, p. 10.
Lacy, C. H.: 1977, Astrophys. J. 212, p. 132.
Limber, D. N.: 1974, Astrophys. J. 192, p. 429.
Limber, D. N.: 1976, in A. Slettebak (ed.), "Be and Shell Stars", IAU
    Symp. 70, p. 371.
Lloyd Evans, T.: 1980, Monthly Notices Roy. Astron. Soc. 192, p. 47.
Luud, L. S.: 1978, Pisma Astron. Zh. 4, p. 454.
MacConnell, D. J.: 1981, Astron. Astrophys. Suppl. (in press).
Mamatkazina, A. K.: 1978, Astrofiz. Inst. Alma-Ata 31, p. 73.
Marlborough, J. M., Snow, T. P., Jr. and Slettebak, A.: 1978, Astrophys.
    J. 224, p. 157.
Martinez, R. E., Muzzio, J. C., and Waldhausen, S.: 1980, Astron.
    Astrophys. Suppl. 42, p. 179.
Metz, K. and Pöllitsch, G.: 1979, ESO Messenger, Sep. 1979.
Peters, G. J.: 1976, in A. Slettebak (ed.), "Be and Shell Stars", IAU
    Symp. 70, p. 436.
Poeckert, R.: 1979, Astrophys. J. 233, L73.
Poeckert, R.: 1979, IAU Circ. No. 3365.
Poeckert, R.: 1981, Dominion Astrophys. Obs. preprint.
Poeckert, R. and Duric, N.: 1980, Publ. Dominion Astrophys. Obs. 15,
    p. 327.
Pöllitsch, G. F.: 1979, IAU Circ. No. 3369.
Reynolds, R. C. and Slettebak, A.: 1980, Publ. Astron. Soc. Pacific 92,
    p. 472.
Saito, M.: 1974, Publ. Astron. Soc. Japan 26, p. 103.
Sanduleak, N.: 1979, Astron. J. 84, p. 1319.
Sanduleak, N. and Bidelman, W. P.: 1979, unpublished.
Sanduleak, N. and Bidelman, W. P.: 1980, Publ. Astron. Soc. Pacific 92,
    p. 72.
Sapargalieva, L. M.: 1978, Tr. Astrofiz. Inst. Alma-Ata 31, p. 61.
Schild, R. and Romanishin, W.: 1976, Astrophys. J. 204, p. 493.
Schoembs, R. and Spannagl, C.: 1976, Astron. Astrophys. Suppl. 26, p. 55.
Slettebak, A.: 1979, Space Science Rev. 23, p. 541.
Slettebak, A.: 1981, (in preparation).
Slettebak, A., Kuzma, T. J., and Collins, G. W., II: 1980, Astrophys.
    J. 242, p. 171.
Slettebak, A. and Reynolds, R. C.: 1978, Astrophys. J. Suppl. 38,
    p. 205.
Slettebak, A. and Snow, T. P., Jr.: 1978, Astrophys. J. 224, L127.
Stephenson, C. B. and Sanduleak, N.: 1977, Astrophys. J. Suppl. 33,
    p. 459.

Struve, O.: 1931, Astrophys. J. 73, p.94
Suzuki, M.: 1980, Publ. Astron. Soc. Japan 32, p. 331.
Turner, D. G., Lyons, R. W., and Bolton, C. T.: 1978, Publ. Astron. Soc.
    Pacific 90, p. 285.
Vega, E. I., Rabolli, M., Muzzio, J. C., and Feinstein, A.: 1980, Astron.
    J. 85, p. 1207.
Vojkhanskaya, N. F.: 1976, Astrofiz. 12, p. 219.
Welin, G.: 1979, Astron. Astrophys. 79, p. 334.

DISCUSSION

Coyne: Is there a best M_V to use for Be stars for determining photometric distances?

Slettebak: No, we cannot specify a single value.

Viotti: First I like to stress the importance to concentrate the
observational and theoretical effort to a number of representative
stars, as you suggested.
Secondly I would like to know how much brighter than MS stars some
Be stars really are since dilution effects in extended atmospheres
of Be stars may affect the luminosity classification. How much could
the luminosity be affected by the assumption of the B.C., being that
of non-emission B stars, and of the intrinsic $(B-V)_o$, which could be
anomalous.

Slettebak: I do not believe that dilution effects affect the luminosity
classification significantly since the sharp absorption lines from the
shells are rather easy to distinguish from the usually broad lines
from the underlying star.
I regret that I cannot answer your second question, which is a photometric one.

Endal: I was surprised to hear that the subgiants and giants have the
same rotational velocities as the main sequence Be stars. If the
luminosity classification is correct and rotation is nearly critical,
I would expect the stars with larger radius to have slower rotational
velocities.

Mermilliod: You did not mention the stars Abt has described as "sn".
He believes that this feature may originate in a weak shell. Can
you comment on them.

Slettebak: The interpretation of the Abt "sn" stars is somewhat controversial. I believe that the broad Helium lines in the spectra of these
stars may have rather sharp cores which are not visible in low resolution spectra, in which case they are not shell stars. But it would be
very desirable to obtain high resolution spectrograms to investigate
this point.

Stalio: In your list of desiderata I would like to add the necessity
to have simultaneous observations from ground and space (UV, X-ray
domain).

Marlborough: (Request by Slettebak concerning comments on role of
magnetic fields): Magnetic fields deduced from the measurement of
circular polarization refer to the line of sight component of the
field averaged over the surface of the star. Very large, small scale
fields can be missed by this technique. Hence magnetic fields may be
important dynamically.

<u>Poeckert</u>: (Question to Sonneborn): 1) Has a "rotating" zero age main
sequence been developed?
2) Can one indicate a band in the HR diagram in which rotating ZAMS
stars can be found?

<u>Sonneborn</u>: To the best of my knowledge, the effects of rotation on
the ZAMS have not been taken into account. Furthermore, a rotational
bias is built into the photometric systems, by using rapidly rotating
stars as standard stars (for example, α Leo). Work is in progress to
determine the rotational speeds of the ZAMS (in a colour-magnitude
diagram) from theoretical atmosphere models.

<u>Mermilliod</u>: There is a lack of observations for faint unevolved stars
in open clusters which can be used to define the ZAMS. Also information on binarity would be needed too, because axial rotation effects
can be confused with binarity effects in the colour-magnitude diagram.

<u>Henrichs</u>: You mentioned that during the $36^{\rm h}$ simultaneous observation
of γ Cas in the H_α and UV region (Slettebak and Snow, 1978) an X-ray
flare was seen when you saw an "event". To my knowledge γ Cas is a
highly variable X-ray source and therefore this apparent covariability
might well be just a coincidence. Perhaps Dr. Peters, who reported
this X-ray observations could comment on that.

<u>Peters</u>: According to Polidan (private communication), prior to the
X-ray flare in early 1977, γ Cas (MX0053+60) was in a low state with
the Copernicus counting rates varying from 1-4 counts (AV: 2.6 cts).
During the pointed observation, the X-ray flux increased to 29 counts!
This was indeed a flare event, not just part of the usual "minute to
minute" fluctuations in X-rays that γ Cas typically displays.

<u>Harmanec</u>: I reply to a remark by Dr. Thomas. I want to mention that
already in 1970 Hardorp and Strittmatter (stellar rotation, ed. by
Slettebak) showed a strong tendency of shell spectra to appear mostly
for stars with highest v sin i observed.
I feel it is the due time to start advertising the binary model. In
the frame work of the binary model, you obtain different luminosity
classes of different Be stars quite naturally - as a consequence of
a) different extent of the Roche lobe around the gainers, b) different
rate of mass transfer.

<u>Sonneborn</u>: It is very important to remember that any theoretical study
of the effects of rotation on observational properties of stars is
independent upon interior models of rotating stars. Our stellar atmosphere work in Ohio State is based on the Sackmann - Arand interiors,
which assume solid body rotation. Different theoretical results might
be obtained if one were to use interior models with differential rotation, for example. The theoretical results are of necessity model-
dependent.

<u>Sareyan</u>: We have to be very careful when we speak of rotation. Because

we are in fact dealing with <u>line broadening</u>, and line broadening can
actually be produced by rotation, but <u>also</u> by many other effects,
like velocity gradients in the atmosphere, for instance. (Line broadening is being observed, and rotation is only <u>one</u> interpretation.)

<u>Thomas</u>: I question several statements:
1. Be stars are defined <u>only</u> as those having at <u>some time</u> shown $H\alpha$
emission. Rotation is an inference, not an observation; <u>broad lines</u>
are a fact.
2. I do see how you assert "shell stars" have a definite limit on
i: especially since shell phase can be transient and non-repetetive
in any fixed time-scale. (Compare 59 Cyg and ρ Oph.)
3. You ask how can a shell arise abruptly, while a wind flows
continously. We observe that winds change equally as rapidly as do
shell phases.
4. You ask how a shell can exist in presence of a wind: a shell <u>is</u>
a phenomenon of the wind.

<u>Sareyan</u>: 1) The distance of Be stars from the main sequence is about
3/4 of a magnitude and it is related to "rotation". The so-called
"classical β Cephei" stars are typically class IV stars, i.e. situated
above the main sequence, like the average Be star. These "classical
β Cephei" are "slow rotators", so we can't expect the shift from the
main sequence to be due to "rotation", at least for those stars.
2) However, in the very middle of the so-called "instability box"
of these 16-18 historical "classical β Cephei", new short period
light variables have been discovered, which are now considered as
β Cep, and which have very large line broadening, i.e. about 300 km/s.

<u>Slettebak</u>: I would not expect the position of β Cephei stars (or any
group of slowly rotating stars) on the H-R diagram to be affected by
gravity darkening effects.

STATISTICAL PROPERTIES OF Be STARS

A.M. Hubert-Delplace[1], M. Jaschek[2], H. Hubert[1], M.Th. Chambon[1]
[1]DEPEG, Observatoire de Paris, 92190 Meudon, France
[2]Observatoire de Strasbourg, 11, rue de l'Université,
67000 Strasbourg, France

Abstract. The emission features of 140 Be stars described in "An Atlas
of Be stars" are briefly reviewed. The time scale of the emission
variations are determined for about 35 stars. For several stars we
estimated, on the Hβ line, the V/R ratio of the two emission components.

I. Emission features of Be stars.

Our sample contains 140 Be stars observed regularly from 1953 to 1976
at the Newton focus of the 120 cm telescope of the Haute Provence
Observatory, see also. "An Atlas of Be stars" (Hubert-Delplace et
Hubert, 1979). During the period 1953–1976, the emission lines of
several stars are not always present, see table 1. The variability of
emission, the shell features, and the percentage of stars which go
from a "B star" to a "Be star" phase and both (notation B $\updownarrow$ Be) are
given. The spectral classification of the stars was taken from
Jaschek et al. (1980).

II. Time scale of "long term" variations of emission lines in some
    Be stars.

"Long term" variations of emission lines are often difficult to estimate because of a lack of continuous observations before 1950 except
for some stars. Some results are given in   table 2. P_0 gives an estimation of the B $\updownarrow$ Be cycle, and P_1 an estimation of the modulation
of the emission during the "Be star" phase. In the case of stars with
a "temporary B phase", the time scale of B $\updownarrow$ Be cycles seems larger
for late Be type stars.

III. Time scale of the V/R variation in some Be stars.

The V/R ratio represents the variation of the V and R emission compo-

125

M. Jaschek and H.-G. Groth (eds.), Be Stars, 125–130.
Copyright © 1982 by the IAU.

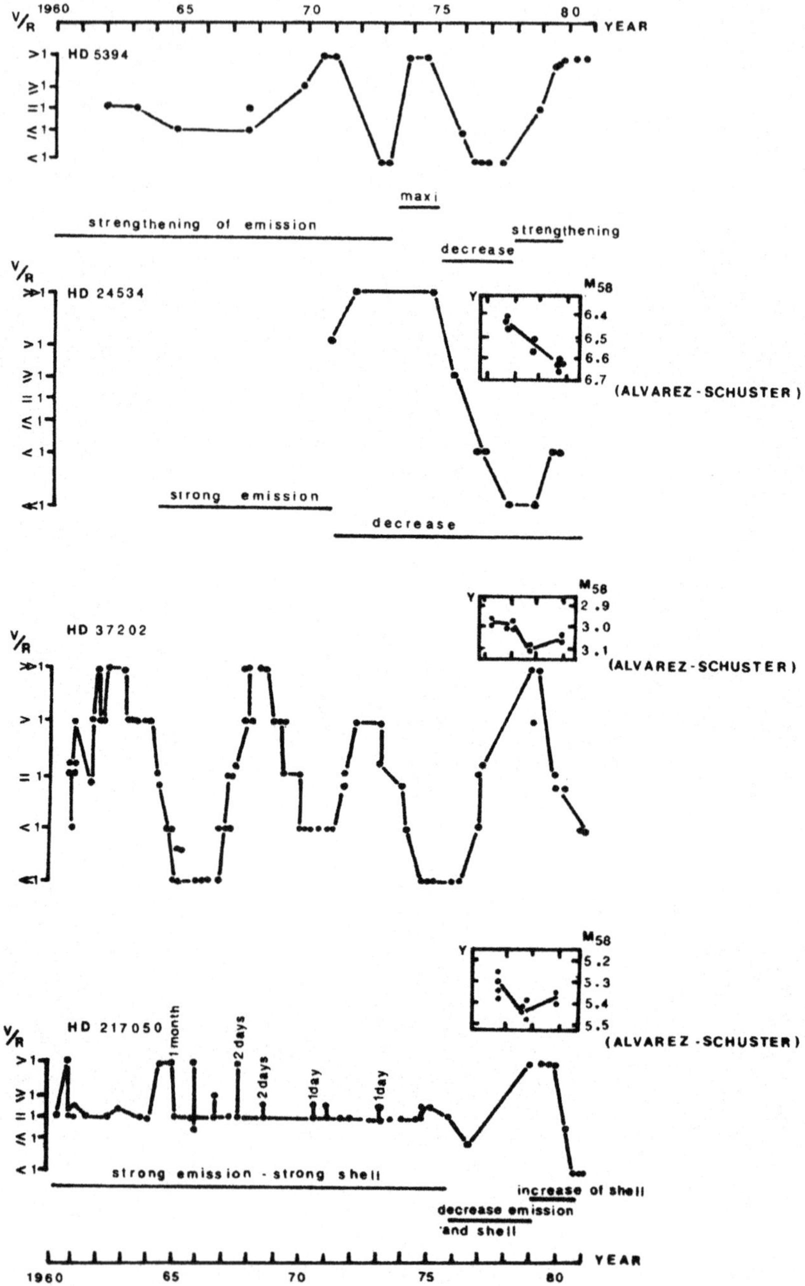

Fig 1. Some examples of V/R variations in Be stars.

nents of the Hβ line profile. For this determination we have used
spectrograms obtained from 1960 to 1980 with the 193 and 152 cm
telescopes (dispersion 9.67 and 12.27 A/mm respectively, Haute-Provence
Observatory). We have searched some correlations between the V/R
variation of the Hβ line profile and the M_{58} variation given by
Alvarez et Schuster (1981) ; some results are given table 3.

Table 1 .- Generalities

	B0-B5e	B6-A0e
A) Permanent emission lines	70,5 %	77,5 %
B) Temporary emission lines	27,0 %	14,0 %
C) No emission lines	2,5 %	8,5 %

A) Permanent emission lines

	B0-B5e	B6-A0e
-strong H emission lines + Fe II emission lines (permanent or temporary)	74,0 %	4,5 %
-variability of emission : strong	27,0 %} 77 %	15,5 %
moderate	50,0 %}	29,0 %
no var.	23,0 %	55,0 %
-shell features (strong or weak , permanent or temporary)	51,0 %	29,0 %

B) Temporary emission lines

	B0-B5e	B6-A0e
-one change B → Be or Be → B	36,0 %	89,0 %
- 2 changes	63,0 %	11,0 %
-shell features during the Be phase	91,0 %	60,0 %

Table 2 .- Time scale of variation of emission lines in some Be stars

A) Stars with a temporary " B phase "

star	sp.t	change	T.S.(years)	lifetime of B phase(years)
HD 35439	B1V	B→Be	P_0 =8,12(var) P_1 =3-4	2
HD 214168	B1V	Be→B	16-20	3
HD 33328	B2IV	B⇄Be	P_0 ν8,12? P_1 ν3-4	> 2
HD 177648	B2V	Be⇄B	ν 10	ν 2
HD 187811	B2V	Be→B	P_0 ν11 P_1 ν 4?	3
HD 191610	B2IV-V	B→Be	> 27	> 6
HD 217543	B2V	Be→B	> 35	> 13
HD 168797	B3V	Be→B	P_0 ν13-16 P_1 ν 4	7
HD 168957	B3V	B⇄Be	> 20	ν 8?
HD 189687	B3IV	B→Be	> 23	ν 13
HD 197419	B3V	B⇄Be	P_0 ν14 P_1 ν 3	5
HD 175863	B4V	B⇄Be	ν 25	ν 5
HD 201733	B4IV	Be B	> 26	4

star	sp.t	change	T.S.(years)	lifetime of B phase(years)
HD 4180	B5III	B⇄Be	P_0 ν17 P_1 ν 4	6
HD 171406	B5V	Be⇄B	13;6(var)	6;3
HD 171480	B6V	B⇄Be	P_0 ν 14 P_1 ν 5?	4
HD 6811	B7III	B→Be	> 32	> 10
HD 22780	B7V	B⇄Be	22<P<36	12<abs<24
HD 142926	B7-8V	B→Be	> 40	14
HD 162732	B7V	Be=B	> 22	
HD 210129	B7V	B→Be	> 50	
HD 175863	B8III	Be→B	> 36	
HD 50658	B8III	B⇄Be	P_0 > 25 P_1 ν 16	
HD 164447	B8V	Be→B	> 45	> 16
HD 23862	B8	Be	> 75	31
HD 144	B9	B→Be	> 45	< 21

B) Stars always in emission

star	sp.t		star	sp.t	
HD 178175	B2V	P ν 10?(lack of observ.)	HD 212076	B2V	P ν 8,var.
HD 205060	B6V	P ν 12-13			

Table 3.- V/R variations of some Be stars

HD	5394 , γCas , B0.5	strong V/R var. during the emission var. Time scale $\sim$4-6 years.
HD	24534 , χPer , O9.5V?	strong V/R var. during the emission decrease. Time scale $\sim$ 10 years? The V mag.(Ferrari-Toniolo et al. 1977)decreases when V/R decreases.M_{58}(Alvarez et Schuster ,1981) increases when V/R increases.
HD	28497 , 228GEri, B2V	strong V/R var. during the emission increase. Time scale $\sim$ 7-9 years.
HD	32991 ,105Tau , B2V	strong V/R var. in 1962-1980. Time scale $\sim$ 10 years.
HD	200120 ,59Cyg, B1V	V/R var. in 1972-1976(2 increases of emission respectively followed by 2 shell phases)and in 1980-1981.
HD	224544 , B6V	V/R var. during the decrease of emission.
HD	224559 , B4V	V/R var. during the decrease of emission. Time scale $\sim$5-7 years.

Be stars with strong shell phases

HD	23862 ,Pléione, B8	weak V/R var. during the decrease of emission (Yilmaz ,1968). Large V/R var. in 1971-1976(strong decrease of emission in 1971-1972,then appearance of a shell phase) The V mag.(Golay,1979) and the B mag.(Sharov et Lyuty,1976) increases when V/R increases.
HD	37202 , ζTau , B2III	strong V/R var. Time scale: 7;4;7 years. Sometimes rapid changes in one day. In1977-1979 M_{58} increases when V/R increases (see Figure 1)
HD	142983 , 48Lib ,B3-4III	strong V/R var. Time scale:9$\rightarrow$12 years.
HD	184279 , B0.5	strong V/R var. Time scale$\sim$4 years .Very rapid changes in some days. In 1977-1979, M_{58} increases when V/R increases.
HD	217050 , EWLac ,B2III	V/R var. in 1976-1980 (see Figure 1)Sometimes very rapid changes.In 1977-1980 M_{58}increases when V/R increases. The same results are obtained with photometric data of Harmanec et al. (1980)

References

Alvarez, M., Schuster, W. : 1981, to be published in Revista Mexicana de Astronomia y Astrophysica.

Ferrari-Toniolo, M., Natali, G., Persi, P., Spada, G. : 1977, Astron. Astrophys. <u>61</u>, 47.

Golay, M. : 1979, Publ. de l'Obs. Astr. de Strasbourg, <u>6</u>, 101.

Harmanec, P., Horn, J., Koubsky, P., Zdársky, F.,Križ, S.,Pavlovski,K.: 1980, Bull. Astron. Inst. Czech. <u>31</u>, 144.

Hubert-Delplace, A.M., Hubert, H. : 1979, An Atlas of Be stars.

Jaschek M., Hubert-Delplace, A.M., Hubert, H., Jaschek, C. : 1980, Astron. Astrophys. Suppl. Ser., <u>42</u>, 103.

Sharov, A.S., Lyuty, V.M. : 1975 IAU Symposium N°70, p. 105.

Yilmaz, N.,:1968, thèse d'Université, Paris, unpublished

DISCUSSION

<u>Divan</u>: The uncertainties in the classification of Be stars due to the
presence of spectral lines originating outside the photosphere can be
quite large if low resolution spectra are used. But MK spectral types
and luminosity classes deduced from spectra with a reasonable resolution are in very good agreement with the BCD classification, and
the presence of Be stars of all luminosity classes from V to III is
confirmed.

<u>Marlborough</u>: For the Be stars which change from Be to B or vice versa,
is there any evidence to suggest that the spectral type is different
when the emission lines are present to times when they are not? If
there are changes in what way do they occur?

<u>Hubert-Delplace</u>: With the MK classification M. Jaschek does not find
significant changes of spectral type when the Be stars change from Be
to B and vice versa. But by using the Herman-Rojas classification
based on $H\gamma$, H_δ, $H\varepsilon$ photospheric lines (not disturbed by emission in
general in the case of stars which present B $\rightleftarrows$ Be cycles), an apparent
change of spectral classification (3 or 4 classes) is observed. (see
also Herman, IAU Symp. 50, 17, 1971).
According to Peton, Astron. Astrophys. 1981, in press, the minimum of
the equivalent width $W\lambda$ ($H\gamma$, $H\delta$, $H\varepsilon$) gives the stellar spectral type
and the maximum of this value gives the influence of the envelope as
a screen effect. This was observed in 66 Oph by Rakotoarigimy and
Herman, Coll. de Liège 1957 and in o And by Peton, Astron. Astrophys.
18, 106 at the beginning of the Be phase. In this case, the <u>apparent</u>
luminosity class is III.

<u>Harmanec</u>: In a paper by Doazan, Harmanec, Koubsky, Kipata and Zdarsky
(submitted to Astron. Astrophys.) we study the behaviour of 88 Her
during the last 20 years. When the star became brighter and bluer and
moved photometrically from B8IV-V to B6 IV-V, the H_α emission almost
disappeared, metallic shell lines completely disappeared, H shell lines
weakened and "photospheric" HeI lines also weakened, behaving thus like
shell lines and indicating a shift to a later spectral type in a formally done spectral classification.

<u>Hirata</u>: In the case of Pleione, the weakening of the Balmer line wings
occured when the brightness decreased. It cannot be explained by the
veiling effect of the envelope, and suggests the variation in the
photospheric level.

<u>Endal</u>: The luminosity classification is important in terms of evoluti-
nary status of these stars. Does the luminosity class change when
there is a Be $\rightarrow$ B transition.

<u>Hubert-Delplace</u>: We have used homogeneous data, spaced out over 23
years of observations at the Haute Provence observatory, to determine
the time scale of the "long term" variations of Be stars, but we know

that these time scales are not periodic, and during the emission phase
we have often observed secondary variations of emission which are
easier to detect with higher dispersion.

RESULTS OF A NEW SURVEY FOR EARLY-TYPE EMISSION STARS

D.J. MacConnell
Centro de Investigación de Astronomía, Mérida, Venezuela.

We discuss the discovery, on objective-prism plates, of 846 stars showing Hα in emission on non-banded spectra. These stars have not previously been reported to have emission and are located primarily along
the southern galactic plane. Twenty-six per cent. of the stars are
known to be of type A0 or earlier.

In MacConnell (1981) we present a catalogue of 731 new stars showing Hα in emission on non-banded spectra. These stars were found on
objective-prism plates (420 A/mm at Hα, widened to 0.3 mm) taken with the
Curtis Schmidt telescope at C.T.I.O. The exposures were of 30 minutes
on IIa-F emulsion behind a RG 610 filter, and the limiting V magnitude
is about 12.5. About 4500 square degrees (220 plate fields) were covered
along the southern galactic plane. The spectral information for these
stars on the plates is very limited as one cannot distinguish spectral
types until about K5 when the TiO bands begin to appear. Nevertheless,
of 202 stars with published spectral information, practically all are
earlier than A0, so we expect that the great majority of the 500-plus
stars with no spectral information are also new Be stars. The prepon-
derance of stars with classifications fall in the range B6-A0, and all
luminosity classes are represented among those with modern classifications. Among HD types B0-B2 no new emission stars were found, and
only 5 stars were found with Houk's HD-revised types in this range; it
is of interest to note that a few emission stars have been classified
as A peculiar by Dr. Houk. Variable Hα emission was noted for several
of the stars which appeared on more than one plate. The number of known
Hα-emission stars rediscovered in the survey is in excess of 2400. Sixteen of the new stars lie in or near eight galactic clusters and are
among the brightest stars in their clusters; two of the clusters are
not previously known to have Be stars.

We report here for the first time an additional 115 new Hα-emission
stars found in a 152-plate extension of the original survey. Spectra
with molecular bands have been eliminated from this group, so it is
similar in all respects to the 731 stars mentioned above except that

M. Jaschek and H.-G. Groth (eds.), Be Stars, 131–133.
Copyright © 1982 by the IAU.

DISTRIBUTION OF 846 Hα-EMISSION
STARS HAVING NON-BANDED SPECTRA
6ʰ
5
8
10
12ʰ
14
16
18ʰ
20
-30°
-60°
-60°
-30°

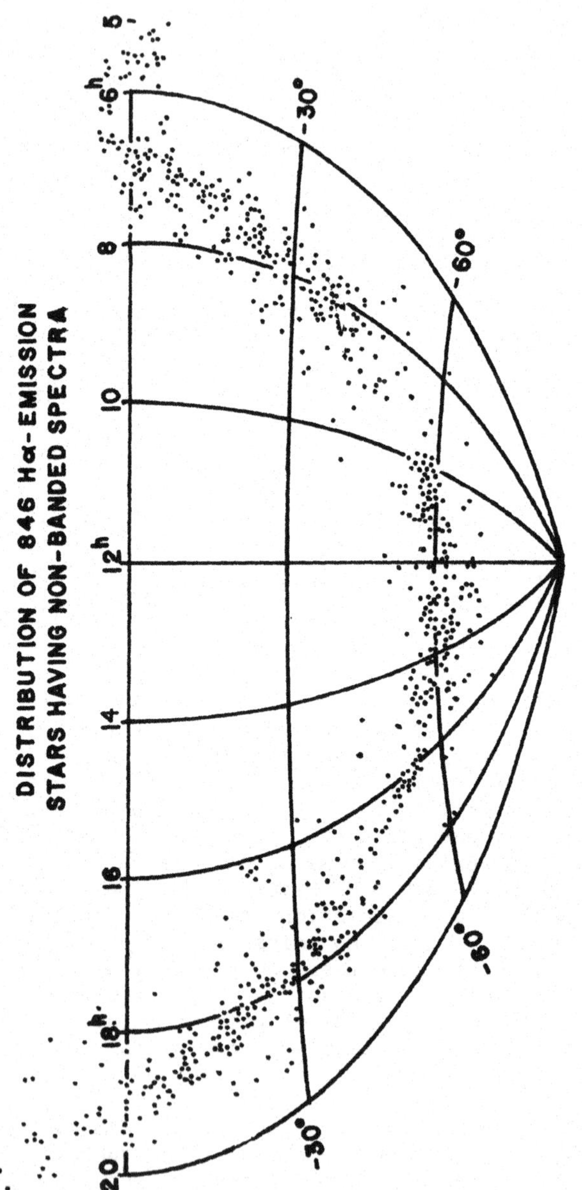

Figure 1

the latter group lies somewhat further from the galactic plane and/or
somewhat north of the celestial equator. The emission stars in the
extension together with other stars of interest found on the same
plates will be prepared shortly for publication in the <u>Astron.Astrophys.
Suppl</u>. Fig. 1 shows the distribution of all the stars discussed here.
The author is presently gathering blue objective-prism plates with the
C.I.D.A. Schmidt in regions of high emission-star density for the purpose of determining approximate spectral types.

Reference:
MacConnell, D.J.: 1981, Astron. Astrophys. Suppl. (in press)

DISCUSSION

<u>Finkenzeller</u>: Are there any H_α emission-line stars which are
associated with reflection nebulosity, and if so, how many?

<u>Mac Connell</u>: My plates do not show any reflection nebulosities that
I am aware of, so I would have to say that the answer is no.

<u>Jaschek</u>: The CDS keeps an updated list of H_α emission-line stars,
with positions, cross identification and other kinds of data.

OBSERVATION DE LA RAIE Hα DANS LES ETOILES Be

Y.Andrillat - Ch.Fehrenbach
Observatoire de Haute Provence

ABSTRACT

Few studies concern the H alpha profile of the Be stars and it is
difficult to do a fine analysis of the line structure, and of its
variations from the results obtained with different resolutions and
accuracies.
Then, it was useful to apply modern technics for the observation of this
line to get an homogeneous material, sensitive enough to show fast
variations of the order of few minutes because an accurate photometry is
available using linear response receivers.

I-INTRODUCTION

Le profil de la raie Hα des étoiles Be a été relativement peu étudié
jusqu'à maintenant et il est difficile de faire une analyse fine de la
structure de cette raie et de ses variations à partir de résultats obtenus avec des résolutions et des précisions différentes.
Il nous a donc paru opportun d'utiliser les techniques modernes pour
l'observation de cette raie afin de disposer d'un matériel homogène,
suffisamment sensible pour mettre en évidence des variations rapides de
l'ordre de quelques minutes et précis car une photométrie à quelques
pourcents est possible en utilisant des récepteurs à réponse linéaire.

II-OBSERVATIONS

Nous présentons un extrait du catalogue des profils de la raie Hα des
étoiles Be dont nous préparons la publication.
Nos observations ont été effectuées à l'aide du spectrographe échelle du
télescope de 152 cm de l'Observatoire de Haute Provence : les spectres
sont enregistrés avec un système de télévision analogique (Adrianzyk et
al 1976).
Nous avons obtenu une centaine de profils de 70 étoiles Be : pour les
plus brillantes, une cinquantaine dont la magnitude est inférieure à 6,
la résolution utilisée est de 10 000. Pour les autres, elle est de 2500
seulement.

135

M. Jaschek and H.-G. Groth (eds.), Be Stars, 135–139.
Copyright © 1982 by the IAU.

Pour quelques étoiles normales (α Lyr, δ UMa), nous avons comparé nos
résultats avec les profils théoriques calculés par Kurucz (1979) :
l'accord est excellent (Andrillat, Fehrenbach, 1981a 1981b).

III-RESULTATS

Variations : La comparaison de nos résultats avec les profils publiées
par différents auteurs montre des variations importantes.
Par exemple, dans le cas de plusieurs étoiles étudiées par Slettebak et al
(1978) (25 Peg, 31 Peg, 66 Oph, $_o$Cas, $_\omega$Ori, ηTau, 120 Tau, κDra, βCMi),
nous décelons une absorption au centre de l'émission Hα , non visible
sur les profils obtenus par ces auteurs avec une résolution inférieure
à la nôtre.
Dans le cas de νGem où la raie Hα est une large émission, Slettebak
observe une absorption centrale très faible alors que sur nos spectres,
elle est intense, son minimum étant nettement au-dessous du continuum.
Il est difficile de savoir si ces variations de profil et d'intensité
sont réelles ou si elles sont dues partiellement ou exclusivement à un
effet de résolution.
L'homogénéité de nos observations effectuées en mars, avril et décembre
1980, nous a permis de mettre en évidence des variations de structure de
la raie Hα dans une vingtaine d'étoiles Be; par exemple :
χPer: en avril l'émission est intense, pointue et montre sur l'aile
violette 2 inflexions tandis qu'en décembre une seule de ces inflexions
subsiste, mais une absorption centrale est apparue (Fig.).
HD 41335: la raie Hα caractérisée en avril par une forte émission avec
une absorption centrale intense, est devenue en décembre une émission
simple avec 2 inflexions sur l'aile violette.
θ CrB: la raie d'enveloppe peu intense visible en mars et avril a
complètement disparu en décembre (Fig.).
κ Dra: l'intensité de l'absorption centrale est sensiblement constante
mais cette raie est nettement dissymétrique, le minimum étant déplacé
tantôt vers le rouge, tantôt vers le violet. Nous avons enregistré un
déplacement correspondant à une cinquantaine de km.s^{-1} dans un intervalle de temps de 24 heures.
25 Ori: l'intensité et le profil de l'absorption centrale ont varié
entre avril et décembre. A cette date, nous observons une absorption peu
intense très large avec un minimum très aplati.
L'étude détaillée d'une telle structure nécessiterait une résolution
supérieure à celle que nous avons utilisée.
Outre les variations de profil, nos spectres montrent des variations
d'intensité de raie, du rapport V/R et de la largeur équivalente. Ce
dernier paramètre étant indépendant de la résolution, nous avons pu
comparer nos résultats avec ceux obtenus notamment par Slettebak et al de
1975 à 1977.

Largeurs de la raie Hα: dans les étoiles Be, les raies d'émission de
l'hydrogène sont généralement très larges. La cause de cet élargisse-
ment n'est pas encore connue (vent stellaire, diffusion par les électrons
dans l'enveloppe...).

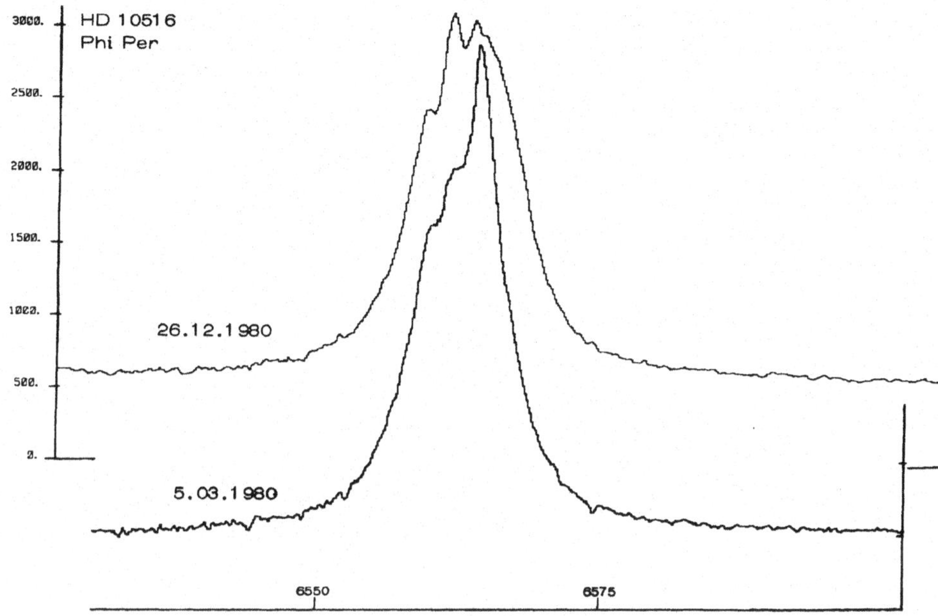
3000.
2500.
2000.
1500.
1000.
500.
0.
HD 10516
Phi Per
26.12.1980
5.03.1980
6550
6575

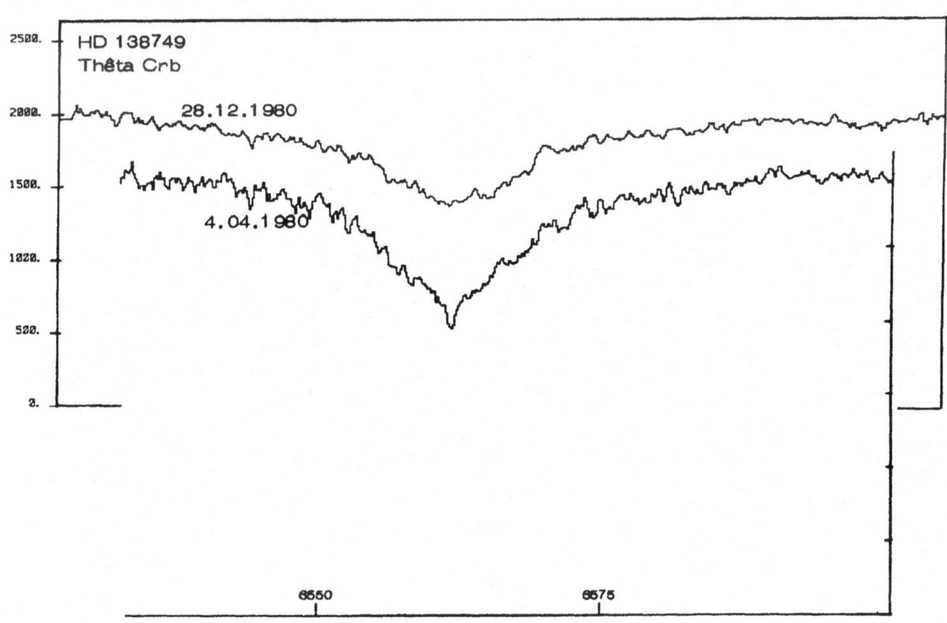
2500.
2000.
1500.
1000.
500.
0.
HD 138749
Thêta Crb
28.12.1980
4.04.1980
6550
6575

Bien que les mesures de largeurs de raies soient très imprécises, nous
avons essayé de déterminer l'extension des ailes de la raie Hα afin de
rechercher une éventuelle corrélation avec divers paramètres (type
spectral, largeur à mi-hauteur, vitesse de rotation).
Nous avons adopté la classification spectrale proposée par Jaschek et al
(1980). Les largeurs sont mesurées à la base de la raie. On constate que
toutes les étoiles du groupe I (exceptées HD 40978 et HD 37657) ont des
largeurs exprimées en km.s^{-1}, supérieures à 800 km.s^{-1}. Par contre, les
étoiles des autres groupes (exceptées 25 Ori et 66 Oph) ont des largeurs
inférieures à 900 km.s^{-1}. Il n'y a pas de distinction visible entre les
groupes II, III, IV et V.

Il s'avère donc que ce sont les étoiles avec de nombreuses et intenses
raies d'émission de H I et Fe II (groupe I) qui ont des ailes de Hα les
plus étendues.
L'émission Hα présente généralement une absorption centrale et la largeur de la raie (mesurée à mi-hauteur) est donc surestimée et ne correspond pas à la véritable largeur de la raie d'émission de l'enveloppe.
Nous avons montré que Hα était bien représenté par une ou plusieurs
fonctions de type gaussien et nous avons déterminé la largeur à mi-
hauteur sur le profil de l'émission pure. Pour les étoiles des groupes
II à V,l'extension des ailes est relativement faible, elle correspond à
la fonction gaussienne qui représente l'émission dans ce cas. Par contre,
une relation grossièrement linéaire semble exister entre ces deux
quantités pour les étoiles du groupe I suggérant que dans ce cas l'élar-
gissement de la raie est d'autant plus grand que la largeur à mi-hauteur
de l'émission pure est plus grande.
En ce qui concerne la vitesse de rotation, elle ne semble pas corrélée
avec l'étendue des ailes. Cependant, pour le groupe I, il semble que
les élargissements importants des raies apparaissent dans les étoiles
ayant une grande vitesse de rotation.

Des mesures plus précises des largeurs, notamment au moyen d'une
représentation gaussienne des profils permettront sans doute de confir-
mer ou d'infirmer l'existence de ces corrélations.

IV-CONCLUSION

L'établissement de ce catalogue des profils de la raie Hα dans les
étoiles Be montre que l'instrumentation utilisée et la résolution choisie
sont bien adaptées à une surveillance suivie des variations spectrales
de ces étoiles.

REFERENCES

Adrianzyk,G., Baietto,J.C., Berger,J.P., Fehrenbach,Ch., Prévot,L., Vin,
        A.:1978,Astron.Astrophys.63,279.
Andrillat,Y., Fehrenbach,Ch.: 1981a,Coll.UAI n°59.
Andrillat,Y., Fehrenbach,Ch.: 1981b, 2° Colloque National du Comité
        Français du Télescope Spatial.

Jaschek,M., Hubert-Delplace,A.M., Hubert,H., Jaschek, C.: 1980,Astron.
        Astrophys.Suppl.Ser.42,113.
Kurucz,L.: 1979,Astrophys.J.Suppl.40,1.
Slettebak,A.,Reynolds,R.C.: Astrophys.J.Suppl.Ser.38,205.

DISCUSSION

Schumann: How many stars will be contained in your catalugue to be
published?

Fehrenbach: 70 stars, mostly stars observed in winter. We are prepared
to observe stars in the Cygnus-Lyra etc. complex.

Peters: 66 Oph is a very interesting star, indeed IUE observations
of this object in April 1980 revealed the presence of highly violet
shifted components to the resonance lines of Si III, Si IV and C IV.
Violet shifts of 700 km s^{-1} were observed. I will discuss the IUE
observations on Thursday.

Snow: I have two questions: First, in some cases where you showed
two profiles of the same star, taken at different times, one spectrum
appeared noisier than the other, why? And second, in how many cases
did you find absorption cores? It is very important to use high
spectral resolution, as you have pointed out, since many stars not
previously suspect of having narrow absorption cores in H_α are found
to have them. This may revise the statistics on frequency of shell
absorption as a function of projected rotational velocity.

Fehrenbach: 1. Some stars observed under bad conditions (near horizon),
the longer exposure time explains the noise.
2. 85% of all stars.

Dachs: (a discussion point raised by Dr. Snow):
At sufficiently high resolution, as shown by Dr. Fehrenbach, central
absorption is very often present in H_α emission line profiles not
only for stars seen equator-on, but also for pole-on stars.

Fehrenbach: I have some difficulties to see the differences between
equator-on and pole-on stars.

Dachs: Broad wings of H_α emission extending to about $\pm$ 25 A from the
line centers can be observed for most early Be-type stars, if the H_α
emission line is strong enough. They probably are due to electron
scattering in the envelope surrounding the star.

Fehrenbach:Many stars with strong emission have broad wings. I think
your explanation is right. We observed some stars with quite strong
lines on an absorption line.

ON THE RADIATION DEFICIENCY OF SHELL STARS IN THE BALMER
CONTINUUM

J.N. Chkhikvadze
Abastumani Astrophysical Observatory of the Academy of Sciences
of Georgian SSR

The present report gives the results of a statistical comparison of
shell stars with normal Be stars to establish the differences in the
parameters of the gaseous envelopes, such as the radiation power in the
Paschen continuum and the optical depth of the envelopes in the Balmer
continuum.

Let us consider Figure 1, showing the dependence of the Q_{BVK} parameter
on the spectral type of shell and normal Be stars ($Q_{BVK}=(B-V)-0.36$ (V-K).
The figure 1 illustrates a well-known property of the emission line B
stars : the radiation power of the gaseous envelope rises when passing
from late to earlier spectral subclasses.

Figure 2 shows the Hα emission line intensity dependence on the Q_{BVK}
parameter. It should be noted that there is a definite correlation. Thus
it can be assumed that the parameter Q_{BVK} can be used as a certain qualitative measure of the radiation power of a gaseous envelope in the
Paschen continuum.

Let us turn to Figure 3, giving the distribution on the parameter Q_{BVK}
for shell and normal stars. According to the criteria of Kolmogorov-
Smirnov and Fisher and Student (Mitropolsky, 1961) both distributions
turned out to be identical at the 95 % confidence level.

Hence the conclusion can be drawn that gaseous envelopes of shell and
normal Be stars do not differ from each other in their radiation power
in the Paschen continuum.

In our previous paper (Chkhikvadze, 1980) we concentrated our attention
on the fact that some of the shell stars are observed to have anomalously high Balmer jumps. In certain cases (EW Lac, Tau, 48 Lib) the anomaly
was expressed as additional absorption jumps. It is easy to understand
that such a phenomenon must be due to an appreciable opacity of the envelopes of such stars in the Blamer continuum.

141

M. Jaschek and H.-G. Groth (eds.), Be Stars, 141–146.
Copyright © 1982 by the IAU.

To solve the problem whether the radiation deficiency is specific of
all shell stars in the Balmer continuum, it would be desirable to take
high resolution spectra of such objects in the Balmer limit région,
although this is a very laborious process. However, as we shell see
later, one can use the following difference ΔQ_{UBV} = Q_{UBV} (observed) −
Q_{UBV} (intrinsic) where Q_{UBV} is Johnson's well known parameter.

Figure 4 borrowed from our previous paper (Chkhikvadze, 1980) displays
the dependence between ΔQ_{UBV} and ΔD, where ΔD is the difference between
the observed and intrinsic Balmer jump. It can be seen that for normal
Be stars, having a small value of Balmer jump compared with normal B
stars of the same subclasses, there is a pronounced correlation between
ΔD and ΔQ_{UBV}, which are, besides, negative. On the other hand for shell
stars with an anomalous jump the differences ΔQ_{UBV} proved to be positive.

Hence, the technique of investigation is based upon the suggestions
as follows :

1. Optically thin gaseous envelope in the Balmer continuun must result
in the radiation excess in the ultraviolet and consequently ΔQ_{UBV}
must be negative.

2. If the optical depht of the envelope is not small then there must
be observed a certain deficiency of the ultraviolet radiation with similar values of the rest of the parameters of the star and the envelope.
That is, for stars with optically thick envelope in the Balmer continuum the difference ΔQ_{UBV} must be algebraically higher than the similar value determined for a star with optically thin envelope.

Let us consider Figure 5 showing the distributions of ΔQ_{UBV} parameters
for shell and normal Be stars.

Estimates of standard statistics of the above mentioned distributions
are given in Table I.

TABLE I.

Statistics	shell	Be
Mean value	+ 0.011	− 0.066
Dispersion	6.68 (−3)	6.22 (−3)
Asymmetry	+ 0.30	− 0.279
Excess	0.80	0.18

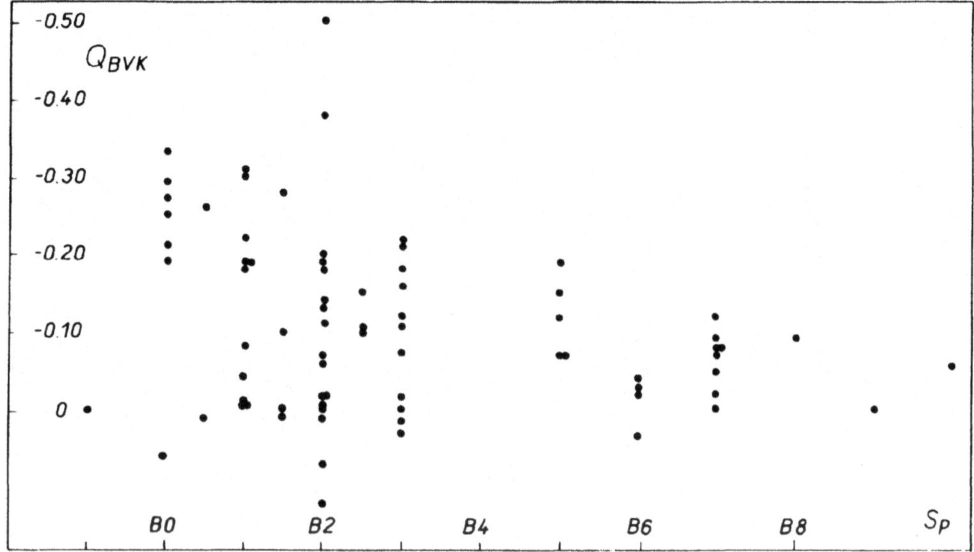

Figure 1.

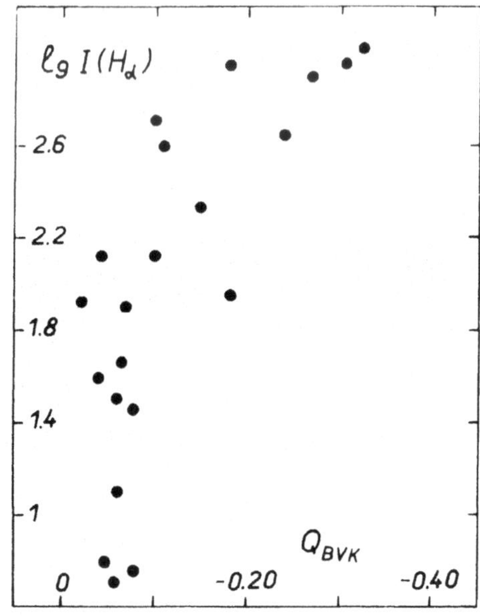

Figure 2.

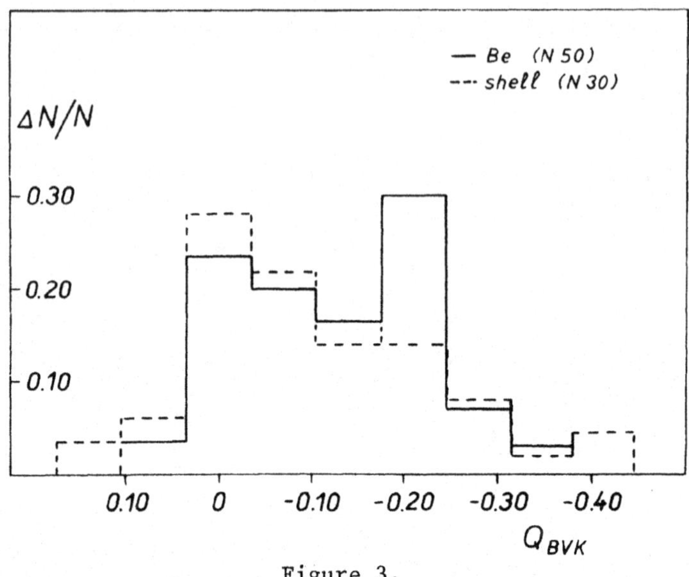

Figure 3.

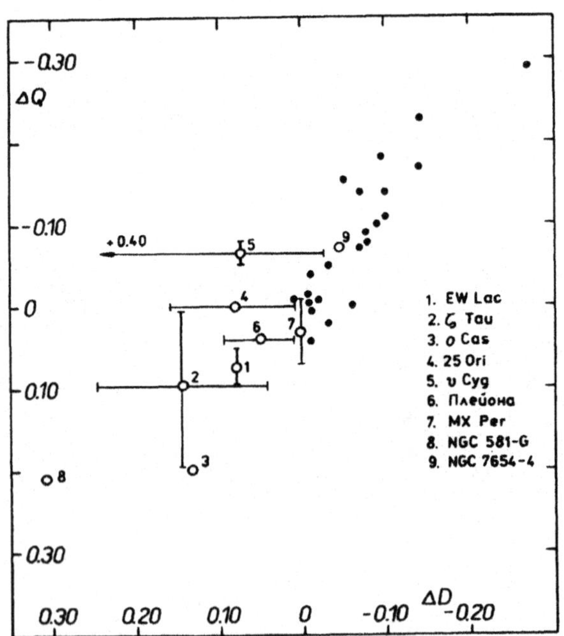

Figure 4.

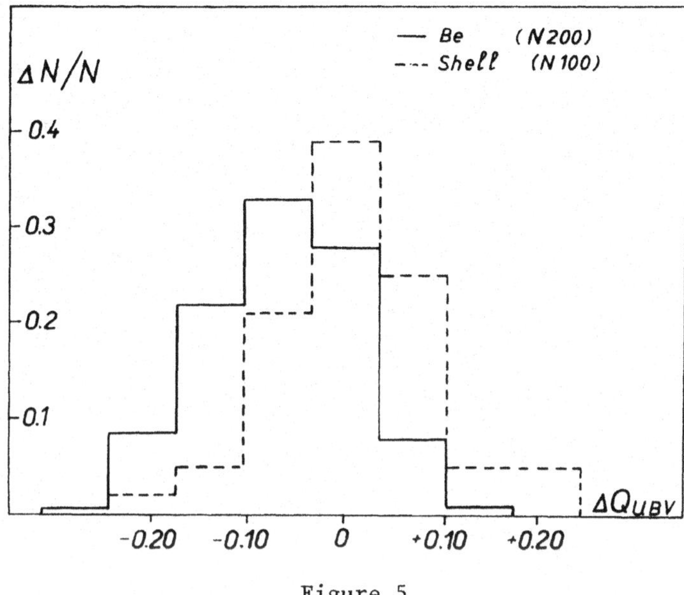

Figure 5.

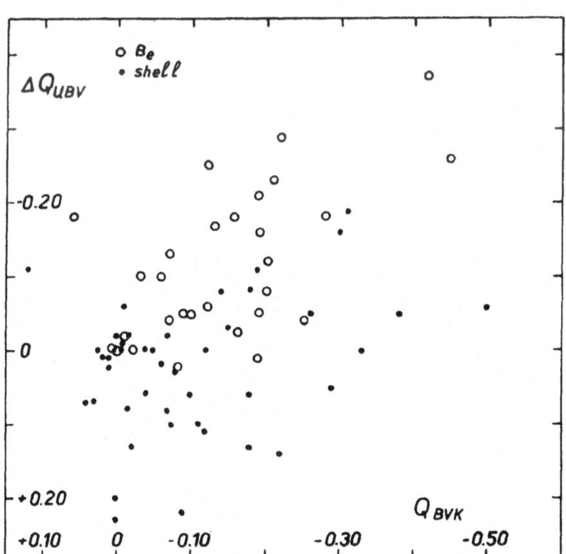

Figure 6.

The calculation showed that at the 95 % confidence level these distributions don't represent samples from one general totality. Hence, the conclusion is drawn that shell stars are characterized by the radiation deficiency in the ultraviolet being due to the fact that the optical depth of the envelopes of shell stars in the Balmer continuum is not small.

The above conclusion is well illustrated in Figure 6 where the shell stars compared with normal Be ones are located otherwise : with the same values of the parameters Q_{BVK} algebraically greater values of ΔQ_{UBV} are specific to shell stars.

REFERENCES

Mitropol'skij, A.A., Tekhnika Statisticheskikh Bychichlenij 1961, Moskva, Fizmatgyz.
Chkhkbadze, Ya. N., Astrofizika 1980, <u>16</u>, no. 4.

INTENSIFIER-DISSECTOR-SCANNER OBSERVATIONS OF THE BRIGHT
NORTHERN Be STARS

Paul K. Barker
The University of Western Ontario

ABSTRACT. Early results are presented from an extensive program of
regular Balmer-line spectroscopy; during eight months in 1980 Hα
profiles were obtained of over 100 northern Be stars. Transient
emission events have been observed in the very early rapid rotator
59 Cygni; sequential variations in Hα are quasi-periodic.

The University of Western Ontario intensifier-dissector-scanner
(IDS) is a 512-channel instrument which when mounted on the Cassegrain
spectrograph of the 1.2m telescope gives a channel center separation of
about 0.7Å at Hα and 0.3Å at the other Balmer lines; resolution is two
or three pixels. The IDS is described by Tomaszewski (1980) and
Tomaszewski et al. (1980). Systematic observations were begun in 1980
April; by the end of November, Hα scans were secured of the program Be
stars as well as of 60 normal B stars and 30 A and Ap stars. Of the Be
stars, one-third were also observed at Hβ and in the near-infrared.
Standard stars are observed on every clear night. This continuing long
term survey work is complemented by intensive observations of selected
active Be stars.

59 Cygni is renowned for its spectral variations--almost ceaseless
since the turn of the century--culminating in the spectacular shell
episode of 1974-75 (Barker 1981). The more recent optical history is
summarized by Hubert-Delplace and Hubert (1979); UV variability and wind
structure is discussed by Doazan et al. (1980) and by Marlborough and
Snow (1980). IDS scans of Hα were obtained on the dates listed in
Figure 1.

The Figure shows that from July onwards the profile undergoes a
regular sequence of variations: an initial asymmetric emission peak,
shallow on the red side, develops a redward shoulder within about 2 days;
a day or so later the strong blue peak subsides, leaving an essentially
flat-topped profile which sometimes shows a weak central reversal. The
flat-topped profile persists for at least 4 days. The April profile
alone has a strong red peak with a blue shoulder.

M. Jaschek and H.-G. Groth (eds.), Be Stars, 147–150.
Copyright © 1982 by the IAU.

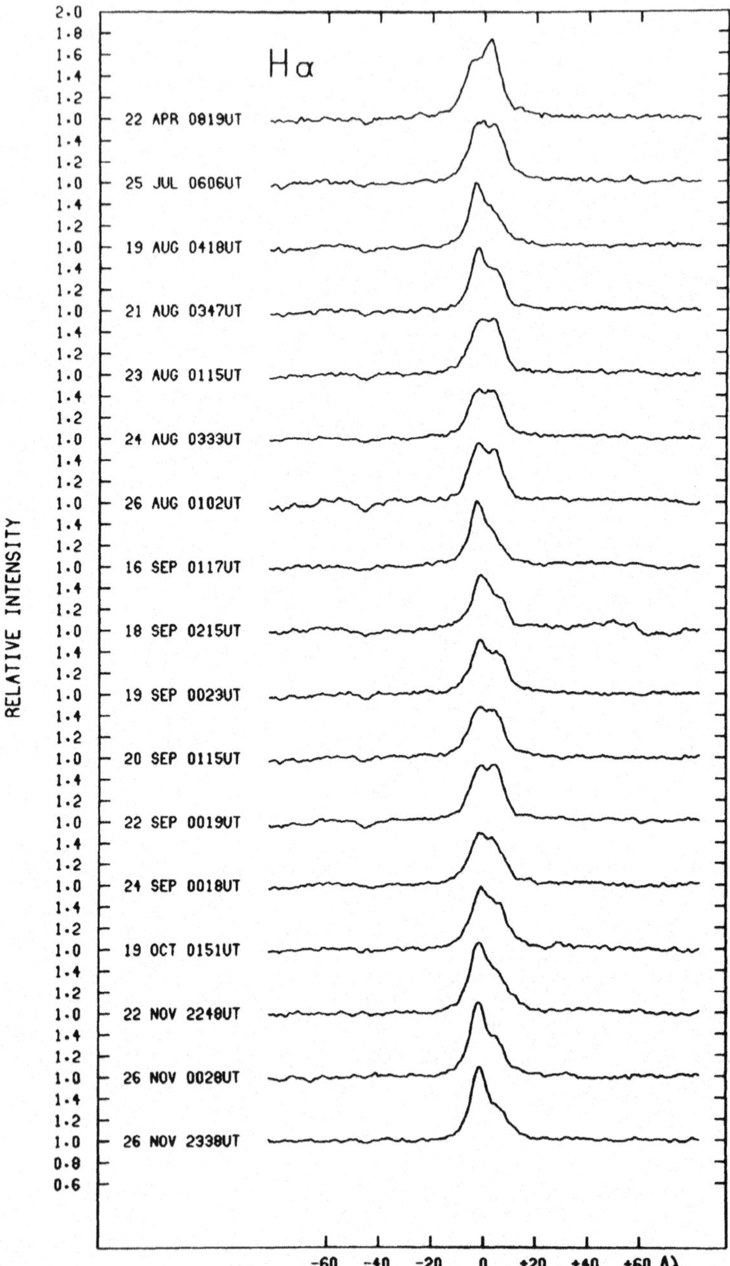

Figure 1.   Hα profiles of 59 Cygni from 1980 April to November.

It is well known that instruments of the IDS breed are subject to electronic variability on a variety of timescales. Therefore repeated observations of presumed non-variable standard stars were made throughout the 59 Cygni sequence. Profiles of Hα in ν Cyg (A5V) and ρ Cyg (G8III) do not vary beyond the 3% noise level typical of the IDS. Thus the variations in the Hα profile of 59 Cygni are undeniably stellar in origin. Further, 17 scans of Hα in 59 Cygni acquired during August 23 and 24 do not show significant changes above the 3% level, so the stellar activity has a characteristic timescale of days and not hours.

The other Balmer lines were observed less frequently. Hβ is largely filled in by emission. At times when Hα is flat-topped, Hβ shows a weak central reversal; when Hα has the strong blue peak, Hβ has a weak blue emission peak. Apparently analogous changes in Hγ are scarcely visible above the IDS noise, while Hδ is more nearly photospheric and unvarying. Between June 23 (when only a flat continuum was seen) and July 25 the Paschen lines P13 to P17 appeared in emission with flat-topped profiles of width 600 km s^{-1} and intensity 1.1 (Barker 1980). No changes were seen in this wavelength region correlated with Hα variations, but a gradual decline of the Paschen lines took place from July to November.

Notice that beginning August 19 and September 16, Hα shows an effectively identical sequence of changes occurring in the same time intervals. This suggests the possibility that these variations may be truly periodic. Any periodicity must be short-lived however, because the November profiles show that a phase shift or other transformation has taken place. Intermittent periodicity has been observed in γ Cas (Hutchings 1970) on a timescale identified as the stellar rotation period. In the case of 59 Cygni, it is interesting that for any realistic choice of stellar radius and rotation velocity, the photospheric rotation period must be less than a day, so the emission variations cannot be associated with photospheric rotation. There is no spectroscopic evidence that the star is a binary. Speculative interpretation of these emission transients is avoided pending the next season of observations; undoubtedly other observers have Hα profiles from the current season which will help elucidate Figure 1.

This work was supported by the National Science and Engineering Research Council of Canada.

REFERENCES
Barker, P.K. 1980, IAU Circ. No. 3498.
Barker, P.K. 1981, preprint.
Doazan, V., Kuhi, L., and Thomas, R.N. 1980, Astrophys.J.Letters,235,L17.
Hubert-Delplace, A.-M., and Hubert, H. 1979, "An Atlas of Be Stars",
          Paris-Meudon Observatory.
Hutchings, J.B. 1970, Monthly Notices Roy. Astron. Soc., 150, p.55.
Marlborough, J.M., and Snow, T.P. 1980, Astrophys. J., 235, p.85.
Tomaszewski, L. 1980, Ph.D. Thesis, The University of Western Ontario.
Tomaszewski, L., Symonds, G.R., and Landstreet, J.D. 1980, Publ. Astron.
          Soc. Pacific, 92, p.518.

DISCUSSION

<u>Metz</u>: Can you exclude 59 Cyg being a binary?

<u>Barker</u>: Photographic spectra obtained from 1974 to 1979 at 37 A/mm
show complex line profile changes during the second shell phase of
59 Cyg, but at times when the lines appear to be photospheric there
is no evidence for any radial velocity variations.

<u>Andrillat</u>: Did you observe the OI line λ 7772 in 59 Cyg and have you
compared its behaviour with respect to λ 8446A?

<u>Barker</u>:The 1-2 A resolution near-infrared spectra show a peculiar
asymmetric emission feature near (but not <u>at</u>) λ 8446, which shares
the general decline of the Paschen emission from July to November 1980.
I have not observed 59 Cyg in the λ 7700 A region.

<u>Fehrenbach</u>: En 59 Cyg les contours pourraient s'expliquer par une
raie d'absorption centrale qui se déplace. Elle est centrale quand
on observe un plateau.

<u>Barker</u>: That is one possible description of the observed variability
However, higher resolution spectra of H_α obtained by Dr. Doazan do
<u>not</u> show significant central absorption when the profile is flatt-
topped.

<u>Harmanec</u>: Have you serious problems with H_2O atmospheric lines in the
neighbourhood of the H_α line?

<u>Barker</u>: Even when the skies at Western Ontario are clear, there are
severe problems with water vapour because we are surrounded by the
Great Lakes; the atmosphere is usually hazy and the humidity is high.
However the repeated observations of H_α in P Cyg show that there are
no effects larger than the instrumental noise level of 3%.

<u>Hubert-Delplace</u>: Have you observed, in the case of 59 Cyg, signifi-
cant variation of H line profiles in one night?

<u>Barker</u>: For none of the observation dates from 1980 April to November
were any variations detected within a single night (above the instrumental noise level of 3%).

OPTICAL SPECTROSCOPY OF HD102567 (4U1145-61)

Eduardo Janot Pacheco
Instituto Astronômico e Geofísico U.S.P./Brasil

C. Chevalier, S.A. Ilovaisky
Observatoire de Paris-Meudon/France

Abstract: Analysis of optical spectra of HD102567 (HEN 715),
the optical counterpart of the X-ray source 4U1145-61 is presented.
Estimates of the star's rotation velocity, inclination angle, distance
and envelope characteristics are given. Some consequences of the
possible existence of a 190^d orbital period for this Be/X-ray sources
are discussed.

I. SPECTROSCOPIC OBSERVATIONS

Ten spectra of HEN 715, the optical counterpart of the X-ray source
4U1145-61 (Sofia et al., 1974) were obtained in March 1976 and March
1977 at the ESO 1.5m telescope with the Echelec spectrograph equipped
with a Lallemand electronographic camera. These spectra were taken
in the first order at dispersions of 62 and 124Å/mm and wavelength
ranges $\lambda\lambda$4000-5000Å and $\lambda\lambda$3600-5500Å, respectively. A detailed analysis
of the observations will be present elsewhere (Janot-Pacheco et al. 1981b).

The spectrum of HEN 715 is that of a rapidly rotating B star. Hβ is
seen in emission with asymmetric profile stronger at the red edge. Hγ
is partially filled in by emission and other Balmer lines are present
in absorption up to H16 (sometimes, up to H25). HeI absorption lines
are prominent. The CIII-NIII-OII blend at $\lambda\lambda$4634-51 is strongly marked
and other strong OII lines are seen mainly around Hγ. Several weak and
often diffuse emission lines were identified with FeII. Many weak
interstellar lines are present throughout the spectrum. They show an
average velocity of 7±1 Km/s. Two Balmer discontinuities seem to be
present at $\lambda\lambda$ 3647 and 3700 although the low signal-to-noise ratio in
this spectral region does not allow a clear conclusion.

Hβ and Hγ show drastic profile changes. Hammerschlag-Hensberge et
al. (1980) reported variations in both the profile and central wavelength of Hβ in timescales of 4 min.

Based on the Ne-Si grid of Walborn (1971) we suggest for HEN 715 a

M. Jaschek and H.-G. Groth (eds.), Be Stars, 151–154.
Copyright © 1982 by the IAU.

spectral classification BO.7-B1 III-V. This classification agrees with
a previous one by Feast et al. (1961). The difficulties of classifying
Be stars should however be kept in mind.

II. ROTATION VELOCITY AND INCLINATION ANGLE

The width of the HeI lines indicate $V \sin i = 250 \pm 29$ Km/s. Some allow-
ance should be given for instrumental profile effects. The width at base
for the Hβ emission line is ∿500 Km/s, similar to that seen in other Be
stars (Janot-Pacheco et al., 1981a).

The absence of shell lines and the shape of Hα, Hβ and Hγ lines
favor i<45° (Poeckert & Marlborough, 1978).

III. VELOCITY VARIATIONS

Radial velocity variation for Hβ emission line and for HeI lines
are shown in Figure 1. Note the variation for Hβ between JD2443212 and
JD2443215, which indicates varying physical conditions in the envelope
(see below).

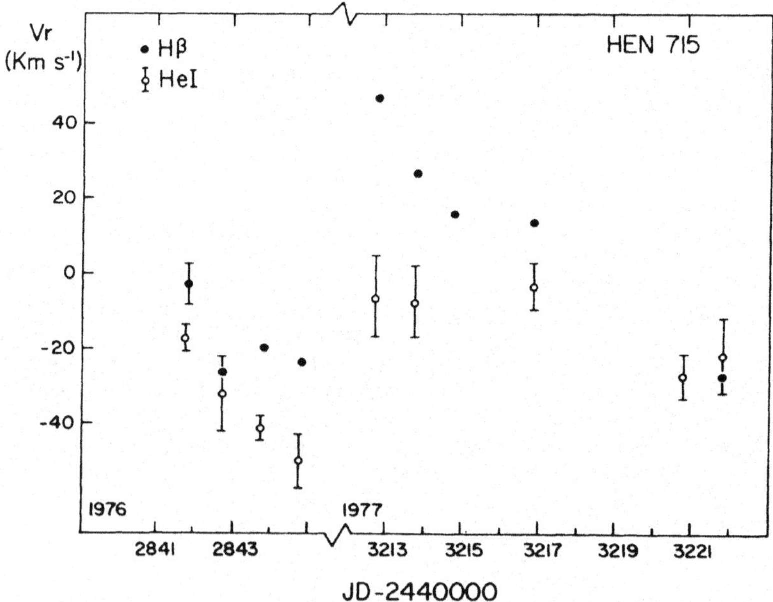

Fig.1 - Velocity variations for the Hβ emission line and HeI
absorption lines of HEN 715. Two epochs are shown.

Watson (1979) suggested a 188^d orbital period for the system 4U1145-61/HEN 715 on the basis of four strong X-ray flux enhancements seen by Ariel V from 1975 to 1978. Our 1976 and 1977 HeI velocities fit in well folded with Watson's ephemeris but they span only from 0.68 to 0.73 in phase.

IV. DISTANCE AND REDDENING

The equivalent width of the CaII K line indicates a distance of 1.3 ± 0.3 Kpc.

Miller (1972) found $A_v<1^m.5$ within the first 2–3 Kpc in the direction of the Carina spiral feature. The position of HEN 715 in the (U-B)X(B-V) diagram in 1964 (Feinstein, 1969) suggests $E_{B-V} \simeq 0^m.30$. This supports the (stellar) color-excess $E_{B-V} = 0^m.25$ derived by Bianchi & Bernacca (1980) from IUE observations. Taking $M_V(B1V)=-2.9$ (Panagia,1973), $V=9.39$ (Feinstein, 1969) we obtain for the star a distance $\sim1.9\pm0.1$ Kpc.

Adopting a distance of 1.6 Kpc, the extreme flux densities of 4U1145-61 (Bradt et al., 1979) correspond to X-ray luminosities $L_X = 6.7\times10^{36}$erg/s and 4.7×10^{34}erg/s, respectively.

V. SOME ENVELOPE CHARACTERISTICS – MASS LOSS

The observations indicate rapidly varying physical conditions in the envelope of HEN 715. Changes in density and/or in expansion velocity can account for line position and profile changes (Poeckert & Marlborough, 1978). The asymmetric $H\beta$ and $H\alpha$ (Watson et al., 1978) profiles stronger at the red edges indicate large expansion velocities in the envelope in the models of Poeckert & Marlborough.

The electron density can be estimated from the quantum number of the last visible Balmer line. This number varies from n=16 to n=25 yielding $12.78 < \mathrm{Log}\ N_e < 14.23$.

The mass loss can be estimated from $dM/dt = 4\Pi R^2\rho(R)V(R)$. Taking $R=6.8\ R_\odot$ (Underhill et al., 1980), V=40 Km/s (the average velocity of the upper Balmer lines) and the densities above we get $3.6\times10^{-8}M_\odot/yr < dM/dt < 8\times10^{-7}M_\odot/yr$.

VI. DISCUSSION

Maraschi et al. (1976) suggested the existence of a class of binary X-ray sources wherein the primary would be a rapidly rotating BVe star. Janot-Pacheco et al. (1981a) argued that direct interaction of a compact object with a typical Be envelope could be important in the production of X-rays. If the system 4U1145-61/HEN 715 has indeed an orbital period of $\sim190^d$, the dimension of the orbit of the compact object will be

$\sim 1.7 \times 10^{13}$ cm for typical masses of the objects involved. This corresponds to a distance of ~ 35 stellar radii from the Be star. If the density in the envelope varies as $r^{-\alpha}$ ($\alpha > 2$), at $r = 35 R_*$ the densities will be of the order $10^7 - 10^9$ cm^{-3}. Accretion onto a neutron star passing through a gas of such densities could explain the 1978 flare of 4U1145-61 (see section IV) for relative velocities of at least ~ 100 Km/s (see Ostriker & Davidson, 1973). Assuming that the envelope is supported by centrifugal forces, such high relative velocities could only be achieved with retrograde motion or in an eccentric orbit.

REFERENCES

Bianchi, L., Bernacca, P.L. 1980, Astron. & Astrophys. 89, 214.
Bradt, H.V. et al. 1978, in IAU/COSPAR Symp. on "X-Ray Astronomy".
Feast, M.W. et al. 1961, Mon. Not. R. astr. Soc. 122, 239.
Feinstein, A. 1969, Mon. Not. R. astr. Soc. 143, 273.
Hammerschlag-Hensberge, G. et al. 1980, Astron. & Astrophys. 85, 119.
Janot-Pacheco, E. et al. 1981a, Astron. & Astrophys. (in press).
─────────────────────────── 1981b, in preparation.
Maraschi, L. et al. 1976, Nature 259, 292.
Miller, E.W. 1972, Astron. J. 77, 216.
Ostriker, J.P., Davidson, K. 1973, IAU Symp. nº 55, "X and γ- Ray Astronomy", Bradt & Giacconi, eds., p. 143.
Panagia, N. 1973, Astron. J. 78, 929.
Poeckert, R., Marlborough, J.M. 1978, Astrophys. J. Suppl. 38, 229.
Sofia, S. 1974, Astrophys. J. 188, L45.
Underhill et al. 1979, Mon. Not. R. astr. Soc. 189, 601.
Walborn, N.R. 1971, Astrophys. J. Suppl. 23, 257.
Watson, M. 1979, private communication.

SEARCH FOR LONG-PERIOD RADIAL VELOCITY VARIATIONS IN SOME
Be STARS

Pastori L.[*], Antonello E.[*], Fracassini M.[** - ***] and
Pasinetti L.E.[* - ***]
*   Osservatorio Astronomico di Milano-Merate
**   Instituto di Astronomia e Geodesia, Milano
***   Dipartimento di Fisica, Università di Milano

INTRODUCTION

Some astronomers have suggested long period phenomena in the Be
stars : Hubert (1971), Delplace and Hubert (1975), Feinstein (1975),
Harmanec et al. (1976), Pustylnik (1976). In particular, Fracassini et
al. (1977) have made a periodogram analysis of the radial velocities
(RV) of the Be star o And from 1900 to 1976, to connect the shell appearance with eventual long-term RV variations. In the present study
all the RV of seven Be stars, found in the literature from the beginning of the century up to now, have been assembled and analysed in the
same way as o And, in order to find out long period phenomena (duplicity,
variability, shell activity, etc...). A brief review on the studied stars
may be found in Harmanec et al. (1980).

METHODS AND RESULTS

The long period analysis of the RV variations has been performed
for EW Lac, 28 Tau, ζ Tau, KX And, CX Dra, 88 Her. Whenever possible,
the hydrogen and metallic lines were considered separately and averaged
for each spectrum and for each month. Therefore in this analysis we
have averaged all the short period variations. The justifications of
this procedure, already adopted in the previous work (Fracassini et al.
1977), are given in a more extensive paper to be published elsewhere.
The reliability of our periodogram analysis was tested with a simulated
RV curve. Only three stars have sufficient data, suitably spaced in the
time, for a periodogram analysis. However, some useful indications were
obtained also for two other stars.

The following conclusions have been drawn :

- EW Lac : the plot of the averaged RV vs date is shown in Fig. 1. In
spite of the dispersion of the points, long-term variations of RV values
are evident. The periodogram analysis gives a period of 40.3 years with

M. Jaschek and H.-G. Groth (eds.), Be Stars, 155–159.
Copyright © 1982 by the IAU.

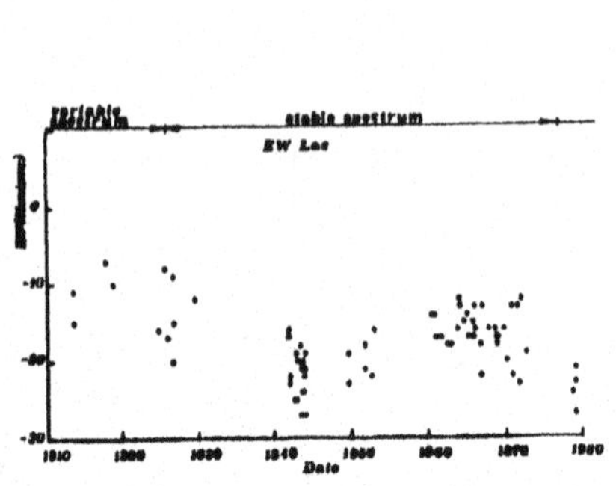

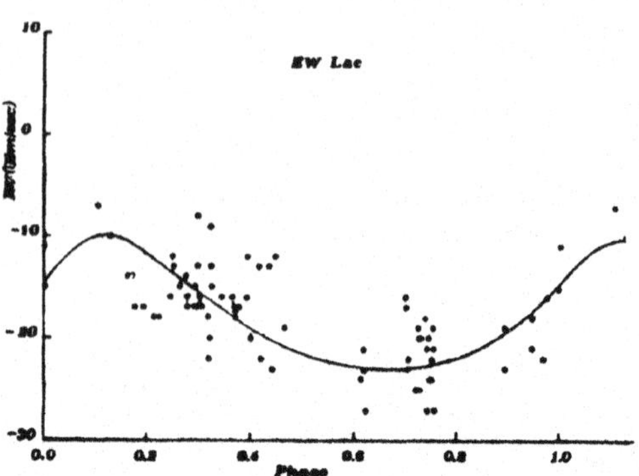

Fig. 1 Averaged RV vs date   Fig. 2 RV vs phases
(P=40.3 years).

a good confidence (Fig. 2). This period could be of dynamical nature
since the envelope of the star seems to have been stable for fifty years,
from 1926 to the end of 1977 (Hadrava et al. 1978 ; Harmanec et al.
1979). This hypothesis may be supported by a result of Scholz (1981)
who suggests a possible existence of a cool companion. In the hypothesis
of a binary system, from the RV curve, we have obtained the following
elements :
ω = 345°, e = 0.22, Epoch = 2421300.38, K = 6.6 Km/s
V_o = -18 km/s, a sin i = 8.68 A.U., f (m) = 0.4 M_o.
Hypothetical parameters : m = 18 M_o, R_* = 12 R_o (allen, 1973)

- 28 Tau : the plot of RV vs date shows long period variations which
could be periodical (Fig. 3). The lowest RV values (1951-54) are ascribed to the asymmetry of the line profiles (Gulliver, 1977). The periodogram analysis gives a period of about 45 years which does not seem
correlated with the spectroscopic behaviour. In the hypothesis of a
periodicity, another decrease of the RV would happen at the end of the
century (probably in the years 1996-99).

- ζ Tau : Fig. 4 reports the plot of the RV of hydrogen lines from
1901 to 1976. Small variations of the averaged RV seem to occur in the
interval 1901-51. In the years 1914-15 the RV are clearly higher than
those of the subsequent years until 1958. We have suggested the occurrence of another activity phase during the interval 1905-21, since the
RV, the V/R ratio and the appearance of the spectrum seem similar to
those subsequent to the year 1951 (Hynek et al. , 1942 ; Hack, 1954).
If the phases of shell activity should be recurrent, the interval between two of them would be of about 37 years. However the duration of the
shell activity would have been shorter than the present one.

- KX And : we have remarked a decrease of the hydrogen RV after the year
1960 and RV variations larger than those of the preceding years. The lack
of data does not allowns to ascertain a periodicity of this phenomenon.

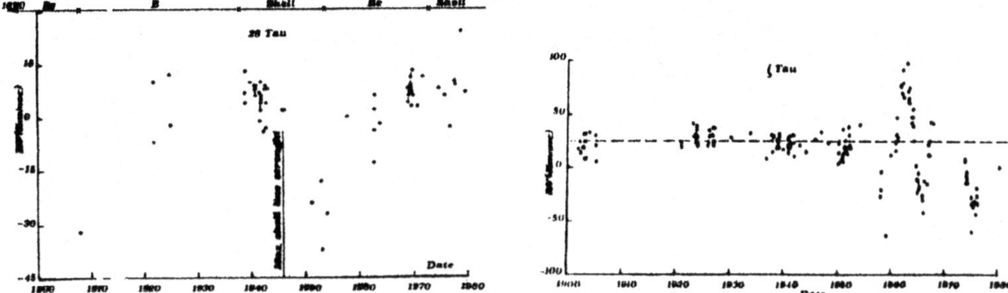

Fig. 3 Averaged RV vs date. Top of the figure :
spectroscopic behaviour according to Hirata et al. (1976).
Fig. 4 Averaged RV of hydrogen lines vs date. The interrupted line is
the mean RV value (+25 km/s) of the star.

- KY And : we have only two groups of RV values for the years 1928 and
1952-56 respectively. The RV ranges and the dispersions of the values
are different in the two groups. The nature of this object being a
spectroscopic binary and an unknown variable is uncertain.

    For 88 Her and CX Dra we cannot attempt any long period analysis
due to the lack of data.

REFERENCES

Allen C.W. : 1973, Astrophysical Quantities, IIIrd ed., Athlone Press,
      London
Delplace A.M. and Hubert H. : 1975, Astron. and Astrophys. 38, pp.75-79
Feinstein A. : 1975, Publ. Astron. Soc. Pacific 87,pp. 603-605.
Fracassini M., Pasinetti L.E. and Pastori L. : 1977, Astrophys. Space
      Sci. 49, pp. 145-167.
Gulliver A.F. : 1977, Astrophys. J. Suppl. Series, 35, pp. 441-459
Hack M. : 1954, Mem. Soc. Astron. It. 25, pp. 281-299
Hadrava P., Harmanec P., Koubsky P., Krpata J. and Zdarsky F. : 1978;
      IAU Circ. N° 3317.
Harmanec P., Koubsky P., Krpata J. and Zdarsky F. : 1976, Bull. Astron.
      Inst. Czech. 27, pp. 47-56.
Harmanec P., Horn J., Koubsky P., Zdarsky F., Kritz S., and Pavlosky :
      1979, IBVS N° 1555.
Harmanec P., Horn J., Koubsky P., Zdarsky F.; Kritz S. and Pavlosky K.
      1980, Bull. Astron. Inst. Czech. 31,pp. 144-159.
Hirata R. and Kogure T. : 1976, Publ. Astron. Soc. Jap. 28, pp. 509-515.
Hynek J.A. and Struve O. : 1942, Astrophys. J. 96, pp. 425-437.
Hubert H. : 1971, Astron. and Astrophys. 11, pp. 100-106.
Pustylnik I. : 1976, Publ. Tartuskai Astrof. Obs. 44, pp. 249-270.
Scholz G. : 1981, Bull. Astron. Inst. Czech. 32, pp. 56-59.

DISCUSSION

<u>Snow</u>: How many Be stars did you examine that did <u>not</u> show any radial velocity variations?

<u>Pastori</u>: Of course none.

<u>Fehrenbach</u>: In the case of ζ Tau you showed very rapid RV variations. Is it possible to reobserve the star to obtain new variations of this type?

<u>Pastori</u>: The variations are irregular.

<u>Bolton</u>: I think the procedure you have adopted is very dangerous and has a high probability of giving incorrect answers because of the inhomogenity of the data you are working with. I am not familiar with the spectra of your individual stars, but in general I think the systematic errors in radial velocity for broad lined and moderately broadened lined stars can approach a substantial fraction of the amplitudes you are finding. There are systematic errors in the older measures due to the use of different wavelengths for the lines, different measurers measuring different sets of lines, different lines having different velocities in the same spectrum, the personal equation for the measurer, unknown weightening of asymmetric line profiles when using microscope measuring engines, and systematic errors between spectrographs. The combination of these can easily exceed 10 km/s for many stars. Averaging bad data will not necessarily produce good data because the errors are systematic. In my opinion the only save approach is to borrow the plates and remeasure them on a modern measuring engine. You should also consider the measures of the radial velocity standard stars at the same time. Otherwise I will have great difficulty believing the results unless you have observed many cycles of variation or else have amplitudes of 50 km/s or more.

<u>Harmanec</u>: (reply to Bolton): We re-analyzed old RV of γ Peg from 1899 – 1903 using new wavelengths. The difference was about 2 km/s only and the β Cep variation with P = .1519 days and semi-amplitude of 3.5 km/s is clearly detectable. Thus, the situation is not so bad.

<u>Pastori</u>: Of course, we had already taken into account all your objections. In our work we have supported a global error (systematic errors of old measurements plus personal equation plus weightening of the asymmetries plus systematic instrumental errors)$\leq$15 km/s. Therefore we have discarded the possibility to find long period variations of this order of amplitude. However, the differences between our V_O and the corresponding RV in the catalogues are equal to 5 km/s (in the average) for all stars and less than 10 km/s for the single stars. Further, the data found in the literature are coming from about ten (in the average) different observers and instruments, for each star. Our declared purpose was to find long

period and large amplitude RV variations. Indeed, our computations
of the orbital elements are only tentative and qualitative ones. We
believe that we have adopted a practical and realistic procedure to
efface the preliminary approach to this problem. Similar, well known,
procedures have been adopted for analogous problems which need the
maximum number of historical informations: the rough determination
of the solar cycle, of the earth's rotation, of the Hubble constant
(by means of the ancient visual photometry of the old supernovae,
etc. Qualitative and quantitative confirmations found by other
A.A. sustain (and do not surprise) our efforts.

<u>Harmanec</u>: Your analysis may be sound for some cases but for KX And
all RV can be reconceiled with a period of 38.9 days.

<u>Pastori</u>: We have looked only for long periodicities. We have not con-
sidered short period binarities or variations as we have averaged all
the data over a month.

<u>Koubsky</u>: You mentioned CX Dra. Do you use RV obtained before 1975?

<u>Pastori</u>: Yes, we have found data for the interval 1919-1929.

SPECTROSCOPIC STUDY OF PLEIONE IN 1977-1979

Ryuko Hirata, Jun'ichi Katahira[*], and Jun Jugaku[†]
DEPEG, Observatoire de Paris, Meudon,
on leave from Department of Astronomy, University of Kyoto.
*Science Education Institute of Sakai City, Osaka, Japan.
†Tokyo Astronomical Observatory, Tokyo, Japan,
guest observer with the International Ultraviolet Explorer
Satellite.

Abstract   Shell spectra of Pleione in 1977-1979 are characterized by further increase in their strengths and by the development of the blue-winged
profiles without noticeable variation of their radial velocities. The MgII
resonance lines at λ2800A have the blue-shifted components with a velocity
of -35 km/s relative to the other shell lines. The development of the shell
structure is derived. The mass loss rate was 7×10^{-11} $M_{\odot}$/yr.

1. INTRODUCTION

We here report the spectroscopic behavior of Pleione(B8e, Vsini=340km/s)
in 1977-1979. Our studies on this star in 1972-1976 are found in Hirata
and Kogure (1977,1978), and Higurashi and Hirata (1978, hereinafter HH).
The reciprocal linear dispersion of our spectrograms is 10A/mm. We also
obtained the IUE spectra in high dispersion mode in October 1978, and
January 1979.

2. GENERAL DESCRIPTION

The strengths of shell lines have increased further in 1977-1979. Their
profiles have still kept their characteristics in 1973-1976, that is, the
higher the ionization potential is, the narrower the line becomes. The
most striking characteristics in 1977-1979 is the development of the blue-
winged profiles without noticeable variation of their radial velocities.
Mean radial velocities of the Balmer shell lines(H6-H20) were +10.7±0.2,
+10.6±0.5, and +8.0±0.5 km/s in October 1977, August and November 1978,
respectively. The CaII K line has still a very broad component.
    The rich shell lines are also seen in the far-ultraviolet region.
Preliminary identification shows these rich shell lines are composed of
the relatively low-ionized metals, just as expected from the visual region. The blue-winged profiles are also seen in the far-ultraviolet region. There is no clear evidence of the existence of CIV and SiIV ions.

M. Jaschek and H.-G. Groth (eds.), Be Stars, 161–165.

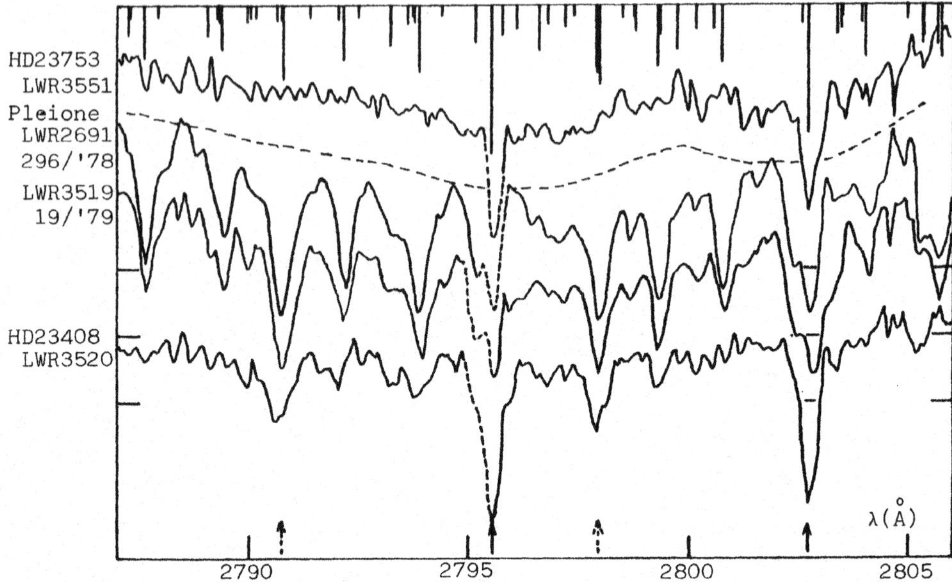

Fig.1. λλ2787-2806 Å region of HD23753, Pleione, and HD2340⟨

The MgII resonance lines at λ2800 A have the blue-shifted components.
Figure 1 shows two spectra in this spectral region, together with those
of HD23753(B8V,Vsini=280 km/s), and HD23408(20 Tau,B7III,Vsini=40 km/s),
all belong to the Pleiades cluster. The spectrum of HD23753 is composed
of the broad photospheric features around the MgII resonance lines at λλ
2795.5, 2802.7A, and sharp interstellar components of these lines, while
HD23408 shows the subordinate lines of the MgII λλ2790.8, 2797.9 A(broke
arrows in the figure), and other photospheric lines of FeII and CrII, in
addition to the MgII resonance lines(solid arrows). Rich shell lines are
seen in the spectra of Pleione. The broken line for LWR2691 is the back
ground photospheric feature implied from the spectrum of HD23753. In the
uppermost part of the figure, we show the expected line depths of the
shell lines in an arbitrary scale, which are approximated by the root of
the caluculated equivalent widths. The equivalent widths of the shell
lines were computed by the curve-of-growth technique for the parameters
derived from the FeII lines in the visual region. Cosmic abundance was
assumed, and the dilution factor was introduced in the ionization equilibrium. We used gf-values of Kurucz and Peytremann (1975). It is seen
that our result of calculation coincides well with the observed spectra.
The MgII resonance lines at λ2802.7A and λ2795.5A have blue-shifted components, whose radial velocities relative to the nearby FeII, CrII shell
lines are -34±1, and -36±1 km/s, respectively. These values were obtaine
from four spectra in October 1978, and two spectra in January 1979. The
value for λ2795.5A should be less weighted, because the line is contaminated by the reseau, and also the non-shifted MnII λ2795.2 line is ex-

pected to contribute appreaciably from our calculation. From the fact
that no blue-shifted component is seen in the MgII subordinate lines, we
conclude that the blue-shifted components are originated in the less
dense part of the envelope.

3. ANALYSIS

We measured the equivalent widths, half half-widths, and central depths
of the metallic shell lines in the spectral region, $\lambda\lambda 3800$-$5020A$,
for six plates obtained in November 1978. The method of analysis is the
same as in HH. We determined the reciprocal excitation temperature θ_{exc},
the column density $\langle Nl\rangle$, and the microscopic velocity ξ from the curve-
of growth. From the half half-widths, we can guess the outer radius of
the absorbing region, r, in units of stellar radius(we assumed the conservation of angular momentum in the envelope). From the central depths,
we get the fractional area of the stellar disk, β, which is screened by
the velocity zone for the line center, and the effective optical thickness of the velocity zone, τ. Figure 2 shows the variation of these parameters in 1973-1978. Typical errors(2σ) are shown in the right bottom

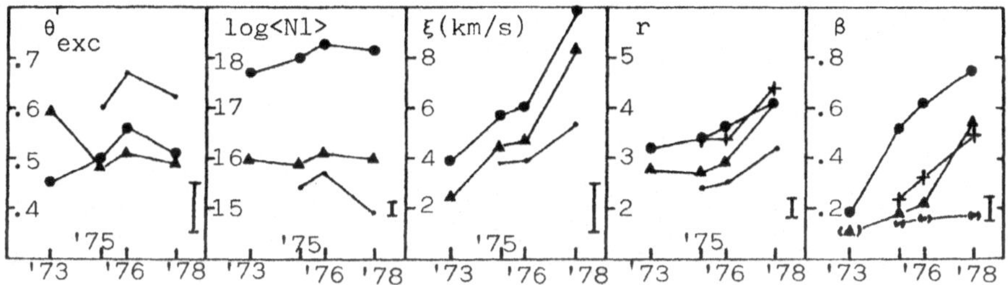

Fig.2 Variation of θ_{exc}, $\langle Nl\rangle$, ξ, r, and β in 1973-1978. The symbols
are FeI($\cdot$), FeII($\bullet$), TiII($\blacktriangle$), and CrII($+$).

corners. The excitation temperatures and column densities of various ions
did not change so greatly, while the radii r and the fractional areas β
have increased further, as a smooth extension in 1973-1976. The structure
of the shell in 1978 is the same as in 1973-1976, i.e., a cold dense core
is surrounded by a hotter, tenuous envelope. The microscopic velocity ξ
increased remarkably. For example, ξ of the FeII lines increased from 6
km/s in 1976 to 10 km/s in 1978. Recalling the development of the blue-
winged profiles in 1977-1979, we can conceive that the outflow became
conspicuous, and this made the apparent values of ξ greater (HH, Mihalas
1979). The increase in ξ affects the widths of shell lines, though we
made no correction for the estimate of r in figure 2. For example, the
outer radius of the FeII formation region in 1978 increases from $4.1R_{*}$
to $4.6R_{*}$ when both ξ- and instrumental broadenings are corrected. This
is an extreme case, and such an effect is not so great in the other ions.
　　Balmer shell lines were analyzed for all plates obtained in October

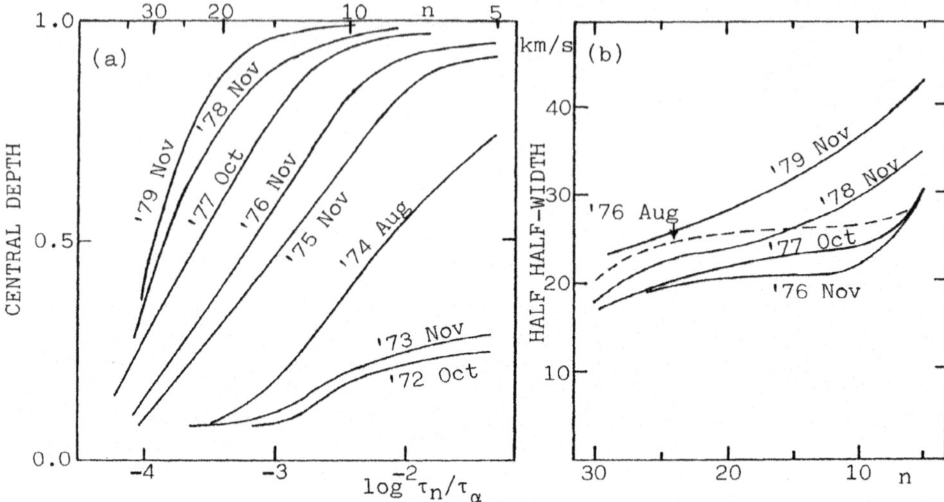

Fig.3 Variation of Balmer shell lines, a) central depths, b) half half-widths.

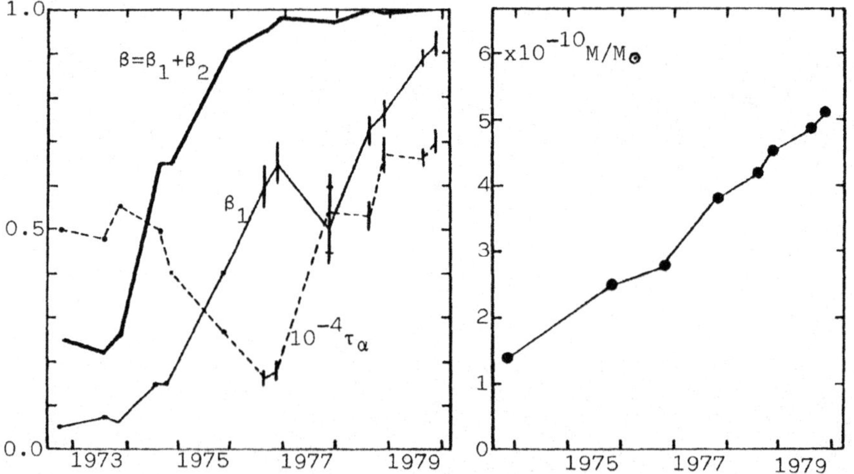

Fig.4 Variation of β and τ_α for the Balmer shell lines.

Fig.5 Variation of the envelope mass.

1977, August, November 1978, and August, November 1979. Figure 3a shows the variation of their central depths in 1972–1979, and figure 3b shows the variation of their half half-widths in 1976–1979. The central depths have increased steadily, and reached to unity even in the intermediate members in 1979, while the half half-widths decreased untill Novemver

1976, then turned to increase since then. The flat part in the H10-H20
lines disappeared at the same time. This behavior of the half half-widths
is in contrast with that of the metallic lines. The equivalent widths of
Balmer shell lines increased steadily in 1972-1979. These observational
facts indicate that the saturation effect influences even in the intermediate members of the Balmer shell lines in 1977-1979. Nevertheless, we
applied our central depth method to the data in 1977-1979. Two-layer approximation was also adopted. The results are shown in figure 4, together
with those in 1972-1976. The fractional area and the optical thickness
in Hα for the first layer are designated by β_1, and τ_α, respectively. It
is seen that the fractional area increased steadily, while the optical
thickness decreased in 1972-1976, and turned to increase since then.
This indicates the growth of dense core, accompaning the outflow, after
the the phase of diffusing-out in 1972-1976. We must note, however, that
the saturation effect in 1977-1979 casts some doubts on the simple application of our central depth method. The mass of the envelope was estimated
from the column density and the outer radius, under the same assumptions
as in HH. We assume the outer radius is $4R_*$ in 1977-1979, as indicated
from FeII, because the half half-widths of Balmer lines give no more the
radius of the hydrogen envelope, due to the saturation effect. Fortunate-
ly, the value of the outer radius does not affect the resultant mass so
seriously. For example, the mass of the envelope increases by 15% for
$5R_*$. Figure 5 shows the variation of envelope mass. The mass loss rate
is estimated as 7×10^{-11} $M_\odot$/yr in 1977-1979, slightly larger than 5×10^{-11}
$M_\odot$/yr in 1973-1976.

We gratefully ackowledge the assistance of the IUE Observatory staff
in acquisition and reduction of the data used in this paper. One of us
(J.J.) would like to express his gratitude for a grant from the Yamada
Science Foundation which made his travel to Greenbelt possible.

REFERENCES

Higurashi,T.,and Hirata,R.: 1978,Publ.Astron.Soc.Japan,30,pp.615-635.
Hirata,R.,and Kogure,T.: 1977,Publ.Astron.Soc.Japan,29,pp.477-495.
Hirata,R.,and Kogure,T.: 1978,Publ.Astron.Soc.Japan,30,pp.601-614.
Kurucz,R.L.,and Peytremann,E.: 1975,Smithsonian Astrophys.Obs.,Special
   Report No.362.
Mihalas,D.: 1979,Monthly Notices Roy.Astron.Soc.189,pp.671-699.

AN UNUSUALLY STABLE AND SHORT SPECTROSCOPIC PERIOD OF THE Be STAR 28 CMa

D. Baade
European Southern Observatory, Karl-Schwarzschild-Straße 2,
D-8046 Garching, Germany

1. OBSERVATIONS

Spectroscopic (1970: ESO, 12 Å/mm, 6 spectra kindly put at my disposal by Prof. A. Van Hoof; 1976: ESO, 12 Å/mm; 1977: Calar Alto Observatory, 42 Å/mm; 1979: ESO, 12 Å/mm) and photometric (1976: ESO and Cerro
Tololo, Hβ, uvby) observations of 28 CMa (B2-3 IV-Ve; $3.52 < m_V < 4.18$,
irregular variations on the time scale of months or years reported;
v_{rot} = 80 km/s) revealed a very complex variability. All observed individual types of variations are known from at least a few other Be stars.
In 28 CMa, however, for the first time a highly significant correlation
between the various variations is established by a stable common period.
The period is 1.365 days which seems to be the shortest stable period
presently known of any Be star. There is no indication that the star's
behaviour changed between 1970 and 1979. Only the equivalent widths of
the emission lines increased noticeably.
  A detailed description of the observations will be given elsewhere
(Baade, 1981). They can be summarized as follows:
a) All stellar absorption lines have asymmetric profiles. The asymmetry
   is variable, line profiles differing in phase by half a period are
   mirror images of one another. The degree of the asymmetry is not the
   same for different lines. In the broad Balmer lines it is barely noticeable, in narrow lines (e.g. Si II) it is very striking.
b) Because of the different effect of the asymmetry on different lines
   a wide range of radial velocity (RV) amplitudes is obtained if RVs
   are derived from measurements of the line minima. The amplitude (2K)
   is the lowest for the H lines (11 km/s), intermediate for the He I
   lines (60 km/s) and the highest for the Si II lines (84 km/s). (RV
   curves for all other lines are ill defined). Within an uncertainty of
   about 5 km/s the line wings are stationary. Thus the RV variations are
   due to a shift of the line cores. The shape of all RV curves is fairly
   sinusoidal. A phase shift between different RV curves is not evident
   from the present observations.
c) The V/R ratio of all emission lines, i.e. the whole Balmer series and
   some Fe II lines, is variable. In Hδ the amplitude is about 70%, averaged over one period V/R=1. The correlation with the stellar absorption

M. Jaschek and H.-G. Groth (eds.), Be Stars, 167–170.

line profiles is that during maximum violet asymmetry (RV minimum)
V/R $<$ 1 whereas during maximum red asymmetry of the absorption lines
(RV maximum) V/R $>$ 1. Within the measuring accuracy, a variation of
the total equivalent widths of the emission lines was not detected.
d) The RVs of all emission lines are variable, those of the whole emission lines as well as those of the violet and red components measured
   individually. The amplitudes are low (2K = 10 km/s), the γ-velocities
   are the same as for the stellar absorption lines, γ = 29 km/s. There
   is however a permanent phase shift of 180 degrees between the RV
   curves of absorption and emission lines. The RV amplitude of the
   central reversals is 10 km/s, the RV curve is in phase with the stellar RV curves. Since the broad stellar absorption lines are very
   little affected by the asymmetry the observed variations of the emission lines cannot be due to variations of the underlying stellar lines
e) Since at the time of the photometric observations no regular short
   term variations were expected and since, furthermore, the comparison
   star turned out to be variable, no useful light curve of 28 CMa was
   obtained which could be compared to the spectroscopic variations.
   The upper limit of the photometric amplitude is about $0^{m}.03$ in uvby.

2. CLASSICAL MODELS

One of the classical models for the V/R variations of Be stars, the
elliptical ring model, can be easily ruled out because the period of 28
CMa is too short. On the other hand, the period is too long to be in
accordance with the observed RV amplitudes and the small brightness
variations if 28 CMa is assumed to pulsate radially. The latter argument
partly also concerns a variable stellar wind model. Since all photospheric lines are affected by the asymmetry, the mass flow had to be considerable. This should have a significant effect on the strengths of the
emission lines which are however observed to be stable on short and
medium time scales.

There are two possibilities to apply a binary model to 28 CMa: The
first is to assume that the asymmetry of the absorption line profiles
is the direct evidence for the presence of a companion, the line profiles being therefore a blend of a broad and of a narrow line component
attributable to two different stars. Since all lines are somewhat asymmetric, both stars must have nearly the same temperature. Because the
luminosities cannot be very different either, both stars must have approximately the same radius. But because the RV amplitude of the line
wings (i.e. the broad line component) is very low, these otherwise almost identical stars would differ in mass by almost a factor of 5! The
second possibility is to assume that the asymmetry reflects only indirectly the presence of a companion. The companion could be the cause
of an asymmetric brightness distribution. However, if this effect is
sufficiently strong to produce the observed asymmetric line profiles,
a light amplitude far higher than the observed one had to be expected.

3. NONRADIAL PULSATIONS

The idea that, because of their proximity in the HRD, there might be a relation between Be and β Cephei stars, is not new. Unfortunately the observations, in particular the lack of real periods, never allowed to pursue this idea in more detail. Another reason is the argument that we heard already yesterday during the discussion of Drs. Jerzykiewicz and Sterken's paper, namely that the observed quasiperiods are often far longer than those known from β Cephei stars (3 - 6 hours). An additional problem is linked to the β Cephei phenomenon. It seems that there is no unique explanation for their variability. Some pulsate radially, in others at least one nonradial mode is excited (c.f. Smith 1980). Since we have excluded radial pulsations from the discussion and since the time is limited, we shall investigate only whether the observed period of 28 CMa is compatible with nonradial pulsations (NRP).

I have tried (Baade, 1981) to obtain a very simple extrapolation of Ledoux's (1951) original relation for the observed periods of nonradially pulsating stars towards fast rotators:

$$\sigma_{k,m} = \sigma_{k,o} - m(1 - C_k)\omega - \frac{\omega^2}{2\sigma_{k,o}} \qquad (1)$$

If it is assumed that the nonrotating star's pulsational frequency, $\sigma_{k,o}$, is of the order of those of β Cephei stars, the observed frequency, $\sigma_{k,m}$, can be smaller than $\sigma_{k,o}$ only if m is not negative. This means that, if m > 0, the pulsational mode is a retrograde one, waves are travelling in the opposite direction to the rotation. Due to the high angular velocity of Be stars, ω, it is then possible that the pulsational period of the star is about the same as in the case of β Cephei stars whereas the observed period is considerably longer.

With the numerical assumptions $\ell = 2$ (this mode has been identified in several β Cephei stars), m = 2, $\sigma_{k,o}$ = 4 d^{-1}, and ω = 1.7 d^{-1}, relation (1) yields a frequency of 0.83 d^{-1}. This may be compared to the observed frequency of 0.73 d^{-1}. In view of the simplicity of the model it is an interesting result that a satisfactory approximation of the observed period of 28 CMa is possible by just taking into account the high rotational velocity of Be stars while all other parameters are kept in the same numerical range as observed in β Cephei stars. That the pulsational mode has to be a retrograde instead of a direct one, may also be due to the fast rotation although this is in contradiction to Hansen's et al. study (1978) of rotating nonradially pulsating stars.

References

Baade, D. 1981. Submitted to Astron. Astrophys.
Hansen, C.J., Cox, J.P. and Carroll, B.W. 1978. Astrophys. J. <u>226</u>, 210
Ledoux, P. 1951. Astrophys. Journ. <u>114</u>, 373
Smith, M.A. 1980. Astrophys. Journ. <u>240</u>, 149.

DISCUSSION

<u>Endal</u>: How did you arrive at the second order term in your equation for the non-radial pulsation periods? How important was this term in determining your $m = +2$ period?

<u>Baade</u>: I included the centrifugal term in my calculations while apart from that I follow Ledoux's approach (1951) and still neglect all other "small" terms. This is certainly not correct, but it was my intention to get just a rough idea whether such a second-order term will turn out to be very important. With the assumptions stated before, <u>this</u> second order term is about one order of magnitude smaller than the linear term.

<u>Souffrin</u>: Can you say something on the stability of the phase – i.e. the life time for phase coherence – of this oscillation?

<u>Baade</u>: Within the known accuracy of the period, I did not find an indication of a phase shift between 1976 and 1979. Between van Hoof's observations in 1970, whose plates I was kindly allowed to remeasure, and my observation in 1976, however, too much time had elapsed to obtain a conclusive result.

<u>Harmanec</u>: Could you make clear the meaning of the angular velocity of your model? Do you assume local conservation of angular momentum or a rigid-body rotation?

<u>Baade</u>: I tried to keep the model as simple as possible. Therefore I assumed a rigid-body rotation.

<u>Henrichs</u>: Is your star known to be an x-ray source?

<u>Baade</u>: No.

<u>Metz</u>: Does your model explain the behaviour of the double emission of H_ε?

<u>Baade</u>: This behaviour does not just concern H_ε but all hydrogen (only H_α is not resolved into two peaks) and FeII (if visible) emission lines. It is easily explained if one assumes that a small additional emission, evoked by the shock effect of the high speed of the travelling waves, moves across the line.

<u>Metz</u>: Is the velocity difference of both the emission peaks consistent with the velocities derived from the system?

<u>Baade</u>: Yes.

THE VARIABLE SHELL PHASE OF HD 184279 BETWEEN 1976 AND 1980

D. Ballereau, A.M. Hubert-Delplace
DEPEG, Observatoire de Meudon, 92190 Meudon.

Abstract. The variable shell (H, HeI, FeIII and once ionized
other metals) BO.5 star HD 184279 presents variations of radial velocity and line profiles with a cycle of about 4 years.
Some correlations between photometric and spectroscopic data
are found. A comparison with 48 Lib and ζ Tau is given.

1. INTRODUCTION

HD 184279, of spectral type BO.5 (Jaschek et al, 1980) and v sin i =
230 kms^{-1} (uesugi, 1978) was first studied in 1929 by Merrill and
Burwell (1932) who noticed dark and narrow hydrogen lines and two very
weak bright edges in Hβ. Swings and Struve (1943) observed broad H and
HeI lines with normal intensity without emission at Hβ in 1942, and
concluded that the shell had completely disappeared. Merrill (1951)
saw extremely weak and diffuse H and HeI lines on a spectrum taken that
year while Lynds (1959) found variations in brightness over 3 months.
Svolopoulos (1975) from observations made in 1971, gave an H$_\alpha$ profile,
some equivalent widths and estimated some shell' parameters. Tempesti
and Patriarca (1976) noted photometric variations from 1968 to 1976,
and we see three distinct phases in the recent life of this star in
the Hubert-Delplace and Hubert' Atlas (1979): 1956-60: B phase, 1963-
70: faint Be phase, 1973-1976: strong Be and shell phase.

2. OUR OBSERVATIONS

Since 1976, 13 spectrogramms of HD 184279 have been taken at 12 Amm^{-1}
in the blue region (Observatoire de Haute Provence). The radial velocities exhibit large variations over four years. The six first spectra
were reduced and traced in intensity to obtain well defined line profiles, equivalent widths and central depths.

3. RADIAL VELOCITIES

The lines of all elements present variations of RV with a cycle of about
4 years. From 1973 to 1977, the shell is essentially seen in H, HeI and

M. Jaschek and H.-G. Groth (eds.), Be Stars, 171–175.

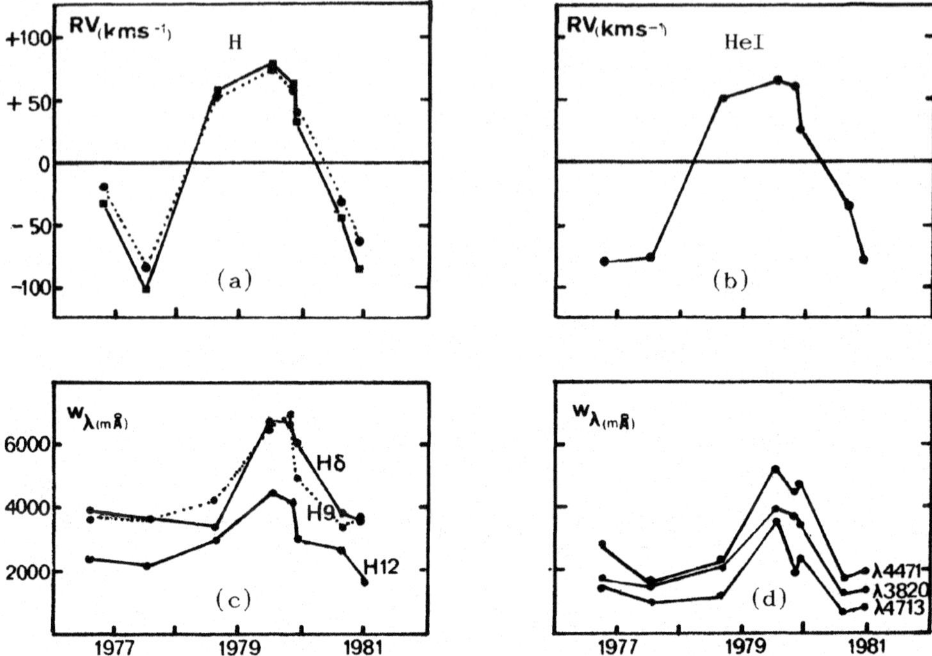

Figure 1-a. Radial velocity of the hydrogen lines between 1976 and 1980 averaged over $H\gamma H\delta H\epsilon$ (solid line) and over H10H11H12 (dashed line).
Figure 1-b. Radial velocity of HeI averaged over the main lines, between 1976 and 1980. Figure 1-c. Equivalent widths of some hydrogen lines (in mA) between 1976 and 1980. Figure 1-d. Equivalent widths of some HeI lines (in mA) between 1976 and 1980.

FeIII, the lines of these elements being strengthened in 1975 when an inverse P Cyg profile was observed in $H\beta$. From 1978 we observe MgII and FeII in addition to the former shell lines; then in 1979-80, MgI, NiII, CaII and SiII. We give in figure 1-a the RV of hydrogen averaged over $H\gamma H\delta H\epsilon$ and over H10H11H12 for eight epochs. It is clear that as for ζ Tau (B2IIIe) and 48 Lib (B3IIIe), the higher Balmer line velocity amplitude is larger than that of the lower members. Also, we note a Balmer progression in 1980, when the RV's are negative. Figure 1-b gives the HeI RV averaged over the main lines. The other elements (FeIII, FeII, MgII, CaII, SiII, NiII) exhibit similar RV variations.

4. BEHAVIOUR OF THE LINE PROFILES

From September 1976 to November 1979, the H and HeI line profiles show important changes. We give in figure 2 the line profiles of $H\beta$ and in figure 3 the line profiles of HeI 4471A and MgII 4481A at the following epochs: 3-9-76, 1-7-77, 1-7-78, 6-6-79, 21-10-79, 28-11-79. We see a P Cyg profile in $H\beta$ in 1976-77 when the RV's of the shell lines are negative, then an inverse P Cyg profile in 1978-79 when the RV are

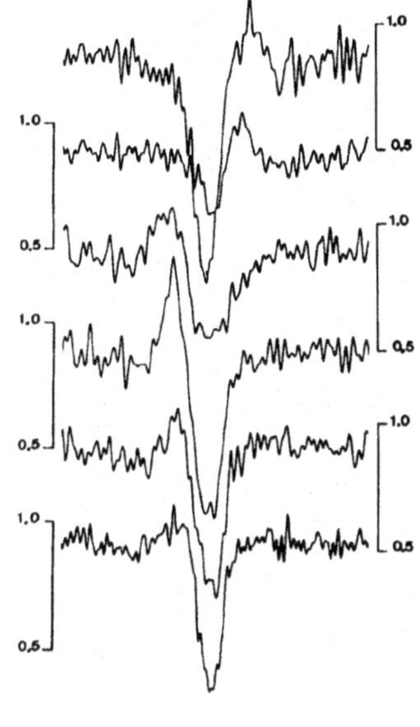

Figure 2. Intensity tracing of the Hβ line profile between 1976 and 1979.

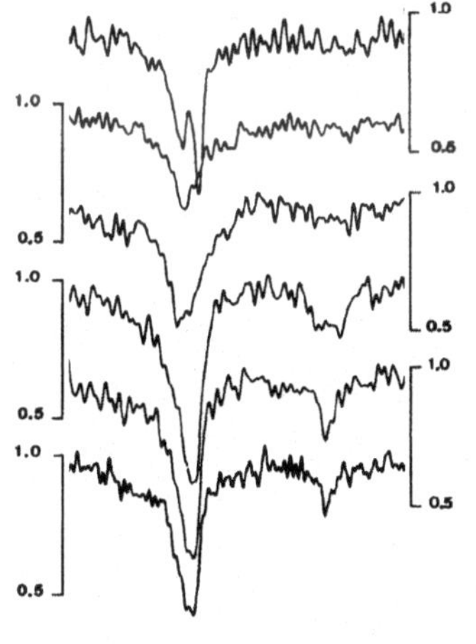

Figure 3. Intensity tracing of the HeI 4471A and MgII 4481A between 1976 and 1979.

positive, with a strong emission peak in July 1979. Later in 1980, Hβ presents a P Cyg profile again. We see a fainter emission component in Hγ which is absent in Hδ. The higher members of the Balmer series are more asymmetric than the lower ones, especially in 1979 when RV are very positive.

The HeI lines originating from the 2^1S, 2^3S, 2^1P^0, 2^3P^0 terms are in very strong and sharp absorption during the whole cycle; λλ4471A, 4026A, 3965A, 3888A, 3820A are particularily enhanced. The profile variations are similar to those of the higher Balmer lines. The FeIII lines (a^5P-z^5P^0 for example) are well observed and exhibit a similar behaviour to the HeI's.

We note that in October 1976 all the hydrogen and helium lines are double. The fainter and the blue shifted secondary component is more visible in the near UV region. This component is very distinct for H9, H10 and HeI 4471A. The difference of RV between these two components is 76 kms^{-1}.

5. EQUIVALENT WIDTHS AND CENTRAL DEPTHS

These two quantities have a similar behaviour. In figure 1-c and 1-d, we see that they decrease from the middle of 1976 to the middle of 1977 to reach minima, when the RV are at minimum. Then, they increase by a factor up to two in the middle of 1979 and after decrease again. Between the middle of 1978 and the middle of 1979, HeI decreases more rapidly than H. The variations of the central depths are similar to those of RV and the equivalent widths.

6. COMPARISON WITH PHOTOMETRIC DATA

Tempesti and Patriarca (1976) have published the photometric variations of this star from 1950. Their measurements show a decrease of brightness between 1968 and 1975 with a pronounced minimum in 1973. At that epoch, we observe a strengthening of emission and the appearance of a shell with H,

HeI and FeIII. In 1975-76, brightness increases again without reaching
the former value.
Alvarez and Schuster (1981) measured a decrease of 0.4 magnitude in
brightness (filter 0.58 µm) between June 1977 and May 1978, to reach
7.2. Between these two epochs, our RV increases from a minimum (-100
kms^{-1}) to strong positive value (+50 kms^{-1}).

7. CONCLUSIONS

a- HD 184279 is an early B type star presenting a temporary HeI shell
as has been observed for stars having the same spectral type (γCas,
59 Cyg).
b- Conspicious correlations between RV, line profiles, equivalent widths
and central depths occur, as for ζ Tau and 48 Lib.
c- The correlation between RV and the magnitude (M58) given by Alvarez
and Schuster (1981) was also noted as in the case of ζ Tau (M58 increases when the RV's are strongly positive).
d- Having a similar period, the amplitude of RV variations is larger
than for ζ Tau' (Delplace et Chambon, 1975).
e- Having a similar amplitude, the period of HD 184279 is much shorter
than that of 48 Lib.
f- The decrease of brightness in 1973 reported for HD 184279 by Tempesti and Patriarca (1976) is to be compared with the decrease of about
0.5 magnitude in the B filter reported by Sharov and Lyuty (1975) in
the case of Pleione for which a shell phase similarly appeared.

star	a	p
HD 184279	90 kms^{-1}	4 years
HD 372 o2	60 kms^{-1}	7 years
HD 142983	60 kms^{-1}	10 years

REFERENCES

Alvarez M. and Schuster W.: 1981, to be published in Revista Mexicana
   de Astronomia y Astrofisica.
Delplace A.M. et Chambon M.T.: 1976, IAU Symposium n° 70, p. 79.
Hubert-Delplace A.M. and Hubert H.: 1979, an Atlas of Be Stars.
Jaschek M., Hubert-Delplace A.M., Hubert H., Jaschek C., A Classifica-
tion of Be Stars, Astron. and Astrophys., Suppl. Ser., 42, 103.
Merrill P.W.: 1951, APJ 115, 47.
Sharov A.S. and Lyuty V.M.: 1975 IAU Symposium n° 70, p. 105.
Svolopoulos S.N.: 1975, Astron. and Astrophys., 41, 199.
Swings J.P. and Struve O. 1943, APJ 97, 194.
Tempesti P. and Patriarca R.: 1976, IBVS n° 1164.
Uesugi A.: 1978, Revised Catalogue of Stellar Rotational Velocities-
   Preliminary Edition.

DISCUSSION

<u>Andrillat</u>: What is the behaviour of the HeI metastable line λ 3889A with respect to other HeI lines (triplets and singulets).

<u>Ballereau</u>: At this dispersion HeI λ 3889A is blended with H8, but the central depth and the width at half intensity of the blend indicate that the contribution of HeI 3889 ($2^3S - 3^3P^o$) is very important. However HeI λ 3965 ($2^1S - 4^1P$) and mainly the triplet lines($2^3P - n^3D$) are also prominent in the spectrum.

<u>Viotti</u>: Did you find any variation of the intensity ratio of singulet and triplet HeI lines?

<u>Ballereau</u>: The triplet lines are always enhanced relative to the singulet lines mainly when the RV's are negative. However, when the RV's are strongly positive all HeI lines are deep and sharp.

A SPECTROGRAPHIC STUDY OF THE SHELL STAR EW Lac

G. Scholz

Zentralinstitut für Astrophysik der Akademie der Wissenschaften der DDR - Potsdam - DDR.

ABSTRACT

38 Zeeman spectrograms of the Be star EW Lac (HD 217050) were obtained in three successive years, 1978, 1979 and 1980. The investigated lines show long-time variations of the radial velocities, line widths, and line intensities. No hints at the occurrence of a global magnetic field larger than 150 Gauss were found.

OBSERVATIONS AND REDUCTIONS

During the new activity of EW Lac (Hadrava et al., 1978 ; Harmanec et al., 1979 ; Poeckert, 1980) spectrograms of this star were obtained with a Zeeman analyzer at the Coudé focus of the 2-m telescope at Tautenburg . The reciprocal linear dispersion is 7.9 Å mm^{-1} the width of the slit was equivalent to 0.16 Å. The Hydrogen line Hβ as well as all unblended lines in the spectral region from about $\lambda\lambda$ 4000 to 4600 Å have been measured with an oscilloscope display machine ; the density data of few special lines were recorded using a Zeiss microdensitometer.

RESULTS

Magnetic field :

The search for an effective magnetic field B_{eff} on EW Lac had a negative result. More exactly, the mean values of $\bar{B}_{eff}$ determined in conventional manner are −22, +34 and +8 Gauss respectively for the years 1978, 1979 and 1980. The maximum absolut values of B_{eff} found for the several years are 260, 210 and 180 Gauss with an accuracy (rms) of nearly ± 250 Gauss for a single plate.

This negative result is supported by missing significant differences in the anticircularly polarized line contours of especially selected lines. Neither a different broadening nor a different asymetry of the profiles could be detected.

M. Jaschek and H.-G. Groth (eds.), Be Stars, 177–179.
Copyright © 1982 by the IAU.

<u>Radial velocities</u> :

- a) <u>Absorption lines</u> :

The result of the observed temporal changes in the radial velocities of the absorption cores of Hβ, Hα, Hδ and the metallic lines,
smoothed by fitting the observations to polynom, is represented in
Fig. 1a.

A SPECTROGRAPHIC STUDY OF THE SHELL STAR EW Lac

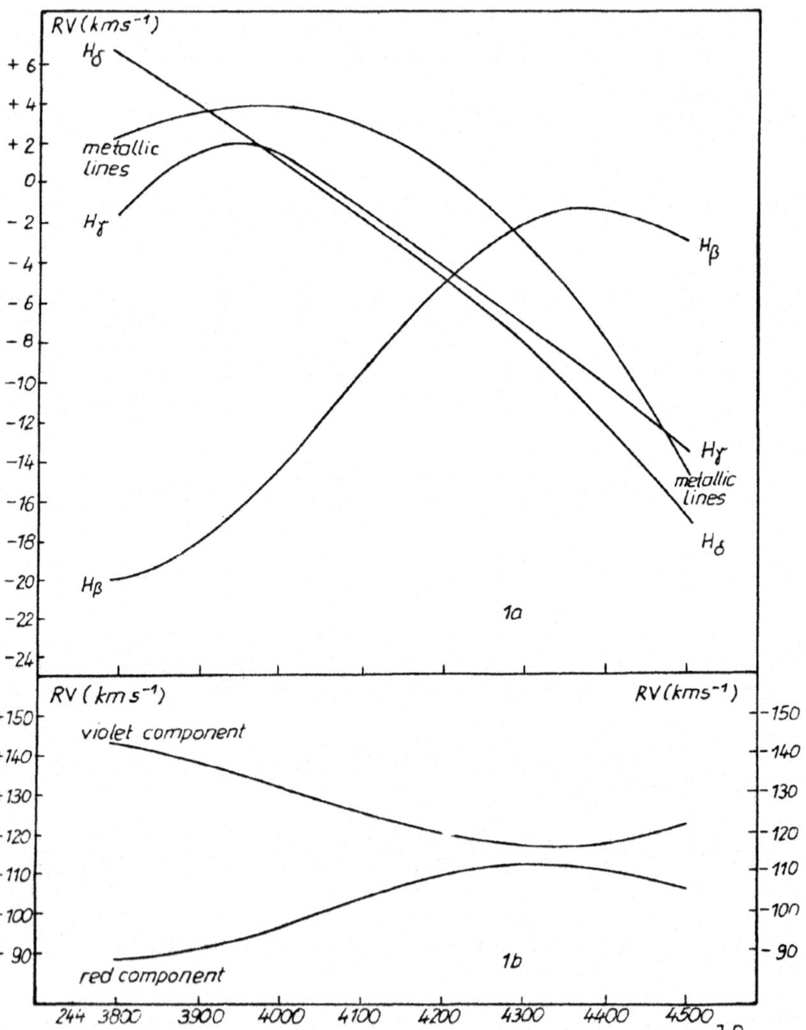

Fig. 1. Radial velocity of the absorption lines (1a) and radial velocity of the emission edges of Hβ (1b) in progress of time.

In 1978 Hβ yields velocities which are practically equal to those of the stable shell phase, that means, about -20 kms^{-1} while the velocity values of Hδ are diminished by nearly 20 kms^{-1}. During 1980 the moving conditions are reversed to those of 1978. Besides the velocity variation, the metallic lines show also changes of the line widths equivalent to rotational velocity variations of 74, 99 and 82 kms^{-1} respectively for the years 1978, 1979 and 1980.

- b) Emission lines :

In Fig. 1b the velocity variations of the red and violet emission component of Hβ are represented. The observations reflect, what is expected by the variation of the absorption core of Hβ. No clear differences exist between the velocity values of Hβ and Hδ, in the case the features of Hδ could be measured. The V/R-ratio determined for some plates in each observational season changes from 2.2 for the year 1978 and 1979 to 1.0 in 1980.

DISCUSSION

The available data suggest that EW Lac is a single star. Obviously, an effect of a global magnetic field larger than about 100 Gauss on the dynamics of the envelope can be excluded. Concerning all observational results the simplest model for EW Lac seems to be an eliptical envelope revolving around the star connected with a strong disturbance crossing successively the inner and outer parts of the shell and indicating with this the presence as well of layers in contraction as in expansion. Furthermore, the existence of a fine structur of the shell is indicated by several peaks in the absorption cores of Hβ and Hδ.

ACKNOWLEDGEMENT

I wish to express my gratitude to the observational staff of the Karl-Schwarzschild-Observatory at Tautenburg.

REFERENCES

Hadrava P., Harmanec P., Koubsky P., Krpata J., and Zdarsky F., 1978 –
     IAU Circ. N° 3317
Harmanec P., Horn J., Koubsky P., and Zdarsky F., 1979 – Inf. Bull.
     Variable Stars, IAU N° 1555
Poeckert R., 1980 – Publ. Dominion Astrophys. Obs. 15, 357.

A PRELIMINARY REPORT ON SIMULTANEOUS ULTRAVIOLET
AND OPTICAL OBSERVATIONS OF LAMBDA ERIDANI

C. T. Bolton
David Dunlap Observatory, University of Toronto
P.O. Box 360, Richmond Hill, Ontario, CANADA L4C 4Y6

Radial velocity observations of λ Eridani between 1971 and 1980 and uvbyHα photometry in 1977 show that the star is both a light and velocity variable with a period of 0.701538 d. Copernicus observations in 1977 are compatible with this result. Both the radial velocity and photometric amplitude are variable, and these variations may be correlated with the hydrogen emission line strength. The character and phasing of the light and velocity curves suggest that λ Eridani is a pulsating variable.

I obtained five 12 $\overset{\circ}{\text{A}}$ mm^{-1} spectrograms of λ Eridani in 1974 after Irvine (1974, 1975) discovered Hα emission in its spectra and suggested that it might be a binary star. No emission was apparent in the Hβ line at that time, and the radial velocity seemed to be constant. I obtained a large number of new spectrograms in 1976 following the report by Percy (1976) that it was varying by $\Delta b \simeq 0\overset{m}{.}06$ (Percy and Lane 1977). There was a broad, double-peaked emission present in Hβ at this time, and the radial velocity measured from the He I lines was varying by about 60 km s^{-1} with a period of either 0.4 or 0.7 d. The velocities of the Hβ emission peaks and central absorption were not variable, but the V/R ratio was correlated with the He I velocities in a way that suggested that the V/R variations were caused by shifts or profile changes of the underlying absorption line.

In order to check the reality of the periodicity, resolve the period ambiguity from the 1976 observations, and determine the nature of the variability I obtained simultaneous ultraviolet and optical spectroscopic and photometric observations of the star during December, 1977. I obtained uvbyHα observations with the University of Toronto's 0.6 m telescope at Las Campanas, Chile from December 5-15, 1977. A total of 233 differential magnitude measurements were obtained in each of the 6 filters on 9 different nights. During a 2.5 day period in the middle of this run I obtained repeated scans of the Fe III lines in the $\lambda\lambda$1114-1135 Å region and the Si IV lines in the $\lambda\lambda$1390-1404 region of the spectrum with the U2 photometer on the Copernicus

181

M. Jaschek and H.-G. Groth (eds.), Be Stars, 181–184.
Copyright © 1982 by the IAU.

satellite. At the same time the U1 photometer was left fixed in a
continuum region near 1200Å. These observations were supplemented
with 12 Å mm^{-1} spectrograms taken during the period November, 1977 to
January, 1978 with the 1.88 m telescope at the David Dunlap Observatory.
I have extended the time coverage of this data by obtaining spectrograms frequently in each succeeding season and by borrowing and re-
measuring KPNO coudé spectrograms obtained by Abt (Abt and Levy 1978)
between 1971 and 1976. This paper is a preliminary report based on
partial reduction and analysis of this material.

There was no emission visible in Hβ during 1977, but Hα was almost
completely filled in. Hβ showed a double-peaked emission again during
the 1978-79 observing season, which was still faintly visible during
part of the 1979-80 season, but there was no obvious Hβ emission during
the past season. Power spectrum analysis of the He I radial velocities
measured from the spectrograms taken during the period 1971-80 indicates that the velocity varies with the period 0.701538 d. There are
no other statistically significant periodic variations in these data,
but the amplitude of the variation does change from less than 30 km s^{-1}
to as much as 56 km s^{-1} on time scales of a year or less. There are
indications that these amplitude changes are correlated with the
strength of the hydrogen emission and changes in the amplitude of the
photometric variations. Examination of the "emission free" Hβ profiles
obtained in 1977 suggests that the absorption line profile varies in a
manner similar to that seen in non-radially pulsating stars (Smith and
McCall 1978), but this must be confirmed by the He I line profiles.

Power spectra of the 1977 _uvby_ observations show the same period
as the radial velocities. There is also significant but much smaller
power at one-half this period. The amplitude of the periodic variation is about $0^{m}_{.}02$. In addition to these variations there was a
(roughly) linear decrease in brightness of $0^{m}_{.}0015$ d^{-1} during the
observing interval. All of the variations have comparable amplitude
in each color, but there is a hint that the amplitude is inversely
proportional to wavelength. The _Copernicus_ data have not been fully
analyzed, but the power spectra of the U1 and V3 data are consistent
with all of the conclusions drawn from the _uvby_ data.

So far I have only examined the _u_ light curve in detail. If phase
$0^{P}_{.}0$ is defined as the time of maximum radial velocity, then the primary
(1Θ) light curve variation has minimum near $0^{P}_{.}25$ and maximum near $0^{P}_{.}75$,
and the secondary (2Θ) variations have minima near $0^{P}_{.}35$ and $0^{P}_{.}85$. The
secondary variations are only significant at about the 3σ level. The
shape of the light curve and its phasing with the radial velocity
curve exclude ellipsoidal and eclipsing variable explanations for the
variations. The variations are consistent with pulsation provided
that the temperature is nearly constant during the cycle. If the
pulsational interpretation is borne out by the complete analysis of
the observations and the correlation between hydrogen emission strength
and amplitude of the variations suggested above is confirmed, we must
consider the possibility that the ejection of emission envelopes in at

least some Be stars may be related to pulsation of the source star.

I would like to thank Helmut Abt for the loan of the KPNO plates used in part of this work and John Percy for numerous helpful conversations regarding the photometric variations. I would also like to thank Ron Lyons and Matthew Bates for their assistance with the data processing. This work was partially supported by the Natural Sciences and Engineering Research Council of Canada.

REFERENCES

Abt, H.A., and Levy, S.G.: 1978, Ap. J. Supp. 36, pp. 241-258.
Irvine, N.: 1974, private communication.
Irvine, N.: 1975, Ap. J. 196, pp. 773-775.
Percy, J.R.: 1976, private communication.
Percy, J.R., and Lane, M.: 1977, A. J. 82, pp. 353-359.
Smith, M., and McCall, M.L.: 1978, Ap. J. 223, pp. 221-233.

DISCUSSION

Jerzykiewicz: The period of $\overset{d}{.}7$ is much too long for a radial
fundamental pulsation mode. If it corresponds to an oscillation, the
mode involved would have to be a g-mode. Such modes are strongly
damped in these stars. Could the $\overset{d}{.}7$ period be due to a spot on the
surface of the star carried around by rotation?

Bolton: I think it is too early in the data analysis to be drawing
conclusions. I suspect that you cannot explain both the light and
velocity amplitudes by a spot. Detailed examination of the HeI line
profiles will probably permit this to be sorted out.

Snow: When you analyzed the power spectrum of the Copernicus
satellite data, did you find periodic behaviour with the orbital
period of the satellite?

Bolton: No, I was surprised that I found nothing because the orbital
period of the satellite changed the position of the stationary U1/V1
tube. Apparently I chose the continuum position well. The orbital
period of the satellite did show up in the window function because
of periodic occultation of the star and the South Atlantic Anomaly.

RADIAL VELOCITY VARIATIONS IN 69 ORIONIS

M. Bossi, G.Guerrero and L. Mantegazza.
Osservatorio Astronomico di Brera, Merate, Italy

69 Ori (B5 V) was discovered as a Be star in 1976 (Doazan et al.,
1977). We observed this object at the Merate Observatory during the
periods Nov. 77. - Jan. 78 - Jan. 78 and Oct. 78 - Feb. 79 obtaining
20 red spectrograms and 25 blue ones with a dispersion of 35 Å/mm. We
performed also B and V photo-electric observations during 12 nights
altogether. The results from the red spectra and the photometric measurements have been discussed by Bossi et al. (1981).

The radial velocities obtained from the blue spectra show a significant difference between the two observational seasons, i.e.
$<V>_{77-78} = 14^{\pm} 6$ km/sec and $<V>_{78-79} = 40^{\pm} 6$ km/sec. This fact can be
due to the presence in the photosphere of rising motions during the
phase of the envelope ejection and/or to a long period binarity. The
radial velocities of the H_α envelope absorption referred to the photospheric one, given as CA in Bossi et al. (1981), also showed significantly different mean values between the two groupes of data.

The only relevant set of radial velocities in literature foregoing
our observations is due to Blaauw and Van Albada (1963). They suggest
a possible binarity with a period of about 19 days.

We have analysed independently our blue and red radial velocities
by means of a least squares periodogram method which took into account
the long-term variation using a third degree polinomial. We can exclude
any significant variability with such a period. On the contary both
sets of our data fit with a period of 1.28 days (Fig. 1 and Fig. 2),
even if we cannot exclude as alternative periods one of about 5 days
or another of about 0.56 days.

With a similar method we computed also the power spectrum resulting from the data of Blaauw and Van Albada (1963) (Fig. 3). The assumed 19 days period results only in aliases corresponding to the peak
that in our data gave the 5 days period. Therefore we believe that also
this last one is an alias. The periods derived from the 1955 data are

M. Jaschek and H.-G. Groth (eds.), Be Stars, 185–188.
Copyright © 1982 by the IAU.

a little shorter (1.13 or 0.53 days) then ours and the connected amplitude of the variations is lower.

If the real period is the longer one, the variability may be attributed or to a contact binarity or to the presence of photospheric spots.
The second hypothesis appears the most reliable, since our photometric
data seems to exclude the presence of eclipses, which would be very
probable in such a close system seen quasi equator-on (Bossi et al,
1981). On the other hand the contemporary increase of period and amplitude, which perhaps happened between 1955 and the present time, does
not follow the Kepler's third law. Finally the different lines cannot
easily fit a binary model.

If, on the contrary, the period of about 0.5 days is the real one,
the variability may be due either to a pulsation or to a rotation of an
object with tri-axial deformation, that would cause a double wave curve
for each stellar rotation. The first hypothesis seems to fails since
such a period is too long for radial pulsations, and the observed amplitudes appear too large for non-radial ones. The second hypothesis deri-
ves from some well known results of classical mechanics, which predict
the tri-axial ellipsoidal configuration for bodies not far from the
rotational instability (Jacobi, 1834 ; Chandrasekar, 1969). In this way
many features of the observed data could be explained. The different
amplitudes in velocity variations for different lines could be connected
with different temperatures in different photospheric regions (where
the coldest zones are the farthest from the barycentre). Moreover, if
the increase in period from 1955 to the present time is real, we would
have a good agreement with the consequent increase in the velocity
amplitude : both phenomena are connected to a growth of the deformation
and hence of the moment of inertia between an inactive period and one
of greater instability.

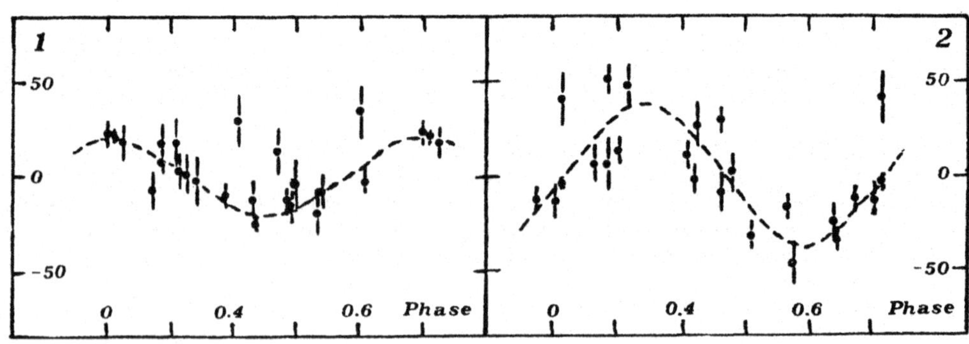

Blue radial velocities (1) and CA velocities (2) in km/sec
phased with a period of 1.28 days.

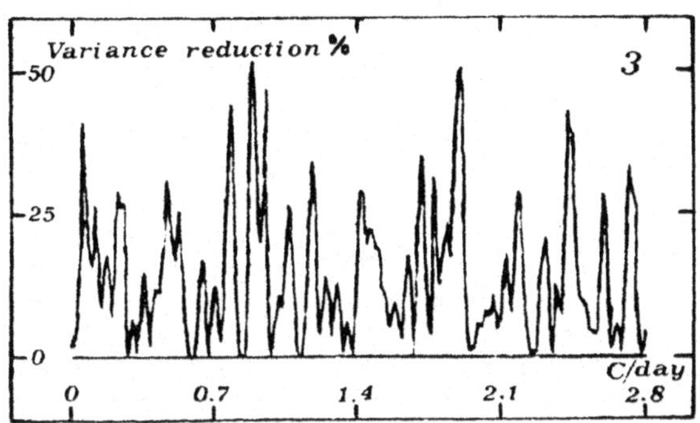

Fig. 3 . Frequency analysis of the 1955 radial velocities.

ACKNOWLEDGEMENTS

We are grateful to Prof. A. Kranjc, who furnished us the program
for radial velocities reductions. We thank also Dr. E. Antonello and
Dr. P. Farinella for their stimulating comments and Mr. M. Scardia for
his technical support.

REFERENCES

Blaauw A., Albada T.S. van, 1963, Astrophys. J. 137, pp 791 - 820
Bossi M., Guerrero G., Mantegazza L., Rusconi L., Scardia M., Sedmak G.,
       1981, Astron. Astrophys. Suppl., to be published.
Chandrasekhar S., 1969, Ellipsoidal figures of equilibrium, Yale Univer-
       sity Press, New Haven.
Doazan V., Bourdonneau B., Leterneur N., 1977, Astron. Astrophys. 56,
       pp 481 - 482.
Jacobi C.G.J., 1934, Poggendorff Annalen der Physik und Chemie, 33,
       pp 229 - 238.

DISCUSSION

Fehrenbach: Pourquoi avez vous rapporté vos VR à celles du shell?
Quel est la raison de votre ajustement polynomial?

Mantegazza: 1. In the red spectra the photospheric H_α absorption is
partially masked by the emission, so it is impossible to measure its
position with an optial comparator. The adopted procedure, based on
the use of a P D S microphotometer, gives more reliable results for
wavelength differences than for absolute wavelengths. The envelope
absorption line was the only unambiguous available reference.
2. Due to the presence of variations in our radial velocities with
time scales larger than our time-base, it was necessary to take them
into account in the analysis for short-term variations. This was
accomplished by computing for each trial frequency a simultaneous
least-square solution for the amplitude and the phase of a sinusoid
and the coefficients of a 3th degree polynomial. This procedure avoids
the introduction of deformations in the power spectrum (see Vanicek,
1971, Astrophys. Space Sci., 12, 10).

Sareyan: Do you think the high dispersion in your radial velocity
curves could be due to the "phasing" scatter, i.e. to the fact that
you put together different cycles that could be not similar in shape
and amplitude?

Mantegazza: May be, especially for the red data, the r.m.s. residual
 of the blue data after the sinusoidal fit is comparable with the
average error bar of the individual measurements.

Harmanec: Be sure that I would like to see periodicity in RV data for
Be stars but I must object you that if you admit to have points
covering almost the whole range of variations in one narrow phase
interval then you can find many such "periods".

Mantegazza: I agree that the fit of the data of Blaauw and van Albada
is quite poor: the error bar of these data is presumably a little
smaller than their amplitude of variation. But if they have a periodic
behaviour the periods suggested by us are the only reliable. Curiously
their values are nearly coincident with the ones determined from our
spectra.

Bolton: Have you calculated the statistical significance of your fit
to the velocity curve? If so, what is it?

Mantegazza: No, we haven't. The correspondence among three periods,
of which two are practically coincident, seems to use sufficiently
significant. They were obtained independently from three different
sets of data.

ON PERIODIC VARIATIONS IN THE SPECTRUM OF THE BOe STAR
X PERSEI ASSOCIATED WITH THE X-RAY SOURCE   3U 0352+30

T.S. GALKINA
Crimean Astrophysical Observatory
U.S.S.R.

The bright BOe star X Persei has been observed for over fifty years,
both as an emission line object and as a variable star.

The discovery that the star is situated very close to the position of a
weak X-ray source renewed the interest on the star and up to the present
it has been under close scrutiny by a number of observers.

In the course of an investigation of all available spectroscopic data,
Hutchings (1974) found evidence for a long period absorption velocity
variation, that may indicate the presence of a massive orbiting companion. A period of about 581 day was found by Hutchings for the broad
Balmer absorption.

Then the 22-hr and 13,9 min periodicities were discovered in the X-ray
range between 0.6 and 7.5 keV. However the search for the 22.4 hr periodicity in the radial velocity data made by Hutchings (1977) was not successful.

At the Crimean Observatory the spectroscopic observation of X Persei
has been started in November 1974. The spectrograms with dispersions
30 and 36 A/mm in the regions Hα and at λλ 4950-3650 A were obtained
from November 1974 to December 1980 ; about 2000 days were covered by
us observations.

The search for the evidence for the periodicities of 581 day and for the
22 hr in the variations of the radial velocity was made by us using the
homogeneous observations from November 1974 to March 1978.

On figure 1 can see that the broad Balmer absorption shows the period
of 581 day in the variations V_r as found by Hutchings, but the velocity
amplitude proved to be about 50 km/sec.

HeI lines revealed variations of V_r with a velocity amplitude of $\pm$ 22
km/sec and a large scatter in the measurements ; furthermore the phase

M. Jaschek and H.-G. Groth (eds.), Be Stars, 189–194.
Copyright © 1982 by the IAU.

is shifted by about one forth of the period, relative to the Balmer
absorption data.

The variations of V_r derived from the violet emission component of the
Hα line have an amplitude K = 30 - 35 km/sec. However, in the phases
$0^P.45 - 0^P.70$ a large scatter of values of the radial velocities (V_r)
was observed.

The radial velocities derived from the V and R emission edges of the
Hα line were analysed for a search of the variations with 22.4 hr
period.

To this purpose our observations from November 1976 to January 1978
were divided into groups, the time intervals used being a short interval of the 580 days period, in view of the superposition of the rapid
rotational and orbiting motions and of the proposed apsidal and other
motions.

In the second figure one can see the variations of the radial velocity
derived from the shifts of emission V and R edges of Hα line with the
phase of the 22.4 hr period (top curve). The zero point has been taken
arbitrarily as the moment of the minimum of the intensity ratio of the
violet to red component (I_V/I_R) of the Hα line emission. The phase dependence of the V/R intensity ratio has been shown on the bottom curve.
After 1978 the spectroscopic observations X Persei were continued.

The observations of last season (1979-1980) have lead to a reexamina-
tion of the 581 day period, because the measurements of radial velocity
revealed a large scatter in the phases of this period.

A search of the new period of the variations was made. According to our
homogeneous spectral observations carried out at the Crimean Astrophysical Observatory from XI.1974 to I.1980, a new period of radial velocities variations has been found, its value being, 308,33 days.

On the third figure one can see the variations of radial velocity from
the broad Balmer absorption line (HI λ 3835), and from the components
of the Hα line in this period. Let us note that the relative intensity
of the V and R components of the Hα line is changing with about the
same period. We assume, therefore that the real orbital period is equal
308,33 days.

For the check the 22,4 hr periodicity the observations of X Persei were
taken in interval of 10 days (from 9 to 20 October 1980). We assume
that this short (10 days) interval, decreases the influence of the orbital motion with large period and large amplitude.

On figure 4 one can see the results of the measurements V_r in the phase
of 22,4 hr period for the R (top) and V (bottom) components of the Hα
line emission. Also the 22,4 hr periodicity in the variations of radial

velocities of the Hα line components, found by us before (Izv. Cr. v. 61, 1980) is confirmed.

This lead us to the evident conclusion about the relation between X Persei and the X ray source 3U 0352 + 30. One notices that if the radius of the B0e component is taken equal to 4,9 solar radü and V sini = 270 km/sec, the period of rotation would be equal to 22,5 hrs.

In figure 5 one can see the variations in I_V/I_R with time (top) and the corresponding variations of the radial velocity of the V_{em} edge and the absorption core of the Hα line (bottom). Three I_V/I_R reversals occur. The time interval between two succesive reversals is about 300 days. The amplitude of I_V/I_R variations seems to differ from cycle to cycle.

On figure 6 one can see the intensity variations of the V (o) and R (x) components with the phase of the period $308^d.3$. We see that the inten- sities of the V and R components of the emission line Hα change in an- tiphase. On the regular variations with a period of $308^d.3$ of V and R are superimposed irregular short time scale pulsations that may be due to irregular matter ejection from the star.

About twenty years ago Boyarchuk showed that the stationary outflow cannot originate from the equatorial region of Be star, but that active processes of irregular nature occur on the stars surface. Our observa- tions confirm this idea.

The ratio I_V/I_R, as Boyarchuk remarked, might be the measure of the expantion velocity of the Be star envelopes. When $I_V > I_R$ the envelope contracts ; when $I_V < I_R$ the envelope extends. But the envelope as a whole can not phase.

Possibly these irregularities are conected with the interaction between the components of the system X Persei/ X Ray 3U 0352+30.

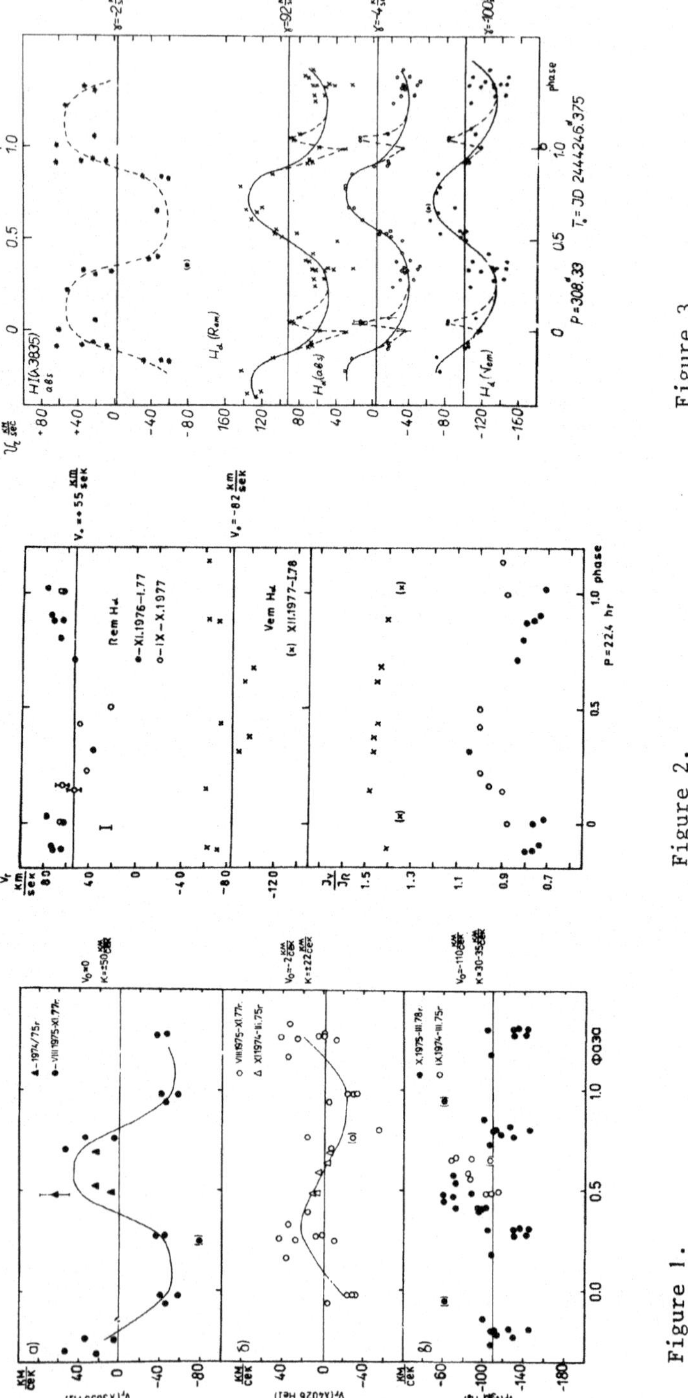

Figure 3.

Figure 2.

Figure 1.

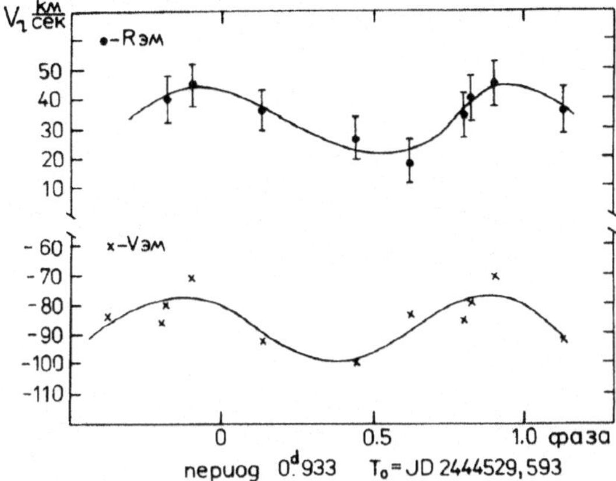

Figure 4.

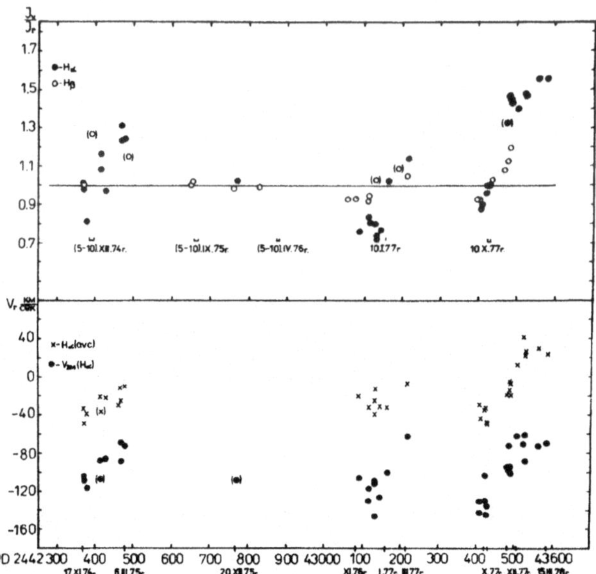

Figure 5.

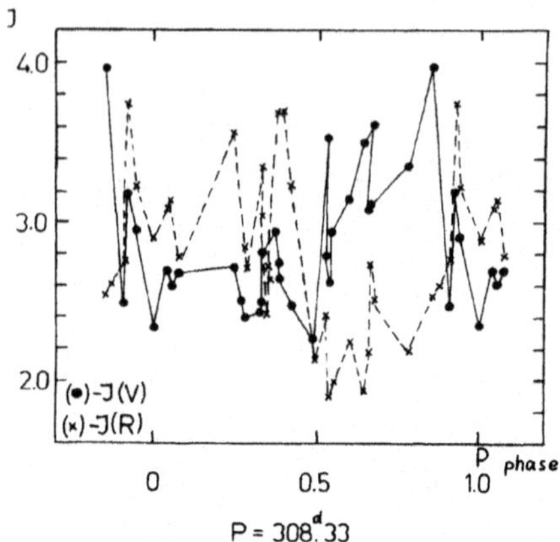

Figure 6.

DISCUSSION

Viotti: Concerning the long-term spectroscopic variations of X Per
it is known that during the last years the star suffered large
luminosity variations with two maxima in 1978 and 1980. It might
be possible that the spectroscopic variations are more or less
correlated to these luminosity variations rather than to any kind
of long-term periodicity.

Galkina: I have made comparisons of the variations of the relative
intensity of the V and R components of the emission line H_α with the
photometric bahaviour of X Per in 1974 - 1977. The spectroscopic and
photometric variations were correlated.

Henrichs: The 22 hour period in the x-ray flux reported by White et al.
(MN 176, 1976) has never been confirmed by any other x-ray measurement.
My question is: would you have found this 22 hour period in your
optical data without knowing it beforehand, or did you only fold the
data with that period? In the latter case one should be very cautious
to call the result a confirmation, especially in the case of such an
irregular varying star like X Per.
Galkina:I have discovered a 22.4 hour periodicity of the variations of
the radial velocities from the emission V and R components of the H_α
line in 1978.
From 1980, October 9 to 20 I have made spectroscopic observations of
X Per. These observations confirmed the 22.4 hr periodicity in the
variations of the radial velocities.

RECENT CHANGES OF THE Be STAR HD 58050

A.M. Hubert-Delplace, H. Hubert, D. Ballereau, M.Th. Chambon
DEPEG, Observatoire de Paris, 92190 Meudon, France

Abstract   Emission line variations in 1961-1981 of the B2 star HD58050
are reported. Brightness variations are recalled. An estimation of
the veiling effect given by continous emission of the envelope, observed in November 1980 in the Balmer lines, is given.

1. INTRODUCTION

This star is classified B2V by Jaschek et al. (1980), and is often
considered as a pole-on star (Kogure, 1970), because of the low value
of its projected rotational velocity, vsini = 140 km/s (Uesugi, 1978).

2. PREVIOUS OBSERVATIONS

A strength ening of emission lines was observed from 1933 to 1952.
According to Tcherg Mao Lin (1946) emission was seen up to H11 in 1945,
according to Burbidge et Burbidge (1953), up to H19 in 1952, emission
being single up to Hδ, Hε, then double with a mean peak separation of
about 145 km/s.
In "An Atlas of Be Stars", Hubert-Delplace et Hubert (1979), this star
exhibited strong H emission lines from 1954 to 1961 . Fe II lines were
also in emission during this period. Then H emission lines slightly
decreased in intensity ; and on medium dispersion spectrograms the
FeII emission lines which were strongest in 1960, decreased also after-
wards and disappeared after 1963.

3. SPECTRAL VARIATIONS SINCE 1961

17 spectrograms have been obtained at the 193 and 152 cm telescopes
of the Haute Provence Observatory (dispersion 9.67 A/mm and 12.27 A/mm
respectively).
From 1961 to 1968 we observe a large decrease of the emission component

M. Jaschek and H.-G. Groth (eds.), Be Stars, 195–199.

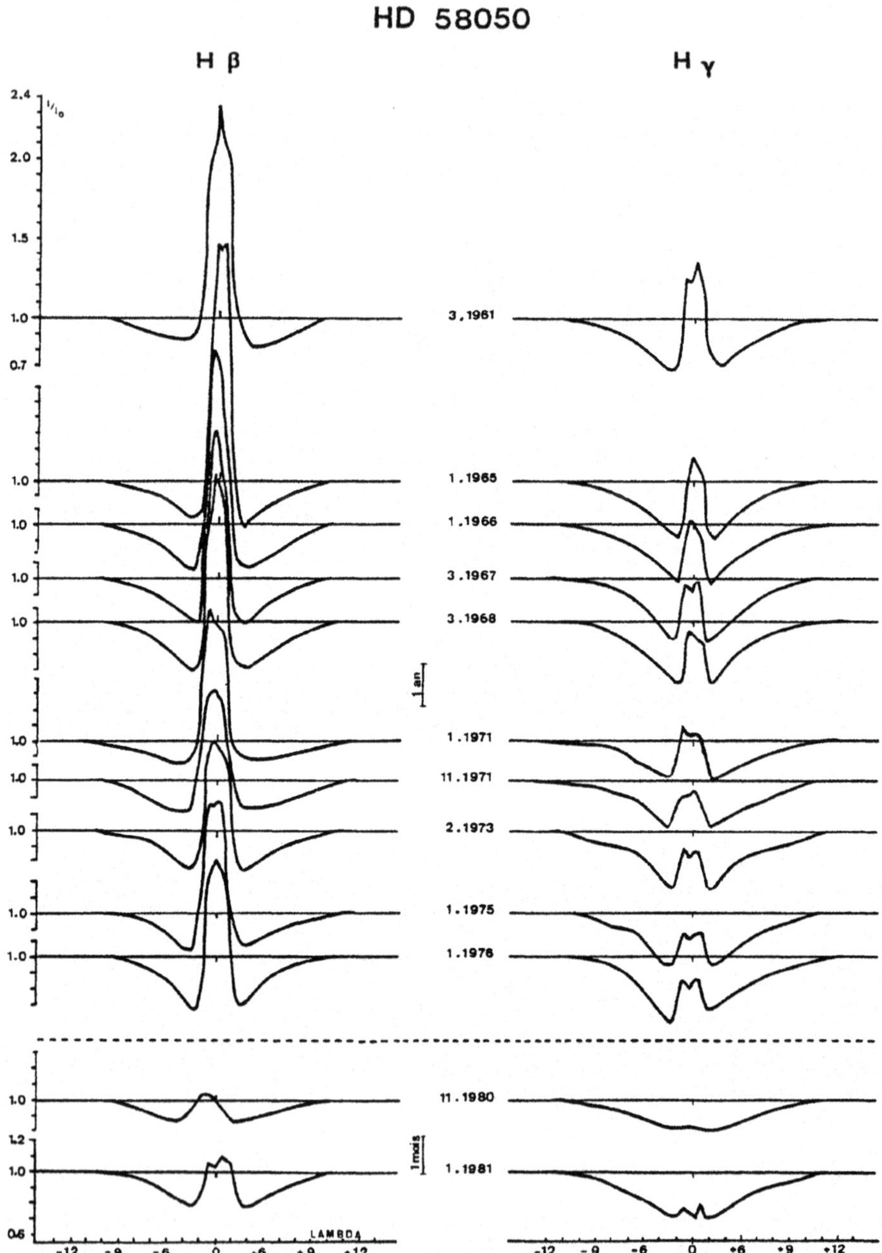

Fig.1. Variations of the Hβ and Hγ line profiles of HD 58050

superimposed in the photospheric lines, fig 1. The last Balmer term
Hn affected by emission is variable (in 1961-1965 n = 19, in 1967-
1968 n = 7) . In 1970-1971 emission again increases and is seen up to
H 19 but without reaching the intensity of 1961, then strongly decreases (end 1971-1976 n = 9, in 1981 n = 7). Rapid fluctuations of the
intensity of the V and R emission components are observed, sometimes
is one day as on January 17 and 18, 1981. These fluctuations could be
produced by propagation of disturbances in the atmosphere.
In November 21, 1980 a spectrogram reveals a weak diffuse emission on
the Hβ line while some traces are also visible in Hγ, Hδ, Hε (emission
is blue-shifted on Hδ, Hε), the other H lines being in diffuse
absorption. A comparison of the higher Balmer profiles of Nov 21, 1980
with those obtained in Feb 1973 and Jan 1981 shows that the equivalent
width and central depth of these lines are always smaller in Nov 1980,
(see Table 1) ; the HeI absorption lines are also fainter.

Table 1

	date	H_{10}	H_{11}	H_{12}	H_{13}
W_λ (Å)	13.2.73	4.2	3.1	2.2	1.5
W_λ	14.1.81	4.2	3.2	2.2	1.4
W_λ	21.11.80	3.0	2.2	1.4	0.8
ω	14.1.81	0.72	0.70	0.64	0.55
R_λ	13.2.73	0.35	0.32	0.27	0.23
R_λ	14.1.81	0.33	0.31	0.25	0.19
R_λ	21.11.80	0.26	0.24	0.17	0.12
ω	14.1.81	0.76	0.76	0.59	0.57

4. BRIGHTNESS VARIATIONS

The Groupe d'Etude et d'Observations Stellaires (GEOS) has observed
this star from the beginning of 1977 to Feb 1981, and found :
a) short time scale brightness variations (period P∼ 3 hours, amplitude
a ∼ 0.15 mag) as in βCMa stars. Variations seem periodic.
b) an increase of brightness in Nov, Dec 1980 of about 0.4 mag.
From 1977 to April 1980, m_V = 6.4 ; in Nov, Dec 1980, m_V = 6.0, and
in Jan, Fev 1981, m_V slightly increases.
Furthermore Dr. Divan and Dr Zorec kindly communicated that the Balmer
discontinuity was seen in emission in Nov 1980 but not in 1977.
So we conclude that the apparent faintness of the H and HeI absorption
lines observed in Nov 1980 could be produced by a "veiling effect"
given by continuous emission of the envelope

5. ESTIMATION OF VEILING EFFECT

In presence of an envelope, the central depth R_λ and the equivalent
width W_λ of an absorption line are given by :

$$R_\lambda = \omega\, R_\lambda^*$$
$$W_\lambda = \omega\, W_\lambda^*$$

$$\omega = \frac{e^{-\tau_e}}{F_e/F_c + e^{-\tau_e}}$$

where R_λ^* and W_λ^* are the photospheric central depth and equivalent width,
   F_c is the stellar continuum flux at the line,
   F_λ the stellar line flux at the wavelength ,
   F_e the envelope continuous emission,
   τ_e the average optical thickness of the continuum,
   ω the veiling factor.
In the approximation $\tau_e = 0$ and $F_e \neq 0$ (true in the case of "pole-on"
stars), for an increase of brightness of 0.4 mag at $\lambda \sim 5500$ Å, we
find
   $$F_e/F_c = 0.45$$
   $$R_\lambda \sim 0.7\, R_\lambda^*$$
   $$W_\lambda \sim 0.7\, W_\lambda^*$$
From our observations, see Table 1, we find
   $$\omega = R\lambda/R_\lambda^* \sim 0.6 - 0.7$$
   $$\omega = W\lambda/W_\lambda^* \sim 0.6 - 0.7$$
In this peculiar case, it is difficult to derive the wavelength
dependence of the veiling effect, because of the presence of a central
emission in the Hβ and Hγ lines, and of the accuracy of our measurements (10-15%) of equivalent widths and central depths.

The authors acknowledge Dr. R. Hirata for his helpful comments, and
Dr. C. Vant'Veer who kindly made 2 spectrograms of this star.

REFERENCES

Burbidge, E.M., Burbidge, G.R. : 1953, Astrophys. J. 118, 262.
Hubert-Delplace, A.M., Hubert, H. : 1979, An Atlas of Be stars.
Jaschek, M., Hubert-Delplace, A.M., Hubert, H., Jaschek, C. : 1980,
      Astron. Astrophys. Suppl. Ser., 42, 103.
Kogure, T. : 1970, Astron. Astrophys. 1, 253.
Tcheng-Mao-Lin.: 1946, Ann. Astron. 9, 120.
Uesugi, A. : 1978, Revised Catalogue of Stellar rotational velocities-
      Preliminary Edition.

DISCUSSION

Sonneborn: I find it misleading to use the term "pole-on" for a star
with line broadening which implies a projected rotational velocity of
150 km/sec. I would reserve the term "pole-on" for stars with sin i
close to zero. The problems of terminology in this field cannot be
emphasized too strongly. I hope there will be more discussions to
these problems during this symposium.

Traving: Can one estimate the effect of veiling (which is in the case
of this star of the order of 50%) on the structure of the photosphere
itself? Such a veiling should change the boundary conditions of the
photosphere and hence affect the formation of absorption lines.

Hubert-Delplace: From the paper by Divan and Zorec it seems possible
in the case of HD58050 to distinguish two kinds of Balmer discontinuities with the BCD classification in Nov. 1980, the first given by
the star itself, the second (at shorter wavelength) by the outer
structure. This is probably a means to determine the effect of veiling
on the stellar continuum. However the veiling effect on the absorption
lines is very difficult to treat and we give here only a rough approach
because we have only one spectrogram. It should be necessary to investigate the wavelength dependence of the veiling factor, if for a
given line there is a variation of this factor over the line.

Sonneborn: I can answer this question for the point of view of model
atmospheres. At Ohio state we have studied model atmospheres with
radiation incident from a second star. In this case the atmospheric
structure is altered and the effective temperature rised. We have not
examined the effect of incident radiation from a circumstellar shell
or disk, although this would be possible with our atmospheric code.
I would expect to find results similar to the incident stellar radiation case.

Sareyan: I would like to comment on the star HD58050. This star seems
to have a reliable 3^h period. When observed spectroscopically it is
a Be star. If its short period light variations had been discovered
by photoelectric photometry through filters placed on the continuum,
it would have been described as a β Cep star. So this is a new member
of a growing intersection between Be and β Cep stars. (HD77320, which
has a 7^h period, according to Burki et al. would also be in this case,
as it is a Be star.) So we don't know really if many Be stars undergo
short period variations, and on the other hand, many β Cep stars could
show - at least irregularly - emission features. (β Cep stars should
be checked periodically for emission, at least at H_α and Be stars
investigated systematically for short period light variations.)

R 81: P CYGNI OF THE LMC

O. STAHL, B. WOLF, Landessternwarte, Königstuhl
D-6900 Heidelberg 1
M.J.H. DE GROOT, Armagh Observatory,
Armagh BT 61 9DG, N. Ireland
C. STERKEN, Astrophysical Institute,
Vrije Universiteit Brussel, Brussel, Belgium

ABSTRACT:

Extensive photometric UBV observations and spectroscopic
high dispersion (20 Å/mm and 38 Å/mm) coudé observations of
the very luminous ($M_V = -8.2$) B2.5 eq supergiant R81 of the
LMC were carried out between 1970 and 1980 at ESO, La Silla.
In addition the IUE satellite was used to obtain a high resolution (0.2 Å) spectrogram in the ultraviolet wavelength
range $1200 < \lambda < 1950$ Å. The most prominent features of the
visual spectrum are P Cygni profiles of the Balmer lines,
indicating a shell with an expansion velocity of about
140 km s^{-1}. The ultraviolet spectrum of R 81 is dominated
by blue-shifted absorption resonance lines (Si II, Si IV,
C II, C IV, Al III etc.) and Fe III absorption lines originating from metastable lower levels. From the UV resonance
lines a very high mass loss rate ($\dot{M} = 5 \quad 10^{-5}\ M_\odot\ yr^{-1}$) was
estimated. The early Balmer lines show very broad shallow
emission wings (total width 40 to 50 Å), attributed to electron scattering. The mass loss is highly variable and presumably occurs in the form of sudden ejections of discrete
shells. Irregular brightness variations of a few tenths of
a magnitude in V on timescales of weeks were found. An absolute bolometric magnitude $M_{bol} \approx -10$ and a photospheric
radius $R \approx 70\ R_\odot$ were estimated. A comparison with theoretical evolutionary tracks indicates a stellar mass $M > 50\ M_\odot$.
The observed spectroscopic properties lead us to suggest
that the LMC star R 81 is a close counterpart of the galactic star P Cyg, representing a short lived transient stage
in the evolution of the most massive stars.

A more detailed presentation of the results has already
been submitted to Astronomy and Astrophysics.

M. Jaschek and H.-G. Groth (eds.), Be Stars, 201–203.
Copyright © 1982 by the IAU.

DISCUSSION

<u>Ballereau</u>: In P Cyg, He I lines have "P Cyg" profiles. Why in your
star only H has P Cyg profiles and not He I?

<u>Stahl</u>: We suppose that the temperature of R81, which seems to be
slightly lower than the temperature of P Cyg is too low to produce
P Cyg profiles of the He I lines.

<u>Viotti</u>: I would like to know more about these absorption lines which
could originate in the deepest layers of the atmosphere of this star,
since they could help us to determine its effective temperature. Are
the radial velocities of excited lines like He I violet shifted or
unshifted with respect to the expected radial velocity of the star
and did you find excited lines, like HeII, NIII, in UV? Finally it
would be important to compare your mass loss rate with that one that
would be determined from the IR.

<u>Stahl</u>: The He I lines are indeed violet shifted with respect to the
system velocity. HeII and NIII lines could not be detected in the UV.
There are no IR observations available for R81. However, we plan such
observations in order to derive the mass loss rate independently from
the IR free-free radiation.

<u>Snow</u>: I have a comment and a question. The comment is that IUE spectra
show very complex interstellar absorption lines towards the Magellanic
Clouds, with numerous velocity components. In particular, highly
ionized species, such as C IV have this appearance. Therefore I am not
convinced that your IUE spectra of this star reveal any evidence for
highly ionized species in the circumstellar envelope; <u>all</u> of the CIV
you see may well be interstellar.
My question has to do with your derivation of the mass-loss rate by
fitting the Castor and Lamers theoretical profiles. If you did not
allow for photospheric contribution to the lines, you may have overestimated $\dot{M}$. On the other hand, this could account for the difficulty
of matching the emission portion of the profiles. How did you treat
this problem?

<u>Stahl</u>: I agree that a great deal of the observed CIV lines may be of
interstellar origin, but it is hard to decide this from our IUE
spectrogram. As far as the fitting of the profiles is concerned: we
did not allow for a photospheric contribution to the lines, because
the photospheric profiles are not known. It is true, that the observed
emission may be reduced by absorption in the photosphere. We think
that the mass-loss rate is not much affected by the photospheric
absorption, as we fitted the shortward wings of the observed line
profiles, which are probably not affected by the photospheric absorption.

<u>Divan</u>: From our spectra taken in Dec. 1971, R81 shows emission in the
Balmer continuum like ordinary Be stars. It is the only supergiant

which shows such emission. The star P Cyg does not show any Balmer
discontinuity, neither in emission nor in absorption. The spectral
classification of R81 is given as B2.5eq. If its absolute magnitude
were not known from its distance, I wonder what luminosity class
would be inferred for its spectrum.

ON THE PROBLEM OF THE CHEMICAL COMPOSITION OF β LYRAE

V. Bahýl'
96201 Zvolenská Slatina 82, Czechoslovakia

Two determinations of helium abundance in the primary
β Lyrae's component existed before 1975. According to Boyarchuk
(1959) helium is 631 times overabundant relative to hydrogen and
according to Hack and Job (1965) the ratio He/H varies from 1 to
2.25 depending on the effective temperature. We started to examine
this probleme with the aim to decide between these extreme results.
In the meantime there were published two papers by Leushin at al.
(1977, 1979) in which there was found, that He/H = 1.5 for T_{eff} =
= 12 000°K and log g = 2.5. The earlier results were based on
equivalent widths analysis with curve of growth method. The authors
of newly published papers have started from convenient models of
the atmosphere, calculated equivalent widths and compared them with
observation. In our paper (Bahýl' 1979) we started from models of
the atmosphere, too. Then we calculated the theoretical profiles.
We have found the best fit for the model Böhm-Vitense (1967),
T_{eff} = 12 900°K, log g = 2 and He/H = 2.72, as shown in Fig. 1.

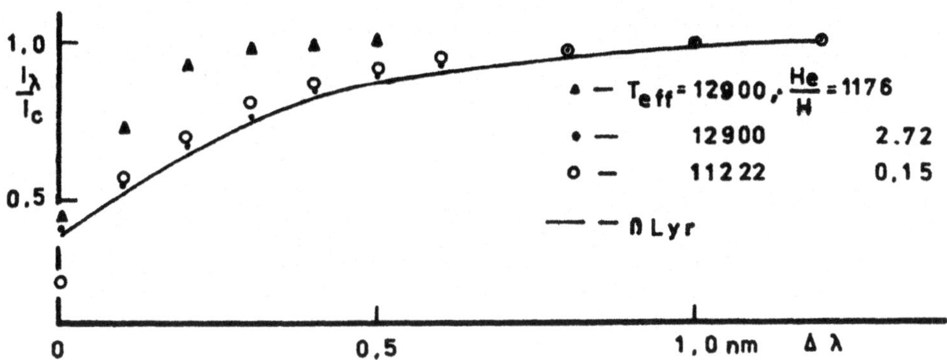

Figure 1. H 6 - some theoretical and the infinum of the observed
profiles.

M. Jaschek and H.-G. Groth (eds.), Be Stars, 205–208.
Copyright © 1982 by the IAU.

The methods designed for spectra which are stable in time
were applied in the cited papers to study ꓐ Lyrae's spectrum. In
this paper we shall try to check out to what extent is this
approach appropriate in the ꓐ Lyrae's case and whether it is not
just the reason of the discrepancies.

The data used in this paper were obtained from 19 high-
dispersion spectrograms (D = 0.8nm/mm) in the spectral range from
365 nm to 490 nm covering the whole light chrve as shown in Fig. 2.

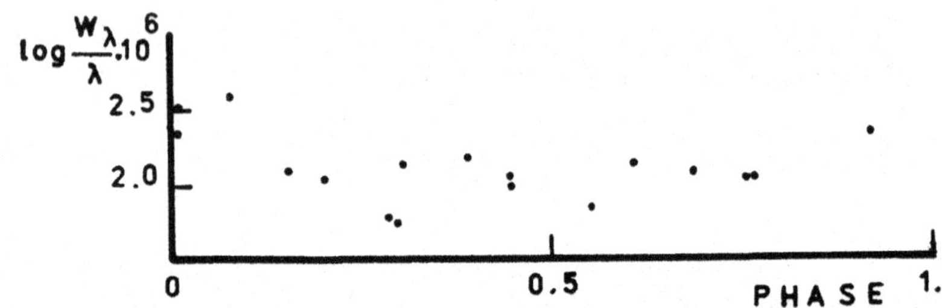

Figure 2. The phase diagram of the equivalent width of the Ca II K
         line.

An important feature was noticed immediately on the
spectral intensity tracings. Marked changes occurs in the absorption
line profiles, and these affect not only the shape of the profiles,
but also the changes of the equivalent widths of the spectral lines
with the phase. The equivalent widths tend to increase around the
minimum phase, as shown for the Ca II K line in Fig. 2.

In the framework of the commonly accepted ꓐ Lyrae's model
as a close binary with an envelope, it is possible to interpret the
spectral changes as a result of the envelope's influence on the
spectrum of the primary. In an effort to work with the profiles as
stable as possible, we directed our attention to the high members
of Balmer hydrogen series. The point is that these come from deeper
layers of the atmosphere, where we can most likely assume conditions
close to LTE, and neglect the influence of the disc-shaped envelope
as a reasonable approximation. We have not reached this aim, as we
were unable to find "stable" lines in ꓐ Lyrae's spectrum; the changes
with phase were observable in the high members of the Balmer series,
too. The conditions are most stable within the phases $0^{p}.2$ to $0^{p}.8$,
but the dispersion is still large. The H 11 profile is plotted in
Fig. 3. The crosses indicate the mean profile from the range $0^{p}.2$
to $0^{p}.8$. The observed mean intensity dispersion is indicated by
vertical bars. The mean profile from the neighbourhood of the
primary minimum $P \in (0.8;1)U(0;0.2)$ is denoted by circle with a dot
and it is given without dispersion, because we have at our disposal
only two goot observations from this period.

Like Leushin at al. (1979) we tried to compare the H 11
profile with the theoretical ones of Klinglesmith (1971). We have
found this theoretical profiles deeper than the observed mean

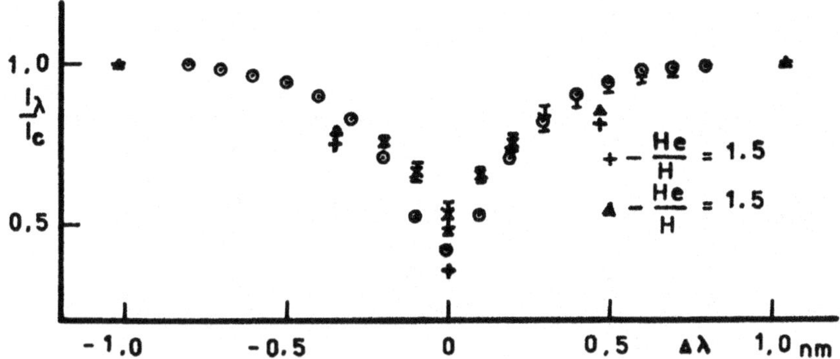

Figure 3. Comparison of the theoretical and observed profiles of
the H 11 line.

profile from the interval $0^p.2 - 0^p.8$. In case of the primary minimum
the profile is indeed deeper, but it is steeper and the agreement
of observation with theory is not improved. The situation is similar
in the other Balmer spectral lines, or the agreement is still worse.
Even then the model with He/H = 0.125 is in better agreement with
observation then the model with He/H = 1.5, as suggested by Leushin
at al. (1979).
Since the application of methods convenient for stable
spectra on the β Lyrae's spectrum may result in different models of
its atmosphere, we finally conclude, that these methods give us the
first approximation to reality only. To find the real abundance
values it would be desirable to compute the model atmosphere taking
into account the common envelope of the whole system and the
processes which take place in its vicinity. Anyway the present
results definitely contradict the helium overabundance in the
β Lyrae's primary component atmosphere as suggested by Boyarchuk
(1959).
My thanks are due to the Ondřejov Observatory for kindly
providing the spectrograms used in this study and to my friends for
stimulating discussions.

References.

Bahýl', V.: 1979, Thesis, Astron. Inst. SAV, Skalnaté Pleso.
Boyarchuk, A. A.: 1959, Astron. Zh. 36,pp.766-776.
Böhm-Vitense, E.: 1967, Astrophys. J. 150,pp.483-500.
Carbon, D. F., Gingerich, O.: 1969, in Theory and Observation of
Normal Stellar Atmospheres, ed. O. Gingerich, M. I. T.
Press, pp.377-472.
Klinglesmith, D. A.: 1971, NASA SP-3065, Wahington DC.
Leushin, V. V., Nevskij, M. Yu., Snezhko, L. I., Sokolov, V. V.:
1977, Astrofiz. Issled. 9,pp.3-15.
Leushin, V. V., Nevskij, M. Yu., Snezhko, L. I.: 1979, Astrofiz.
Issled. 11,pp.40-50.
Mihalas, D.: 1966, Astrophys. J., Suppl. Ser. 114,pp.1-30.

DISCUSSION

<u>Sonneborn</u>: Before you include the effects of circumstellar material,
the effects of rotation must be taken into account. For a star such
as β Lyr (v sin i = 240 km/s) the temperature variation over the
surface will be important for the hydrogen lines. Using a model
atmosphere with a single effective temperature can only produce
misleading results, especially in abundance determinations.

IV. INFRARED OBSERVATIONS

SPECTROSCOPIC OBSERVATIONS OF Be STARS
ESPECIALLY IN THE INFRARED

L. Houziaux
Institut d'Astrophysique, Université de Liège
Avenue de Cointe, 5, B-4200 Liège (Belgium)

Y. Andrillat
Observatoire de Haute Provence, C.N.R.S.
F-04870 St. Michel l'Observatoire, France

Since the time of I.A.U. Colloquium n° 70, new data
have been obtained in the infrared, far less however than
in the ultraviolet or in the visible. Few line spectra have
been recorded but numerous results have been published in
the field of spectrometry of the continuum. It is however
neither easy nor very useful to report on the infrared
wavelength range alone, and therefore we shall consider
when appropriate other spectral regions in this review.
After all, the astrophysicist nowadays has the advantage to
seek for information in a spectrum which encompasses a
fantastic frequency range of about 10 dex, from $\nu = 10^8 s^{-1}$
up to $\nu = 10^{18} s^{-1}$. This means that a contemporary astronomer
will have to become familiar with a somewhat larger span of
phenomena than with the rather narrow one he has been
accustomed to for the last 40 or 50 years.

The fields of photometry and polarimetry in the
infrared have been reviewed earlier during this symposium
respectively by E.E. Mendoza and G.V. Coyne.

Among recent works on line spectroscopy in the
infrared, we would like first to mention the stellar atlas
published by the late H.L. Johnson (1977) and based on
spectra obtained via Fourier transform spectroscopy. This
atlas contains namely the spectra of γ Cas, P Cygni and
φ Per, from about Hβ to 10,300 A. The fluxes have been
normalised to give a constant value per unit frequency
interval. The reduction procedure allows a resolution of
1.93 A at 5000 A, but the instrument is capable of a much
higher resolution (0.13 A at 5000 A).

Work at Haute Provence and Asiago allowed to observe several lines up to 11,200 A, revealing in peculiar Be stars the Paschen lines up to P6, the He I 10830 and 10913 A lines, the C I multiplet around 10690 A, the [S II] lines at 10284, 10318 and 10336 A, [S III] lines at 9069 and 9532 A, [Fe III] lines at 8838, 9960, 10433 and 10504 A. Several lines remain however unidentified.

The infrared spectrum of ζ Tau has been obtained between 1.5 and 4.7 µm by Smith, Thronson, Larson and Fink (1979) using Fourier spectroscopy. The outstanding features in this range are the Brackett lines. Bα is in emission at 4.05 µm but from B(4-10) to B(4-25), around 1.6 µm, these lines appear in absorption. This behaviour is unlike that of the Brackett series in γ Cas, in which all lines are in emission. In addition to the Brackett lines, the Pfund line Pβ(5-7) and some Humphreys lines (6-11 and 6-10) are detected in emission. Typical line width is around 130 km/s, i.e. about 40 A at 4 µm, which is much less than v sin i $\sim$ 310 km/s measured from emission features in the visible. The equivalent width of the Brackett α line is about 33 A, while its width (FWMH) is 47.2 A. It is suggested by the authors that therefore Bα is optically thick and originates in a small rapid rotation hot region close to the star, while the absorption lines arise from a cooler and more extended atmosphere.

It has been known since 1924 that the bright O I line at 8446 A should be due to a fluorescence mechanism pumping Ly β photons. However the line at 11287 A, which precedes 8446 A in the cascade towards the ground level has not been reported in emission. More curious still is the fact that the resonance multiplet at 1302 A has not been observed either in several Be stars where the 8446 A line appears as a strong emission (as reported by Oegerle, Polidan and Peters (1979)). In HD 50138, where λ 8446 is the strongest feature in the near infrared, we have been unable to detect any emission at λ 1302 on high dispersion I.U.E. spectra. Of course, it is well known that strong interstellar absorption is present in this line but a broad and strong emission should be easily detectable.

We shall come back on infrared lines later but we would like to report now about important observations of the infrared continuum. Garrison (1978) has extended his observations of Herbig Ae/Be stars to the near infrared, covering the λ 3400 to λ 8300 range.

The most remarkable characteristic feature is the heavy reddening of these objects and a much smaller Balmer discontinuity than that observed in stars with an analogous spectral type. This is shown on fig. 1, where the spectrum is however dereddened using model atmosphere fluxes for stars of similar effective temperatures. It is also apparent that some stars show an infrared excess with respect to the model fluxes.

Garrison explains the small Balmer jump by free-bound emission and establishes a means of computing the "Balmer excess" for an optically thin model

$$\Delta D_B = -2.5 \log \frac{4\pi R_*^2 \pi F_\nu}{4\pi R_*^2 \pi F_\nu + 4\pi \int_{V_s} j_{\nu_{bf}} dV_s}$$

where

F_ν is the flux from the star shortward of $1/\lambda = 2.7 \ \mu m^{-1}$.

$j_{\nu_{bf}}$ is the volume emission for the free-bound H transitions,

V_s being the volume of the emitting region.

$$j_\nu = 1.88 \ 10^{-34} \ T_s^{-3/2} \ N_e N_i \ cgs$$

where T_s is the shell temperature; then an idea of the shell emission called the "emission measure" and defined as

$$\varepsilon = \frac{\int_V N_e N_i \ dV}{R_*^2}$$

can be computed.

But

$$\int_V j_{\nu_{bf}} dV = 4\pi R_*^2 \ \pi F_\nu \ (10^{0.4 \ \Delta D_B} - 1)$$

hence

$$1.88 \ 10^{-34} \ T_s^{-3/2} \int_V N_e N_i \ dV = 4\pi R_*^2 \pi F_\nu \ (10^{0.4\Delta D_B} - 1)$$

and

$$\varepsilon = \frac{4\pi^2 F_\nu}{1.88} (10^{0.4\Delta D_B} - 1) \; 10^{34} \; T_s^{3/2}$$

can be deduced from ΔD_B, T_s and F_ν taken from a model
atmosphere.

The emission measure can also be used to predict the
Paschen excess due to Paschen free-bound and free-free
emission. The excess in magnitude is given by

$$\Delta m_{ir} = -2.5 \; \log \frac{4\pi R_*^2 \; \pi F_\nu'}{4\pi R_*^2 \pi F_\nu' + 4\pi \int j_{\nu_{bf+ff}} \; dV_s}$$

F_ν' meaning here the theoretical stellar flux at some point
in the Paschen continuum. The emissivity for free-free
transitions is

$$4\pi j_{\nu_{ff}} = 7.39 \; 10^{-38} \; T^{-1/2} e^{(-1.44 \; 10^4 \lambda^{-1}/T_s)} \; N_e N_i$$

and Δm_{ir} may then be computed for various T_s in order to
give a best fit with the observed energy distribution. These
temperatures are given on fig. 1.

The reliability of these models is based on the
assumption that the shell is optically thin in the
continuum. The optical depth for free-free absorption,
bound-free absorption and electron scattering are discussed
by Garrison taking into account a model where the density
drops off at a rate

$$N = N_o \left(\frac{r_o}{r}\right)^n$$

for realistic values of r_o and N_o at the stellar surface.
The conclusion is that the shell is optically thin for free-
free absorption and electron scattering, while this might
not be true for bound-free absorption at the Balmer
continuum. Garrison concludes also that the Paschen
reemission alters only very slightly the flux below 5000 A,
so that the dereddening procedure is not affected. This in
our opinion is a very important point that should be
examined with caution in each case since it affects all the
results. The dereddening procedure has thus to be confined
in a small region between 5000 and say 4000 A where the
reddening is assumed to be truly interstellar.

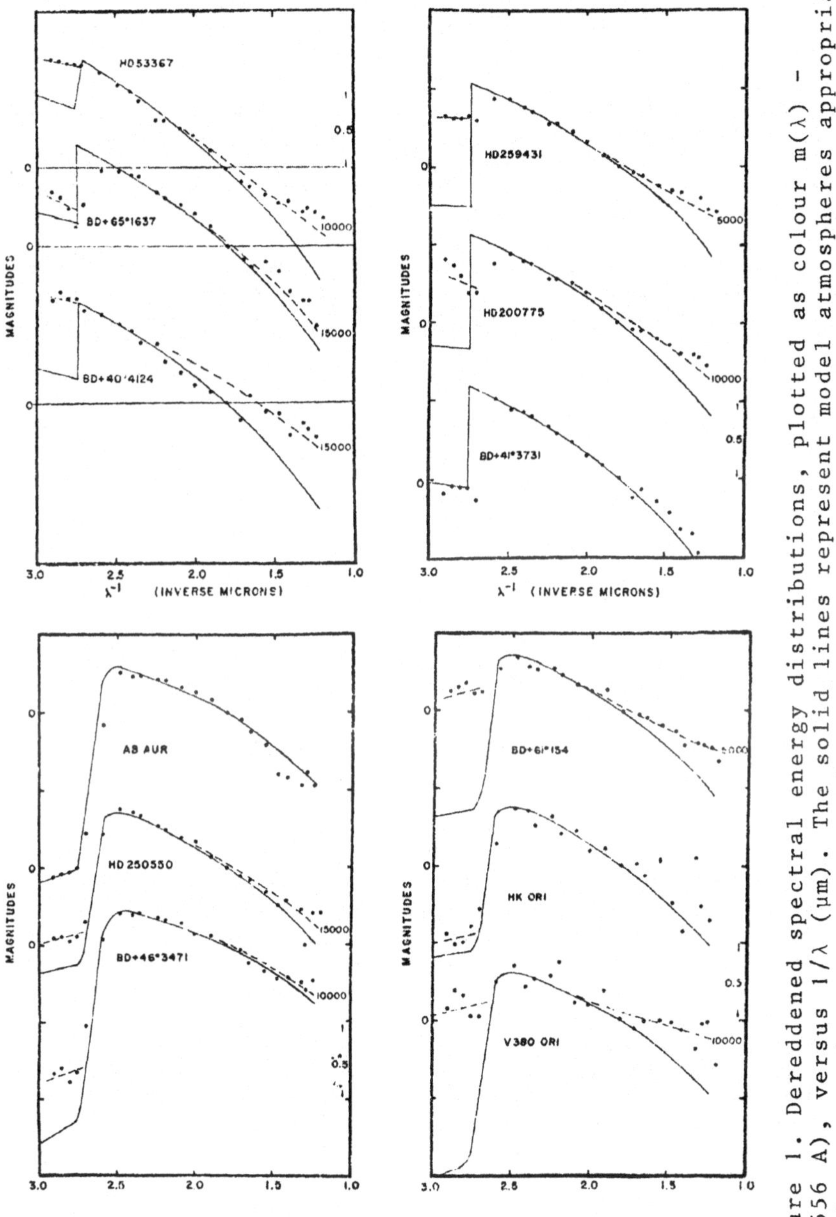

Figure 1. Dereddened spectral energy distributions, plotted as colour $m(\lambda)$ – $m(5556\ \mathrm{A})$, versus $1/\lambda$ (μm). The solid lines represent model atmospheres appropriate to the spectral types. Dashed lines indicate model fluxes plus optically thin emission from a circumstellar shell. Numbers to the right of each curve give the temperature of the shell yielding the best fit with the observations (L.M. Garrison, Jr. The Astrophysical Journal, 224, pp. 540–541, 1978).

The infrared excess up to 5 μ has been computed in the same way as for the Paschen excess, using Johnson's results for the reddening curve below 1 μ. The results show that all the IR excess cannot be attributed to free-free and bound-free emission including n = 4 and 5 levels.

A thermal component is thus necessary also to explain a peak in the blackbody function around 3 μ, hence linked to temperatures of T = 2,890/λ(μm) or about 1000°K. These results are also in agreement with the emission measures derived from the Balmer excess.

Using the characteristic velocity derived from the H_α profile and integrated intensities from the Balmer excess, the mass flow rate is derived.

For spherical symmetry and constant density

$$\epsilon = \frac{4\pi R_s^3 \, N_e^2}{3R_*^2}$$

assuming complete ionization and taking no account of the star's volume and occultation. R_s is the radius of the shell. Then

$$\frac{dM}{dt} = \dot{M} = 4\pi R_s^2 \, N_H \mu \, m_H \, \frac{dR}{dt}$$

If Hα and the Balmer excess are formed at about the same R_S,

$$\dot{M} = (3\epsilon)^{2/3} R_*^{4/3} \, 4\pi \, \mu m_H \, v_s \, N_e^{-1/3}$$

Thus to know the massflow rate when the velocity v_s and ϵ are known, one needs to estimate the electron density and the stellar radius. Massflow rates are of the order of 10^{-6} to 10^{-7} $M_\odot y^{-1}$. General properties of the shells are deduced from all available data.

Scargle, Erickson, Witteborn and Strecker (1978) have studied the infrared spectrum of γ Cas. Observations, carried out aboard the Kuiper Airbone Observatory cover a region from 1.12 to 4.1 μm. Main results are summarised on fig. 2. Various stellar models are proposed by the authors. The flux F_ν is decomposed into star and shell contributions

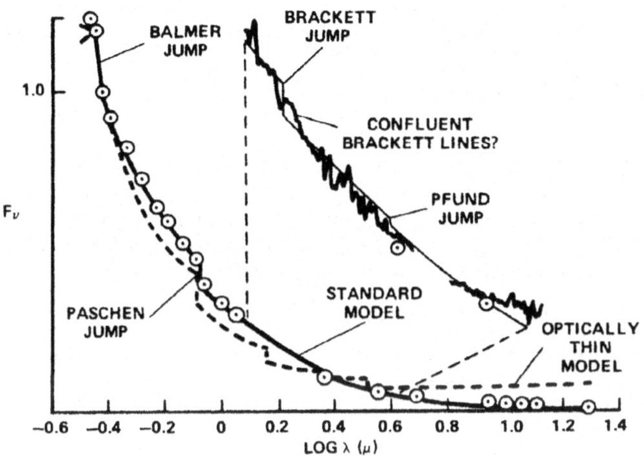

Figure 2. The spectrum of γ Cas, on a linear scale, corrected for a reddening corresponding to E(B−V) = 0.1. The points ⊝ are broad band data. The smooth solid line is the flux for a slab geometry model with T_e = 28,000°K and T_{shell} = 17,540°K. The dashed line represents the flux of an optically thin model which is clearly inadequate. The inset shows the dereddened data obtained by Scargle, Erickson, Witteborn and Strecker (from the Astrophysical J., 224, p. 528, 1978).

$$F = (1 - f)\psi_\nu + f\phi_\nu$$

$$= (1 - f) F_\nu^{*} \exp(-\tau_\nu^{*}) + f\phi_\nu(\tau_\nu^{s})$$

where F_ν is the stellar flux, attenuated by an optical thickness τ_ν^{*}. For the isothermal slab model, the function ϕ_ν, which represents the flux emitted by the shell, is

$$\phi_\nu(\tau) = B_\nu(T_s)(1 - e^{-\tau})$$

The parameter f represents the fraction of the total radiation contributed by the shell at the long wavelength side of the Balmer discontinuity; it is depending on the size of the shell relative to the stellar photosphere. f is determined by forcing the model Balmer jump to equal the observed one.

It is found that the IR excess in the case of γ Cas is due mostly to free-free and bound-free emission from a shell of gas at T_s = 18,000°K, with an optical depth $\tau(1\ \mu m)$ = 0.5. Disk models for the shell give the best fit with N_e = 10^{12}cm^{-3} and a size of 2.10^{12}cm.

The authors attract attention on the fact that the shell adds radiation to that of the star and also substracts radiation from the star. In certain situations, the net effect may be zero and shell is invisible at some wavelengths in the continuum.

A study of HD 200775 has been carried out by Altamore et al. (1980) using data available in all frequency ranges. In the infrared, new photometric observations have been made from 2.3 to 12.6 μm. No variability in the IR magnitudes has been found. Extinction due to interstellar matter has been estimated at 0.25 m, in contrast to E(B-V) = 0.57 - 0.70 as derived from ground based photometry. The stellar continuum fits a 16,000°, log g = 4 atmosphere, especially in the UV. Also in the case of this star, a discontinuity of -0.36 magnitude for the shell is found. The authors ascribe most of the IR excess to thermal radiation while an outflow of material is found to be

$$\frac{\dot{M}}{v_{es}} = 2.3\ 10^{-9}\ M_\odot\ yr^{-1}/km\ s^{-1}$$

where v_{es} is the escape velocity.

On the photometric side, Whittet and van Breda (1980)

B0e   B1e
Ne + Ar
HD 181308
HD 24534
HD 12856
HD 206773
HD 187350
HD 193089
HD 13851
HD 39680
HD 216411
HD 33461
HD 142926

B2e
Ne + Ar
HD 85875
HD 35345
HD 198944
HD 224055
HD 50083
HD 12882
HD 11806
HD 198895
BD + 61° 623
HD 21212
HD 13669
HD 184279
Hα
6678   HeI   7065
7772
CaII   PI2
8446

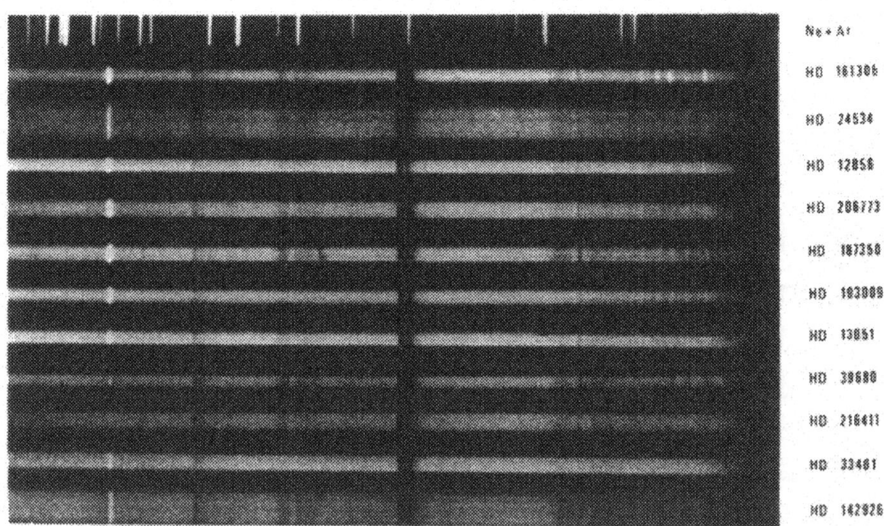

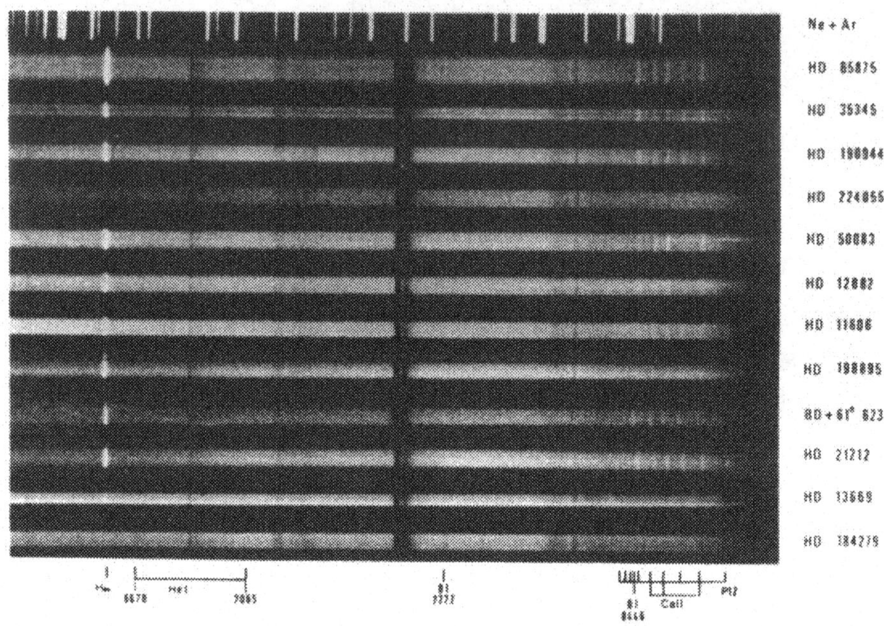

Plate 1

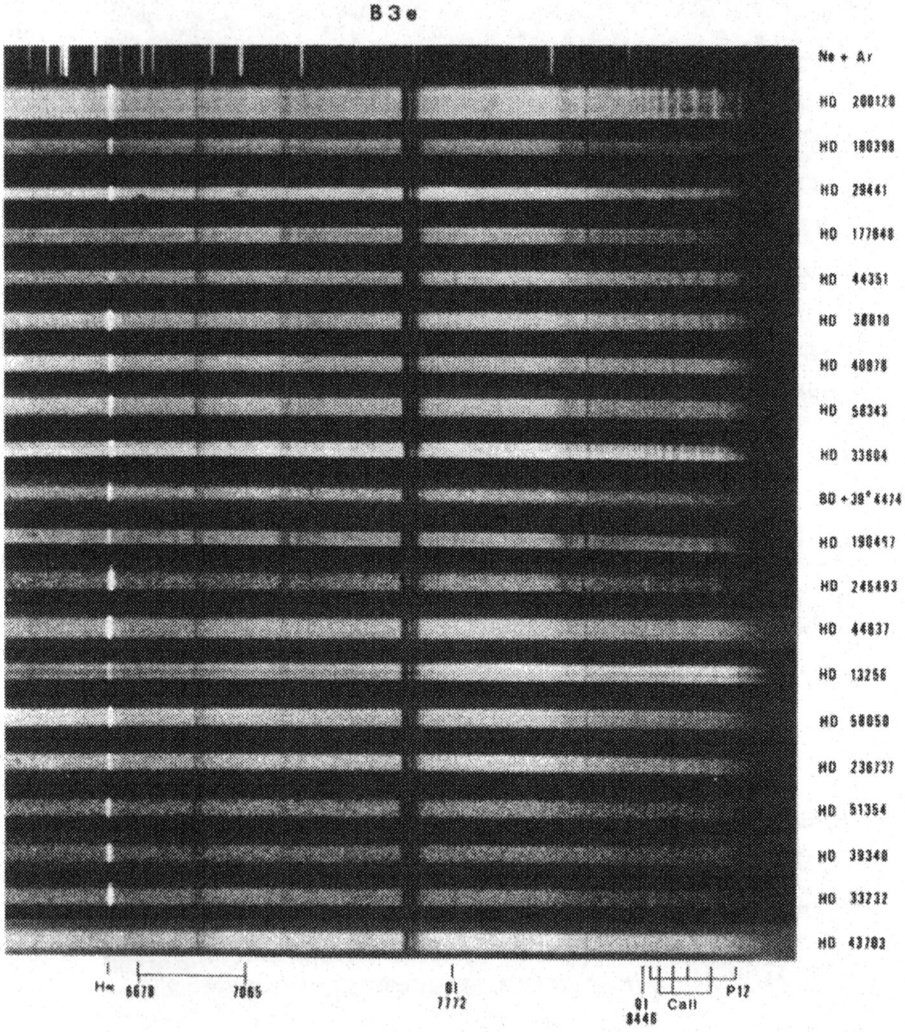
B3e
Ne + Ar
HD 200120
HD 180398
HD 29441
HD 177648
HD 44351
HD 38010
HD 40978
HD 58343
HD 33604
BD +39° 4474
HD 198457
HD 245493
HD 44637
HD 13256
HD 58050
HD 236737
HD 51354
HD 39340
HD 33232
HD 43703
Hα
6678
7065
OI
7772
OI
8446
CaII
P12

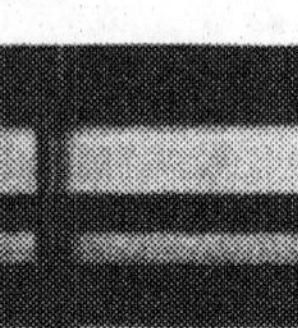

Plate 2

B 4e   B 5e   B 6e

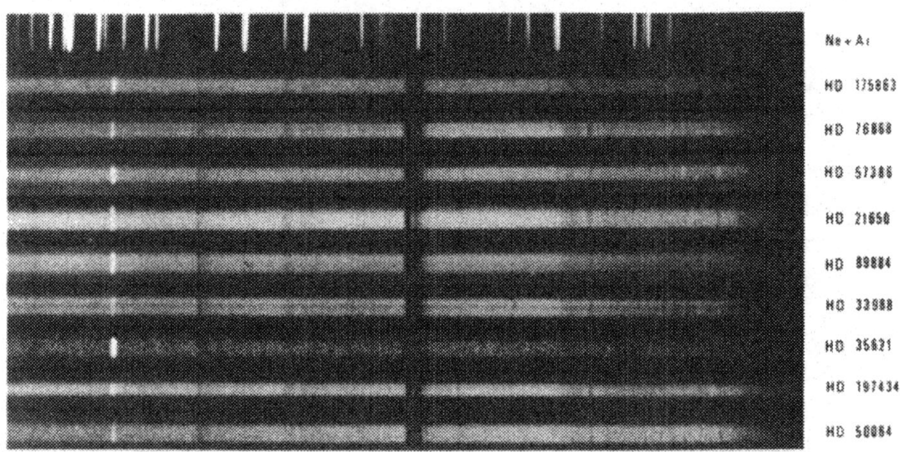

B 7e   B 8e   B 9e

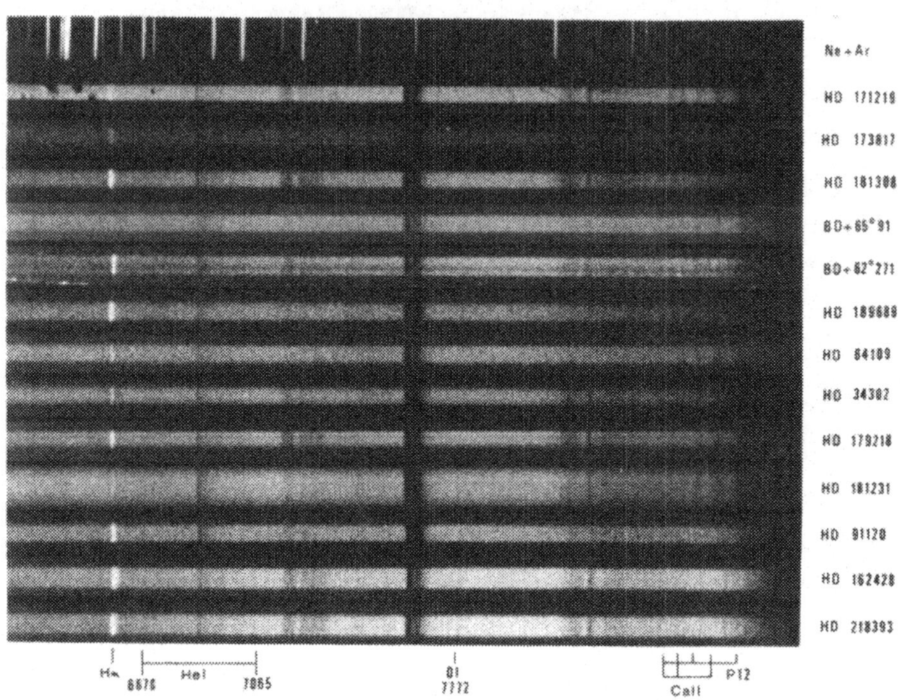

Plate 3

L. HOUZIAUX AND Y. ANDRILLAT

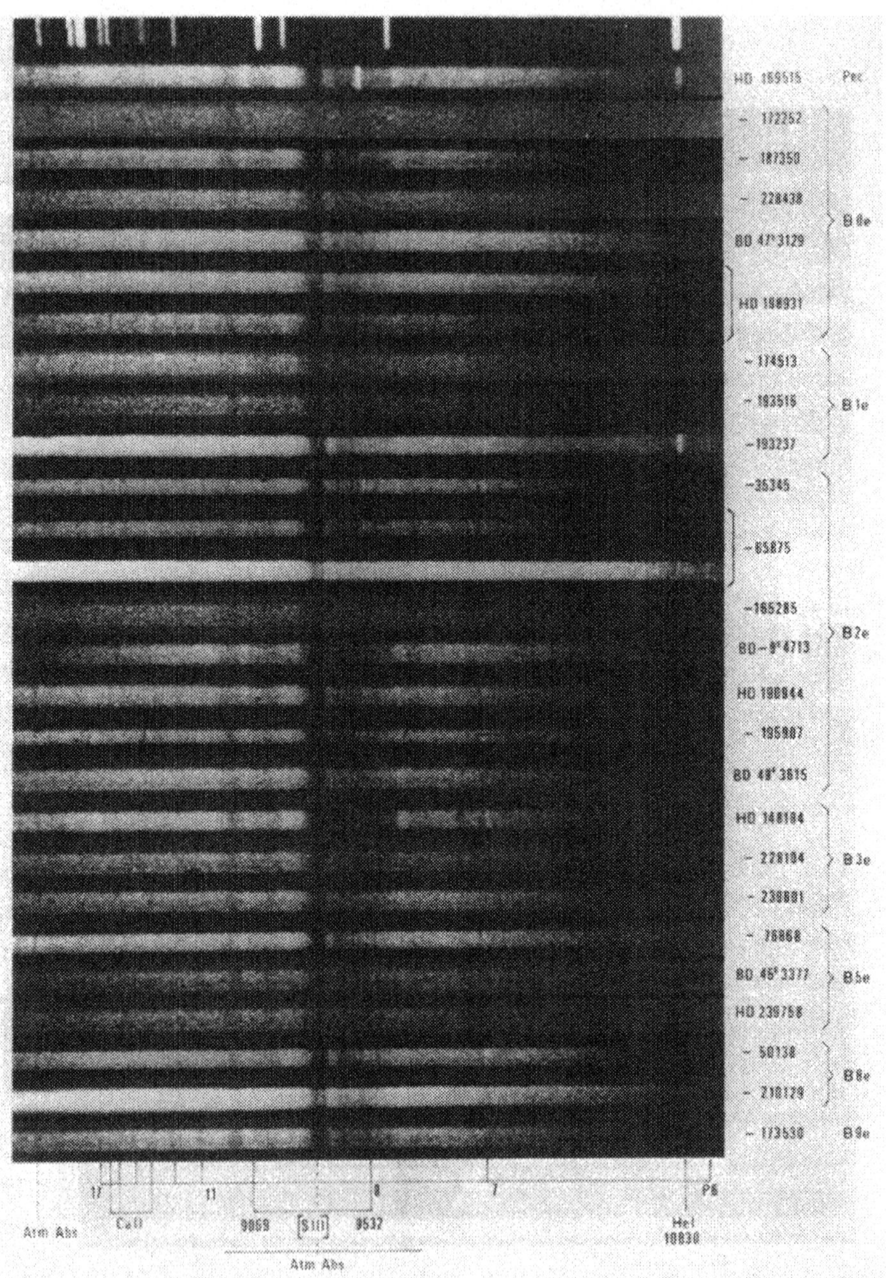

Plate 4

have measured JHKL (1.2 to 3.5 μm) magnitudes of southern
early-type stars. Intrinsic visual-infrared indices have
been determined and R is found to be 3.12 $\pm$ 0.05, thus close
close to the classical value for field stars.

At this occasion, colour indices revealed infrared
excess in some of the shell stars. E(V-K) may be unusually
large with respect to E(B-V), which is itself larger than
in normal stars. Their study shows also that high rotation
Be stars and nebular shell stars exhibit an infrared
excess, (with two stars showing extreme values), because
normal or dark clouds reddening corrections cannot place
the star in the domain of unreddened stars.

In the H-K, K-L diagram the slope and normal reddening
points almost in the direction of a black-body so that the
dereddening is not as unambiguous than in the previous
case.

Infrared observations of 27 CMa have led Danks and
Houziaux (1978) to propose an extension of 17 stellar radii
and an electron density of 6.10^{11} cm^{-3} for the shell
surrounding this object.

In 1967, we published infrared spectra of bright Be
stars. The work has been continued in order to increase the
number of objects, on hypersensitised IN plates using the
120-cm telescope at Haute-Provence and with the Roucas
image tube spectrograph at the Cassegrain focus of the
193-cm telescope. In both cases the reciprocal dispersion
is 230 A mm^{-1} and a projected slit width reaches 7 A.

The spectra are shown on plates 1 to 3. The
classification has been made according to Mount Wilson
Catalogue, all these stars belonging to the Merrill-Burwell
Catalogue.

As one can see, the H$_\alpha$ emission has in some stars
seriously weakened or disappeared. The minimum detectable
equivalent width is about 0.5 A. In about half of the stars
O I at 8446 A is present and the proportion is somewhat
less for high Paschen lines. The highest proportion for O I
and H lines emissions is the Mount Wilson class B2. This is
also true for the Ca II triplet.

When looking further into the infrared (plate 4) the
quality of the spectrum is somewhat degraded. In some of
the stars one sees P7 and P6 in emission and He I λ 10830
but the spectrum is often underexposed above 1 μm.

Although these objects have been identified as bright
H_α stars for several decades, one can find only very little
about them in the literature. In looking in the files of
the S2/68 ultraviolet spectrometer units, we found a fair
proportion of the stars observed during the sky survey and
typical results between 1350 and 2740 A are shown on fig. 3.
There is nothing spectacular about these spectra where the
resolution is only 35 A. The noise is fairly high since the
V magnitudes of the objects are in the 7-9 range and no
emission appears under such conditions. The "line" spectrum
resembles that of ordinary B stars with noticeable λ 2200
interstellar absorption in fainter objects (as HD 43703
(B 1 IV p) V = 8.62 and 206773 (B 0 V p) V = 6.92). The
noise has been computed for flux averages over 100 A and
then the monochromatic magnitudes

$$m_\lambda = -2.5 \log F_\lambda - 21.17$$

have been computed from 1400 A to 2500 A and at 2740 A. In
order to obtain spectral characteristics in the visual,
spectra have been obtained with the CNRS-Liège
objective prism Schmidt telescope on IIaJ and 098-02 plates.

Certain stars show clearly variations but very few have
been observed for a sufficient period of time to state
anything on periodicity.

In dereddening these stars in the B-V, U-B diagramme,
the representative points reach the unreddened sequence at
places corresponding to an earlier spectral type than the
MK type determined chiefly from inspection of He I lines,
supposed to be of photospheric origin. Hence there appears
differences in colour excesses $\Delta E(B-V)$ and $\Delta E(U-B)$, which
are most likely due to shell reemission in the Paschen and
Balmer continua. In all but few cases, the colour excess
E(B-V) derived by the Q method from U, B, V magnitudes is
larger than the E(B-V) deduced from ultraviolet colours
(Thompson et al., 1978). Starting from model atmospheres by
Kurucz, Peytremann and Avrett (1977), Delcroix (1979) has
computed ultraviolet colours $(m_\lambda - V)_0$ averaged over 50 A
wavebands in the range λ 1400 to λ 2500. These colours have
been reddened for values of E(B-V) from 0.1 (0.1) to 0.4.
Furthermore, in summing up all the fluxes in the range
λ 1400 to λ 2500, Delcroix has defined an index

$$TD1 - V = -2.5 \log (F_{1565} + F_{1965} + F_{2365}) - V + C$$

where the fluxes F_{1565}, F_{1965} and F_{2365} are taken from
Thompson et al. (1978). The constant C is such that TD1 - V

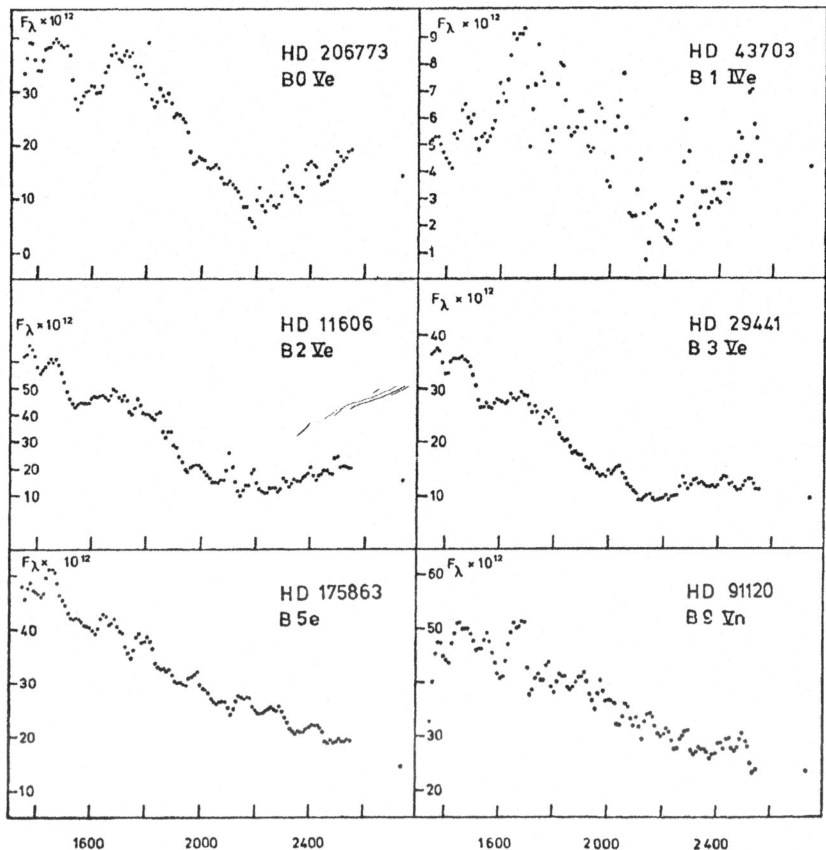

Figure 3. Sample of spectra of Be stars obtained with the
S2/68 ultraviolet scanning spectrometer. Abscissa is
wavelength in Angströms. Ordinates give fluxes in
10^{-12} ergs $cm^{-2}s^{-1}°A^{-1}$. The resolution of 35 A permits
to distinguish a few spectral features, notably at λ 1400–
1420 (Si IV in early B and Si II in late B stars), λ 1475
(Si II, late B), λ 1550 (C IV), $\lambda\lambda$ 1850, 1960, 2110 (Fe
Fe III, early B stars), $\lambda\lambda$ 2240, 2350, 2450 (Fe II and
ionised metals). The interstellar band at λ 2200 is quite
conspicuous in the intrinsically bright objects (B0, B1).
Noise varies according to the V magnitude (V = 7.03 for
HD 11606 and V = 8.62 for HD 43703).

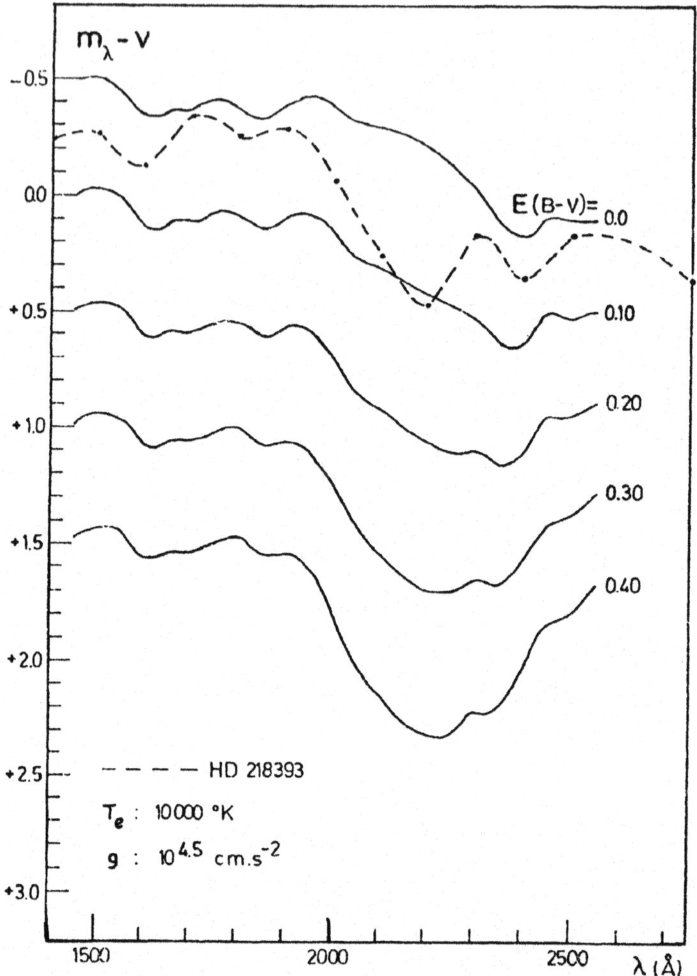

Figure 4. Colour indices computed from a model atmosphere including a great number of lines and averaged over 50 A intervals. The four lower lines show how the colour indices are modified by interstellar absorption for four values of E(B-V). Points represent the colour indices of the star HD 218393, derived from S2/68 data averaged over 100 A wavebands. Note that the shape of the observations does not fit the model, although the range in colour index is appropriate to a slightly reddened 10,000°K model atmosphere.

equals zero for an average AOV star. The index TD1 - V has
also been computed in the same way at the $(m_\lambda - V)$ colours
mentioned above, and reddened for several values of E(B-V).
When E(B-V) is determined from the observed colours
m(1565) - m(2740) and m(1565) - m(2365), as well as its
mean error, it is possible to assign from the observed
TD1-V index, values of T_e and log g to a given star. This
procedure minimises the errors on the ultraviolet fluxes
due to noise. It is also possible to compare the observed
$(m_\lambda - V)$ indices to $(m_\lambda - V)$ for model atmospheres
appropriately reddened. Such a comparison leads generally
to a very satisfactory fit, except in a few particular
cases. We mention as an example on fig. 4 the $(m_\lambda - V)$
indices of the star HD 218393. As seen in plate 3, this
star shows a bright Hα line but no other emission in the
near-infrared. It has been erroneously classified as Ave in
the Mount Wilson Catalogue. The total ultraviolet flux
leads, via the TD1 - V procedure to an effective temperature
of 10,100°K. However, as seen on fig. 4, the colours
indicate that this star cannot have such a low temperature.
It is rather a hot object with an appreciable amount of
reddening and/or abnormally strong absorption features in
the ultraviolet.

Near infrared spectra of these stars led us to extend
the observations to other spectral regions. Furthermore,
infrared lines carry important information, especially the
Paschen and Brackett lines of hydrogen. Emission lines of
chemically abundant atoms are more frequently seen at
infrared wavelengths than in visible or ultraviolet regions,
since the radiation emitted from the shell has to compete
with relatively weak photospheric emission.

We have to hope that observatories will develop and
provide new facilities for visiting astronomers, so that
both photometry and spectrometry in the infrared will retain
the attention of many Be stars observers.

REFERENCES

Altamore, A., Baratta, G.B., Casatella, A., Grasdalen, G.L.,
    Persi, P. and Viotti, R. : 1980, Astron. and Astrophys.,
    90, p. 290.
Andrillat, Y. et Houziaux, L. : 1967, J. Observateurs, 50,
    p. 107.
Danks, A. and Houziaux, L. : 1978, Publ. Astron. Soc. Pac.,
    90, p. 453.
Delcroix, A. : 1979, private communication.
Garrison Jr, M.L. : 1978, Astrophys. J., 224, p. 535.

Johnson, H.L.: 1977, Rev. Mexicana Astron. y Astrofisica, 2, p. 71.
Kurucz, R.L., Peytremann, E. and Avrett, E.H.: 1974, Blanketed Model
    Atmospheres for Early-Type Stars, Smithsonian Institution.
Oegerle, W., Polidan, R. and Peters, G.: 1979, B.A.A.S., 11, p. 682
Scargle, D.S., Erickson, E.F., Witteborn, F.C. and Strecker, D.W.:
    1978, Astrophys. J., 224, p. 527
Smith, H.A., Thronson, H.A., Larson, H.P. and Fink, U.: 1979, B.A.A.S.,
    11, p. 668
Thompson, G.I., Nandy, K., Jamar, C., Monfils, A. Houziaux, L.,
    Carnochan, D.J. and Wilson, R.: 1978, Catalogue of Ultraviolet
    Stellar Fluxes, The Science Research Council, London.
Whittet, D.C.B., van Breda, I.G.: 1980, Monthly Notices R. Astron.
    Soc., 192, 467

DISCUSSION

Coyne: Does the curvature in the reddening curve shown by models in
the far UV imply any curvature, at least theoretically, in the B-V,
U-B reddening law?

Houziaux: The curvature in the reddening line in the UV does not imply
a curvature in the B-V, U-B diagram, since the shape of the absorption
curve is not the same in the ultraviolet and in the visible.
However, because of the width of the U,B,V filters such a curvature
may well appear in the visible, especially if one considers large
values of E(B-V).

Thomas: I caution against diagnosing these very essential IR data
only in terms of uniform shells, as you have done. In particular,
in stressing agreement between data and computations based on shells
of T_e ~20000 K you ignore the contributions of those chromospheric-
wind regions which we know exist. For example the computations by
Casinelli should be mentioned. Also, we know now the great uncertainty
in deriving mass flux velocities from H_α profiles alone (cf. Doazan
et al this colloquium).

Houziaux: I quite agree that the compatibility between mass flux
derivations from infrared data and from computations based on shells
of T_e 20000 K is not a proof of the consistency of the model,
especially when velocities used in calculations based on infrared
observations are derived from H_α profiles, the width of which may
not be due to motion effects.

LE SPECTRE DES ETOILES Oe DANS LE ROUGE ET LE PROCHE INFRAROUGE

Y.Andrillat (1) - J.M.Vreux (2) - M. Dennefeld (3)
(1) Observatoire de Haute Provence
(2) Institut d'Astrophysique de Liège
(3) Institut d'Astrophysique de Paris

ABSTRACT

Among the O stars, Conti and Leep (1974) define the Oe class as those
stars showing emission in the hydrogen lines but without emission in
N III or 4686 He II.
In the blue-red region, aside from the presence of Lambda 4542 in
absorption, the overall appearance is similar to Be stars. So far no
Oe star has been observed until 1.1 µ: the aim of this paper is to fill
this gap. We present observations of Oe stars in the near infra-red
region as well as new observations of Hα profiles.

I-INTRODUCTION

Parmi les étoiles de type O, Conti et Leep (1974) distinguent
celles dont le spectre est caractérisé par de l'émission dans les raies
de l'hydrogène et par l'absence d'émission due à N III 4630-40-41 A et
à He II 4686 A (classe Oe). Dans la région bleu rouge, les spectres de
ces étoiles sont semblables à ceux des étoiles Be (Frost et Conti 1976).
A notre connaissance, aucune étoile Oe n'a été observée jusqu'à 1,1µ.
Il nous a paru intéressant de prolonger la comparaison des spectres de
ces étoiles à ceux des Be jusqu'à la région du proche infrarouge. Nous
avons également observé les profils de la raie Hα.

II-OBSERVATIONS ET RESULTATS

HD 155806 : O7,5V|n|pe - O7,5IIIe. La raie Hα observée par Frost et
Conti (1976) est une forte émission, 4 fois plus intense que le continu
sous-jacent et légèrement variable de 1972 à 1974.
2 spectres de HD 155806 ont été obtenus dans la région 0,6-1,1µ . Ils
sont caractérisés par un continu intense sur lequel se superposent de
fortes et nombreuses émissions de H I, He I, O I, Ca II, Fe II (Fig.1).
H I est bien représenté par Hα dont l'intensité est sensiblement la
même qu'en 1972-1974 et par de nombreuses raies de la série de Paschen :
P7 (W=1,7 A), P11 (2,1), P12 (1,9), P13 (2,5), P14 (2,2), P15 (2,3),

M. Jaschek and H.-G. Groth (eds.), Be Stars, 229–233.
Copyright © 1982 by the IAU.

P16 (2,2). Les mesures de P8, P9 et P10 sont rendues impossibles par la
présence de bandes de la vapeur d'eau. La comparaison des diverses inten
-sités de P11 à P16 suggère la contamimation de P13, P15 et P16 par les
raies de Ca II (8662, 8542 et 8498 A).
He I est également bien développé : 10830 A intense (W=5,7 A), 7065 A
(W=0,5 A) et 6678 A faible. Frost et Conti ont observé en absorption les
raies de la série des triplets et celles de la série des singulets. Sur
nos spectres obtenus en 1979, 6678A est une émission nette tandis qu'en
1972-1974, la raie était une faible et large absorption.
O I: les émissions 8446 et 7772 A sont très intenses (W=2,2 et 1,0 A).
Fe II est également présent à 7515, 7712 et 9997 A.
Dans l'intervalle 6500-8750 A, le spectre de HD 155806 est semblable à
celui des étoiles Be caractérisé par les raies de l'hydrogène de la série
de Paschen, de O I, de He I et du triplet de Ca II.
Les émissions de Fe II, notamment 7712 A, sont visibles dans les étoiles
Be qui dans le bleu-violet, montrent également Fe II en émission.
Polidan et Peters (1976) constatent que 20 % des étoiles Be montrent à
la fois une émission de O I à 7772 et de Fe II à 7712 A, ce qui est le
cas pour HD 155806.
Le spectre de cette étoile est analogue à celui de HD 202904 et HD 32343
de type B3e (Andrillat, Houziaux 1967).

χ Per: OBe. Hα est une émission intense (Fig.2) : I/I_C=2, W=12,74 A. La
raie est pointue et l'aile violette montre une inflexion, suggérant une
structure double visible peut-être avec une meilleure résolution. La
largeur à la base correspond à 750 km.s^{-1} environ.
He I 6678 est une faible émission (W=2,21 A): on devine une structure
double.

HD 60848: 08V:pe. Le profil de la raie Hα que nous observons en février
1981 (Fig.2) est analogue à celui obtenu par Frost et Conti (1976) en
1972-1974. Il est caractérisé par une émission avec une faible absorption
centrale (cas de nombreuses étoiles Be). L'intensité de Hα n'a pas varié:
I/I_C=1,5 et W=16,13 A.
He I 6678 A est une émission faible : son intensité est environ le 1/10e
de celle du continu et W=1,05 A. Elle est large et on soupçonne une
structure double nettement visible sur le spectre publié par Frost et
Conti.
Dans le domaine 0,8-1,1μ le spectre que nous avons observé en mars 1980
(télescope 193cm de l'OHP, spectrographe ROUCAS + RETICON - dispersion
230 A.mm^{-1}) montre les émissions des raies de l'hydrogène de la série de
Paschen jusqu'à P16, ainsi que celle intense de He I à 10830.
Comme dans le cas de HD 155806, l'émission est visible dans les termes
élevés de la série de Paschen tandis que, pour la série de Balmer, elle
est limitée aux premières raies : Hα, Hβ. Nous avons constaté ce compor-
tement différent des raies homologues des 2 séries dans de nombreuses
étoiles Be (Andrillat, Houziaux 1967).

HD 39680: 06V|n|pe var. Sur notre spectre obtenu en février 1981 (Fig.2),
Hα est une émission avec une très faible absorption centrale que l'on
soupçonne sur le profil observé en 1972-1974 par Frost et Conti. L'inten-

Fig.1 : HD 155806
Télescope 3m60 ESO
Spectro Boller Chivens + RETICON
Dispersion 228 A.mm⁻¹
Hα
HeI 6678
HeI 7065
FeII 7515
FeII 7712
OI 7772
OI 8446
P 16
11
10
P9
P8
FeII 9999
P7
3000
2000
1000
0
5840
7208
8576
9944

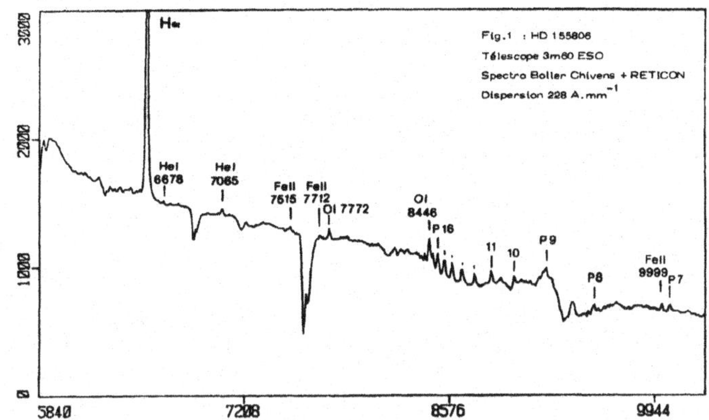

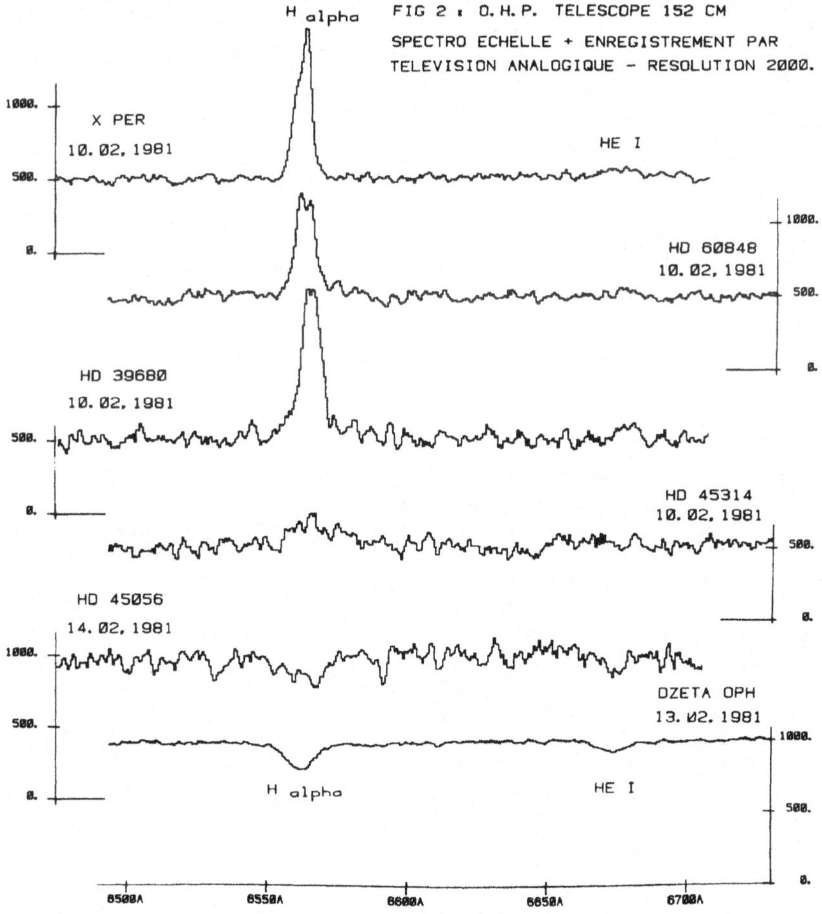

FIG 2 : O.H.P. TELESCOPE 152 CM
SPECTRO ECHELLE + ENREGISTREMENT PAR
TELEVISION ANALOGIQUE - RESOLUTION 2000.
H alpha
X PER
10.02, 1981
HE I
HD 60848
10.02, 1981
HD 39680
10.02, 1981
HD 45314
10.02, 1981
HD 45056
14.02, 1981
DZETA OPH
13.02. 1981
H alpha
HE I
1000.
500.
0.
6500A
6550A
6600A
6650A
6700A

sité n'a pas varié : nous mesurons I/I_C=2,22, W=22,75 A.
Le contour de He I à 6678 A présente également un profil identique. Cette
émission est faible : I/I_C=0,2 environ, W=1,71 A. En 1972-1974, la structure de cette raie et son intensité étaient semblables.
Alors que,dans la région bleu rouge, le spectre de HD 39680 ressemble
à celui de HD 60848, il n'en est pas de même dans la région 0,8-1,1µ où
aucune émission n'est visible dans HD 39680. On note seulement la présence des absorptions intenses dues aux raies de l'hydrogène de la série de
Paschen visibles jusqu'à P15.

HD 45314: OBe. En février 1981, la raie H_α est une émission, large, peu
intense et l'on soupçonne une absorption centrale: I/I_C=0,5, W=7,35 A
(Fig.2). Une très faible absorption pourrait être due à He I 6678 A.
Dans la région 0,8-1,1µ le spectre observé en mars 1980 ne montre aucune
raie nettement visible. La série de Paschen est peut-être présente en
absorption. On distingue les raies de P12,P13,P14.

HD 46056: 08V(e). Dans cette étoile, l'émission est intermittente. Sur
le spectre obtenu en février 1981, la raie H_α est faiblement en absorption
avec un profil nettement dissymétrique, l'aile violette étant plus étendue :I/I_C=0,29, W=-3-04 A. La raie He I à 6678 A est en absorption:I/I_C
=0,20, W=1,41 A. Le profil est dissymétrique comme celui de H_α (Fig.2).

ζ Oph:08V - 07,5III:n((f)). Comme pour HD 46505, la raie H_α apparaît
en émission de façon intermittente. Elle est en absorption sur notre
spectre obtenu en février 1981: I/I_C=0,81 W=-2,48 A. La raie He I6678 A
est également en absorption : I/I_C=0,08 W=-0,89 A (Fig.2).
Dans le domaine 0,6-1,1µ les absorptions sont dues aux raies de la série
de Paschen de P7 à P16 (W=-3,02 pour P7, -1,95 pour P11, -2,13 pour P12,
-1,18 pour P13 et 0,38 pour P14). Les absorptions de He I à 7065 A
(W=-2,13 A) et 6678 A (W=-1,24 A) sont également présentes.

III-CONCLUSION

Dans la région du proche IR, les éléments identifiés dans les étoiles
Oe sont les mêmes que ceux trouvés dans les Be.
HD 155806 s'apparente aux étoiles Be des premiers types qui possèdent
une enveloppe à forte métallicité visible quand OI 8446 et Fe II 7712 A
sont en émission et H_α, très interne, sans structure.
Une étude relative à 68 étoiles Be (Andrillat, Houziaux, communication
précédente) montre que la série de Paschen et OI 8446 A apparaissent en
émission dans les étoiles B0e, B1e...B5e mais principalement dans les
B2e. Ces émissions sont présentes dans HD 155806 et HD 60848.
La raie 6678 A de He I est visible en absorption seulement dans les
classes B0e...B3e mais pas au-delà. Elle est présente dans tous nos
spectres d'étoiles Oe et en émission dans 4 d'entre elles.
Nos observations dans le rouge et le proche infrarouge confirment donc
la ressemblance des Oe et des Be et permettent de préciser que les
étoiles Oe s'apparentent aux étoiles Be des premiers types probablement
aux B2e ou B3e.

REFERENCES

Andrillat,Y. and Houziaux,L.: 1967,J.Obs.50,107=Publ.OHP 9,n°11.
Conti,P.S. and Leep,E.M.:1974,Astrophys.J.193,113.
Frost,S.A. and Conti,P.S.:1976,in A.Slettebak(ed.),Be and shell stars,
      D.Reidel Publ.Co.,Dordrecht-Holland,139.
Polidan,R.S. and Peters,G.J.:1976,in A.Slettebak(ed.),Be and shell stars,
      D.Reidel Publ.Co.,Dordrecht-Holland,59.

DISCUSSION

Henrichs: Because X Per is associated with an x-ray source I wonder if any peculiar feature is found in the near IR of this star. Does it just look like an "normal" Oe star?

Andrillat: In our spectra of X Per (Sept. 28, 1972), H_α is in moderate and narrow emission, there is a weak absorption in HeIλ 7065. The Paschen lines seem to show P Cyg profiles. There is no emission either in the CaII triplet nor in OI λ 8446.

INFRARED PHOTOMETRY OF Be STARS

Alejandro Feinstein
Observatorio Astronómico,
Universidad Nacional de La Plata
1900 La Plata, Argentina.

A photometric study of Be stars in the JHKL system is reported here. Early papers (see Feinstein and Marraco, 1980) about bright southern Be stars observed photoelectrically in the UBVRI system and in the Balmer-lines: Hα, Hβ and Hγ, permit us to derive several conclusions about their light variability, the amount of emission in the α and β indices, and some correlations about these data. On the other hand, we recently found (Feinstein and Marraco, 1981) that the emission strengths of the hydrogen lines are well correlated to the amount of excess radiation in the near-infrared of the RI system.

It is already known that Be stars exhibit an infrared excess at wavelengths longer than 1 micron as compared with the radiation of normal B-type stars. This is interpreted as the contribution of free-free radiation from a circumstellar shell surrounding the star. Therefore, for the same bright southern Be stars cited above, it should be interesting to look for the JHKL data already published, which we completed with measurements for some stars observed by us. The latter were carried out with a PbS detector cooled with liquid nitrogen and standard filters, attached to the 83-cm reflector at La Plata. A mechanical choper was used to switch beams. Our measures are referred to the list of standard stars given by Glass (1974) having an error estimated of $\pm$ 0.05 for the J, H and K bands, while at L it may reach about $\pm$ 0.08.

The analysis of the combined data permit us to derive several conclusions: two-color diagrams have been used to locate the Be stars with respect to the normal B main sequence stars (Figures 1 and 2).

The deviations of the stars' positions from the main sequence are the result of the infrared excess plus some interstellar reddening. The only significant influence is due to the former, which we have computed for each star as the difference in the K-L index between its actual location and the main sequence. We assumed that the effect takes place only in this K-L index, which we called E_K . One star with an unusual high K-L value is HD 178175. Perhaps a large error in the L magnitude could be responsible for this extreme value.

M. Jaschek and H.-G. Groth (eds.), Be Stars, 235–240.
Copyright © 1982 by the IAU.

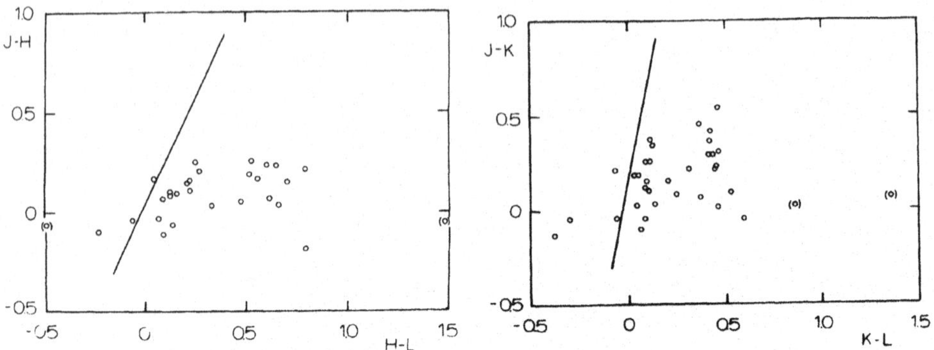

Figure 1. Left: The (J-H,H-L) array for Be stars. Figure 2. Right: The
(J-K,K-L) array for Be stars. In both diagrams the solid line corresponds
to the normal main sequence stars.

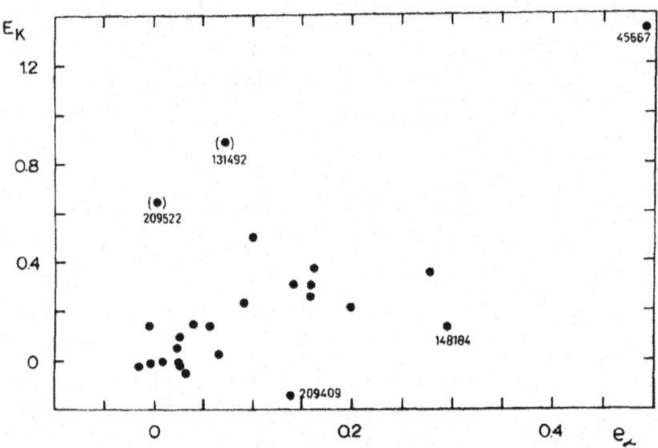

Figure 3. The infrared excess E_K versus the amount
of the emission in the Hα-
line given by the e_α index.

Then, we compared the infrared excesses with the amount of emission
in the Hα-line, which we displayed in Figure 3. With the exception of
some stars, slightly far from the mean relation and plotted with their
HD numbers, there is a quite good relation between the two sets of values.

On the other hand we looked for a correlation of the E_K values with
the ultraviolet excess through the use of the Q values, as defined by
Johnson. But, in Figure 4, which gives the $(E_K, \Delta Q)$ array, no correlation
at all appears between infrared and ultraviolet excesses. However, it is
interesting to notice that a few stars which are far from the mean relation in the (E_K, e_α) diagram are also discordant in Figure 3.

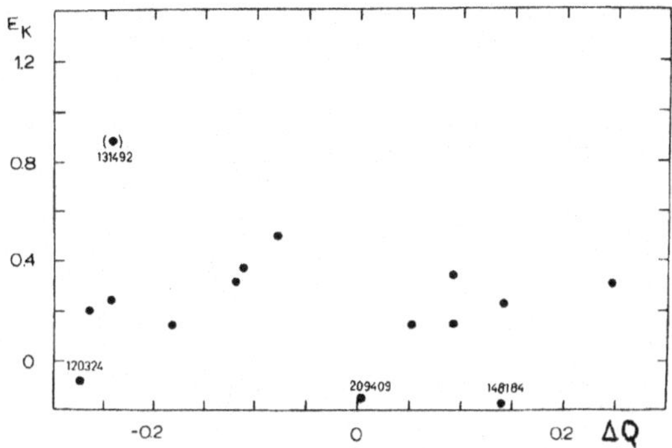

Figure 4. The infrared excess E_K versus the ultraviolet excess given by ΔQ.

Evidence that a few of the Be stars may suffer variations in the infrared bands is provided by the comparison of measures obtained at different times and by different observers. The results are summarized in Table 1, where only those stars with significant changes are listed. Syste-

Table 1. Variable Be stars in the infrared.

	J	H	K	L	n	
HD 45677						
1971/72	–	6.49	4.62	2.15	–	Allen (1973)
Dec 1975	6.95	6.03	4.74	2.16	3	Feinstein
Feb 1981	7.19	–	4.93	2.19	5	Feinstein
HD 120324						
1971/72	–	3.72	3.61	3.20	–	Allen (1973)
May 1975	4.01	4.03	4.15	4.5:	1	Feinstein
Mar 1981	3.93	–	3.96	3.55	1	Feinstein
HD 148184						
Jul 1974	3.69	3.37	3.19	2.79	3	Feinstein
1976/77	3.50	3.29	3.04	2.71	2	Whittet and van Breda (1980)
Aug 1976	3.46	3.26	2.98	2.64	4	Feinstein
May 1980	3.46	–	2.89	2.28	1	Daminelli
HD 158427						
Aug 1976	2.99	2.98	2.76	2.30	5	Feinstein
Aug 1980	3.11	–	2.94	2.42	3	Feinstein
HD 178175						
1974/75	5.69	5.74	5.66	–	5	Feinstein
1977	5.31	5.20	5.16	4.96	1	Whittet and van Breda (1980)
Aug 1980	5.71	–	5.45	–	1	Feinstein

matic differences among observers may, however, affect these conclusions,
but it becomes difficult to judge how large they are, because these are
the only available data we have at hand, and not very many stars are included.

In conclusion, our results confirm the idea that the same region of
the envelope is responsible for the infrared excess and the Balmer line
emission.

An observational program to measure more Be stars in the infrared
is in course.

REFERENCES

Allen, D. A. 1973, Mon. Not. R. astr. Soc. 161, 145.
Feinstein, A. and Marraco, H.G. 1980, Astron. J. 84, 1713.
Feinstein, A. and Marraco, H.G. 1981, Publ. Astron. Soc. Pacific, in
          press.
Glass, I.S. 1974, Mon. Notes R. Astr. Soc. Sth. Afr. 33,53.
Whittet, D. C. B. and van Breda, I. G. 1980, Mon. Not. R. astr. Soc.
          192, 467.

DISCUSSION

<u>Metz</u>: Would you briefly comment the surprising fact that there is no correlation between IR excess and UV excess.

<u>Feinstein</u>: The data suggest that no relation appears between IR and UV excess. Perhaps it would be necessary to add another information to clear this point.

<u>Houziaux</u>: 1. What is the typical rms error on the H,K,L magnitudes for the stars observed, i.e. at which level can you talk about variability?
2. What do you think about the band pass of H,K,L filters of different observers? Can they been considered as identical?
3. What is the limiting magnitude you can reach with your instrument?

<u>Feinstein</u>: 1. The rms error depends on the magnitudes of the star. For the B band the rms error is about $^{m}05$ in a 5^{m} star.
2. There may be some differences, but I think these differences larger than .1 or .2 must be considered as variabilities.
3. In the L band the limiting magnitude is about $5^{m}5$.

<u>Coyne</u>: To determine IR and UV excesses is the normal colour–colour curve in the IR determined from the same group of stars as for the colour–colour curve in the UV?

<u>Feinstein</u>: It is assumed that these are the same stars or at least stars with the same characteristics.

<u>Poeckert</u>: What do you call an excess? What is the excess relative to?

<u>Feinstein</u>: The (colour) excess in K–L is the difference between the observed colour index K–L minus the K–L of the standard relation for main sequence stars in the J–K, K–L diagram.

<u>Snow</u>: I don't know much about IR photometry, but my understanding is that due to a variety of causes (difficulties of calibration, variable atmospheric conditions, differences between observing systems) there can be large uncertainties in IR magnitudes. You have shown data taken from different observers at different times, without showing error bars, and indicated from this that the star is variable, at levels as small as only a few tenths of a magnitude. In view of the uncertainties, how confident are you that the variations are real?

<u>Feinstein</u>: All the stars which I show are bright in the IR, so assumed errors about $^{m}05$ mean that differences larger than a few tenths of a magnitude must be variations.

<u>Viotti</u>: I would like to make a general comment about the problem of IR excess in Be stars. Dr. Thomas told us that one needs non-uniform models to understand the IR system, but I think that no models can be

built up without knowing what happens at the "base" of the atmospheres.
Since in the middle UV the continuum opacity is smaller, one has more
chances to go deeper in the atmosphere and to have a more realistic
value of the classical stellar parameters, effective temperature,
gravity, etc. So I strongly suggest IR astronomers not to restrict
themselves to start from scanner and high resolution optical spectra
used to discribe the "underlying" star, but to use IUE high resolution
spectra to look at the continuum spectra and the excited lines to
derive the "stellar" parameters, and not only the resonance lines
which are formed at higher levels. In other words, don't make a model
without making its feet before.

SEARCH FOR VARIABILITY IN NEAR INFRARED FLUXES
OF PECULIAR EMISSION-LINE OBJECTS[*]

P. Bouchet,
European Southern Observatory,

J.P. Swings
European Southern Observatory and
Institut d'Astrophysique, Université de Liège

Thirty-five peculiar emission line objects
(essentially B[e] stars) were observed at J(1.2 μ),
H(1.6 μ), K(2.2 μ), L(3.5 μ), and M(4.8 μ) at the E.S.O.
1 m telescope where the photometer is equipped with an
InSb detector. The results of the observations performed
during several nights and days in Feb. 1979 and March 1980
are listed in Table 1.

Our K magnitudes and H-K color indices are compared
in Table 2 with those obtained in the early 1970's and
given e.g. in Allen and Swings (1976, Astron. Astrophys.
47, 293). Only one case (M1-11) is definitely variable at
K, as already reported by Allen (1973, M.N.R.A.S. 161, 145);
six other objects are marginally variable in H-K.

It therefore appears that the dust surrounding these
peculiar emission-line objects, although patchy, does
exist in sufficient quantities and at relatively constant
temperature(s) so that repeated observations in the near
infrared do not reveal any really meaningful (> .1 or .2 m)
variation.

---

[*]Based on data collected at the European Southern
Observatory (La Silla, Chile).

TABLE 1

NEAR INFRARED MAGNITUDES OF B[e] OBJECTS

Object	Date (Yr, Mo, Day)	V	J		H		K		L		M	
HD 45677	79:02:17	8.5	7.79±	.10	6.49±	.08	4.83±	.10	2.13±	.15	1.07±	.16
	79 02 20		7.70	.03	6.44	.05	4.63	.10	1.94	.06	1.00	.11
	80 03 08		7.93	.05	6.76	.04	4.81	.03	2.11	.05	1.30	.07
	80 03 09		7.91	.04	6.79	.04	4.86	.03	2.22	.05	1.35	.07
MWC 819	80 03 09	14.9	12.55	.79	10.76	.13	·8.88	.08	6.17	.06	5.23	.54
HD 50138	80 03 10	6.7	5.86	.02	5.10	.02	4.18	.03	2.77	.04	2.16	.08
HD 51585	80 03 09	11.2	–		–		–		6.41	.09	5.83	.40
M 1-11	80 03 10	13.5	11.04	.40	10.09	.14	8.98	.09	7.11	.18	6.18	.56
CD-24°5721	80 03 09	11.4	9.28	.07	8.53	.08	7.33	.03	5.62	.06	4.88	.20
(RX Puppis	80 03 10	∼8.5	5.14	.02	4.00	.02	2.79	.02	1.08	.04	0.48	.06)
Ve 27	80 03 09	13.8	11.30	.40	10.28	.17	8.54	.05	5.66	.06	4.34	.13
Hen 230	80 03 09	12.4	9.11	.09	7.79	.04	6.51	.03	4.91	.05	4.26	.10
He 2-34	80 03 10	15.6	8.19	.05	6.58	.02	5.14	.02	3.48	.03	2.94	.07
HD 87643	80 03 08	8.7	6.20	.04	5.04	.04	3.75	.03	2.08	.05	1.46	.08
Hen 373	80 03 10	13.5	9.33	.10	8.19	.03	6.92	.03	5.41	.10	4.69	.20
η Car	80 03 10	6.2	3.61	.02	2.63	.03	1.12	.01	−1.77	.06	−3.13	.05
Hen 485	80 03 10	11.7	8.40	.05	7.24	.02	6.02	.02	4.52	.04	4.07	.17
GG Car	80:03:08	8.7	6.84±	.04	6.09±	.05	5.00±	.03	3.69±	.05	3.09±	.08
	80 03 09		6.87	.04	6.01	.05	4.98	.03	3.56	.05	2.99	.08
	80 03 10		6.86	.03	6.03	.03	4.99	.01	3.58	.04	3.03	.05
He 2-79	80 03 08	14.1	11.76	.86	10.06	.10	7.89	.04	5.42	.05	4.81	.16
He 2-80	80 03 08	13	10.00	.14	8.81	.07	7.44	.03	5.67	.06	4.92	.15
Hen 782	80 03 10	Var.	8.17	.04	6.44	.03	4.90	.01	3.17	.04	2.58	.06
He 2-90	80 03 08	13.3	11.08	.38	10.24	.14	7.89	.04	4.86	.06	3.61	.13
He 2-91	79 02 20	14.4	9.18	.06	7.70	.05	6.36	.09	4.34	.06	3.44	.11
	80 03 08		9.72	.10	8.00	.05	6.44	.03	4.53	.06	3.87	.10
Hen 938	80 03 10	13.3	9.08	.08	7.82	.03	6.23	.02	4.33	.04	3.67	.11
He 2-101	79 02 21	16.9	11.57	.59	9.72	.14	8.33	.05	6.49	.03	5.57	.35
	80 03 09		12.9	.80	10.00	.09	8.46	.05	6.46	.10	5.80	.07
He 2-139	80 03 10	16.8	8.76	.05	7.49	.04	6.38	.02	5.00	.03	4.73	.11
CPD-52°9243	79 02 18	10.3	6.03	.07	5.14	.07	4.06	.25	2.53	.11	1.79	.01
	80 03 08		6.45	.05	5.35	.04	4.08	.03	2.66	.05	2.07	.07
He 2-147	80 03 08	15	6.23	.07	6.13	.04	5.21	.03	4.23	.05	4.15	.12
Mz 3	79 02 18	14	9.30	.07	7.35	.07	5.58	.25	3.04	.11	2.02	.01
	80 03 08		9.32	.07	7.54	.04	5.62	.03	3.19	.05	2.33	.08
Pe 2-9	80 03 10	17.0	11.43	1.00	9.43	.08	7.17	.02	4.67	.04	3.85	.05
Hen 1191	79 02 19	13.8	9.88	.15	8.25	.05	6.25	.03	3.80	.04	2.84	.08
	80 03 08		10.35	.20	8.44	.05	6.43	.03	3.90	.05	3.05	.08
M2-9(core)	79 02 18	14.7	10.91	.15	9.28	.08	7.08	.25	3.95	.11	2.64	.02
	80 03 08		11.32	.44	9.41	.07	7.03	.03	3.90	.05	2.71	.07
AS 222	79:02:20	12.5	7.83±	.02	6.79±	.05	5.68±	.09	4.01±	.06	3.23±	.11
H1-25	80 03 09	16.5	10.55	.18	8.70	.05	6.86	.03	4.61	.05	3.78	.09
HD 316248	80 03 09	12.1	9.93	.40	9.30	.16	8.52	.13	7.32	.52	7.14	1.00
HD 163296	80 03 09	6.8	6.34	.04	5.54	.03	4.72	.02	3.55	.04	3.14	.08
MWC 922	79 02 18	13.9	8.86	.07	7.38	.07	5.66	.25	2.78	.11	1.38	.01
	80 03 09		8.99	.18	7.49	.04	5.59	.08	2.75	.04	1.51	.08
MWC 300	79 02 18	11.6	9.08	.11	8.14	.08	6.19	.25	3.06	.11	1.74	.02
	80 03 09		9.16	.22	8.11	.06	6.12	.02	3.02	.04	1.83	.07
MWC 939	79 02 18	12.4	10.45	.27	8.98	.08	7.23	.25	4.93	.11	4.05	.04
	80 03 09		10.07	.39	8.72	.13	7.01	.03	4.86	.06	4.03	.15
HD 190073	79 02 17	7.9	8.20	.15	7.16	.15	6.18	.03	4.80	.05	4.59	.19

TABLE 2 : SEARCH FOR INFRARED VARIABILITY

Name	Group	K Ref. 1976	K 1979	K 1980	H-K Ref. 1976	H-K 1979	H-K 1980	Notes
HD 45677	2	4.6	4.6	4.8	1.87	1.81	1.93	
MWC 819	2	8.3	−	8.9	1.80	−	1.9	
HD 50138	1	4.2	−	4.2	0.91	−	0.92	
M 1-11	3	7.5	−	9.0	1.15	−	1.11	(1)
CD −24°5721	2	7.2	−	7.3	1.02	−	1.20	
RX Pup	3	3.5	−	2.8	1.11	−	1.22	(2)
Ve 27	2	8.5	−	8.5	1.6	−	1.74	
Hen 230	2	6.5	−	6.5	1.23	−	1.29	
He 2-34	3	5.8	−	5.1	1.45	−	1.44	
HD 87643	2	3.6	−	3.7	1.19	−	1.28	
Hen 373	1	6.7	−	6.9	1.14	−	1.27	
η Car	2	1.2	−	1.1	1.51	−	1.51	
Hen 485	1	6.0	−	6.0	1.17	−	1.23	
GG Car	2	4.9	−	5.0	0.97	−	1.04	
He 2-79	2	8.0	−	7.9	2.3	−	2.16	
He 2-80	2	7.3	−	7.4	1.20	−	1.37	
Hen 782	1	4.7	−	4.9	1.16	−	1.54	
He 2-90	3	7.8	−	7.9	1.9	−	2.36	
He 2-91	2	6.3	6.4	6.4	1.55	1.43	1.56	(1)
Hen 938	1	6.1	−	6.2	1.54	−	1.56	
He 2-101	3	8.4	8.3	8.5	1.7	1.39	1.54	
He 2-139	3	5.3	−	6.4	1.0	−	1.11	
CPD −52°9243	1	4.2	4.1	4.1	1.24	1.10	1.26	
Mz 3	3	5.4	5.6	5.6	1.88	1.80	1.92	
Pe 2-9	1	7.0	−	7.2	1.76	−	2.26	
Hen 1191	2	6.3	6.3	6.4	1.95	2.00	2.00	
M 2-9	3	7.0	7.1	7.0	2.24	2.22	2.38	
AS 222	1	5.6	5.7	5.7	1.25	1.19	1.31	
H 1-25	3	7.3	6.6:	6.9	2.1	1.81	1.85	
HD 316248	3	8.1	−	8.5	1.01	−	0.78	
HD 163296	1	4.5	−	4.7	0.73	−	0.82	
MWC 922	2	5.6	5.7	5.6	1.79	1.75	1.90	
MWC 300	2	5.9	6.2	6.1	2.07	1.97	1.99	
MWC 939	2	6.8	7.2	7.0	1.55	1.78	1.71	
HD 190073	1	5.8	6.2	−	0.79	0.91	−	

(1) possibly variable (Allen, 1973)

(2) has mags. and colors varying as for LPV

DISCUSSION

<u>Selvelli</u>: 1. I think you cannot rule out the possible presence of a
cool companion in a star with IR excess by saying that no radial
velocity variation has been observed. A not neglegible percentage
of objects might be viewed pole-on.
2. Is it straightforward to distinguish between an IR excess produced
by a possible cool companion from that produced by a dust envelope?

<u>Swings</u>: I agree; however, the only case I was talking about concerned
HD45677 which, as you know, is most probably seen equator-on.
2. I think so because the colours will be very different, especially
in the I,J,H band areas.

<u>Coyne</u>: What size temperature change in the dust would give changes in
J,H,K,L,M,N of the order of .5 mag?

<u>Swings</u>: It is of course the colours that are important; if one looks
for example, at an H-K/K-L diagram, the decrease of K-L of $\approx$ .5 mag.
would change the dust temperature by +200-300K.

<u>Chkhikvadze</u>: It is well known that IR excess is correlated with the
presence of some forbidden lines ([FeII] and others). On the other
hand there are many such peculiar Be stars (MWC 374, XX Oph, αSco (B),
HR3164 etc.), but they have no IR (dust) excess. What do you think
about it?

<u>Swings</u>: What I said is that for B [e] stars there exists a strong
correlation between the existence of an infrared excess and the
presence of e.g. [OI], [FeII] , [SII] in their spectra. The cases you
mention are different: for XX Oph, see Swings and Allen (PASP <u>84</u>, 523,
1972); I do not think αSco B has been measured alone, i.e. without the
"contamination" of the M supergiant. I am talking here about VV Cephei
stars or symbitics.

<u>Sterken</u>: 1. How did you calculate the reddening correction?
2. How sensitive are the temperatures in the dust shells on the
accuracy of the method of the dereddening?

<u>Swings</u>: 1. The reduction was performed by one of us (P.B.) on La Silla,
using Wamsteker's calibration (ESO preprint No 132, in press in A&A).
The contribution of reddening for dust shells is very small, and if it
affects very slightly the magnitudes, I do not think, it will change
the IR colours by values greater than the observational errors.
2. See e.g. Allen's graph (fig. 2., MNRAS <u>161</u>, p 157, 1973): a few tens
of degree, may be, which is probably again below the accuracy of the
measurements.

<u>Andrillat</u>: In order to decide whether a cool companion is present,
perhaps it is possible to study the behaviour of the IR triplet of

CaII, because in the Be stars the intensities of these three lines are different.

<u>Swings</u>: I agree; one could actually get one's inspiration from a recent paper by G. Herbig on T Tauri stars where that specific problem is examined.

<u>Persi</u>: What is the typical Signal/Noise ratio of your IR observations at 5μ ? To be sure of the presence of dust surrounding your observed Be star it is necessary to have a small statistical error at 5μ and to extend the observations at larger wavelengths.

<u>Swings</u>: I don't have the count rates with me here, but most of the objects observed were much brighter at 5μ than the limit of the system $(m \approx 6)$.
Having the H,K,L map is enough to be sure about the presence of dust; having data at longer wavelengths allows one to have an idea about the temperature distribution of the dust graines.

INFRARED EMISSION FROM FOUR Be STARS OPTICAL COUNTERPARTS OF
GALACTIC X-RAY SOURCES

P.Persi,M.Ferrari-Toniolo
Istituto Astrofisica Spaziale CNR
Frascati,Italy

G.L.Grasdalen
University of Wyoming
Laramie,USA

ABSTRACT. Preliminary results of our infrared observations from 2.3
up to 10 and 20 microns of the Be-X-ray stars X Per, γ Cas and
HDE 245770,indicate the presence of an ionized circumstellar disk with
an electron density law of the type $n_e \propto r^{-3.5}$. X Per and γ Cas show
besides,variable infrared excess at 10μ suggesting variability in the
stellar wind. LS I+65°010 presents an anomalous infrared energy
distribution for a Be star.

1. INTRODUCTION.

In the last years,many galactic X-ray sources have been identified
with B-emission line stars (Bradt et al.1978). Maraschi et al.(1976)
propose the Be stars as a new class of X-ray transient in which the
sudden variations in mass-loss rate from the star in the presence of
a compact companion,could produce transient X-ray emission.
High-dispersion IUE spectrograms of several Be stars show
asymmetric or violet-shifted ultraviolet resonance lines,indicating
the presence of moderately strong stellar wind with terminal velocities
ranging from 700 to 1100 Km s^{-1} (Dachs,1980). The observed variabilities
in different spectral regions (ultraviolet,optical and infrared) of
some Be stars (Doazan et al.1980; Henrichs et al.1980; Slettebak et al.
1979;Ferrari-Toniolo et al.1978 and Elias et al.1978) suggests a
variability in the characteristics of the stellar wind.
In order to determine the wind parameters of the Be stars,and to
look at the infrared variability,four Be stars optical counterparts
of galactic X-ray sources have been observed during 1978-1980. The

247

M. Jaschek and H.-G. Groth (eds.), Be Stars, 247–251.
Copyright © 1982 by the IAU.

infrared observations from 2.3 up to 10 and 20 microns were carried
out at the 2.3 m infrared telescope of the University of Wyoming,
using a Ge-bolometer. Table I reports the log of the IR observations
and the optical and X-ray characteristics of the observed Be stars.

Table I : Log of IR observations

Star	X-Ray	S.P.	L_x/L_*	IR	N	Date
HDE 245770	A 0535+262	B0(II-V)e	$8\ 10^{-2}$	$2.3-10\mu$	3	78Oct.–80Sept.
LS I+65010	2S0114+650	B0.5IIIe	1.510^{-4}	$2.3-4.9\mu$	1	80Sept.
X Per	4U0352+309	B0(III-V)e	$2\ 10^{-5}$	$2.3-10\mu$	9	78July–80Sept.
γ Cas	4U0053+604	B0.5IVe	$6\ 10^{-6}$	$2.3-19.5\mu$	11	78Jun.–80Sept.

2. DISCUSSION.

During 1978-80,the Be stars γ Cas and X Per have been observed in
different occasions.Infraed variability has been found especially at
longer wavelengths.The IR light-curves of X Per seem to be correlated
to the visual light-curve and will be discussed in a forthcoming paper.
HDE 245770 has been observed during an X-ray quiescent phase and very
close (Sept.25) to the recent X-ray flare-up of Oct.2 (Oda,1980).

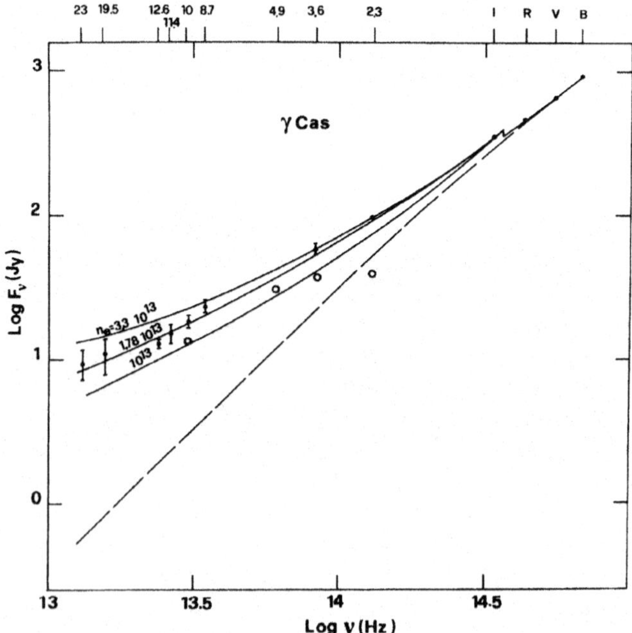

Figure 1. IR energy distribution of γ Cas.

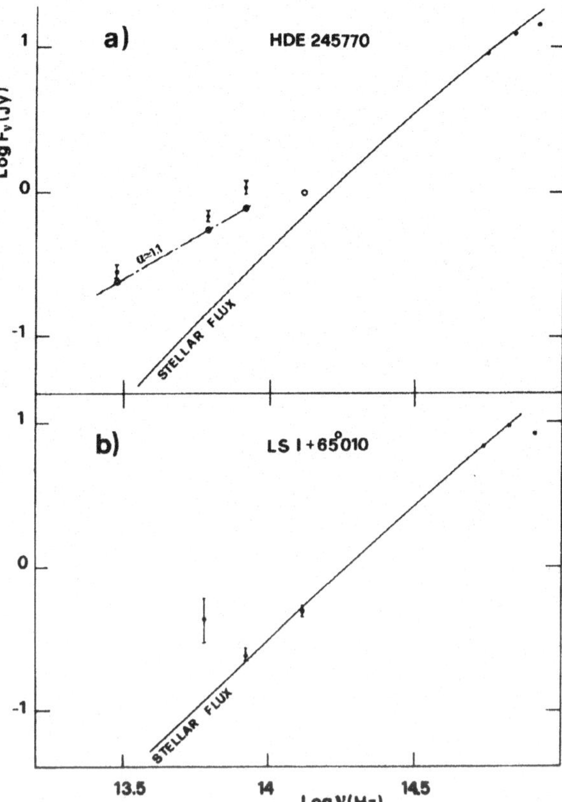

Figure 2a),b). IR energy distribution of HDE 245770 and
LS I+65010.

No significative infrared variations has been found during these X-ray
phases.

The IR energy distribution for our observed stars,were obtained
correcting the measured magnitudes for the standard i.s. reddening law
and the visual extintions reported in Table II. The continuum IR
spectrum of γ Cas relative to the period 1980 Aug.–Sept.,was compared to
the model described by Poeckert and Marlborough(1978).The best fit
results for an electron density of the envelope $n_e(R*)=1.78\ 10^{13}cm^{-3}$.
The IR excess (difference between the observed flux and the Kurucz model
of the star) (open circle in the figures)shows a power spectrum law
$S_\nu \propto \nu^\alpha$ between 5 and 10 microns with a spectral index $\alpha \sim 1.0$ for X Per,
γ Cas and HDE 245770. This observed spectral index,different from a ff+bf
spectrum with electron density $n_e \propto r^{-2}$ $(\alpha \sim 0.65)$,indicates a steeper
electron density law in the envelope surrounding our Be stars.

Using the spectral slope obtained by Hartmann (1978) for a thin isotherm: disk,we derive $n_e \propto r^{-3.5}$ implying acceleration of the wind in the region where originates the 5 -10 microns radiation. From our measured IR exce: at 10μ and using the spherical ff+bf models with a wind velocity law of the type $v(r) = v_\infty (0.01+0.99(1-R*/r)^\beta$)developed by Persi et al.(1980), we obtain a value of mass-loss rate for γ Cas,X Per and HDE 245770. These values of $\dot{M}$,reported in Table II, and obtained for an exponential velocity β =1-2,could be overestimated because of the spherical symmetry of the envelope here adopted.

Table II: Mass-Loss Rate.

Star	Log T* (K)	Log g	A_V	R* ($R_\odot$)	D (Kpc)	$S_{10\mu}^{ff+bf}$ (Jy)	v_∞ (Kms^{-1})	$\dot{M}$(M$_\odot$y^{-1}) (10^{-7})
HDE 245770	4.48	3.5	2.31	10	1.80	0.24±0.03	300	29
γ Cas	4.40	3.5	0.31	11	0.25	15.6±2.0	860	76
X Per	4.48	4.0	1.08	5	0.35	0.34±0.06	650	4.9

The very reddened star LS I+65°010 (A_V=4.1) shows an anomalous infrared energy distribution with respect to other Be star (see Fig.2b). The flux at 5 micron could suggest the presence of dust surrounding the star.Further IR observations especially at longer wavelengths need to confirm the presence of dust in LS I +65°010.

REFERENCES

Bradt H.V.,Doxsey R.E.,and Jernigan J.G. 1978,IAU/CoSPAR Symp.,CSR-P-78-5
Dachs J. 1980,Second European IUE Conference,ESA SP-157,139
Doazan V.,Kuhi L.V., and Thomas R.N. 1980, Ap.J.(Lett.),235,L17
Elias J.,Lanning H.,and Neugebauer G. 1978,P.A.S.P.,84,697
Ferrari-Toniolo M.,Persi P., and Viotti R. 1978,M.N.R.A.S.,185,841
Hartmann L. 1978,Ap.J.,224,520
Henrichs H.F.,Hammerschlag-Hensberge G.,and Lamers H.J.G.L.M. 1980,
    Second European IUE Conference ESA SP-157,147
Maraschi,L.,Treves A.,and Van den Heuvel E.P.J. 1976,Nature,259,292
Oda M., 1980, IAU Circular No. 3527
Persi P., Ferrari-Toniolo M., and Grasdalen G.L. 1980, Astron. Astrophys.
    92,238
Poeckert R., and Marlborough J.M. 1978, Ap.J.Suppl. Series,38,229
Slettebak,A.,and Snow T.P. 1978,Ap.J.(Lett.),224,L127

DISCUSSION

<u>Poeckert</u>: Your data tells you that there is an acceleration zone
where the IR emission is produced, but you take the UV resonance lines
as indicative of the absolute velocity. You can have an accelerating
zone without having large velocities. Your $\dot{M}$ may be much lower.

<u>Persi</u>: This is a very important point. Our IR observations give us a
correct value of the ratio $\dot{M}/v_\infty$, but I don't believe that our choice
of the terminal velocity could strongly modify the values of $\dot{M}$.

<u>Henrichs</u>: The fact that your mass-loss rate value derived from the
IR excess exceeds the value for UV measurements by a large factor
does not surprise me. The reason is that the UV mass-loss rate value
is obtained from the <u>absorption</u> column between the observer and the
star, whereas from the IR <u>emission</u> you consider the projection of the
whole (presumably) non-spheric envelope. Thus only in perfect spherical
symmetry you would expect the same answer.

<u>Editorial remark</u>: See also discussion following the paper by Guarnieri
et. al.

A PRELIMINARY DIGITAL ANALYSIS OF THE SPECTRUM OF β LYRAE

Eugenio E. Mendoza V.
Instituto de Astronomía
Universidad Nacional Autónoma de México

Abstract. This paper describes preliminary results from a digital analysis of the Fourier transform spectroscopy of β Lyrae. The spectra are photometrically precise in the 4800-10200 Å range with a spectral resolution of 3.85/cm. The most remarkable spectral features are the emission strength of some neutral Helium and Hydrogen lines, and their variability with time. Other fainter emission lines disappear, now and then.

An excellent review of the complex system of β Lyrae has been written by Sahade (1980) with an emphasis of the problems posed by this interesting eclipsing binary system. Sahade also gives extensive references to previous works.

This paper is based upon three Fourier-Transform Spectra of β Lyr obtained by Johnson on June 14, June 16 and October 10, 1976 with a Michelson Spectrophotometer. The system yields photometrically precise spectra (Johnson: 1977a, 1977b, 1978) in the 4800-10200 Å range. The spectral resolution is 3.85 cm^{-1}. The scan time of the interferometer was set to 2.8 seconds and it was necessary to add together many interferograms to obtain spectra of high quality. The probable error of noise ripple, expressed as a percentage of the stellar flux is around one percent (Mendoza 1981); except on both ends where it is near the two percent level. A compressed spectrum of β Lyrae is shown in Figure 1.

A preliminary analysis has been made in three directions: (1) wavelength measures for the purpose of line identification, (2) equivalent width measures of all the lines stronger than three times the probable error (Johnson and Mendoza, 1980), and (3) a study of line variability. The accuracy of measurement is limited by line blending; part of this blending is due to instrumental blending and part to blending of the stellar line themselves, and by profile and line strength variability. It is prohibitive to publish here all the measurements that we have made.

253

M. Jaschek and H.-G. Groth (eds.), Be Stars, 253–255.

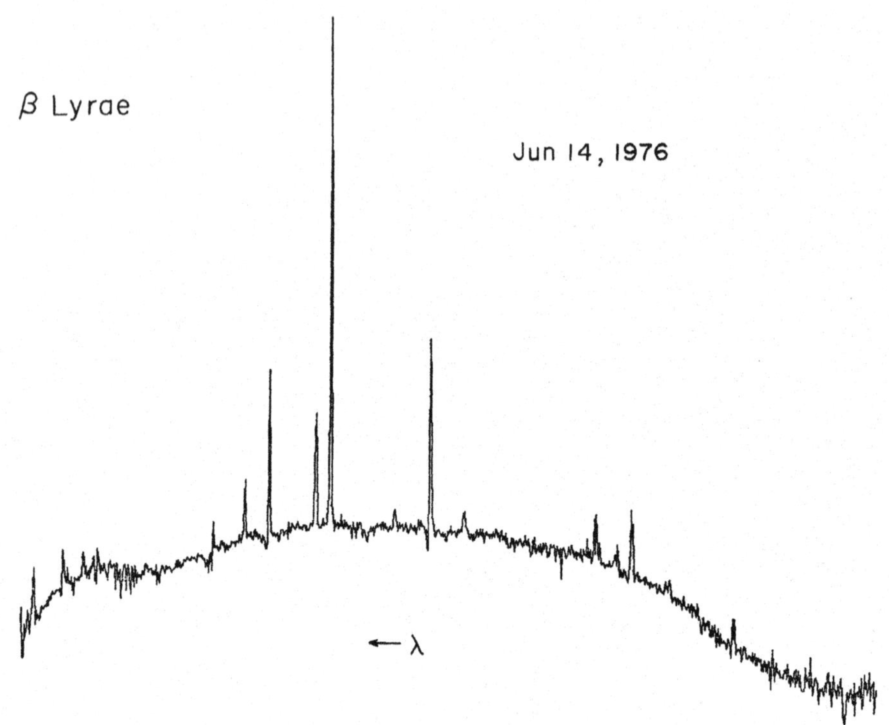

Fig. 1.- The compressed spectrum of β Lyrae from 4000 Å (right) to
λ10300 Å (left). The strongest emission line is Hα. The ab-
scissae are frequencies (cm⁻¹) and the ordinates relative intensity.

A great enhancement by emission in some neutral Helium and Hydrogen
lines are among the most remarkable spectral features seen in our three
spectra. Furthermore these lines are double, with the red component
stronger than the blue one, in general but not always. These lines are
variable in profile and strength. The changes are dramatic sometimes.
Other fainter emission lines appear and disappear with time. We show in
Figure 2 the spectral behavior of the N II(λ6329 Å) and the He I
(λ7281 Å) lines, to illustrate the above.

Hα is the strongest emission line in our spectra of β Lyrae (see
Figure 1). However, the higher Paschen lines are very little contami-

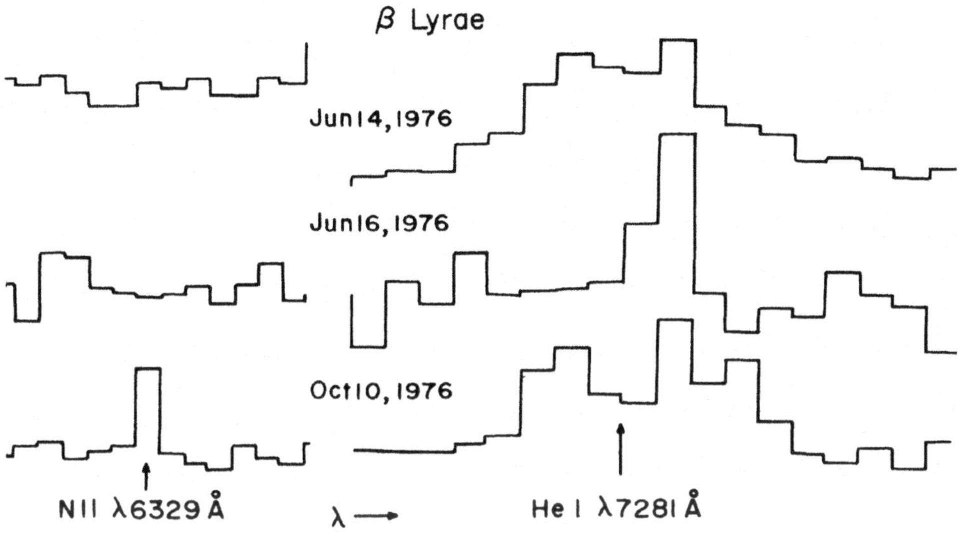

Fig. 2.- The spectral behavior of N II and He I lines in β Lyrae.

nated by emission, if any.

We have measured 600 absorption and 60 emission lines, in round figures. We find, in a first approximation, the same multiplets found in γ Cas, φ Per and P Cyg by Johnson et al. (1978), including the Mg II doublet (λλ9218, 9244 Å).

We are indebted to the IBM Scientific Center (Mexico) for extended computing facilities.

REFERENCES

Johnson, H.L.: 1977a, Rev. Mex. Astron. Astrof., Vol. 2, pp. 71-170.
Johnson, H.L.: 1977b, Rev. Mex. Astron. Astrof., Vol. 2, pp. 219-230.
Johnson, H.L.: 1978, Rev. Mex. Astron. Astrof., Vol. 4, pp. 3-201.
Johnson, H.L. and Mendoza, E.E.: 1980, Applications of Digital Image
    Processing to Astronomy, Proc. SPIE, Vol. 264, pp. 230-235.
Johnson, H.L., Wiśniewski, W.Z., and Fäy, T.D.: 1978, Rev. Mex. Astron.
    Astrof., Vol. 2, pp. 273-308.
Mendoza, E.E.: 1981, Rev. Mex. Astron. Astrof., Vol. 6, in press.
Sahade, J.: 1980, Space Science Reviews, Vol. 26, pp. 349-389. D. Rei-
    del Publishing Co., Dordrecht-Holland.

BIBLIOGRAPHY AND OBSERVING CAMPAIGNS

INTRODUCTORY TALK : BIBLIOGRAPHY OF Be STARS

Pavel Koubský
Ondřejov Observatory
251 65 Ondřejov, Czechoslovakia

As you know the first list of bibliography on Be stars was published
in the second Be Star Newsletter. Six contributors are trying to cover
the production of papers in the field. No doubt that the bibliography
is very useful and I am very glad that Dr. M. Jaschek has offered the
Newsletter for this purpose and that she is acting as the editor of the
bibliography.

Since the start of this service I was thinking how to make the
bibliography even more useful with reasonable amount of additional work.
The first question is the scope of the bibliography : Be stars are not
isolated in the Universe. When studying the Be phenomenon, people have
found similtarities with early type supergiants, beta Cep stars, and
even symbiotic stars. Several components of the Be binaries are WR stars
or even X-ray sources. Very broad spectral region (from X-rays to ratio
waves) and different techniques including speckle interferometry are used
to obtain data on Be stars. I think that the contributors should include
also important papers from these fields in the Be star bibliography.

The second question is the bibliographical system. In a letter sent
last May to Dr. M. Jaschek I proposed a system which is a modification
of the system used by Commission 42 (See table 1). In fact, there are
three categories of bibliographical entries : papers dealing with one
or several stars, papers describing properties of tens of stars and
general papers. The format for the first group should be the name of the
star (HD number obligatorily if available), reference and code (Table 1)
The format for the second group should be reference, code and file of
all stars the paper is dealing with. Dr. L. Pastori has tried to process
the bibliography with the proposed system and I think the result was
very good.

So far, there are only few announcements of currents programs, so
it is not worthwhile to include them as a permanent part of the bibliography. But I think people should use more the possibility offered by
Dr. M. Jaschek and contribute to the Newsletter.

M. Jaschek and H.-G. Groth (eds.), Be Stars, 259–260.
Copyright © 1982 by the IAU.

TABLE 1 – Be STAR BIBLIOGRAPHY – CLASSIFICATION SCHEMA

1. Conferences, symposia and monographs

2. Catalogues, discoveries
   a) catalogues
   b) discoveries of new Be stars
   c) identifications of optical counterparts with X, IR and radio
      sources

3. Observational data
   a) spectrometric (including spectrographic)
   b) radial velocities
   c) photometric
   d) polarimetric
   e) optocal region
   f) UV
   g) X
   h) IR
   i) radio

4. Time scale
   a) several points
   b) long series

5. Theoretical investigations
   a) models of envelopes
   b) comparison with observational data
   c) origin of the envelopes
   d) evolution of Be stars

6. Statistical data

7. Mescellaneous

I do realize that the proposed version of the bibliography will
require from the contributors and the editor an extra work, but I
hope better system will be the reward of all the people involved.

Harmanec : There is still no general agreement on the nature of the Be
phenomenon. Consequently, I vote for the system in which all early-
type emission-lin objects are considered.

A CATALOGUE OF Be STARS

M. Jaschek and D. Egret
Observatoire de Strasbourg

We are presently undertaking, at the Stellar Data Center, a large
effort to produce the *Catalogue of Stellar Groups*, which is a catalogue
listing all the kinds of stars with spectral peculiarities existing up
to now. The work is in progress and presently about fifty groups (Ap,
Am, Be, WR, H and K emission line stars, etc.) are defined, gathering
more than 25000 stars. The catalogue was started with the aim to set up
lists defined by homogeneous criteria, because up to now the available
information is scattered through the literature. We hope that the catalogue will be useful for formulating observing programs and for further
detailed research on each of the groups.

For this symposium we have compiled a list of Be stars and prepared
a comprehensive catalogue of Be stars on microfiche.

1. DEFINITION

The definition of a Be star we accepted is that of a non-supergiant
B star which showed emission in one Balmer line at least once. For this
reason we have excluded, with one exception, all stars for which no MK
classification exists. The exception is that of Bidelman and MacConnell
(1973) who used the same definition but provided no MK type. (Note that
only 132 stars come from this list).

Besides supergiants, also planetary nebulae and spectroscopic binaries were excluded. The B[e] stars are listed as a separate table.
These stars are defined as being those showing proper emissions of Be
stars plus emission of some forbidden line.

The catalogue contains about 1100 stars. From confrontation with
the number of stars in our catalogue group "Hα emission stars", which
includes about 5000 stars, it is evident that we have made a strong
selection. On the contrary, one can be reasonably sure that the stars
in our catalogue are authentic Be stars.

Please notice that MK types are established in a wavelength region
in which Hα is not included. Therefore many stars having Hα in emission
are classified as normal stars, being called for instance B3 V and not
B3 V e.

M. Jaschek and H.-G. Groth (eds.), Be Stars, 261–263.
Copyright © 1982 by the IAU.

2. THE DATA

The final catalogue can be considered as the intersection of three main files: the lists of emission-line stars, the compilations or lists of MK classifications and the data base of the Stellar Data Center.

Our major source for emission line stars was the *"Catalogue of Stars Whose Spectra have shown Emission Lines"* (Wackerling, 1970). We also used the list by Bidelman and MacConnell (1973), and some additional stars were found in the lists of MK classifications.

Our sources for the MK classifications are the *"Catalogue of Selected Spectral Types"* (Jaschek, 1978), the Michigan Spectral Survey for HD stars, vol. I and II (Houk and Cowley, 1975, Houk, 1978), the compilation by Uesugi (1976) and the last edition of the MK Extension (Kennedy, 1980). Other classifications by Lesh (1968), Graham (1970), Henize (1976), Andersen and Nordström (1977, 1979) and Jaschek et al. (1980) are used.

In a final step we have searched for each star the fundamental data and measures available in the *Catalogue of Stellar Identifications* (Ochsenbein et al., 1981). The following data are listed in the microfiche edition of the catalogue: DM and HD identification, number in the main lists of emission line stars (MWC, AS, Henize), coordinates, V magnitude and UBV photometry (when available from Nicolet's compilation, 1978), adopted MK classification, radial velocity, rotational velocity; an asterisk indicates when the star is known as a visual binary or a spectroscopic binary, and if the star is measured in other photometric systems (uvby, infrared and ultraviolet); if the star is known as a variable star, the name is given.

The list of B[e] stars is given in a separate table.

A bibliographical file for the same stars is also in preparation: we have used the *Bibliographical Star Index* (Spite et al., 1980) to obtain a list of the papers (complete reference and title) in which the star is quoted, in the last ten years (since 1970, date of the preparation of the bibliographical catalogue by Jaschek et al., 1971). This list will be published as a separate microfiche.

3. FINAL REMARKS

A first analysis of this catalogue is presented in the first session of this symposium (Egret, 1982).

The microfiche can be requested from the Centre de Données Stellaires (Observatoire de Strasbourg).

REFERENCES

Andersen,J. and Nordström,B.: 1977, Astron.Astrophys.Suppl.Ser. 29,309.
Andersen,J. and Nordström,B.: 1979, Inf.Bull.C.D.S. 15,39.
Bidelman,W.P. and MacConnell,D.,J.: 1973, Astron.J. 78,687.
Egret,D.: 1982, IAU Symp.98 *Be Stars*, Jaschek,M. and Groth,H.G. eds., 23
Graham,J.,A.: 1970, Astron.J. 75,703.

Henize,K.: 1976, Astrophys.J.Suppl.Ser. 30,491.
Houk,N. and Cowley,A.,P.: 1975, *Michigan Catalogue of 2-dimensional Spectral Types for the HD Stars*, vol.I, Univ. Michigan.
Houk,N.: 1978, *Michigan Catalogue of 2-dimensional Spectral Types for the HD Stars*, vol.II, Univ. Michigan.
Jaschek,M.: 1978, Inf.Bull.C.D.S. 15,121.
Jaschek,C.,Ferrer,L.,Jaschek,M.: 1971, La Plata Obs., Ser.Astron.,XXXVII.
Jaschek,M.,Hubert-Delplace,A.-M.,Hubert,H.,Jaschek,C.: 1980, Astron. Astrophys.Suppl.Ser. 42,103.
Kennedy,P.,M.: 1980, *MK Classification Extension*, MtStromlo Obs.
Lesh,J.,R.: 1968, Astrophys.J.Suppl.Ser. 17,371.
Nicolet,B.: 1978, Astron.Astrophys.Suppl.Ser. 34,1.
Ochsenbein,F.,Bischoff,M.,Egret,D.: 1981, Astron.Astrophys.Suppl.Ser. 43,259.
Spite,F.,Ochsenbein,F.,Kirchner,S.,Lahmek,R.: 1980, Inf.Bull.C.D.S. 18, 89.
Uesugi,A.: 1976, *A Revised Catalogue of Stellar Rotational Velocities*, preliminary edition, preprint.
Wackerling,L.,R.: 1970, Mem.Roy.Astron.Soc. 73,153.

DISCUSSION

Andrillat: Dans la définition que vous avez adoptée pour les étoiles Be, vous indiquez qu'il s'agit d'étoiles ayant montré une fois H_α en émission : à partir de quelle date?

Egret: The oldest observations of H_α in emission we used are those compiled by Wackerling (1970) (including the Mt. Wilson catalogue, etc.).

A CATALOGUE OF Hα OBSERVATIONS

J.R. Ducati [+]
Centre de Données Stellaires
11, rue de l'Université - 67000 Strasbourg - France

ABSTRACT. A catalogue of quantitative observations of Hα published between 1949 and 1980 has been prepared. It contains 4095 measurements concerning 2700 stars.

We have compiled a catalogue of quantitative Hα observations made since 1949, by means of a survey in the literature that revealed the existence of 37 publications in which measurements are given. This catalogue is available at the Stellar Data Center at Strasbourg.

The first part of the catalogue consists in the original measurements, as published by their authors, amounting to 4095 Hα magnitudes, ratios, equivalent widths or other forms of quantifying the Hα line intensity ; as there are stars that were observed by more than one author, or even by the same author, several times, these 4095 observations refer to some 2700 different stars. In this first part of the catalogue, the following data are provided : star identification, spectral type, m_v, the Hα measurement, the error in Hα, the number of observations, remarks concerning duplicity, variability, emission, etc..., and the reference. Some of these data may not be present, if not provided in the original paper.

The second part of the catalogue represents a tentative to furnish an unified, homogeneous Hα system. As a justification to this effort, one must remark that it does not exist, until our days, an accepted system for Hα. For the Hβ case, on the other hand, there is the well-known and widely accepted system developed by Crawford and his co-workers since almost twenty years ago.

It is not always possible to transform a photometric system to another,

[+] In leave of absence from the Departamento de Astronomia do Instituto de Fisica da Universidade Federal do Rio Grande do Sul, Porto Alegre, Brasil

M. Jaschek and H.-G. Groth (eds.), Be Stars, 265–268.
Copyright © 1982 by the IAU.

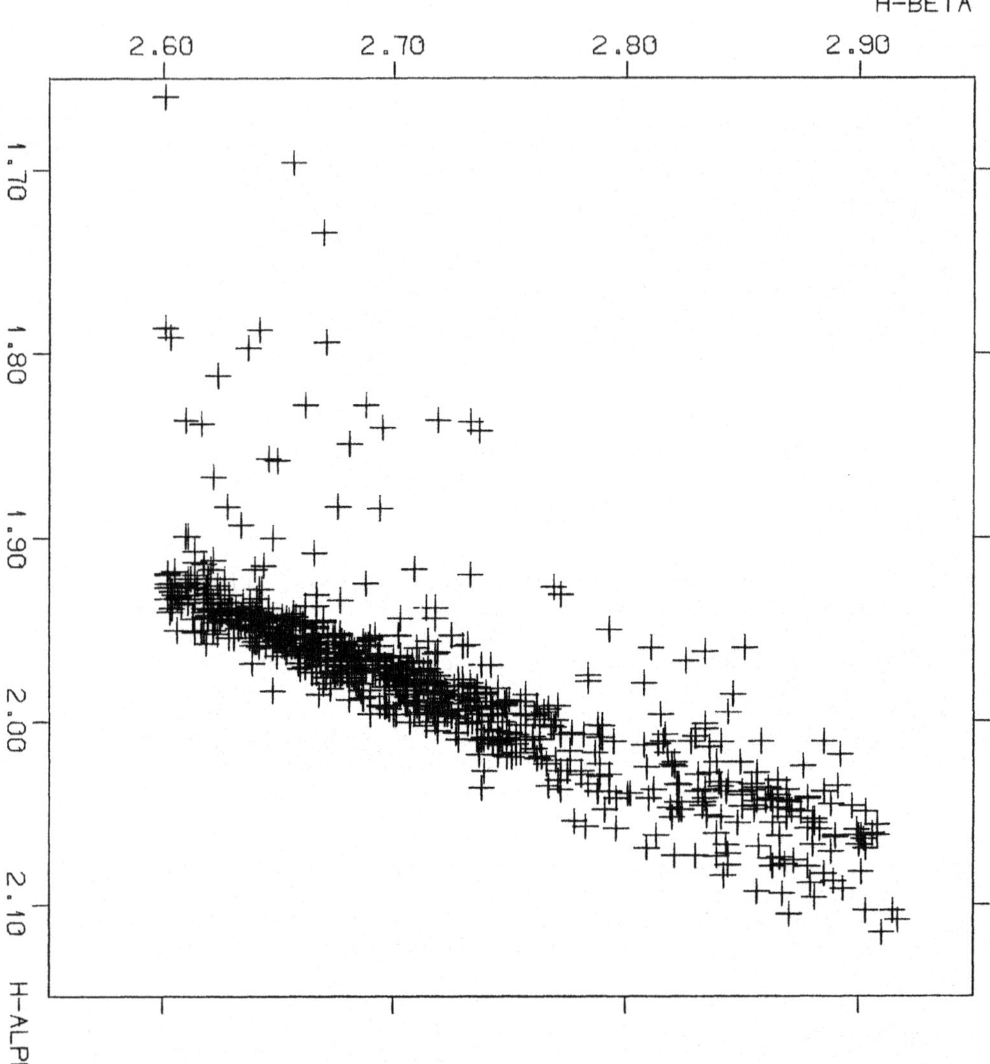

Fig. 1. Hα vs. Hβ for 700 stars which have both lines observed. The α-scale is in the Strauss-Ducati system, the β-scale in the standard Crawford system.

since different observing techniques may prevent any meaningful attach-
ment. Studying the first part of this catalogue and looking for suitable systems or lists of observations, we found twelwe papers, by
seven authors, that have enough data to allow a transformation, in linear
form, to another system, at least for a certain range in the respective
domains of the Hα indices. These twelwe papers are : Feinstein (1974,
1978), Dachs and Schmidt-Kaler (1975), Strauss and Ducati (1981),
Mendoza (1976 a, b, c, 1977, 1979), Peat (1964, 1966), Andrews (1968),
and Cester et al (1977).

Merging these systems, we arrived at a catalogue of 2300 stars with
unique α indices. Only in a few cases these indices are means made from
more than one author, the general case being that the index in the catalogue comes from a single reference, having undergone a transformation.

The unified indices in the catalogue are given in a system chosen arbitrarily between those that form the catalogue ; nevertheless transformation from this system to another one is possible by using equations
given in the third part.

As a check for the quality of this transformation, we made a comparison
of the Hα and Hβ indices for the stars that have both observed. As figure 1 shows, the correlation for 700 stars is good, taken in account
the spread in the observations that occurs generally. The stars that
are scattered in the left of the diagram, are in their majority supergiants or emission stars, which classically are separated in this kind
of comparison.

REFERENCES

Andrews, P.J. 1968, Mem. R.A.S. 72,35
Cester, B., Giuricin, G., Mardirossian, F., Pucillo, M., Castelli, F.,
     Flora, U. 1977, Astron. Astrophys. Suppl. 30,1
Dachs, J. Schmidt-Kaler, Th. 1975, Astron. Astrophys. Suppl. 21,81
Feinstein, A. 1974, Monthly Notices R.A.S. 169,171
Feinstein, A. 1978, Rev. Mex. Astron. Astrof. 2,331
Mendoza, E.E. 1976a, Rev. Mex. Astron. Astrof. 1,363
Mendoza, E.E. 1976b, Rev. Mex. Astron. Astrof. 2,69
Mendoza, E.E. 1976c, Rev. Mex. Astron. Astrof. 2,33
Mendoza, E.E. 1977, Rev. Mex. Astron. Astrof. 2,259
Mendoza, E.E. 1979, Astron. Astrophys. 71,147
Peat, D.W. 1964, Monthly Notices R.A.S. 128,435
Peat, D.W. 1966, Monthly Notices R.A.S. 131,467
Strauss, F.M., Ducati, J.R. 1981, accepted for publication in Astron.
     Astrophys. Suppl.

DISCUSSION

<u>Mermilliod</u>: The transformation of the H_α measurement into an unique
system seems to be rather problematic since the number of common
stars between different systems is generally small and the slope of
the relation is unfavourable to perform a good transformation. Furthermore, transformation can be made only for normal stars and unfortunately not for the Be stars, for which it would be most interesting.

<u>Ducati</u>: Taking into account these problems we excluded two thirds of
the work used to make the general catalogue, forming the unique system
only with sufficiently correlated observations if the slope of the
correlation equation was great enough to avoid an amplification of the
errors. Nevertheless, an important proportion of Be stars was really
excluded and this cannot be avoided, since each observer drives an H_α
system for his specific purpose, which is not the case for H_β.

<u>Mermilliod</u>: Mermilliod and Mermilliod published a compilation of
Balmer-line photometry $(\alpha,\beta,\gamma,\delta)$, made in the frame of the collabora-
tion of the Lausanne Institute of Astronomy with the Stellar Data Center.
The description has recently been published in the Bulletin of the CDS.

<u>Mendoza</u>: Why did you not include in your catalugue other hydrogen
lines?

<u>Ducati</u>: Catalogues of H_β and H_γ already exist and are available in the
Stellar Data Center in Strassbourg. For H_β data, especially, there is
the uvbyβ Catalogue by Hauck and Mermilliod (1980).

AN OBSERVING CAMPAIGN FOR SYSTEMATIC PHOTOELECTRIC OBSERVATIONS
OF BRIGHT Be STARS

P.Harmanec, J.Horn, P.Koubský
Astronomical Institute of the Czechoslovak Academy
of Sciences, 251 65 Ondřejov, Czechoslovakia

Most of our present-day knowledge of the photometric behaviour
of Be stars still comes from various surveys only. Little attention
was paid to systematic studies of possible photometric variability
of particular objects on different time scales. Nevertheless, just
this kind of information is necessary if one wants to study the
physical relationship and/or differences between Be stars and other
types of B stars like Beta Cepheids, supergiants, helium-rich stars
etc., and to test various models of the Be phenomenon. Long systematic photoelectric observations do exist only for several Be stars.
They invariably indicate that the long-term variability - if present
- is always the most pronounced one. In some cases, this long-term
variability masks the periodic light variations occuring on shorter
time scales (see, e.g., the case of CX Dra, Koubský et al 1980).

All this suggests that the series of systematic (differentially
measured) photoelectric observations of individual Be stars secured
over many years would be of a great value. Consequently, an appeal
was made by Harmanec et al (1980a) and by Harmanec (1980) to organize
an international cooperation on systematic photoelectric observations
of a large but defined group of bright Be stars over a period of at
least ten years.

After a positive reaction from a number of colleagues all over
the world we prepared a detailed observing programme, selecting
comparison and check stars etc. for about 140 bright emission-line
B objects (including several supergiants) - roughly up to the
magnitude limit of the Bright Star Catalogue and north of -20°
declination. Detailed instructions concerning the practicalities
of the campaign have been published in the second issue of the
Be Star Newsletter which is available on request from Dr.M.Jaschek
(Harmanec et al 1980b).

Important points to be discussed here are:

1. All the measurements should be performed differentially, with

M. Jaschek and H.-G. Groth (eds.), Be Stars, 269–274.
Copyright © 1982 by the IAU.

respect to comparison stars which, after some optimization of their
choice, should become obligatory for all the participants.

2. A careful transformation to standard photometric systems must
be performed. Colour coefficients should be determined for each
observing season.

3. To ensure convincing detection of possible rapid variations
of a small amplitude and to give an idea about confidence of a
particular set of measurements, each participant should observe
the check stars <u>as frequently as</u> variables and to publish the
results of these measurements, too.

4. Which would be the most suitable form of publishing and storing
the data obtained in the course of the campaign is a point which
we should agree on here.

REFERENCES

Harmanec, P.: 1980, Be Star Newsletter (ed.M.Jaschek) 1, 2
Harmanec, P., Horn, J., Koubský, P., Žďárský, F., Kříž, S.,
      Pavlovski, K.: 1980a, Bull. Astron. Inst. Czech. 31, 144
Harmanec, P., Horn, J., Koubský, P.: 1980b,
      Be Star Newsletter (ed. M.Jaschek) 2, 3
Koubský, P., Harmanec, P., Horn, J., Jerzykiewicz, M., Kříž, S.,
      Papoušek, J., Pavlovski, K., Žďárský, F.: 1980,
      Bull. Astron. Inst. Czech. 31, 75

HARMANEC: Let me to summarize the main ideas from our discussion as follows:

1. As the campaign is planned for 10 or 20 years, we should afford to devote say one or two years to a search for optimal comparison stars in particular cases.

2. Individual observations from all participants should be collected in the data files which already exist at the Ondřejov Observatory Computing Centre. They should be made available to anybody on request.

3. All participants should feel free to publish their results in usual astronomical journals. However before submitting their papers to particular journals they should send them to the Ondřejov Campaign Centre which should play the role of a first referee to check whether the papers contain all basic pieces of information on which we agreed here.

4. The Campaign Centre should extend the observing programme also to southern Be stars.

5. New detailed instructions concerning the organization of the campaign, amount of obligatory information to be published, etc. will be prepared for the next issue of the Be Star Newsletter. In particular, a list of suitable standard stars with obligatory standard values will be prepared for convenience of participants and in an effort to homogenize the data as much as possible.

DISCUSSION

<u>C. Jaschek</u>: Is a regular journal going to accept to publish observational data in small bits?

<u>Sterken</u>: 1. IBVS will not object against publishing numerous short reports. The publication is fast and there are no page charges.
2. Are you willing to act as a kind of referee when you get the progress reports? Do you plan to inform people if they are not observing in the correct way? This would be very desirable and would insure the homogenity and the credibility of the results.

<u>Harmanec</u>: 1. Yes, IBVS is a very good possibility for rapid publishing.
2. Also yes, we agree to do that. We shall prepare some suggestion along this line for the next Be Newsletters.

<u>Sterken</u>: 1. I have a comment on the use of small or medium passbands: since amateur observers will be involved, and since many of them work with apertures of 30 to 35 cm, it might be a problem to obtain the same accuracy when using narrow bands.
2. I agree that it is very important to observe the same comparison stars all the time, even if they are relatively distant to the program star. But near-by comparison stars are absolute needed in order to carry out accurate photometry (especially if short time variations are studied). I would therefore suggest that every observer selects one comparison star in the very neighbourhood of the program star, and that he observes them <u>together</u> with the official comparison star. After two or three years the promoters of the campaign will be able to single out the good (constant) nearby comparison star, after which they could be kept as definite comparison stars. On a project which is meant to be executed over 10 or 20 years, it would really pay to look first for good comparison stars.

<u>Harmanec</u>: Well, I hope that it will be possible to distribute the objects in such a way that amateur observers will observe mainly the brightest program stars. (As to sec. question, see my reply to Dr. Sareyan)

<u>Sareyan</u>: I strongly suggest that we use no longer the UBV system, even if it is a "realistic" choice just now, it's a very bad one on the long-term scale. Any narrow band pass filter system is easier to reduce than UBV, which has numerous draw-backs: the U filter has a red leak, the flux in which depends on the actual multiplier's cathode; the V filter is cut off on its red side only by the photomultiplier cathode's cut-off. So these two limits are PMT dependent. Moreover it would be really useful (and necessary) that each observation is given with the PMT temperature, and its stability during the night, <u>even</u> if the observations are made differentially (a 1% drift is given by a 1°C drift in flux of the photocathode, and this cannot be completely removed when we observe with large bandpass filters some variable and comparison stars that have very different spectral types).

2. It appears that very short period variations ($<$ 1 day) cannot be investigated as exactly as longer periods. For example, the comparison star in rapid differential photometry has to be very close to the variable; so they probably won't be photometric standards, and no reduction "to the system" can be done that particular night, in order to increase the measurement's precision _and_ the time resolution.

Harmanec: Starting with the first appeal for photometric observations of Be stars we always stressed that the ubvyβ system should be preferred over the Johnson system. In fact, we ordered the Strömgren filters from the Kitt Peak and in future we shall use them in Ondrejov and at Hvar. The same did Dr. Muminović in Sarajevo. Yet, I am afraid that for most of the data we will have to live with the UBV data. I believe that if the choice is to have no photometric information or to have the UBV data only, than the latter possibility is to be preferred. It is certainly possible to use different comparison stars for various types of observations according to the special needs of the program, but in all such cases, at least few differential observations of this new comparison star with respect to the obligatory one should be performed during each observing night. Then you do not lose the information for possible long-term variability.

Sterken: Could you consider to extend your program to the southern hemisphere? The southern hemisphere offers many possibilities for long observing runs under good observing conditions, and especially in Australia and New Zealand there are many amateurs who are able to perform professionally, and who are surely willing to participate such a campaign.

Harmanec: Well, we shall do it, but I shall send you a preliminary version of the program to check the suitibility of selected comparison stars.

Mermilliod: ubvy photometry is homogenous only as long as small groups are working in this system. If a large number of people are to work with it, the same will happen as happened to the UBV system: systematic differences and a large scatter will appear for stars observed by various observers.

Harmanec: Thank you for reminding me this point. I would like to ask the audience whether it would be useful to prepare an _obligatory_ list of standard stars and ask all the participants to use just these stars for deriving the colour transformation coefficients of a given photometric system. - yes? It seems that you all agree.

Bolton: We should not worry about transformations between Johnson U and Strömgren u because the bandpass of U crosses the Balmer jump. This will make the transformation for Be stars very difficult and makes the U band useless for physical interpretations.

Harmanec: But you probably agree that at least the V and B-V data can be reconciled with the corresponding Strömgren values. Then we can save at least some brightness and colour information about the long-term variability.

SPECTROSCOPIC OBSERVING CAMPAIGN

Paul K. Barker
The University of Western Ontario

With organization of the Photometric Observing Campaign well under
way, there is an exciting opportunity for collaboration amongst the
spectroscopists. I am proposing a program of regular ground-based
spectroscopy of the 100 or so brightest northern Be stars, to be con-
temporaneous with the photometric program.

Of fundamental importance to dynamical theories for Be stars is an
observational understanding of their time-dependent behavior, yet even
though previous long-term studies of spectral variability have been
undertaken (as well as intensive studies of individual stars) the
tendency is for each observer to concentrate on particular favorite
objects. Systematic observations of all the bright northern Be stars
would give, without prejudice, extensive information on the nature and
timescale of the spectroscopic variations. Spectroscopy concurrent
with the photometric program can only enhance the value of the data.

Obviously, several groups already have well established observational programs of both a long term and a concentrated nature; one
program has been organized to combine observations at many diverse
wavelength regions, for a few selected stars. The point of this proposal is that, in the effort to better understand Be variability in a
larger more general group of stars, the scientific aims might be well
served by a cooperative venture in which the observational burden is
shared. Even in the outstanding "Atlas of Be Stars" for example, only
some of the stars are observed more than once a year. Further, a
special advantage of a cooperative campaign is the facility to detect,
at the earliest possible stage, developing activity in previously
quiescent stars. Rapid communication of the incipient activity to all
collaborators would enable intensive study of the active phase by both
photometrists and spectroscopists. Even isolated sporadic observations
could contribute significantly when collated with other spectra.

Such a program of regular spectroscopy is in progress with an
intensifier-dissector-scanner at The University of Western Ontario 1.2m
telescope. The scanner is a 512-channel instrument with present

M. Jaschek and H.-G. Groth (eds.), Be Stars, 275–276.
Copyright © 1982 by the IAU.

resolution of about 0.7Å in the blue and 1-2Å in the red and near-
infrared. The capabilities of the scanner are demonstrated by the
obseryations described elsewhere at this Symposium. The scanner is
sufficiently fast that from 1980 April to November, Hα scans were
secured of all the program Be stars, along with Hβ and near-infrared
scans of one-third of the stars, as well as complete and repeated
spectral coverage of selected active stars (ζ Oph, o And, θ CrB, and
59 Cyg that year for example). Resolution for the 1981 Be season will
be doubled by a new grating, and it is quite realistic to expect comparable observational productivity in the future.

     On a practical level, one must consider seriously the mechanics
required for cooperative spectroscopy. A specific obstacle to successful collaboration might arise from the difficulties which attend inter-
comparison of spectral data obtained with diverse equipment and instrumental profiles. The labor involved in effective juxtaposition of data
could well prove to be overwhelming. In this event the only sensible
mode for collaboration might be to simply exchange information on
active stars quickly enough that all observers have the opportunity to
follow the active episodes.

     One way to evaluate the prospects for true interchange of data
would be for all participants to observe a small number of agreed
standards, and to compare the profiles from the different observatories,
before embarking on such a project. Even if this test gives discour-
aging results, there are still occasions when knowledge of the qualitative profile at a crucial phase of activity would be valuable.
Comments and suggestions (for example, how should the star list be
subdivided?) are sought from observers interested in a possible
campaign of this general nature.

V. ROTATION AND BINARITY

ROTATION, EXPANSION AND DUPLICITY OF Be STARS

Petr Harmanec
Astronomical Institute of the Czechoslovak Academy
of Sciences, 251 65 Ondřejov, Czechoslovakia

1. INTRODUCTION

In recent years there is a general tendency to divide Be stars
into different groups. Bidelman's (1976) division on supergiants,
rapidly rotating single stars, interacting binaries, early type
nebular variables and quasi (or young) planetary nebulae is often
quoted. Another example is the classification by Lesh (1968) or a
very recent classification by Jaschek et al (1980), limited a priori
to "normal" Be stars i.e. stars of luminosity classes V to III. I do
not think such an approach is recommendable. There is still no
general agreement as to the nature of the Be phenomenon and all such
classifications must be more or less descriptive. The possibility
that the hydrogen emission observed in the spectra of apparently very
different objects has always _the same_ physical cause, and that the
differences between these objects are caused only by continuously
varying physical and/or geometrical parameters should seriously be
considered, along with all other concepts. At the moment, I find it
more promising to study Be stars in context of all early type emission objects, including supergiants, P Cyg stars, Beta Cep objects,
WR stars, symbiotic stars and novae. Such a viewpoint is not new -
- it was expressed already by Struve (1942).

Let us consider the case of supergiants in particular. Little
is known about their long-term spectral variability. However their
RV and photometric variations do seem to be rather similar to what
is observed for Be III-V stars (c.f., e.g. Sterken 1977, Sterken and
Wolf 1978 or Rufener et al 1978). Excluding the supergiants from
considerations leads from time to time even to inconsistencies. In
many catalogues, the c prefix of the spectral type is quoted from
older sources and may refer to the shell spectrum of a Be III-V
object. Thus, considering the above-mentioned facts, I prefer to let
the subject of this study somewhat "undefined".

Until now, many different models have been suggested, or even
computed, to explain various specific aspects of the Be phenomenon.

M. Jaschek and H.-G. Groth (eds.), Be Stars, 279–297.
Copyright © 1982 by the IAU.

Yet, as far as I know, there are only three general conceptions
attempting to explain the Be phenomenon in its complexity:
1. The rotational hypothesis proposed originally by Struve (1931),
2. The hypothesis of radial outflow of matter, first suggested
   by Gerasimovič (1934), and
3. The binary hypothesis, formulated in a general way by Kříž and
   Harmanec (1975).

Here, I shall try to outline the basic principles and to
evaluate successes and pitfalls of these three competing conceptions
in their relation to the available observational data. Inevitably,
this evaluation will reflects my personal knowledge and interests
and will be neither complete nor the only possible.

2. THE ROTATIONAL MODEL (RM)

This model starts from the observed correlation between rotational velocity and width and shape of emission lines in Be stars.
It is assumed that Be stars are rotationally unstable at equator and
eject matter, which forms envelopes and gives rise to the emission
lines. Different shapes of the lines are easily explained by the
aspect effect: Be stars with single-peaked narrow emission lines have
also sharp absorption lines and are understood as rapidly rotating
objects seen roughly pole-on. Also the observed proportionality of
the width of the hydrogen emission lines to their wavelength, first
recognized by Curtiss (1923), agrees well with RM. RM is thus based
on assumption that a rapid axial rotation is a common property of all
Be stars, a property which these stars gained during their contractic
towards the ZAMS and during subsequent evolution. Sackmann and Anand
(1970) showed that, on assumption of a rigid rotation, a B star
evolving from the ZAMS becomes rotationally unstable at the equator
very early and still long before the end of its main-sequence stage
– unless the initial rotation is very slow. Also Strittmatter et al
(1970) computed evolutionary models of uniformly rotating B stars,
considering also the equatoreal mass loss which they found to be
between 3×10^{-9} and 4×10^{-7} $M_\odot$/year in their particular case.
Kippenhahn et al (1970) considered more complicated models in which
local angular momentum conservation was assumed in regions of varying chemical composition and overall conservation of angular momentum
and solid body rotation in all chemically homogeneous regions.
Assuming the critical rotation at the ZAMS, they obtained equatoreal
mass loss at the end of the main-sequence stage. On the other hand,
models assuming local conservation of angular momentum in radiative
regions and solid body rotation with overall conservation of angular
momentum in convective layers, avoided the mass loss during the main-
-sequence evolution. Meyer-Hofmeister and Thomas (1971) repeated
these computations taking the mass loss into account. They estimated
the mass-loss rate to be of the order of 10^{-9} $M_\odot$/year. They also
computed theoretical distribution of rotational velocities and
compared it with the distribution observed for Be stars. Unfortunate-

ly, too many assumptions involved in this comparison do not allow to
make any firm conclusion. Ostriker (1970) has shown that a contract-
ing, rotating star will never shed mass at the equator unless also
viscous and/or magnetic forces are taken into consideration. Sum-
marizing the theoretical results, we can say that the available
predictions of rotational instability are model dependent and therefore somewhat inconclusive.

It seems that the observational evidence is more specific on
this point. The diagrams showing the distribution of observed projected rotational velocities of Be stars (v.sin i) versus spectral type,
with the line of critical rotation drawn, clearly show that the
question of rotational instability does not represent any serious
obstacle for RM (Slettebak 1976, 1979). Doazan (1970), measuring
profiles of 26 Be stars, concluded that no correlation exists
between the width of the H I emission lines and rotational velocities
of underlying stars, casting thus some doubts on one of Struve's
(1931) basic arguments. Nevertheless, Slettebak (1976) and Slettebak
and Reynolds (1978) presented convincing evidence of this correlation, for a large group of Be stars. Hardorp and Strittmatter (1970)
demonstrated a strong tendency of shell spectra to be observed
mainly among most rapidly rotating Be stars. Poeckert and Marlborough
(1976) observed a correlation between polarization of Be stars and
v.sin i. All these pieces of evidence strongly support the view that
the envelopes of Be stars are to a great extent axisymmetric, rather
than spherically symmetric, and that their structure <u>must strongly
depend on rotation</u>. This <u>need not mean</u> however that the high rotation <u>causes</u> the Be phenomenon.

Using Jaschek's et al (1980) homogenized data for 140 bright
northern Be stars, I constructed histograms of v.sin i for B2e-B5e
and B6e-B9e stars. They are shown in the upper panel of Fig.1. One
interesting difference of these histograms is that while there is
almost no qualitative difference between distribution of early and
late type non-emission B stars, there exists a striking deficiency
of slowly rotating B6e-B9e stars in comparison to B2e-B5e stars. In
my opinion, one possible interpretation of this effect is that in
fact many B6e-B9e stars are B2e-B5e stars <u>seen roughly equator-on</u>
and that the corresponding parts of their envelopes simulate a later
spectral type. (In the section devoted to binary models, I present
further evidence supporting this view.) The whole problem may be
further complicated by Collin's (1974) finding that rapid rotation
strongly affects strengths of He I 447.2 and Mg II 448.1 lines in
the sense that a rapidly rotating star appears to have a spectral
type 2 or 3 subclasses too late in terms of that appropriate for its
mass, radius and luminosity. According to him, high rotation of B
stars systematically increases the number of B7-B9 stars at the
expense of B3-B5 objects (i.e. in the region, where the He I/Mg II
ratio is used as the main classification criterium). Due to above-
-mentioned facts, our knowledge of the distribution of true rotat-

ional velocities (which several authors derived using either the method of Chandrasekhar and Münch 1950 or a different approach of Bernacca 1970) need not be reliable.

A great disadvantage of the original RM was that it had offered no clear explanation of the variations observed. It could, in principle, explain light and spectral variations with typical cycles

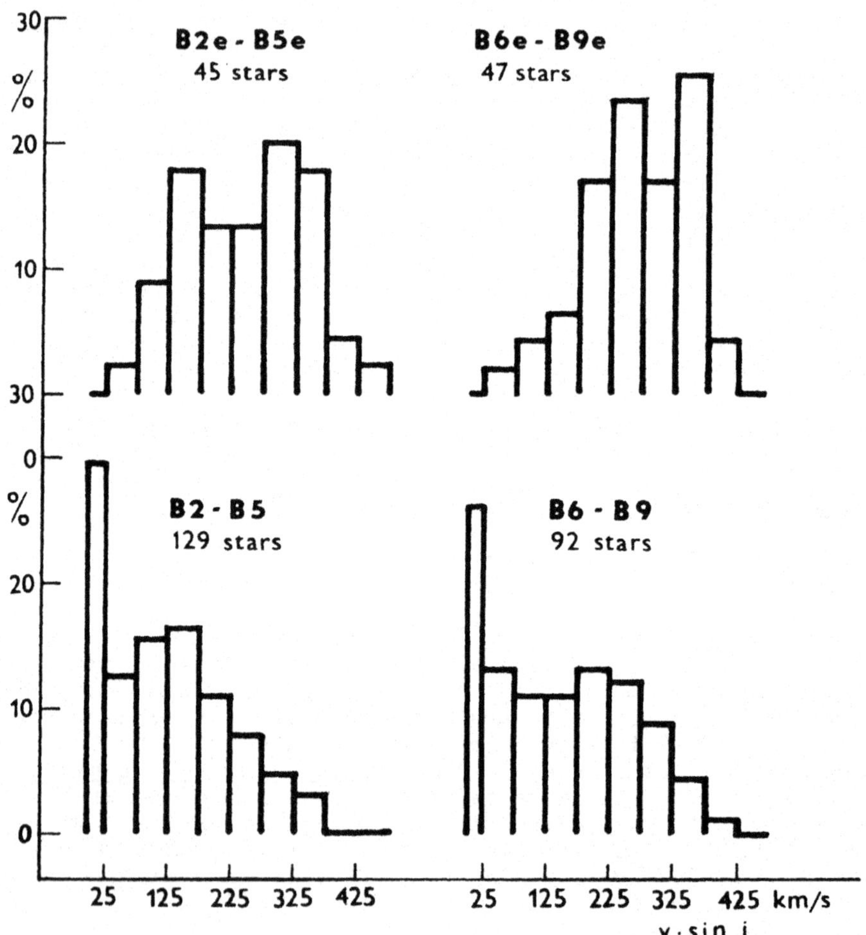

Fig. 1. Distribution of v.sin i for bright Be and B stars (the data for non-emission stars were adopted from Hardorp and Strittmatter (1970). The observed velocities are subdivided into intervals of 50 km/s, with the exception of the lowest interval which is only 25 km/s. The number of stars in this last group has been doubled in order to represent each star by the same area in the histogram.

around 1 day. However, to explain long-term E/C and RV and V/R
variations of emission lines, some other mechanism had to be sought.
Already Struve (1931) suggested a possibility to explain the cyclic
V/R variations by assumption that the envelope has a form of an
elliptical ring the line of apsides of which rotates slowly in space.
McLaughlin (1961) formulated this model mathematically and showed
that it gives correct amplitudes of RV variations. Huang (1973a)
revived McLaughlin's ideas and computed also emission profiles of
the ring, assuming that the ring is optically thin for line radiation.
Although he showed that also the V/R changes are reproduced well by
the model, the idea of elliptical ring was not generally accepted.
It was not clear how such a narrow ring, arbitrarily tilted with
respect to the observer, could produce the deep shell lines (see
Marlborough 1976). However Kříž (1976, 1979a,b) using more realistic
physical assumptions and assuming a geometrically thick elliptical
envelope, was able to reproduce the observed profiles as well as
their V/R and RV variations quite reasonably. The model thus seems
to be very promising. The long-term E/C variations are much less
understood and it is possible that several different types of these
variations exist. Some ideas how to explain them within the framework
of RM were proposed, for example, by Limber (1970,1976) or by Huang
(1973b, 1976). Nevertheless, no quantitative theory is available to
show whether the suggested mechanisms can work properly.

A new development of RM offers the study by Baade (1981). He
suggests to interpret Be stars as analogs of Beta Cep objects subjected to a rapid axial rotation. Developing a crude theory of non-radial
pulsations of rapidly rotating stars, he succeeded to explain the
1.37-day periodicity of RV and V/R variations of 28 CMa as a non-
-radial pulsation with waves travelling opposite to the sense of
rotation. Baade pointed out that this retrograde mode is very sensitive to small changes in the angular velocity of rotation and may
potentially be responsible for long-term V/R and RV variations of Be
stars and their instability. Baade's idea is certainly quite promising but it needs further quantitative development. For example:
assuming the pulsational period of a non-rotating star to be 0.25^d,
Baade obtained a theoretical period of 1.2^d for 28 CMa, which he
considers a reasonable agreement with the observed value of 1.37^d.
However using 0.19^d (a more typical "average" period of Beta Cep
stars), one obtains 0.46^d - a less satisfactory result. The question
then arises whether such modelling is more than a "fitting of free
parameters". One would like to see more independent theoretical
predictions which could be confronted with observations. For example,
if a long- and short-term variability is simultaneously observed in
one star, can both these variations be identified with specific
pulsational modes for the same value of angular velocity?

3. THE MODEL OF RADIAL OUTFLOW OF MATTER (OM)

Gerasimovič (1934,1935) was the first who introduced the concept
of radial outflow of matter as an alternative how to explain presence

of emission lines in the spectra of Be stars. He argued that in
luminous stars like supergiants and P Cyg stars, the acceleration
due to radiation pressure must outweight gravitation for hydrogen,
resulting thus in the formation of a steady expanding envelope which
he called a "non-static chromosphere". In less luminous rapidly
rotating Be stars this radiative dissipation is facilitated by stellar rotation. Gerasimovič supposed that the gravitation does not
allow the formation of a permanent non-static chromosphere. He
expected that the outflow proceeds until the line optical thickness
of the expanding chromosphere is sufficiently large to stop the outflow. Only after this chromosphere has sufficiently dispersed into
space as a detached expanding shell, a new cycle of events begins.
Supposing that the envelope is expanding with decreasing velocity,
Gerasimovič was able to explain qualitatively central absorptions in
double-peaked emissions. Gerasimovič (1935) outlined also some preliminary considerations how to explain even the long-term V/R changes
as consequences of varying optical thickness of the expanding
envelope.

   It is beyond the scope of this paper to review all the work
dealing with line formation in expanding atmospheres (readers may
refer to Hummer's 1976 review). Here, I only want to mention a modification of OM represented by recent papers by Doazan et al (1980a,
b,c). In some respect, their model represents a revival of old
Gerasimovič's ideas. They suppose that the Be phenomenon is only an
enhancement in the chromosphere-corona complex which should exist
around every star. However instead of assuming, like most other
advocates of OM, that the mass loss from massive stars is due to the
radiatively-driven stellar wind, they accept the concept of Thomas
(1973) and Cannon and Thomas (1977) that the flux of matter is caused
by subatmospheric nonthermal storage modes. Their model of a radially
expanding envelope of Gamma Cas (Doazan et al 1980c) predicts correct
ly the presence of highly ionized lines in the UV spectrum and of
the X-ray emission, which should originate in the high-temperature
regions close to the star. The H I emission should originate in
cooler outer layers of the envelope, where the expansion velocity is
decreasing outwards, as in Gerasimovič's model. The authors mention
however that the quantitative agreement of their model with the
observations in optical region is not satisfactory since the densities in outer layers are too low to give rise to the observed
hydrogen emission.

   Another modification of OM was suggested, with some hesitation,
by Snow et al (1979). They assumed the radiatively driven stellar
wind model and a velocity field with expansion velocity increasing
outwards. In their model, the hydrogen emission originates near the
stellar surface, where the velocity is low; the UV shell lines, like
Fe III, with expansion velocities around 100 km/s are formed above
these layers. Finally, the high-velocity Si IV and N V lines, observed in some cases, originate in the outer layers of the envelope,
where both, the acceleration of the expanding matter, and the degree
of ionization, have increased substantially.

The trouble is that all available theoretical profiles of the
H I lines originating in expanding atmospheres use the theory of
radiatively driven stellar wind (e.g. Kunacz 1980). The resulting
profiles differs substantially from the profiles observed in Be
stars. As far as I know, no modern theoretical hydrogen-line profiles
originating in an expanding atmosphere with the velocity fields
assumed by Gerasimovič or by Doazan et al were published. On the
other hand, the profiles computed on assumption of RM give satisfactory agreement with observed profiles (see, e.g., Poeckert and
Marlborough 1978 or Kříž 1979a,b). From this point of view, the
general validity of OM is open to further investigation. Similarly,
one would like to see a quantitative explanation of different variations observed in Be stars within the framework of OM. As far as
I know, this important problem was not tackled by the proponents of
OM since the time of Gerasimovič's preliminary considerations.

4. THE BINARY MODEL (BM)

In a general form, BM was outlined by us in two papers six years
ago (Kříž and Harmanec 1975, Harmanec and Kříž 1976). These papers
contain also more or less complete references to all earlier work in
this direction. Now, I take this opportunity to re-formulate the
basic ideas of BM. A drawback of our original formulation was that it
was too model-dependent. Many of the ideas have been demonstrated
using the results of the theory of mass exchange in close binaries.
Quite naturally, any prediction made on the assumption that a
particular Be star is a case B mass-exchanging binary could be only
as good (or, as wrong) as the theory of mass exchange itself. The
observational data accumulated during recent years represent serious
warning that this theory is still rather oversimplified and cannot
provide us with reliable quantitative predictions. Indeed, some of
the objections against BM were rather objections against the predictions of the theory of mass exchange. For this reason, I prefer to
use analogy with really observed interacting binaries instead of
comparison with some particular theoretical mode of mass exchange.

What is the essence of BM?
A. The Be envelope is formed by the matter transferred to the
B star from the other component of the binary system. The most
probable situation (let us call it mode 1) is that such an envelope
is formed within the corresponding Roche lobe. To provide space
enough for the envelopes with dimensions several times larger than
the radius of a B star, the orbital periods of such systems must be
relatively long. In one important parameter we thus have a simple
geometrical sequence: short-period interacting systems (periods up to
a few days) will appear as Algol binaries. Because of limited space
around their mass-gaining stars (gainers), these systems exhibit
mainly the absorption lines of gas streams but only moderate or no
H emission lines (see Plavec and Polidan 1976). For longer periods
(up to several hundreds of days or more) we observe a Be star
(Peters 1980 found the H$_\alpha$ emission in almost all Algol systems with

orbital period longer than 6 days). The other end of this sequence
for very long orbital periods may be represented by symbiotic and
VV Cep stars.

No doubt, the period is not the only important parameter. The
masses of both components must also play an important role, and
especially the mass ratio which may control the rate and the type
of mass exchange. The appearance of real systems depends also on
which of both stars dominates in the optical spectrum. Thus, the
above-mentioned simple picture can be almost arbitrarily complicated.
For example: if the mass transfer is strong enough, the emission
may be associated with hot spots, especially in extremely short-
-periodic systems such as dwarf novae. Let us denote this situation
as mode 2. Yet, it seems that most of Be stars with their double
emission lines do not originate in this manner. Another consequence
of a rapid mass transfer can be a substantial mass loss from the
system. In principle, such material can form an outer envelope
around the whole system (let us call it mode 3). Then, no apparent
restriction of the orbital period of such binary can be predicted.
Because the mass loss will be probably connected also with a substantial loss of angular momentum, a rather short orbital period is quite
probable. With some fantasy, one could re-interpret Baade's observations of 28 CMa as such a configuration of two B stars seen nearly
pole-on. Some emission may originate even in clouds corresponding to
stable periodic orbits around the Lagrangian points L_4 and L_5 (see
Wu 1975). In real systems, combinations of several modes are quite
probable - as a further complication.

Concerning the nature of the mass-losing components (losers),
I stress that we never claimed that it must be a late type star, as
it is often quoted in literature. Already in our first paper we
pointed out that the losers may be of very different spectral types
in particular cases.

B. In any of the three modes considered, original Plavec's
(1970) argument that the excess angular momentum brought to the
gainer must increase its rotational velocity, is still valid. As the
deviations of the equipotential surfaces from the spherical shape
are significant only near the critical lobe, it is clear that
especially for modes 1 and 3 there must be a high degree of axial
symmetry, and that the models of single rotating stars with rotationally supported envelopes, are quite appropriate to describe such a
configuration, at least as a good first-order approximation. Further,
because the outer equipotentials enclosing a binary become more and
more spherical with increasing distance and because it seems from
the UV observations (Plavec 1980a,b) that the high-excitation
emission lines are not associated with neither star in the observed
systems, it is not surprising that OM gives good predictions for
these lines. It is thus possible that one day all three (now competing) conceptions will join into one complex model.

C. Besides the formation of the envelope, the other component
may be responsible for the long-term RV and V/R variations observed
for many Be stars. Computations by Kříž and Harmanec (1975) showed
that the envelope around the gainer produced by a variable mass
transfer may have a form of an elliptical ring whose line of apsides
slowly rotates due to the perturbing force of the secondary component.
Thus, BM can explain the origin of the elliptical envelope discussed
already in section 2. The excellent <u>quantitative</u> agreement of Kříž's
and Harmanec's computations with the observed variations of ζ Tau is
probably one of the greatest goals of BM. The long-term RV and V/R
variations are far the most pronounced changes observed in Be stars
and no other theory has offered their quantitative explanation in
such a consistent way. Similar computations were further developed
by Castle (1977), who discussed a whole grid of models, including
those corresponding to what I call mode 3. Castle confirmed the
applicability of the model to Be stars and pointed out that elliptical
disks can produce radial velocity-excitation gradients of either sign.
To be honest, I must say that this model will probably work even if
the elliptical envelope would be formed by another process than by
mass transfer from the other star. Yet, it seems that the role of the
secondary, as the perturbing agent forcing the line of apsides of the
envelope to rotate, is substantial. Even Marlborough et al (1978) had
to suppose an invisible secondary for explaining the observed V/R
changes of γ Cas, though they had not accepted some other aspects of
BM.

   D. From analogy with other types of binaries it seems that the
secondary may be responsible even for the long-term E/C variations of
Be stars. Outbursts of reccurent novae are also cyclic, but not
periodic phenomena, similarly as the less pronounced changes of many
other binaries. Olson (1980a,b), observing U Cep, found that a bright-
ening of the loser always preceded formation of an accretion disk
around the B gainer and appearance of the H emission in the spectrum.
Olson (1981, private comm.) can show that for several Algol binaries
a size increase of the loser preceded a mass-transfer event. If a
similar correlation could be found in some known Be binary, it would
be a strong argument in favour of BM.

   E. If Be stars are really binaries, they should manifest themselves by periodic variations due to the orbital motion. One should
observe RV curves of both components, periodic V/R variations of
double emission lines, additional absorptions from gas streams at
certain orbital phases etc. Some Be stars should appear as eclipsing
binaries and others could still manifest themselves by periodic light
variations due to ellipticity or eclipses by gaseous streams and/or
disks. Though <u>all</u> the above-mentioned kinds of variations were really
found in particular cases, the search for them was so far negative in
others. Some variations on a time scale of weeks or months were usually detected but it was impossible to find a periodicity for them.

   F. The light and spectral variations (often pseudoperiodic)

taking place on a time scale around one day or shorter may have
various causes. Some possibilities to be mentioned are:
 i.Effects of differential rotation coupled with unhomogeneous
distribution of surface brightness.
 ii.Analogy of flickering of dwarf novae in larger geometrical scale.
 iii.A pulsation of the B star (see the considerations by Percy 1979,
1980 and by Baade 1980).
 iv.Effects of binary motion in a short-periodic system, the most
probably mode 3 binary. I suggest that the peculiar eclipsing binary
EM Cep (P=0.806^d) may be an example (see Breinhorst and Karimie 1980
and references herein). The duplicity of EM Cep is still questionable,
its light variations may well be a more regular case of the changes
observed for EW Lac or V923 Aql, but Tremko (1981,private comm.)
believes that he has detected lines of two components in the spectrum.
 v.Effects of binary motion in a long-periodic binary, important only
in some orbital phases. The short-living shell phase of HR 2142,
observed every 80.86 days, may serve as a good example (see Peters
1976). The whole secondary shell phase, with very pronounced spectral
variations, lasts one day only. An accidental observer would probably
conclude that he has observed rapid variability of the envelope of
HR 2142 on a time scale of <u>hours</u>, having no apparent periodicity!

     Two main objections were raised against BM. One is the negative
result of the search of duplicity in particular Be stars. The other
(by Plavec 1976) is: if all Be stars were semidetached binaries, we
should observe more eclipsing binaries among them.

     Let us consider the first problem. In the majority of cases,
when we dot not observe the secondary directly, the detection of
duplicity of a Be star is uneasy and possible only after accumulation
of long series of observations, or impossible in less favourable
cases. For mode 1 binaries, a typical semi-amplitude of the RV curve
of the Be component must be of the order of 10 km/s or less. For
mode 3 binaries, with opaque outer envelope, the situation may be
even worse because the envelope lines reflect only a small fraction
of the orbital motion of underlying stars (Kříž and Harmanec 1975).
Castle (1977) pointed out that also the presence of long-term RV
variations can delay or advance maxima and minima of orbital RV
changes for more than one tenth of the period making thus period
finding very difficult. Rapid (probably irregular) variations of the
envelopes very complicate period finding, too. Olson (1980a,b) observed photometric changes in the disk of U Cep on a time scale as short
as four orbital periods. Strong interference of orbital and non-orbit-
al variations was found also by Kříž et al (1980) in RX Cas or by
Koubský et al (1980) in CX Dra. Gulliver (1977) was not able to detect
duplicity of Pleione. I compiled most of published RV's of Pleione and
averaged them over about 100 days. It resulted in a smooth RV curve,
with a possible period of about 13000 days, which is in phase with
appearance of shell phases. This RV curve perhaps represent some slow
atmospheric motions but it is tempting to speculate whether Pleione
is not a long-periodic binary and whether its shell phases are not

analogies of shell phases of HR 2142, occuring in larger dimensions.
Also o And (P=10000^d?), 59 Cyg or γ Cas may be candidates of such
interpretation. Speckle interferometry may help to answer. So far,
only φ And, η Ori, ν Gem, λ Cyg (a triple system) and o And (a triple
system?) are known speckle-interferometric binaries with a Be
component (McAlister 1981, private comm.).

Tentatively, one can conclude that the absence of clear periodic
variations need not a priori exclude the binary nature of a particular
Be star. Clearly, it is necessary to seek other ways how to confirm
or clearly deny the binary nature of individual Be stars.

Let us proceed to the problem of "missing" Be eclipsing binaries.
Poeckert (1979,1981) and Suzuki (1976,1980) published very important
studies of φ Per. Poeckert, using new high-dispersion spectrograms,
detected He II 468.6 emission, apparently associated with the secondary. Measuring also the RV curve of the primary, defined by broad
absorption wings of H I and He I, and by Fe II emission, he arrived
at component masses $m_1 \sin^3 i = 21.1 \pm 5.6$ M$_\odot$ and $m_2 \sin^3 i = 3.4 \pm 0.8$ M$_\odot$.
Shell lines of H I and He I follow the RV curve of the primary, with
the exception of a "bump" on the descending branch. Weaker "secondary"
shell lines of He I follow the RV curve of the secondary. Apparently,
the secondary is <u>well detached</u> from its Roche lobe, otherwise it
would be much brighter than the primary. Poeckert (1981) suggests a
model consisting of a Be primary enclosed by a nearly circular disk
emitting Fe II and most of H I emission, and of a small hot secondary
with a disk emitting He II emission. φ Per thus seems to consist of
a Be primary and a WR secondary. Poeckert's (1981) data clearly show
an increase in the strength of some shell lines around the conjunction
with the secondary in front, followed in later phases by an increase
in their RV's. There are at least two possible explanations of this
effect. Either, we suppose (like Struve 1941) that we observe a
projection of a gas stream flowing from the secondary to the primary,
or we have to postulate (like Suzuki 1976,1980) that the motion of
the outer parts of the disk around the primary corresponds to stable
periodic orbits of the restricted problem of three bodies. The first
possibility needs a special geometry of the stream because the
maximum velocity of the shell lines is observed <u>after</u>, not before the
conjunction as it is usual for Algol binaries. Kříž (1981, private
comm.) computed some trajectories of particles flowing from the
rotating disk around the secondary and obtained indeed a stream with
needed geometry. Yet, before his continuing computations will convince
me, I see further objections against the gas-stream interpretation.
The observed v.sin i of the primary, 450 km/s, indicates that the
value of sin i can hardly be smaller than, say, 0.75. The velocity
of the stream in the vicinity of the gainer should be of the order of
several hundreds of km/s. Why, then, should we observe only a small
fraction of this velocity, of the order of several km/s, projected
against the disk of the primary? The same applies also to other Be
binaries showing "bumps" on their shell-velocity curves (ζ Tau, 4 Her,
KX And). In all cases, the secondary maximum of the shell velocity

represents only a fraction of projected orbital velocity of the star
in question. I thus prefer Suzuki's interpretation of the effect.
Suzuki applied his model of stable orbits to Hynek's (1940,1944)
measurements of φ Per. Assuming sin i = 1, he arrived at the values
of m_1=20 $M_\odot$ and m_2=4 $M_\odot$ for the masses of both stars, in excellent
agreement with the above-mentioned result of Poeckert (1981)!
Suzuki's model does not explain the strengthening of the shell lines
around the conjunction. Stable periodic orbits are symmetric with
respect to the line joining the stars and one would expect that a
similar strengthening should occur also a half period later, which is
not the case. Tentatively, I suggest that the observed strengthening
is caused by absorption in an outer part of the disk around the
secondary, projected against the disk of the primary in phases near
conjunction. Such geometry needs the orbital inclination to be close
to, but differ from 90°, which is quite probable. If my explanation is
a correct one, Suzuki's analysis remains sound, because the velocity
of additional absorption should be close to the systemic velocity,
in agreement with the velocity of the "normal" shell lines.

The new development in our understanding of φ Per is important
also in that sense that φ Per is a prototype of a Be binary in which
mass transfer between components - if present at all - is driven by
some other mechanism than the Roche-lobe overflow, and in which the
secondary is well detached and quite small. If a non-negligible
fraction of Be stars are binaries similar to φ Per, then the problem
of "missing" eclipsing binaries among Be stars does not longer exist.

In conclusion, I shall discuss some promises of a statistical
approach to the problem: Recently Dr. Kříž called my attention to the
results of Burki and Maeder (1977) who found that the percentage of
Be stars and binaries among B0-B4 stars both exhibit the same
dependence on galactic longitude, reaching a maximum between 30° and
90°. Such a finding is certainly interesting from the viewpoint of BM.

Kogure (1981) presented a statistical study of the relations
between Be/Ae binaries and their non-emission counterparts. He
restricted himself to stars of luminosity classes III to V and studied
period distributions for B and A stars separately. He found a double-
-peaked distribution for Be binaries, in contrast to a single-peaked
distribution for all B binaries, and he concluded that the Be binaries
may be divided into two groups: short-periodic binaries (P < 30 days),
which are essentially Algol-type interacting binaries, and long-
-periodic binaries, which have statistically higher v.sin i and in
which the Be phenomenon is due to a high rotation, as in single Be
stars. I do not agree with this interpretation. The "short-periodic"
group consists almost exclusively from eclipsing binaries and reflects
thus probably selection effects due to decreasing probability of
discovery of eclipses for long-periodic systems and due to exclusion
of "high-luminosity" systems. Kogure himself remarks that after
inclusion of the systems classified I, II or c, the short- and long-
periodic groups become almost unseparable. But I realized another

interesting outcome of Kogure's statistics: with the exception of
17 Lep, all the Ae binaries are _eclipsing_ binaries! Together with
the histograms mentioned when discussing RM, it _may_ indicate that
all these systems in fact contain a B star which excites the H I
emission , and the spectrum of which is masked by the accretion disk,
seen equator-on and simulating a later spectral type. This view is
even strengthened by Plavec's (1980a,b) results. Plavec showed that
the UV spectra of the emission-line binaries SX Cas and W Ser contain
hot continua corresponding to B stars. These continua are eclipsed
in primary minima and are therefore associated with the hotter
components of the systems, which however are classified A6 IIIe and
F5 II-IIIe, respectively, from the optical spectra! Plavec interprets
these continua as originating from accretion of matter in the inner
parts of narrow equatoreal disks in these systems subjected to a
heavy mass transfer. (In systems with low rates of transfer, he
found no contradiction between optical and UV spectral types.) I
suggest an alternative explanation - to suppose that the main role of
the heavy mass transfer is to produce a flattened, optically thick
envelope, which - seen roughly equator-on - almost screens the star.
In my interpretation both, SX Cas and W Ser do contain B stars whose
radiation from less screened polar regions is detected in the UV
spectrum. The cooler continua, seen in the optical spectra, originate
in "photospheres" of the accretion envelopes. As a speculation, I
suggest an effect not considered so far. The reservoir of the thermal
energy contained in gas flow from loser must be smaller when the
matter is flowing from a cool star then when it comes from a hot
object. May it not be, at least partly, a consequence of this effect
that we observe cooler optical continua in SX Cas and W Ser, but not
in β Lyr (in spite of its even higher rate of mass transfer) in which
the loser is a B8 star?

      I wish to thank Dr. S. Kříž for many stimulating discussions
and for a critical reading of the manuscript.

REFERENCES

Stellar Rotation: IAU Colloq.4 (ed.A.Slettebak/Reidel)
Be and Shell Stars: IAU Symp.70 (ed.A.Slettebak/Reidel)
Close Binary Stars: Observations and Interpretation: IAU Symp.88
      (ed.M.J.Plavec, D.M.Popper, and R.K.Ulrich/Reidel)

Baade, D.: 1981, Astron.Astrophys. (in press)
Bernacca, P.L.: 1970, Stellar Rotation, 227
Bidelman, W.P.: 1976, Be and Shell Stars, 453
Breinhorst, R.A., Karimie, M.T.: 1980, Publ.Astr.Soc.Pacific 92,432
Burki, G., Maeder, A.: 1977, Astron.Astrophys. 57,401
Cannon, C.J., Thomas, R.N.: 1977, Astrophys.J. 211,910
Castle, K.G.: 1977, Publ.Astr.Soc.Pacific 89,862
Chandrasekhar, S., Münch, G.: 1950, Astrophys.J. 111,142
Collins, G.W., II: 1974, Astrophys.J. 191,157

Curtis, R.H.: 1923, Publ.Obs.Univ.Michigan 3,1
Doazan, V.: 1970, Astron.Astrophys. 8,148
Doazan, V., Kuhi, L.V., Thomas, R.N.: 1980a, Astrophys.J. 235,L17
Doazan, V., Selvelli, P., Stalio, R., Thomas, R.N.: 1980b, Proc.
    Sec.Europ.IUE-ESA Conf., Tübingen, p.145
Doazan, V., Stalio, R., Thomas, R.N.: 1980c, NASA IUE Symp.,
    Greenbelt, May 1980
Gerasimovič, B.P.: 1934, Monthly Not.Roy.Astr.Soc. 94,737
Gerasimovič, B.P.: 1935, Observatory 58,115
Gulliver, A.F.: 1977, Astrophys.J.Suppl. 35,441
Hardorp, J., Strittmatter, P.A.: 1970, Stellar Rotation, 48
Harmanec, P., Kříž, S.: 1976, Be and Shell Stars, 385
Huang, S.-S.: 1973a, Astrophys.J. 183,541
Huang, S.-S.: 1973b, in "Extended Atmospheres and Circumstellar
    Matter in Spectroscopic Binary Systems" (ed.A.H.Batten/Reidel),22
Huang, S.-S.: 1976, Publ.Astr.Soc.Pacific 88,448
Hummer, D.G.: 1976, Be and Shell Stars, 281
Hynek, J.A.: 1940, Contr.Perkins Obs. No.14
Hynek, J.A.: 1944, Astrophys.J. 100,151
Jaschek, M., Hubert-Delplace, A.-M., Hubert, H., Jaschek, C.: 1980,
    Astron.Astrophys.Suppl. 42,103
Kippenhahn, R., Meyer-Hofmeister, E., Thomas, H.C.: 1970,
    Astron.Astrophys. 5,155
Kogure, T.: 1981, Publ.Astr.Soc.Japan (in press)
Koubský, P., Harmanec, P., Horn, J., Jerzykiewicz, M., Kříž, S.,
    Papoušek, J., Pavlovski, K., Žďárský, F.: 1980,
    Bull.Astr.Inst.Czech. 31,75
Kříž, S.: 1976, Bull.Astr.Inst.Czech. 27,321
Kříž, S.: 1979a, Bull.Astr.Inst.Czech. 30,83
Kříž, S.: 1979b, Bull.Astr.Inst.Czech. 30,95
Kříž, S., Harmanec, P.: 1975, Bull.Astr.Inst.Czech. 26,65
Kříž, S., Arsenijevič, J., Grygar, J., Harmanec, P., Horn, J.,
    Koubský, P., Pavlovski, K., Zverko, J., Žďárský, F.: 1980,
    Bull.Astr.Inst.Czech. 31,284
Kunacz, P.B.: 1980, Astrophys.J. 237,819
Lesh, J.R.: 1968, Astrophys.J.Suppl. 17,371
Limber, D.N.: 1970, Stellar Rotation, 274
Limber, D.N.: 1976, Be and Shell Stars, 371
Marlborough, J.M.: 1976, Be and Shell Stars, 335
Marlborough, J.M., Snow, T.P., Slettebak, A.: 1978, Astrophys.J.224,
    157
McLaughlin, D.B.: 1961, J.Roy.Astr.Soc.Can. 55,13 and 73
Meyer-Hofmeister, E., Thomas, H.-C.: 1971, Max Planck Inst.Prepr.
    Astro 41, July 1971
Olson, E.C.: 1980a, Close Binary Stars, 243
Olson, E.C.: 1980b, Astrophys.J. 237,496
Ostriker, J.P.: 1970, Stellar Rotation, 19
Percy, J.R.: 1979, Inf.Bull.Var.Stars No.1530
Percy, J.R.: 1980, preprint
Peters, G.J.: 1976, Be and Shell Stars, 417
Peters, G.J.: 1980, Close Binary Stars, 287

Plavec, M.: 1970, Stellar Rotation, 133
Plavec, M.: 1976, Be and Shell Stars, 439
Plavec, M.J.: 1980a, Invited Review, NASA-IUE Symp., Greenbelt,May
Plavec, M.J.: 1980b, Close Binary Stars, 251
Plavec, M.J., Polidan, R.S.: 1976, in "Structure and Evolution of
     Close Binary Systems" (ed.P.Eggleton, S.Mitton, and J.Whelan/
     Reidel), IAU Symp.73, p.289
Poeckert, R.: 1979, Astrophys.J. 233,L73
Poeckert, R.: 1981, Publ.Astr.Soc.Pacific (in press)
Poeckert, R., Marlborough, J.M.: 1976, Astrophys.J. 206,182
Poeckert, R., Marlborough, J.M.: 1978, Astrophys.J.Suppl. 38,229
Rufener, F., Maeder, A., Burki, G.: 1978, Astron.Astrophys.Suppl.31,
     179
Sackmann, I.J., Anand, S.P.S.: 1970, Stellar Rotation, 63
Slettebak, A.: 1976, Be and Shell Stars, 123
Slettebak, A.: 1979, Space Sci.Reviews 23,541
Slettebak, A., Reynolds, R.C.: 1978, Astrophys.J.Suppl. 38,205
Snow, T.P., Peters, G.J., Mathieu, R.D.: 1979, Astrophys.J.Suppl.
     39,359
Sterken, C.: 1977, Astron.Astrophys. 57,361
Sterken, C., Wolf, B.: 1978, Astron.Astrophys. 70,641
Strittmatter, P.A., Robertson, J.W., Faulkner, D.J.: 1970,
     Astron.Astrophys. 5,426
Struve, O.: 1931, Astrophys.J. 73,94
Struve, O.: 1941, Popular Astr. 49,129
Struve, O.: 1942, Astrophys.J. 95,134
Suzuki, M.: 1976, Astrophys.Spa.Sci. 39,495
Suzuki, M.: 1980, Publ.Astr.Soc.Japan 32,331
Thomas, R.N.: 1973, Astron.Astrophys. 29,297
Wu, C.-C.: 1975, Astrophys.Spa.Sci. 36,407

DISCUSSION

<u>Slettebak</u>: In the histograms showing the number of Be stars versus
v sin i, the small number of Be stars with low v sin i is precisely
what would be expected if all the stars in that group rotate with
nearly the same (critical) equatorial velocity and have their rotation
axes randomly distributed in space. Therefore, I do not see the need
for your suggestion that Be stars of late type may actually have
early-type stars inside their envelopes.

<u>Harmanec</u>: My argument was the following: While the distributions for
early and late type non-emission B stars are very similar to each
other, there is a notable difference between late- and early-type Be
stars in the sense that late-type Be stars have a peak for high
v sin i while the distribution of early Be stars is much flatter.
What is important is that I used a homogenous sample of objects. I am
not going to say that all late Be stars are early Be stars seen
roughly equator-on. Rather, I guess that the sample of late-type
Be stars can be contaminated by such objects.

<u>Sonneborn</u>: The recent work by Slettebak et al. on the effect of
rotation on spectral classification shows that the shift to later
types can be only as large as 2 subtypes. It is not possible for
a B2 star to appear spectroscopically as a B8 or B9 star because of
rotational effects alone.

<u>Harmanec</u>: Well, I speculate that the equatorial parts of the <u>envelope</u>
must (or better: could) simulate a later spectral type even in continuum.

<u>Selvelli</u>: I would like to anticipate some results about 17 Lep that
will be presented tomorrow. 17 Lep is seen almost pole-on. From IUE
observations we have found the presence of a very thick shell <u>fully</u>
covering the system. From the shell lines we have derive a temperature
around 8000 K in agreement with that derived from the UV continuum.
It seems likely, therefore, that the observed continuum is formed in
the thick shell surrounding the primary.

<u>Peters</u>:1.The list of higher temperature Be stars mimicing stars of
later spectral type is endless. I will cite two examples:
TT Hya is usually classified as an A 2 star but inspection of the
"blue spectrum" reveals the presence of a B7 - B8 star. RZ Oph (P=262^d)
has been classified as an F giant, but whereas the Fe II and similar
lines are sharp, the Mg II 4481 line is broad, indicating that the
rotational velocity at the photosphere is relatively high.
2. We agree that hot spots exist (probably in the region where the
gas stream impacts the disk) because we observe strong (variable) N V
in some objects.
3. In his analysis of the IR Ca II triplet emission in Be stars,
Polidan concluded that a relatively cool (6000 K) dense external disk
was required to explain the existence of the optically thick triplet

lines. Since a higher percentage of the calcium triplet emitters are
confirmed to be binaries, we approve of your suggestion that semidetached systems can be surrounded by a circum system disk.
4. Suzuki's calculations pertained only to the orbital plane. It is
not clear what the situation is at the higher latitudes through which
our lines of sight presumably pass.

Harmanec: Thank you for your comments. To your fourth remark I can
only comment that Dr. Suzuki is probably doing something to generalize
his model because recently I received a preprint of his new study in
which he tries to interpret our observations of 4 Her by his model
assuming an inclination of 50°. I cannot tell you more details,
unfortunately, because the paper is written in Japanese language.

Hubert-Delplace: In the case of ζ Tau, you mentioned that the 7 year
period is well explained by the elliptic ring, but after the 7 year
cycle observed in 1960-1967, we have observed a 4 year, then again a
7 year cycle. How do you explain the change of period, what physical
parameter change in the model that you propose?

Harmanec: In the model we suggested the elliptical envelope is formed
by a short-living mass transfer event lasting only a fraction of the
133-day orbital period. It appears quite natural to me that - in
absence of further mass transfer activity - viscous and/or magnetic
forces in the discs will tend to destroy the original structure of
the elliptical envelope and to force it to rotate with the orbital
period. Thus, the period of the long-term variation should decrease
and after a few cycles disappear completely. It should probably
convert gradually into an envelope of Suzuki's type which maintains
stable configuration with respect to the line joining the binary
components exhibiting thus only variations with orbital period. It
may happen, however, that a new mass transfer event will occur
causing a new long-term variability.

Baade: M. Smith has made a survey of bright narrow-lined B stars of
the northern hemisphere. He found that an extremely high percentage
of these objects shows variable line profiles. This suggests that
some type of pulsational instability is quite common in this region
of the HRD. If we think that there is no observational evidence for
pulsations of Be stars, we shall have to explain such a difference
between B and Be stars. So, regardless which result we expect, we
must study the problem of oscillations of Be stars.

Bolton: Equatorial pulsation in a Cepheid type atmosphere is unlikely
as both, Baade and I, see variations in the He I lines. We should
be careful in talking about pulsation in Be stars until we have a
better understanding of pulsation in ordinary B stars.

Thomas: You questioned our model, I resume:
A) 1. Our model, being empirical, can give large atmospheric densities
if F_M is large. E.g. to get 3×10^{13} at base of atmosphere we need.

F_M lo^{-5}M /y.
2. Our velocity first increases outward, to v(max), fixed observationally; then eventually decreases to reach the $\gtrsim$100 km/s required by
H and Fe II <u>displacements</u>.
3. Agreed - we produce Fe II and Balmer lines <u>outside</u> the superionized
lines; I do not see, with the <u>observed</u> data, you can do otherwise.
B) I would be cautious of eclipsing binary location at atmopsheric
features. I think only the Sun, V444 Cyg, and ζ Aur (+ 31 Cyg etc.)
have been sufficiently analyzed to date to be sure of what occurs
where.
4. I do know of computations from <u>any</u> model which gives the observed
range of e.g. 59 Cyg Hα profiles.

<u>Harmanec</u>: Thank you for your explanatory comments. But could you
comment on my argument concerning SX Cas in particular? The emission-
line component of this star is being totally eclipsed by the star
every 36 days. During this eclipse, H emission exhibit strong variations, while the UV emission lines of N V, C IV etc. remain completely
unchanged (see Plavec 1980 a,b). For me it speaks in favour of the
hypothesis that the H emission originates closer to the star than the
UV lines, in this particular star, of course.

<u>Thomas</u>: I have never seen the data on SX Cas; and never do I draw
conclusions before looking <u>myself</u> at the data.

<u>Giovanelli</u>: About RX Cas, would you like to point out the reason why
it is possible to look at the orbital period in one colour and not
in the other ones.

<u>Harmanec</u>: You clearly see that while the V curve reproduces quite well
from one cycle to another, the light level in the U varies, but the
time scale in these variations is longer than one orbital period
because the curve in each particular cycle <u>is</u> defined. The U curve
seems to reflect effects of gas streaming, detectable also spectroscopically. In contrary, in KX And we observe stable U curve and
larger scatter in the V curve.

<u>de Loore</u>: Do you have ideas about the progenitor system for Be-binaries,
I mean mass ranges, mass ratios leading to such systems?
A second point is, how will these systems evolve later on?

Harmanec: Six years ago I had. Now I am much less sure. Essentially,
Be binaries should originate from detached MS binaries with periods
of a few days. If such a system survives the first, probably the most
rapid phase of mass exchange without losing too much mass and angular
momentum, its period must increase substantially, to the range of
period giving sufficient space for the Be envelope to be formed within
the Roche lobe. As the mass ratio is usually more than reversed by this
process, you may obtain a B star even from a binary composed originally
from two A stars, say. This effect is able to explain the observed high
percentage of Be stars among B stars (see Kriz and Harmanec, 1975).

After the Helium burning or the electron degeneracy stops, the mass
transfer from the loser will probably shrink rapidly. Consequently,
its rotation must increase substantially and you should obtain something like a WR object, hot small Helium star with a rapidly rotating
envelope. As already mentioned here by Dr. Poeckert, ϕ Per may be
just in this evolutionary phase. However, to predict original masses
may be uneasy because we still do not know how much of mass and angular
momentum can be lost from the system. That at least some fraction is
lost seems to be well established now - see the evidence of an outer
envelope around β Lyr and other interacting binaries or the UV observation of the mass loss.

THE EVOLUTION OF RAPIDLY ROTATING B/BE STARS

A. S. Endal
Dept. of Physics and Astronomy, Louisiana State University,
   U.S.A.

ABSTRACT

   Rotation can significantly change the moment-of-inertia of a main
sequence star. As a result, the ZAMS rotation rate need only be within
~30% of the critical value in order to reach critical rotation during
the hydrogen burning stage. Calculations of the evolution of rotating
stars show that the Be stars result from a normal (Maxwellian) distribution of B-star rotation velocities.

1. INTRODUCTION

   The classical model for a Be star involves a single star rotating
so rapidly that material is ejected from its surface, forming a disk in
the equatorial plane. The high incidence of Be stars (~10% of all B
stars) requires that the critical velocity be reached during the long-
lived main sequence stage. The locations of the Be stars relative to
the ZAMS (Bond, 1973; Schild and Romanishin 1976) suggest that many of
these stars arrive on the ZAMS with subcritical velocities and reach
critical rotation as they evolve away from the ZAMS.

2. EFFECTS OF ROTATION ON MAIN SEQUENCE STARS

   Sackmann and Anand (1970) computed the evolution of a rigidly-
rotating $10M_\odot$ star and found that a moderate initial rotation (~½ of
critical) is sufficient to reach critical rotation during the main
sequence stage. This differs from the results of Hardorp and Strittmatter (1970), based on the moments of inertia of nonrotating models.
They concluded that the initial rotation rate must be very close to
critical in order to produce critical rotation from core contraction
during the main sequence stage. The difference between these results
can be understood in terms of the effect of rigid-body rotation on
stellar structure.

299

M. Jaschek and H.-G. Groth (eds.), Be Stars, 299–302.
Copyright © 1982 by the IAU.

If a star rotates rigidly, the centripetal acceleration is greatest
at the surface. As a result, the fraction of the gravitational force
which must be supported by the pressure gradient increases inward, i.e.,
the pressure gradient is more centrally concentrated in the rotating
star than in the nonrotating case. This can be achieved by changing
either the density stratification or the temperature stratification (or
both), but the temperature structure is largely fixed by the requirement
of thermal equilibrium. (Rigid-body rotation does not significantly
affect the core luminosity.) Thus, rotation tends to increase the density
gradient and, thereby, decrease the moment of inertia. If, during the
main sequence stage, the rotation increases due to core contraction, the
above effect tends to amplify the contraction. This amplification effect
would not be present in nonrotating models.

Figure 1 shows the main sequence evolution of four sequences of
rotating models (M = 5M$_\odot$) computed by Endal and Sofia (1979). The ZAMS
surface rotation velocities of sequences A to D are: 178, 239, 324, and
418 km/s, respectively. By comparison, the ZAMS critical velocity at
this mass is 580 km/s. The quantity λ is the ratio of centripetal
acceleration to gravity, at the surface and in the equatorial plane.

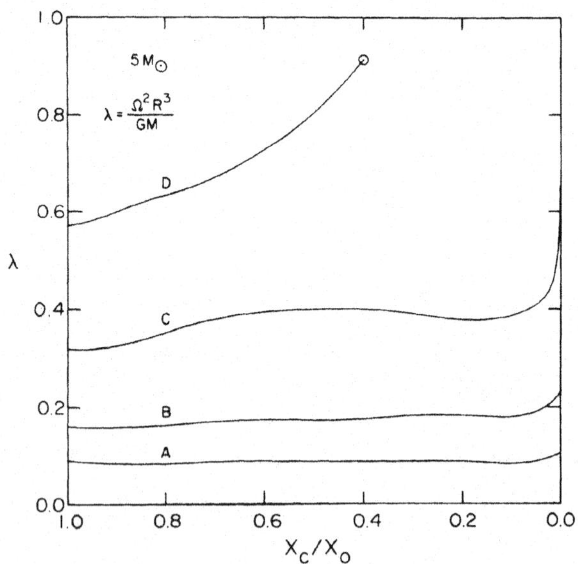

Fig. 1 - The evolution of the rotation parameter λ as a function of core
        hydrogen abundance.

A reasonable analytic fit to the increase (with age) of the
rotational parameter λ is given by

$$\lambda(\lambda_o, X_c/X_o) = \lambda_o[1+(2.57\lambda_o-0.31)(1-X_c/X_o)],\qquad(1)$$

where subscript o refers to ZAMS values. The minimum λ_o required to reach critical rotation ($\lambda=1$) prior to hydrogen exhaustion is $\lambda_o = 0.50$, which corresponds to 70% of the critical rotation rate. This is higher than the value of Sackmann and Anand (1970) because a $5M_\odot$ star contracts less in response to hydrogen depletion than a $10M_\odot$ star.

3. COMPARISON TO OBSERVATIONS

Equation (1) was combined with an analytic fit to the time dependence of X_c/X_o and the result convolved with a Maxwellian distribution of ZAMS velocities (characterized by the mean velocity $\langle v \rangle$). This gives the fraction of B stars which will be Be stars for a random distribution of ages. The results are given in Table 1. The percentages have been multiplied by 0.75 to account for close binaries ($\sim$25%), where synchronization will prevent critical velocities; $\langle v \rangle$ has been corrected for close binaries and evolutionary effects.

Table 1. Predicted frequency of Be stars

$\langle v \rangle$	% Be	$\langle v \rangle$	% Be	$\langle v \rangle$	% Be
169	2	204	7	231	13
185	4	218	10	243	16

According to Bernacca and Perinotto (1974), the mean rotation rate at $5M_\odot$ is 215 km/s, so our models predict that 10% of these stars should be Be stars. Massa (1975) finds that $\sim$20% of the B stars brighter than $m_V = 5.5$ have shown emission at some time. However, the Be stars are, on the average, ~ 0.5 magnitude brighter than normal B stars (Slettebak, 1979) and this implies a Malmquist correction factor of 2, due to the larger volume sampled for the brighter objects. Thus, the correct number of Be stars is closer to 10%, in agreement with the prediction.

This research was supported by the National Science Foundation grant AST 79-19688.

REFERENCES

Bernacca, P. L., and Perinotto, M.: 1974, Astron. Astrophys., 33,
     pp. 443-450.
Bond, H. E.: 1973, Pub. Astron. Soc. Pac., 85, pp. 405-407.
Endal, A. S., and Sofia, S.: 1979, Astrophys. J., 232, pp. 279-290.
Hardorp, J., and Strittmatter, P. A.: 1970, in "Stellar Rotation"
     ed. A. Slettebak (Dordrecht: Reidel) pp. 48-59.

Massa, D.: 1975, Pub. Astron. Soc. Pac., 87, pp. 777-784.
Sackmann, I.-J., and Anand, S.P.S.: 1970, Astrophys. J., 162, pp. 105-124.
Schild, R., and Romanishin, W.: 1976, Astrophys. J., 204, pp. 493-501.
Slettebak, A.: 1979, Space Sci. Rev., 23, pp. 541-580.

DISCUSSION

Thomas: Suppose you use your imagination and lift the quasistatic
corrections: leaving free the possibility of $F_M \neq 0$ at the outer boundary of your interior: Do you think that this nonstatic interior could
possibly give F_M up to 10^{-4}, independent of any atmospheric modeling?

Endal: The stellar structure equations are highly nonlinear, especially
for rotating stars where you have shear instabilities, etc. contributing additional nonlinear effects. For this reason, it is difficult to
predict (without detailed calculations) how the interior will react
to a change in boundary conditions. My guess is that the effect would
be quite small, but this is purely a guess and should be checked by
real calculations. My experience is that the response of a stellar
interior model to a given change is very difficult to predict _a priori_.

Marlborough: Is the neglect of the effect of circulation currents on
the structure of the outer regions of your models a reasonable
assumption?

Endal: In terms of the global structure of the star, this is certainly
a good assumption. Circulation currents have velocities many orders-
of-magnitude smaller than the rotational velocities. Departures from
hydrostatic equilibrium may, however, have a substantial feedback
effect on the circulation itself.

Andrillat: What catalogue do you use for the values of v sin i?

Endal: I used the results of Bernaca and Perinotto (1974), based on
their catalogue.

Peters: The published rotational velocities for eclipsing (Algol)
binaries are systematically low because they are based upon the
widths of lines formed in the circumstellar disk. R. Polidan and I
are analyzing the Mg II line (λ 4481) in Algol systems, which usually
is not formed in the disk, and find considerably larger values of v sin
i.

DETERMINATION OF THE INCLINATION OF ROTATIONAL AXES AND
ROTATIONAL VELOCITY FROM THE LINE PROFILES OF ROTATING STARS

M. RUUSALEPP
Tartu Astrophysical Observatory, Estonia, USSR

For a grid of rotating hot stars the computed line
widths at the half maximum depth for the lines HeI λ 4471
and MgII λ 4481 are given. Based on the correspondent
observational data the inclination of the rotational axes i
and the reduced rotational velocity w have been found for
19 B and Be stars.
Key words: stellar rotation - line profiles.

To make it possible to specify the rotational state of
a star we must start by calculating some line profiles for
a grid of stars, which are differently oriented and rotate
at different angular velocities. As the fit features for
such a task we have chosen the HeI line at λ 4471 $\overset{o}{A}$ and
MgII line at 4481 $\overset{o}{A}$.

Calculated model line profiles for those lines we
have taken from a paper by Stoeckley and Mihalas (1973),
where a grid of intensity profiles for nonrotating stars
is given for six effective temperatures, T_{eff}, from 15000 K
to 27500 K and for ten values of the surface projection
cosines, μ, from 1.0 to 0.1. Taking the limb darkening into
account a grid of rotationally broadened profiles for these
two lines was computed assuming that the shape of the
stellar surface is given by the Roche model, the von Zeipel
theorem for the effective temperature distribution holds and
the star rotates as a solid body.

For a nonrotating star we have taken the surface
gravity, g, to be 10^4 cm/s^2, and the effective temperature
values 27500 K and 22500 K. Model profiles were calculated
for 4 values of the inclination of the rotational axes i
and for 8 values of the reduced angular velocity w, defined

303

M. Jaschek and H.-G. Groth (eds.), Be Stars, 303–310.
Copyright © 1982 by the IAU.

as the ratio of the angular velocity to the critical angular
velocity. The calculated half-widths (the line width at half
maximum depth) are given in Table 1 and Fig. 1.

For both spectral lines used by us each value of the
half-width measured from observations gives us a dependence
of i versus w as a curve and an intersection of these curves
defines the values of i and w. Such intersecting curves are
given in Fig. 2 for nine stars. The observed values for
half-widths have been taken from a paper by A. Slettebak
et al. (1975). The error box for i and w has been singled
out by taking the error of the half-widths to be ± 0.1 Å.
The extreme values for i and w thus obtained are given in
Table 2, where we have compiled the main results of our
study for 19 B and Be stars. An example of finding the
error box for β Lup is given in Fig. 3.

In Fig. 4 the mean values of i and w are given for the
stars under study. These values do not show any clustering
or any peculiar distribution which might indicate systematic
errors in the measurements analysis.

In our study we have followed the paper by Hutchings
et al. (1977), where it was shown that i and w can be determined for rotating stars by finding the ratio of the half-
widths of the UV and the visual spectral lines. Among those
14 stars, unfortunately, none coincided with the stars
studied by us.

If i, w and V_c (the critical velocity at the stellar
equations) are known, we can determine V sin i by V sin i =
V_c w sin i. The value obtained depends on the atopted value
of V_c. The values of V_c used by us were taken from a hand-
book by Allen (1973). The calculated values of V sin i are
given in Table 2, where also the values of V sin i, found
by Boyarchuk and Kopylov (1964), by Uesugi and Fukuda (1970)
and by Slettebak et al. (1975) are presented. The results
of the last two catalogues are compared with our results on
Fig. 5. There is a tendency for our values to run slightly
higher than those of Slettebak et al. In addition, the
results of Stoeckley (1968), namely that for η U Ma holds
i = 50° ÷ 90° and for 48 Per holds i = 30° ÷ 50° overlap
with our results. To use in full the rotationally distorted
spectral lines for finding the values of i, w and V_c, more
spectral line profiles, including those of UV region, are
to be used.

I express sincere thanks to Dr. A. Sapar for useful
suggestions, comments and help in formulation of the
English version of present report.

Table 1: Computed half-widths of MgII λ 4481 and
         HeI λ 4471 lines

w	Lines	i			
		$0°$	$30°$	$60°$	$90°$
		T_{eff} = $27500°$	log g = 4.0		
0.2	HeI	1.40	1.50	1.80	2.00
	MgII	0.45	0.65	1.00	1.25
0.3	HeI	1.40	2.40	3.00	3.20
	MgII	0.45	1.10	1.50	1.75
0.4	HeI	1.40	3.00	3.60	4.00
	MgII	0.45	1.42	2.10	2.50
0.5	HeI	1.40	3.40	4.25	4.80
	MgII	0.45	1.80	2.75	3.35
0.6	HeI	1.40	3.70	4.70	5.40
	MgII	0.45	2.30	3.65	4.25
0.7	HeI	1.30	4.00	5.20	6.00
	MgII	0.50	2.75	4.35	5.00
0.8	HeI	1.30	4.20	5.60	6.40
	MgII	0.50	3.35	5.00	5.60
0.9	HeI	1.30	4.40	5.80	6.60
	MgII	0.50	3.95	5.50	6.10
		T_{eff} " $22500°$	log g = 4.0		
0.2	HeI	1.00	1.20	1.80	2.25
	MgII	0.50	0.60	0.95	1.20
0.3	HeI	1.00	1.60	2.80	3.20
	MgII	0.50	0.90	1.35	1.80
0.4	HeI	1.00	2.20	3.40	3.85
	MgII	0.50	1.30	2.10	2.50
0.5	HeI	1.00	2.80	3.80	4.25
	MgII	0.50	1.85	2.90	3.45
0.6	HeI	1.00	3.20	4.20	4.75
	MgII	0.50	2.45	3.70	4.30
0.7	HeI	1.00	3.40	4.60	5.20
	MgII	0.50	3.30	4.50	5.10
0.8	HeI	1.00	3.60	4.80	5.60
	MgII	0.50	4.25	5.30	5.90
0.9	HeI	1.00	3.90	5.10	5.90
	MgII	0.50	5.10	6.00	6.60

Table 2: i, w and V sin i values for program stars

N°.	Star	HD	Sp.	HeI (Å)	MgII (Å)	V sin i (km/s)[*]				i	w
						S.	B.	U.	R.		
1.	HR 3237	68980	B1.5IIIe	3.37	2.10	115	166	167	90-113	$14^{\circ}-25^{\circ}$	0.58-0.85
2.	β^2 Sco	144218	B2 V	2.62	1.21	50	85	84	81- 96	$21^{\circ}-28^{\circ}$	0.36-0.40
3.	β Lup	132058	B2 V	3.34	3.00	100	130	130	167-200	$27^{\circ}-34^{\circ}$	0.63-0.70
4.	ν Cen	120307	B2 V	2.70	1.75	70	170	94	130-155	$27^{\circ}-40^{\circ}$	0.40-0.50
5.	χ Car	65575	B2 V	1.85	1.31	50	100	98	38- 63	$5^{\circ}-10^{\circ}$	0.60-0.76
6.	α Mus	109668	B2 IV-V	4.14	3.51	110	215	198	191-233	$22^{\circ}-30^{\circ}$	0.80-0.90
7.	ν^2 Cen	121790	B2 IV-V	4.18	3.26	125	150	153	203-231	$25^{\circ}-35^{\circ}$	0.70-0.85
8.	ϕ Cen	121743	B2 IV-V	3.08	2.03	80	180	115	154-183	$40^{\circ}-54^{\circ}$	0.41-0.47
9.	ω Ori	37490	B2 IIIe	4.43	4.14	160	204	195	192-206	$54^{\circ}-71^{\circ}$	0.62-0.68
10.	γ Ori	35468	B2 III	2.27	1.30	50	61	64	64- 78	$23^{\circ}-39^{\circ}$	0.35-0.47
11.	HR 6143	148703	B2 III	3.23	1.55	75	77	83	93:	50°	0.35
12.	ν Ori	41753	B3 V	1.99	1.03	40	40	42	94-104	$30^{\circ}-34^{\circ}$	0.35-0.36
13.	η Hya	74280	B3 V	3.55	2.33	100	130	132	134-146	$20^{\circ}-23^{\circ}$	0.67-0.80
14.	HR 3206	68243	B3 V	3.82	3.72	90	158	160	215-243	$34^{\circ}-42^{\circ}$	0.68-0.72
15.	η UMa	120315	B3 V	4.85	4.00	150	215	216	261-300	$41^{\circ}-60^{\circ}$	0.60-0.81
16.	HR 6087	116087	B3 V	5.75	4.89	190	–	–	366-389	$66^{\circ}-76^{\circ}$	0.71-0.77
17.	HR 4537	102776	B3 Ve	5.93	5.43	205	–	270	412-436	$60^{\circ}-73^{\circ}$	0.82-0.90
18.	48 Per	25940	B3 Vpe	4.71	4.03	250	210	217	254-277	$34^{\circ}-40^{\circ}$	0.76-0.89
19.	λ CMa	45813	B4 V	3.60	2.86	110	–	138	183-215	$36^{\circ}-49^{\circ}$	0.52-0.60

[*]S.- Slettebak et al., B.- Boyarchuk and Kopylov, U.- Uesugi and Fukuda, R.- Ruusalepp

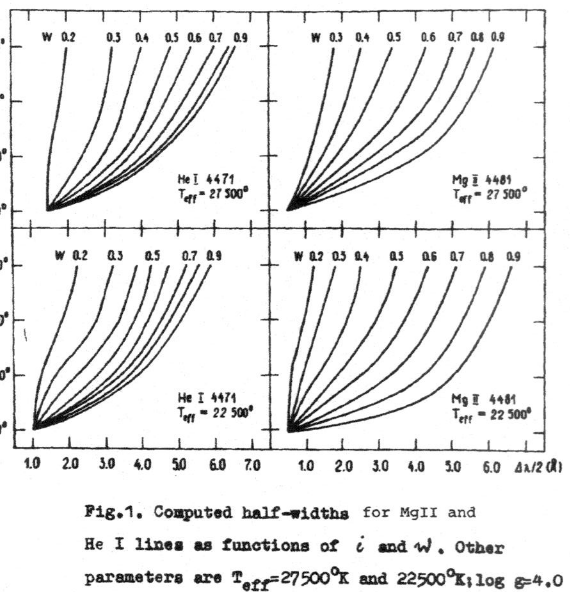

Fig.1. Computed half-widths for MgII and He I lines as functions of i and W. Other parameters are T_{eff}=27500°K and 22500°K; log g=4.0

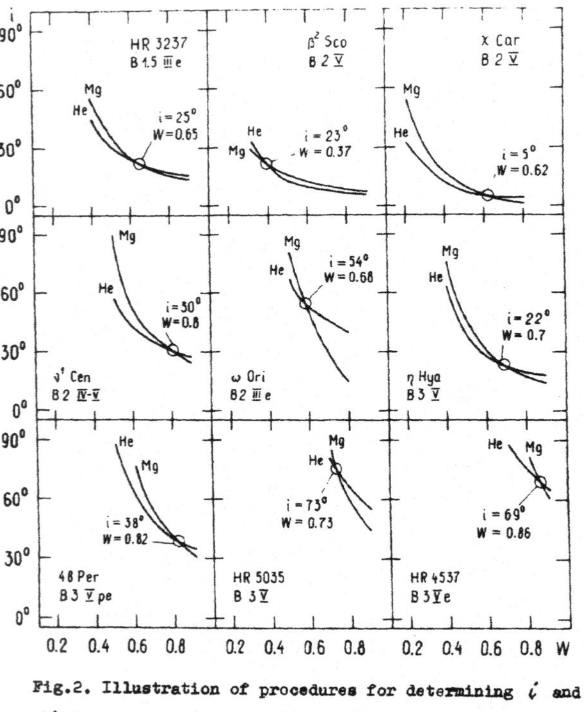

Fig.2. Illustration of procedures for determining i and W from measured line half-widths for a sample of program stars.

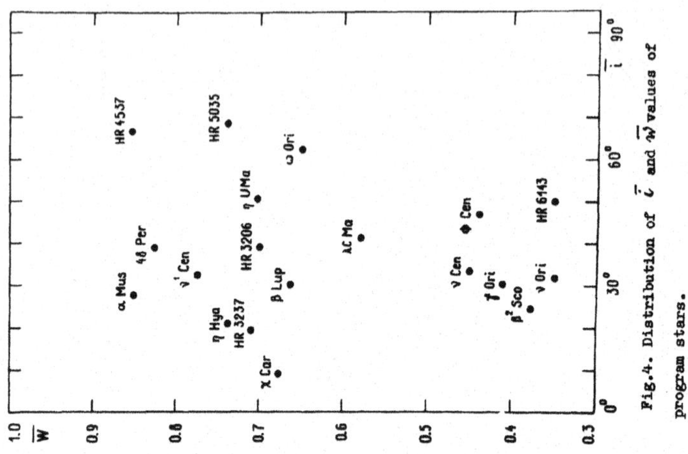

Fig.4. Distribution of $\bar{i}$ and $\bar{W}$ values of program stars.

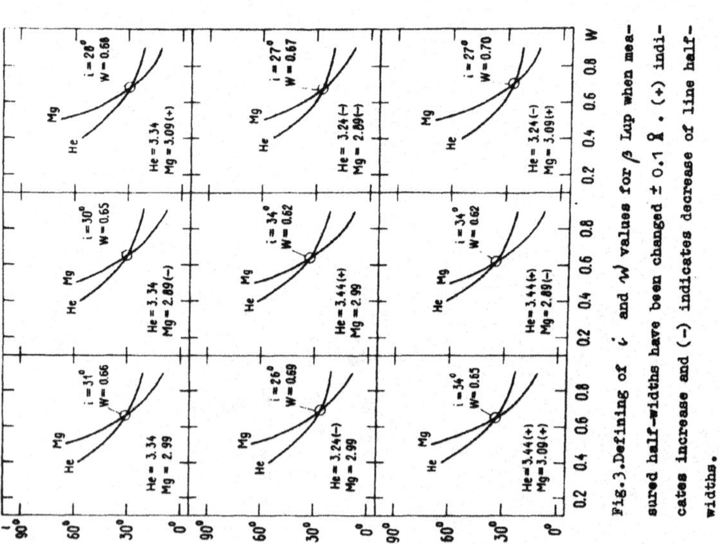

Fig.3. Defining of $\bar{i}$ and $\bar{W}$ values for β Lup when measured half-widths have been changed $\pm$ 0.1 Å. (+) indicates increase and (−) indicates decrease of line half-widths.

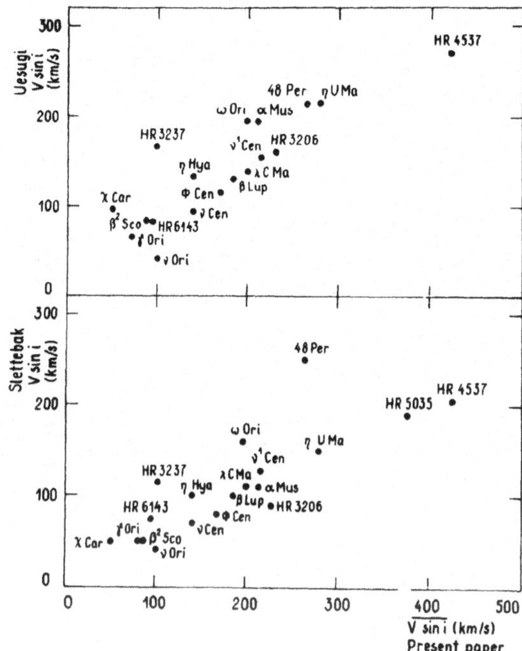

Fig.5. A comparison of rotational velocities
given by Slettebak and by Uesugi with those given
in the present paper.

REFERENCES:

Allen, C.W.: 1973, Astrophysical Quantities

Boyarchuk, A.A. and Kopylov, I.M.: 1964, Publ. Crimean Ap.
    Obs. 31,44

Hutchings, J.B. and Stoeckley, T.R.: 1977, PASP, 89,19

Slettebak, A., Collins, G.W. II, Boyce, P.B., White, N.M.
    and Parkinson, T.D.: 1975, Ap. J. Suppl. 29,137

Stoeckley, T.R.: 1968, Mon. Not. R. Astr. Soc. 140,121

Stoeckley, T.R. and Mihalas, D.: 1973, NCAR-TN/STR 84

Uesugi, A. and Fukuda, I.: 1970, Contr. Inst. Ap. and
    Kwasan Obs., Univ. of Kyoto, No. 189

DISCUSSION

<u>Sonneborn</u>: What are the errors in your determinations of i and ω?

<u>Ruusalepp</u>: The error is dependent on the star and on the observed half-widths of the spectral lines. The characteristic values of the errors are $\pm$ 10° for i and $\pm$ .1 for ω/ω_c.

<u>Endal</u>: From the viewpoint of someone who models rotating stars, the indeterminacy of the sin i factor in observations is very aggravating. I believe that any investigation into measuring v and sin i separately should be strongly encouraged.

<u>Marlborough</u>: Have you compared your results for i and ω/ω_c with those published by Hutchings and collaborators?

<u>Ruusalepp</u>: Among the sample of stars under examination there were no stars which coincide with stars studied in that paper.

Be STARS AS INTERACTING BINARIES

Geraldine J. Peters
Department of Astronomy
University of Southern California
Los Angeles, CA   90007   USA

At the last IAU Symposium on Be stars held in Bass River, MA in
1975, several of us (including P. Harmanec, M. Plavec, R.S. Polidan,
and myself) suggested that Be stars are mass transfer binaries and represent the higher mass counterparts to the familiar Algol systems.
Although a number of alternatives to the rotational hypothesis have
been presented at this symposium (including "chromospheric" activity
and non-radial pulsation) and some of these may well explain certain
objects, I strongly believe that binary mass transfer is responsible
(either directly or indirectly) for the circumstellar envelopes observed in a large percentage of Be stars.

Why is it reasonable to consider the binary hypothesis? Four
replies come immediately to mind. First, the frequency of spectroscopic binaries among B stars is high, 36% (Abt and Levy 1978) and,
according to Plavec (1970) most of the unevolved, detached spectroscopic binary systems observed today should undergo mass exchange at
some point during their post main sequence evolution. Second, the mass
losing secondary (MLS) is the logical source of some of the matter in
the envelope since, for early B stars, V_{eq}/V_{cr} appears to be less than
0.85 (Slettebak 1979). Third, the MLS is the logical source of the high
angular momentum which Be stars, as a class, seem to display. This high
angular momentum undoubtedly aids in the establishment and/or support
of the envelope. Finally, some Be stars which usually do not look
unusual in the visible portion of the spectrum are confirmed spectroscopic binaries (e.g. HR 2142, HR 7084). Alternatively, some Algol
systems are spectroscopically similar to Be stars (e.g. TT Hya, AU Mon).

The primary reason that many researchers have been reluctant to
accept the binary hypothesis is that few Be stars have been confirmed
to be spectroscopic binaries. But several factors conspire against us
to make it quite difficult to detect binary motion. These include the
fact that we expect the semi-amplitude for the radial velocity variation
to be small and, more often than not, the spectral lines are broad complex features containing both absorption and weak emission components
which render the measurements more uncertain than those for sharper

311

M. Jaschek and H.-G. Groth (eds.), Be Stars, 311–314.
Copyright © 1982 by the IAU.

symmetrical lines. Uncertainties in the radial velocities from such
complex features are usually about 10 - 15 km s^{-1}. Therefore, if the
semi-amplitude, K, for the radial velocity curve is less than the
latter value, it is indeed a challenge to find a solution for the orbit

Now, we expect the mass ratio, q, for an interacting binary Be
star to be less than 0.3 and the systemic mass to be in the range
5 - 15 M$_\odot$. The maximum expected value of K (assuming that the system
is viewed equator-on) as a function of orbital period, computed for
q = 0.1 and 0.3 and ΣM = 5 and 15M$_\odot$, is shown in Figure 1. As one can
readily see from the curves, for all but the high mass systems containing more equal components, only the binaries of relatively short
period (<50^d) can be easily detected. Since a q of 0.1 is probably
more representative of the interacting binary Be stars than the larger
values considered, it appears reasonable to assume that there are a
large number of intermediate to long period systems which have not yet
been discovered.

Orbits have recently been determined for two relatively long
period interacting binary Be stars, HR 2142 (Peters 1981) and ϕ Per
(Poeckert 1981). In each case, the semi-amplitude of the velocity
curve is small (9.4 and 16.8 km s^{-1} for HR 2142 and ϕ Per, respectively). Despite the low values of K, orbits for both stars were

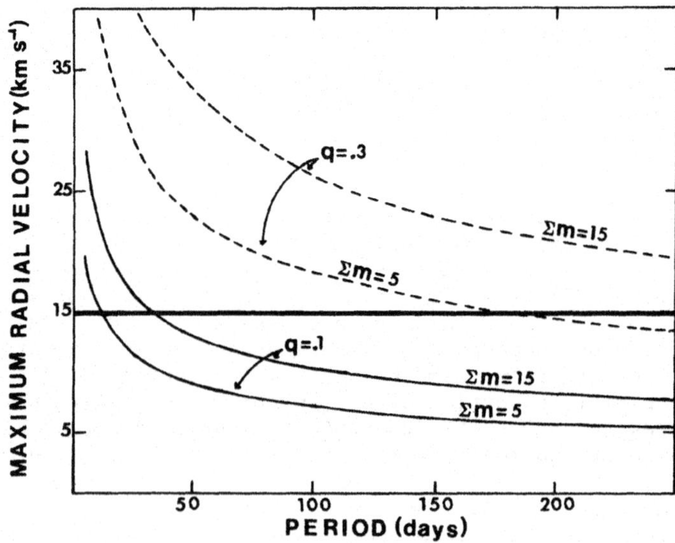

Fig. 1 - The maximum expected semi-amplitude for the radial velocity
curve versus orbital period. Computations for two pertinent mass
ratios and values of the systemic mass are presented. The bold,
horizontal line separates the domain of easy detection (above) from
that in which detection is more difficult.

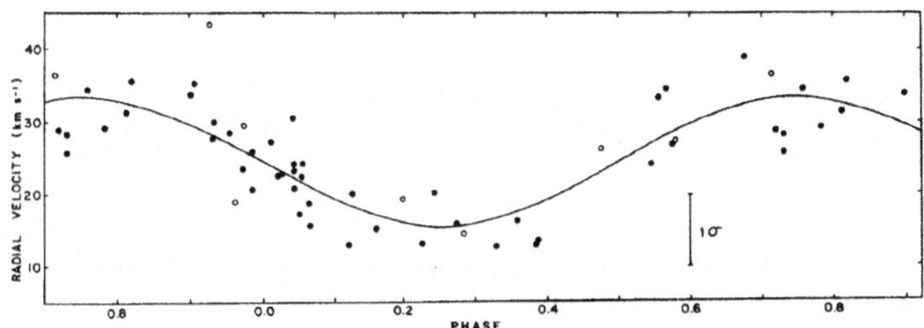

Fig. 2 – Radial velocity curve for HR 2142. Parameters from the orbital solution (indicated by the solid curve) are: $P = 80\overset{d}{.}860$, $T = 2441990.5$, $V_0 = 24.1$ km s^{-1}, $K = 9.4$ km s^{-1}, $f(m) = 0.007$.

obtained with a minimum of difficulty because the periods for these systems were known apriori. In the case of HR 2142, the period had already been determined from the strict periodicity of the short-termed, conspicuous shell phases (Peters 1976). The orbital solution for HR 2142, shown in Figure 2, is discussed in more detail in Peters (1981).

Since it is clearly advantageous to have some idea of the period when one searches for Be binaries, I offer the following suggestion. Polidan (1979) found a remarkably good correlation between the intensity of the infrared Ca II triplet emission (present in about 20% of the Be stars, a high percentage of which are confirmed binaries) and the period of the system. He determined that $I\alpha P^{4/3}$ (log I/log P = 1.20±.20, I in energy units). One could, therefore, use Polidan's results along with a measurement of the IR Ca II triplet emission strength to obtain an initial estimate for the period. This combined with good spectral data and careful measurements should lead to the discovery of many more interacting binary Be stars.

At this symposium much emphasis has been placed on the fact that Be stars lose mass, or at least develop circumstellar envelopes, episodically. Such behavior _is_ compatible with the binary model as changes in the circumstellar disk are a result of variable mass transfer which is often observed in Algol binaries. The discovery of variable N V and C IV in Be stars has been offered as evidence that the Be phenomenon is a "chromospheric" one. However, it should be mentioned that N V and C IV have been observed in the interacting binary Be stars AU Mon (B5V) and HR 7084 (B3V) and their presence is most assuredly associated with the gas streams in these systems (Peters and Polidan 1982).

Ultimately, the success of the binary model for Be stars will depend upon the number of spectroscopic binaries that are found and confirmed. It is clear what the objective of our future endeavors should be.

REFERENCES

Abt, H.A., Levy, S.G.: 1978, Astrophys.J.Suppl 36,pp.241-258.
Peters, G.J.: 1976, in A. Slettebak (ed.), 'Be and Shell Stars',
     IAU Symp. 70, pp.417-428.
Peters, G.J.: 1981, in preparation.
Peters, G.J., and Polidan, R.S.: 1982, this volume, pp.405-409.
Plavec, M.: 1970, Publ. Astron.Soc.Pacific 82,pp.957-995
Poeckert, R.: 1981, Publ.Astron.Soc.Pacific 93, in press.
Polidan, R.S.: 1979, Ph.D. dissertation, UCLA.
Slettebak, A.: 1979, Space Science Reviews 23, pp.541-580.

Koubsky: Has the Calcium emission been detected in systems of small
period?

Peters: The systems of short period do not appear to display IR CaII
triplet emission.

Giovannelli: What is the confidence level of the fit with the sinus
of the HR2142 data you have shown?

Peters: The uncertainties in K, v_0 and T are .9, .6, and 1.1,
respectively.

Endal: I would like to make 3 comments about your statement that
observed rotation velocities are inevitably less than 85% of the
critical value.
1. For rapid rotators, the observed velocities may contain substantial errors.
2. The critical velocity is usually based on nonrotating, unevolved
models. Rotation and evolution can substantially reduce the critical
velocity.
3. It is not at all clear that $v/v_{crit}=1$ is necessary for instability.

Peters: Investigations published to date suggest a minimum of .85 for
v/v_{crit}. I'm sure that the value of this ratio will be debated for
years to come.

RADIAL-VELOCITY AND PHOTOMETRIC VARIATIONS OF o And:
CRITICAL EVALUATION OF POSSIBLE PERIODS

J. Horn[1], P. Koubský[1], J. Arsenijević[3], J. Grygar[2],
P. Harmanec[1], J. Krpata[1], S. Kříž[1], and K. Pavlovski[4]
1 Ondřejov Observatory, Czechoslovakia
2 Institute of Physics, Řež, Czechoslovakia
3 Beograd Observatory, Yugoslavia
4 Hvar Observatory, Yugoslavia

The bright Be star o And has been observed for over 90 years.
Throughout this time numerous observers detected highly variable
spectrum, photometric changes and substantial range of radial velocity of this star. Much effort has been devoted for period searching
in these data. In the present study we collected all available (to
our knowledge) radial velocities and photometric measurements in the
visible region and tried to evaluate the reported periodicities.

We have found 418 radial velocities of o And in the literature
and have measured RV for additional 80 plates (see Table 1).

Table 1: Sources of radial velocities for o And

File	Number of RV		Dates of obs.	Reference
B	70		1906 - 1912	Beardsley (1969)
W	50		1906 - 1907	Harper (1915)
P	105		1936 - 1938	Tremblot (1938)
IA	27		1961 - 1967	Galeotti, Pasinetti (1968)
K	28		1965	this paper, Crimea
IB	69		1967 - 1975	Fracassini et al. (1977)
O	52	H shell	1975 - 1976	this paper, Ondřejov
G	64	lines	1975 - 1977	Gulliver, Bolton (1978)
Misc.	33		1900 - 1980	

The histograms of some files are presented in Fig. 1. The scatter
is quite important. In the phases without shell the scatter must be
caused mainly by measuring errors (especially when oscilloscopic de-
vice is not used). On the other hand, good measurements of sharp cores are much more reliable but they reveal real variation ± 10 km/s.
The velocities from files IA, IB are affected by even larger scatter
caused probably by the instabilities of the spectrograph. Thus, only
files B, W, P, K, O and G have been used for the study of long-term
variations. In order to reduce the scatter, the velocities were con-

315

M. Jaschek and H.-G. Groth (eds.), Be Stars, 315–318.

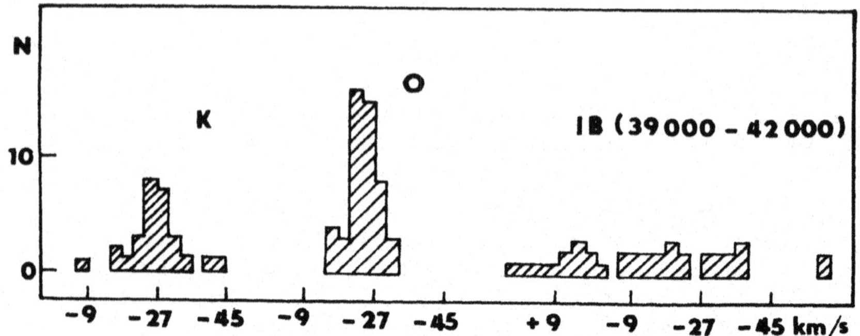

Figure 1. Frequency distribution of the RV of o And (see Table 1).

densed into 29 points, each representing the mean over intervals
50 - 100 days. A search for periods in the range 5 to 140 years was
then made. The least squares so-
lution gave the following element:
$P=25.9$ years, $e=0.41$, $\omega=60^{\circ}$,
$K=9.4$ km/s, $\gamma=-21.6$ km/s. Fig. 2
gives the phase diagram. The result is close to the period reported by Fracassini et al.(1977)
even though they used different
material (files B, W, IA, IB).

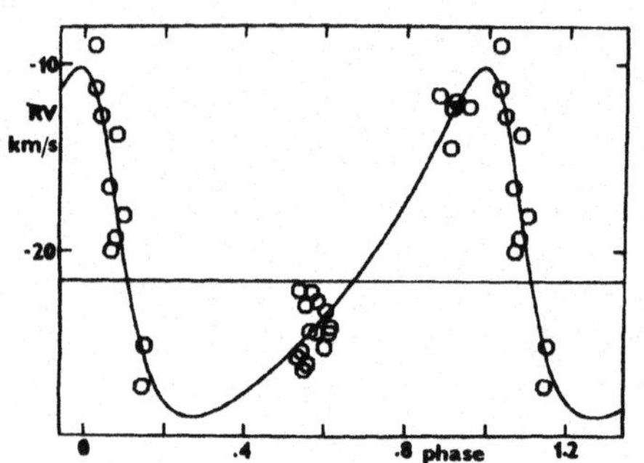

Figure 2. RV curve of o And.

A search for periods in the
range 0.5 to 1000 days was made
for velocities of files O and G.
With the exception of two quasi-
periodicities 3.5 and 1.2 d the
search was unsuccessful.

We have analysed 879 UBV measurements of o And and 201 points
in the instrumental system. Details are given in Table 2. It is clear
that long-term variations are present in the data (see Fig. 3).
There are also short-term changes but their amplitude is not constant
This complicated behaviour of the photometric data causes that any

Table 2: Sources of photometry for o And.

No. of obs.	Type	Dates of obs.	Reference
201	Inst.	1956 - 1958	Schmidt (1959)
3	UBV	1964	Johnson et al. (1966)
253	UBV	1966	Olsen (1972)
270	UBV	1975 - 1979	this paper, Hvar
161	UBV	1975 - 1979	Bossi et al. (1980)
192	UBV	1976 - 1977	Padalia (1979)

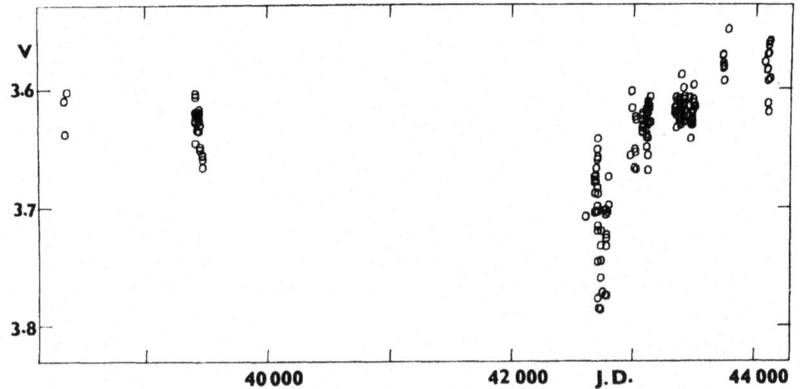

Figure 3. V magnitude versus time for o And.

attempt to find simple period must fail. When the data were segment-
ed it was possible to disclose periods in several segments but we
were unable to find any period common to all segments.

The only positive result of the analysis of radial velocities
is the 25.9 year period. But the coverage is not good and one should
look for reliable RV in the time of expected minima. All ,shorter
periods found in the data are not convincing and it is necessary to
analyse data from at least two shell events. It was not possible to
detect the period of about 1.5 days in good RV measurements from the
last shell episode.

No short-term period was found in the UBV data even after remo-
val of long-term variation. However, it seems that there is a correlation between amplitude of short-term changes and the shell activity.

REFERENCES

Bossi,M.,Guerrero,G.,Mantegazza,L.,and Scardia,M.: 1980,Astrophys.
        Space Sci.
Fracassini,M.,Pasinetti,L.E.,and Pastori,L.: 1977,Astrophys. Space
        Sci. 49, 145
Galeotti,P.,and Pasinetti,L.E.: 1968, Rend. Acc. Naz. Lincei 45, 39
Gulliver,A.F., and Bolton,C.T.: 1978, Publ. Astron. Soc. Pacific
        90, 732
Harper,W.E.: 1915, Report to the Chief Astronomer for 1911, Ottawa,
        p. 154
Johnson,H.L.,Mitchell,R.I.,Iriarte,B.,and Wiśniewski,W.Z.: 1966,
        Comm. LPL 4, 99
Olsen,E.H.: 1972, Astron. Astrophys. 20, 167
Padalia,T.D.: 1979, private communication
Schmidt,H.: 1959, Z. Astrophys. 48, 249
Tremblot, F.: 1938, Bull. astron. 11, 401

DISCUSSION

<u>Harmanec</u>: It should be made clear that the only possible real period
in RV we believe is the 25.9 year period. As three independent groups
of observers confirmed this star to be a speckle-interferometric
binary, it is quite reasonable to believe in such a long period.

<u>Fracassini</u>: I wish to point out that the orbital elements you have
found are quite near to those found by us, with our method and
measurements, with a period of 30 years. The mass function value found
by you is near to the particular case with $i \simeq 90^{\circ}$ and $M_1 = M_2$. Therefore, it should be possible (and very interesting to observe some
eclipsing phenomena.

<u>Mantegazza</u>: Looking at the large gap in the R.V. data phased with
the 26 years period, I think that there may be a large number of
periods that can equally well phase the data. (The PDM method tends
to overestimate the long periods when the data distribution contains
large gaps.)
I have also tried to analyze the several sets of photometric data
existing in the literature in the search for short-term periodicities.
My conclusions are the following: It doesn't exist any clear periodicity, or, if it exists, it is hidden by some types of stellar and/or
shell noise. The only thing that can be told is that the variations
have a time scale around one day. I believe that the existing sets of
data are inadequate for an accurate study of such variations. In fact,
if you generate a synthetic pseudo-periodic random time series and if
you sample it with the times of the real observations, from this
analysis you obtain the same inconclusive results as with the real
observational data.

<u>Koubsky</u>: The 26 year period is a result of a period search in the
range 5-140 years, using programs developed by Morbey.
Both, the mean variance and significance were far the best for this
period. On the other hand, the fit for the period derived by Fracassini et al. (24 years) was much worse.

ROTATIONAL VELOCITY VERSUS MASS LOSS IN Be STARS.[*]

V. Doazan, M.L. Franco, R. Stalio, R.N. Thomas
Observatoire de Paris, France; Scuola Internazionale Superiore
di Studi Avanzati (SISSA), Trieste, Italy; Osservatorio Astronomico and SISSA, Trieste, Italy; Institut d'Astrophysique,
Paris, France.

ABSTRACT

C IV and Si IV resonance line profiles of 21 Be, B-shell and normal
stars are studied with the aim of detecting evidences for mass loss. We
found that almost all our sampled stars are loosing mass. A relation
between an estimated lower limit for the rate of mass loss and the observed rotational velocity was searched but not found.

I. INTRODUCTION

A large portion of the literature on Be stars is characterized by
attempts to relate their rotational velocities to a number of observed
phenomena and models. The reason for this lies in the fact that Be stars
have statistically larger v sin i than their "normal" counterparts.
Thus one asks if the origin of the extended atmospheres, where emission
lines are formed, is due to rotation. More specifically one asks if the
origin of the displacements of the superionized lines observed in the
UV in terms of the stellar wind produced by various mechanisms, like
radiation pressure, could become more efficient in a rapidly rotating
star where the effective gravity is lower. One also asks if it is possible to explain the observed variability of lines, continua, polarization, etc in terms of a rotationally oblated shape of the outer atmosphere of the star.
These questions persist despite increasing evidence from UV, X-ray,
high-dispersion visual, IR, and radio data that most of these phenomena
are related to the non-thermal structure of the atmospheres and sub-

[*] Based on observations by the International Ultraviolet Explorer (IUE)
collected at the Villafranca Satellite Tracking Station of the European
Space Agency. This work was partly supported by a CNR-Italy contract,
and a SISSA contribution.

M. Jaschek and H.-G. Groth (eds.), Be Stars, 319–324.
Copyright © 1982 by the IAU.

TABLE 1.

STAR	SP. TYPE	V sini	OBSERV.	IONS
X Per	O9.5e	150	4/4/78	Si IV, C IV
ζ Oph	O9.5Vnn	380	3/8/78	Si IV
τ Sco	BOV	25	Copernicus	Si IV
HR 2678	BO.5III	155	23/1/79	Si IV
γ Cas	BO.5IVe	300	18/10/79	Si IV
κ CMa	B1.5IVne	200	23/1/80	Si IV, C IV
χ Oph	B1.5Ve	120	23/1/80	Si IV
216 Pup	B1.5Vp	275	3/6/78	Si IV, C IV
59 Cyg	B1.5Ve	375	12/12/78	Si IV, C IV
HR 5223	B2IIIe	85*	23/1/80	Si IV
δ Cen	B2IVne	180	23/1/80	Si IV
HR 4009	B2IVpne	360	23/1/80	Si IV, C IV
HR 2142	B2Ve	415	21/8/78	Si IV, C IV
ι Her	B3IV	8	Copernicus	Si IV
ζ Tau	B4IIIp	310	21/8/78	Si IV
τ Ori	B5III	35	6/1/79	Si IV
δ Per	B5III	259	11/9/78	Si IV
48 Lib	B5IIIp	390	23/1/78	Si IV
κ Dra	B6IIIp	250	27/3/79	Si IV
θ CrB	B6Vnn	395	20/5/80	Si IV, C IV
η Tau	B7III	215	4/1/80	Si IV

* Dachs et al. (1981).

atmospheres of stars in all the HR diagram. In a subsequent paper Doazan,
Stalio and Thomas will present a proposed gross structural pattern for
the atmospheres of Be Stars. Here we only want to emphasize in a qualitative way the lack of any observed relation between mass loss and
rotational velocities for a group of Be, B-shell and B normal stars.

II. OBSERVATIONS AND DATA ANALYSIS

IUE high resolution spectra taken in the short wavelength region
of 19 Be, B-shell and normal B stars of spectral type between O9.5 and
B 7 and classes luminosity III to V, and two Copernicus spectra of τ Sco
(BOV) and ι Her (B3IV) have been used in order to detect mass loss from
asymmetric and/or displaced Si IV and C IV resonance absorption lines.
The list of stars, their spectral type taken from Lesh (1968,1972),
v sini values taken from Uesugi (1976) and the day of observation are
presented in Table 1. Note that only one observation per star has been
considered, thus ignoring the fact that most of our stars may have

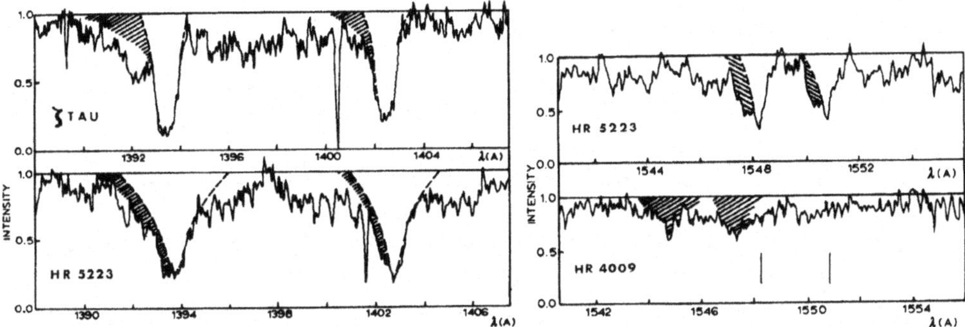

Figure 1.: Si IV and C IV resonance profiles. The shaded areas are the estimated wind contributions.

variable amount of mass loss; for our purposes this is enough in view of the qualitative character of our considerations.

In order to decide whether a star loses mass or not we have estimated lower limits for the mass loss rate using the simple curve of growth approach. The assumptions we made are numerous; some of them act in the sense we require, i.e. of establishing lower limits for the mass loss rates.

a) From asymmetric profiles with violet extended wings we have substracted a symmetric profile created by transferring the long wavelength wing to the violet side. We have then assumed that the result represents the contribution of the stellar wind to the total profile; the position of the minimum absorption gives the velocity regions where there is the maximum concentration of the ions considered. In the case of multiple components (most frequently two), which are found often, we have taken only those at maximum blue displacement, thus neglecting the contribution to the mass loss rate from the low velocity components. In Figure 1 we present as examples the Si IV profiles and the resulting wind components of ζ Tau and HR 5223; the same has been done for the C IV profiles of HR 5223 and HR 4009.

b) The mass loss rates has been estimated by assuming the following:
- all wind components have equivalent widths falling in the linear part of the curve of growth,
- the wind velocity is measured at the minimum absorption of the wind profiles,
- the photospheric radius of all stars has been taken to be 6.3 $R_\odot$,
- solar Si and C abundances have been adopted.

Under these assumptions, the equivalent widths give us column densities and by adopting the equation of mass continuity we derive the following expression for the mass loss rate:

$$\dot{M} = 2.347 \times 10^{-12} \times \frac{W_\lambda\, v_o}{f\lambda^2 \left(\frac{n_{ion}}{n_{el}}\right)\left(\frac{n_{el}}{n_H}\right)} \quad M_\odot\, yr^{-1}$$

with W_λ, the equivalent width, measured in A, v_o in km s^{-1}, λ in A. (n_{el}/n_H) is the element abundance and (n_{ion}/n_{el}) the ionization fraction of Si IV and C IV in the wind.

c) We then have adopted an ionization fraction $(n_{Si\ IV}/n_{Si}) = 1$; this value gives very conservative rates of mass loss. For example the rate of mass loss we obtained from τ Sco is 1.7 10^{-11} $M_\odot$ yr^{-1}. If instead we use the Si IV ionization fraction of 0.01 derived by Lamers and Rogerson (1978) we obtain a rate 100 times langer which is much close to their 5-8 10^{-9} $M_\odot$ yr^{-1}. The ionization fraction of C IV has been scaled to that of Si IV by adopting the same ratio $(n_{CIV}/n_C)\big/(n_{SiIV}/n_{Si}) = 0.25$ as for τ Sco in Lamers and Rogerson analysis. Three stars ζ Tau, 48 Lib and HR 5223 require a much smaller ratio in order to give rates consistent with those derived from Si IV. For these stars we have adopted the rates derived from the Si IV profiles.

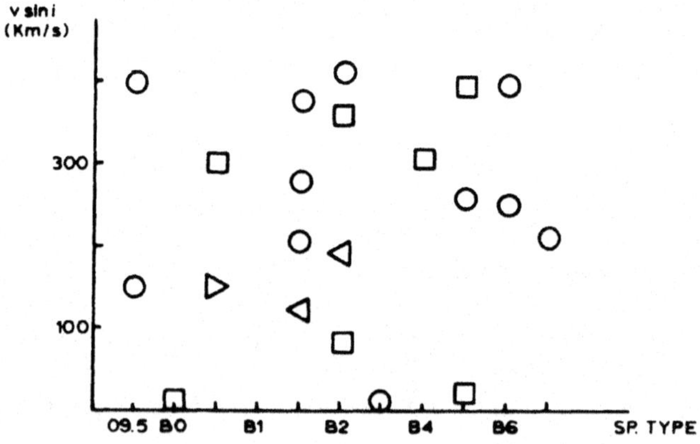

Figure 2. : v sin i versus spectral type diagram. The symbols indicate values of $\dot{M} \leq 10^{-12}$ $M_\odot$ yr^{-1} ($\triangleleft$), $10^{-12} < \dot{M} \leq 10^{-11}$ ($\circ$), $10^{-11} < \dot{M} \leq 10^{-10}$ ($\square$), $10^{-10} < \dot{M}$ ($\triangleright$). $\dot{M}$ has been estimated as described in the text.

III. RESULTS

The results of our study are shown in Figure 2, where in a diagram
v sin i versus spectral type we have plotted all stars using different
symbols for different ranges of mass loss rates.

There appears to be a complete lack of any correlation between
our lower limits of mass loss rates and the rotational velocity. If
instead of lower limits we would be able to give a real estimate of the
rates, the situation would not be expected to change very much: the same
large scattering is present also in the more restricted range of spectral types from B1.5 to B2.

This result seems to be in contradiction with a similar diagram
presented at the previous IAU Symposium on Be Stars by Marlborough and
Snow (1976), in which however they used only Si IV. In that diagram
mass loss seemed to occur from the hotter and more rapidly rotating
stars. The apparent contradiction arises from the fact that: (1) our
sample of objects is larger than theirs, (2) we measure the C IV doublet
in addition to Si IV.

REFERENCES

Dachs, J., Eichendorf, W., Schleicher, H., Schmidt – Kaler, Th., Stift,
    M. and Tug, H.: 1981, Astron. & Astrophys. Suppl. 43, 427.
Lamers, H.J.G.L.M. and Rogerson, J.B. 1978: Astron. & Astrophys. 66,
    417.
Lesh, J.R.: 1968, Astrophys. J. Suppl. 17, 371.
Lesh, J.R.: 1972, Astron. & Astrophys. Suppl. 5, 129.
Marlborough, J.M. and Snow, T.P.: 1976, in I.A.U. Symp. 70, Be and shell
    Stars, A. Slettebak (ed.), p. 179.
Uesugi, A.: 1976, Revised catalogue of Stellar Rotation velocities,
    preliminary edition, University of Kyoto, Japan.

DISCUSSION

<u>Snow</u>: There are at least two very uncertain assumptions in this.
First, you assumed that the equivalent widths lie on the linear part
of the curve of growth, yet in the figure you showed, this is obviously not the case; the two C IV lines in the figure had about equal
equivalent widths, indicating substantial saturation. Second, you
have assumed that the ionization balance in the wind is the same for
all your stars, from O9 to B7. This is almost certainly not correct.
Both of these uncertainties, and others, will by themselves create
a lack of correlation between M and v sin i.

<u>Franco</u>: It is true that for some of the stars in our sample the
equivalent widths are not in the linear part of the curve of growth
and it is certainly true that the ionization balance in the wind is
not the same for all the stars. We have assumed that all the Si is
present in the wind in the form of Si IV; this is a very conservative
estimate: in τ Sco this fraction is 1/1oo; in all stars of our sample
that show C IV profiles formed in the wind (that are the most) the
ionization fraction of Si IV showed not to be different from τ Sco.
This, in our opinion, makes us confident that we are estimating lower
limits of mass loss rates; this is our goal. From lower limits we can
calculate that there are stars with low v sin i values losing mass as
well as stars with high v sin i values.

VI. X-RAY OBSERVATIONS

X-RAY OBSERVATIONS OF Be STARS

Saul RAPPAPORT
Massachusetts Institute of Technology
Edward P.J. VAN DEN HEUVEL
University of Amsterdam

Be star binaries with neutron star companions are shown to constitute
a major class of X-ray sources. Some general observational and inter-
pretive techniques of X-ray astronomy are reviewed. Data for 12 Be/X-ray
binary systems are summarized. The Be/X-ray binaries are found to be
systematically wider systems, with lower-mass primaries, and with significantly more transient behavior than the "standard" massive X-ray
binaries such as Cen X-3 and SMC X-1. The difference between the two
types of X-ray binaries is explained in the context of slightly different evolutionary scenarios for the progenitor binaries. The "standard"
massive X-ray binaries result from wind-mass-loss dominated evolution
of very massive close binaries, while Be/X-ray binaries probably
result from mass-transfer dominated evolution of systems with primary
masses $\lesssim 20$ $M_\odot$. The implications of the X-ray observations of Be/X-ray
binaries for Be stars in general are discussed.

A. INTRODUCTION

X-ray and optical observations of binary systems containing an accreting
neutron star can provide a powerful astrophysical probe of the system
parameters. Through such studies it is possible to determine the mass
and radius of the optical star, the mass of the neutron star and an
estimate of the stellar wind intensity; mass transfer mechanisms and
evolutionary histories may also be investigated. There are presently
~ 75 bright galactic X-ray sources [with $F_X \geq 2 \times 10^{-10}$ ergs/cm$^2 \cdot$s,
in the energy range 1-20 keV; see, e.g., Forman et al.(1978)], many
of which are believed to be binary systems containing an accreting neutron star. About 20 of these binary X-ray systems are identified with
early-type companion stars (see e.g., Bradt, Doxsey and Jernigan 1979).
It was noted several years ago that 4 of the sources were identified
with Be star companions (Maraschi, Treves and van den Heuvel 1976).
Presently, there are at least 12 such X-ray systems identified with
Be stars (including three in the Magellanic Clouds).

In this review we describe what can be learned about Be star binary
systems through the study of X-ray emission from a compact companion

327

M. Jaschek and H.-G. Groth (eds.), Be Stars, 327–346.
Copyright © 1982 by the IAU.

star (e.g., a neutron star). In § B we review briefly how observations
of X-ray binaries, in general, can yield the system parameters. To date,
for reasons that will become apparent, more information has been obtained from observations of X-ray binaries with supergiant companions
than for systems containing Be stars. Section B is therefore included
to establish the credibility of the observational techniques and the
theoretical interpretation in those systems for which more information
is available. In § C we summarize the state of X-ray observations of
Be star binary systems. In § D we present a simple model to explain the
observations and describe a plausible evolutionary scenario for Be star
binaries with neutron star companions. Finally, we draw a number of
conclusions about Be star/X-ray binaries, some of which may be applied
to Be stars in general.

B. DIAGNOSTICS OF X-RAY BINARY SYSTEMS

Most of the bright galactic X-ray sources are luminous X-ray emitters
with $L_x \sim 10^{35}$ to 10^{38} ergs/s. These objects are widely interpreted as
binary systems powered by accretion from a "normal" optical star into
the deep gravitational potential well of a companion neutron star or
other collapsed star. The inferred mass accretion rates are in the range
$\sim 10^{-11}$ to 10^{-8} $M_\odot$/yr. Much of our understanding of X-ray binaries
comes from investigations of a relatively small sample of well studied
sources. This latter group consists mostly of X-ray pulsars.

There are presently $\sim$ 18 known X-ray pulsars (see, e.g., Rappaport and
Joss 1981). The X-ray pulse profiles for 14 of these sources are collected in Figure 1. Note that the pulses have a distinctly larger duty cycle
than those of radio pulsars (Manchester and Taylor 1977). The X-ray beam
pattern is determined by the external magnetic field of the neutron star,
the resultant funneling of the accretion flow, and the complex radiative
transfer processes for X-rays propagating from the neutron-star surface
through regions of accreting magnetized plasma (see e.g., Lamb, Pethick
and Pines 1973). The pulsations result from an X-ray beam pattern that
is misaligned with the rotation axis of the neutron star and viewed from
different directions as the star rotates.

The pulse periods of a number of the X-ray pulsars have been carefully
studied over the past decade. The pulse period histories for 8 X-ray
pulsars are shown in Figure 2. Note the trend in most of the X-ray pulsars for a secular decrease in pulse period with time (see, e.g., Schreie
1977). This is in sharp contrast with the radio pulsars which are powered
by their store of rotational kinetic energy. This "spin-up" behavior is
interpreted as due to torques on the neutron star applied by the accretin
matter (Pringle and Rees 1972; Lamb, Pethick and Pines 1973; Rappaport
and Joss 1977, 1981; Mason 1977).

The theory of accretion torques on neutron stars (see, e.g., Ghosh and
Lamb 1979) predicts that, under many conditions, the fractional rate of
change in the pulse period can be expressed as

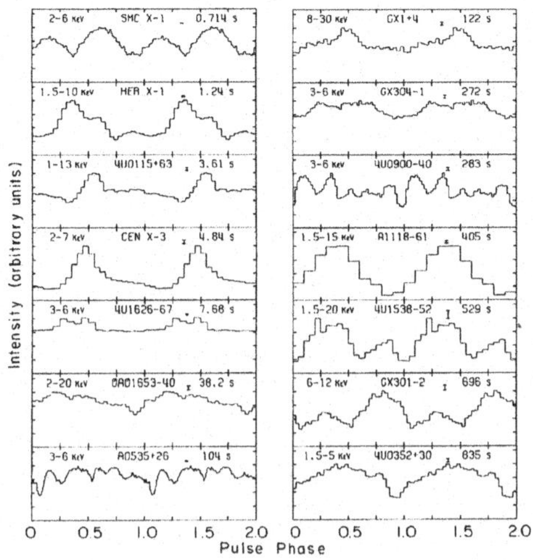

<u>Figure 1</u>: Sample pulse profiles for 14 binary X-ray pulsars (from Rappaport and Joss 1981). In each case, the data are folded modulo the pulse period and plotted against pulse phase for two complete cycles. The approximate pulse periods and energy intervals are indicated for each pulsar. Non-source background counting rates have been subtracted. A typical ± 1 σ error bar, derived from photon counting statistics, is indicated for each pulse profile.

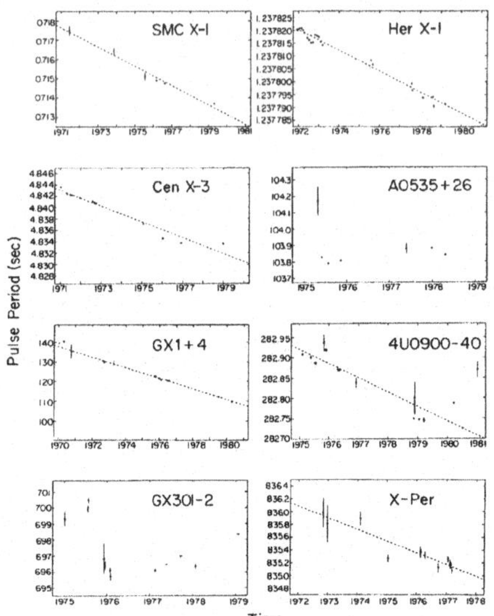

<u>Figure 2</u>: Pulse period histories for 8 binary X-ray pulsars (from Rappaport and Joss 1981;1982). The heavy dots are individual measurements of pulse period; the vertical bars represent the 1 σ uncertainties in the period determination. The data are from the Uhuru, Copernicus, Ariel 5, SAS-3, OSO-8, HEAO-1, Hakucho, and Einstein satellites, the Apollo-Soyuz Test Project, and a balloon and sounding rocket experiment. The dashed lines are minimum chi-squared fits of a straight line to the data points.

$$\dot{P}/P \simeq -3 \times 10^{-5} \, f \left(\frac{P}{1s}\right) \left(\frac{L_x}{10^{37} \text{ ergs s}^{-1}}\right)^{6/7} , \qquad (1)$$

(Rappaport and Joss 1977) where the dimensionless function $\underline{f}$ is expected
to be of order unity for a neutron star with mass, radius and magnetic
moment equal to 1 $M_\odot$, 10 km and 10^{30} G cm^3 , respectively. For an accreting white dwarf $\underline{f}$ will be about two orders of magnitude smaller (Rappaport and Joss 1977). The observed spin-up rates for the X-ray pulsars
seem to fit equation (1) quite well given the uncertainties in $\underline{f}$ and
L_x. The data are clearly inconsistent with the expected spin-up rates
for accreting white dwarfs.

Observations of X-ray pulsations can also enable one to measure orbital
motion. The method of determining orbits from the Doppler delays of
X-ray pulse arrival times (see e.g., Gursky and Schreier 1975) is analogous to classical optical Doppler measurements, and is illustrated sche-
matically in Figure 3. For a perfect clock moving with uniform velocity,
a plot of pulse arrival time vs. pulse number yields a simple linear
relation. If the intrinsic rate of the clock increases with time (as is
the case for the spin-up of a neutron star; see above) the same type
of plot yields a curved line such as the one shown in Figure 3. If,
furthermore, the clock is in a Keplerian orbit there will be superposed
periodic systematic Doppler delays due to the time-of-flight of the
pulses across the orbit. For the case where the curvature due to orbital
motion is much greater than that due to changes in the intrinsic pulse
period (or where the measurement interval contains a number of orbital
cycles) the orbit can be determined by subtracting a simple polynomial
from the arrival-time plot. Complications arise when the reverse of the
above condition obtains, and such orbits are difficult to determine.
This is the case for slow pulsars (with their concomitantly large values
of $|\dot{P}|$) in long-period orbits.

An example of orbital determination from a measurement of Doppler delays
of X-ray pulses is shown for the case of 4U0115+63 in Figure 4. These
data were obtained with SAS-3 in observations that spanned an interval
of 26 days (Cominsky et al. 1978; Rappaport et al. 1978). The Doppler
delays in pulse arrival time (Fig. 4) were obtained by subtracting off
the expected pulse delays from a constant pulse period. Analyses of these
data yielded an orbital period, P_{orb} = 24.3 days, a projected semi-major
axis, $a_x \sin i$ = 140 1t-sec = 60 $R_\odot$, and a moderate eccentricity, e = 0.34
(Rappaport et al. 1978). The orbital elements are all determined with a
precision of ~ 1 part in 10^3 .

Doppler velocity curves have now been measured for the X-ray star in 7
binary systems (see e.g., Rappaport and Joss 1981). In 5 of these systems
it has also been possible to measure the Doppler velocity curve of the
optical companion star (see, e.g., Rappaport and Joss 1981 and references
therein). As these systems also exhibit X-ray eclipses they can be said
to be "double-line spectroscopic eclipsing binaries". With this type of

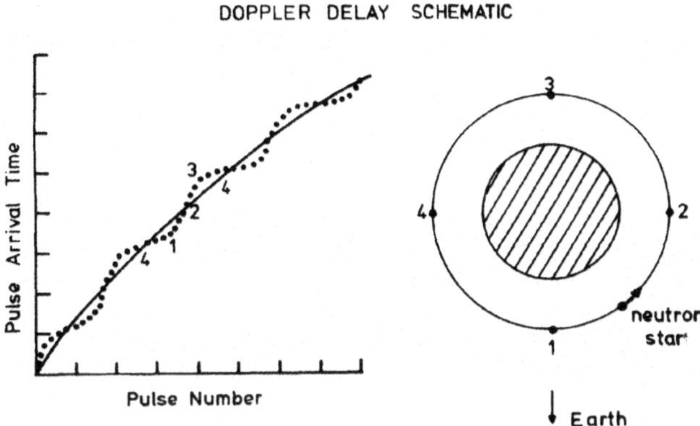

Figure 3: Schematic of the Doppler delays in X-ray pulse arrival times. Such measurements are used to determine the orbital elements in binary X-ray pulsar systems (see text).

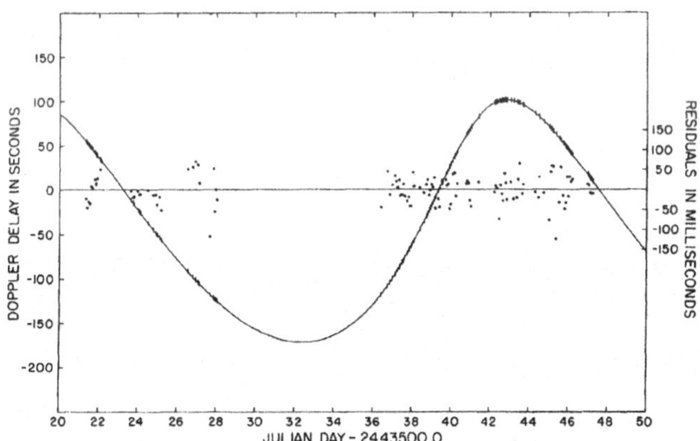

Figure 4: Doppler delay data for 4U0115+63 (from Rappaport et al. 1978). The vertical bars are the measured delays in pulse arrival time (left-hand scale); the length of each bar is con- siderably greater than the uncertainty in the measurement. The solid curve represents the expected delays for a Keplerian orbit with the best-fit orbital parameters (see text). Small circles (right-hand scale) indicate the residual differences between the measured delays and the best-fit curve.

information, many of the binary system parameters for these 5 sources
can be determined; in particular,the masses of the neutron stars have
been obtained. The most probable masses of the individual neutron stars
and the corresponding uncertainties are summarized in Figure 5. Most
of the neutron stars are seen to have a mass consistent with ~ 1.5 $M_\odot$
(Joss and Rappaport 1976; Bahcall 1978; Rappaport and Joss 1981).

C. X-RAY OBSERVATIONS OF Be STARS

Hard X-rays (E $\gtrsim$ 3 keV) have been observed from the vicinity of at least
12 Be stars. The observational results are summarized in Table I. In
general, the identification of the X-ray source with the Be star (or
Be star binary system) is based on a positional coincidence (the X-ray
positions have typical uncertainties of ~ 30"). The identifications
of most of these Be stars are discussed extensively by Bradt, Doxsey
and Jernigan (1979).

A priori, we must consider the possibilities that the X-ray emission
could come from either the Be star or from an unseen companion (e.g.,
a neutron star). Binarity can be established in one of three ways. First,
a direct measurement of orbital motion may be possible. Though there
have been a number of searches, the only orbital motion detected from
X-ray observations has been for the case of 4U0115+63 (discussed in § B).
Secondly, the existence of rapid X-ray pulsations (P $\lesssim$ 1000 s) directly
implies the presence of a compact companion star. Because X-ray pulsations are observed from the first 6 sources listed in Table I we can conclude that these are Be star binaries with a compact companion. Finally,
we note that X-ray luminosities in excess of ~10^{34} ergs/s are apparently
too large to come from the stellar corona of an isolated Be star. Recent
results from the Einstein observatory (Pallavicini et al. 1981) indicate
that the coronal X-ray emission (at least for E $\leq$ 4 keV) from early type
stars is ~ (1.4 ± 0.3) x 10^{-7} times the bolometric luminosity. This high
optical luminosity criterion applies to all of the non-X-ray pulsars in
Table I except possibly for γ Cas (Jernigan 1976). We note, however,
that if anything the X-ray fluxes from Be stars may be slightly lower than
predicted by the above relation (Rosner 1981). Thus, we conclude that
at least 11 and possibly all of the Be stars listed in Table I are binaries with a compact companion star.

Also of interest is the nature of the compact companion star in these
Be-binary systems. For the 6 X-ray pulsars listed in Table I only white
dwarfs or neutron stars are serious candidates for the compact objects.
The spin-up rates for X Per (White, Mason and Sanford 1977) and
4U0115+63 (Rappaport et al. 1978) are too large to be associated with
a white dwarf companion (see § B). The X-ray luminosities of A0535+26
and A1118-61 (at peak brightness; Rosenberg et al. 1975; Ives, Sanford
and Bell-Burnell 1974) are too large to be produced by accretion onto
a white dwarf (Katz 1977; Kylafis and Lamb 1979). From the fact that
the X-ray properties (e.g., pulse profiles and energy spectra) of GX304-1
(McClintock et al. 1977) and 4U1145-61 (White et al. 1980) are very

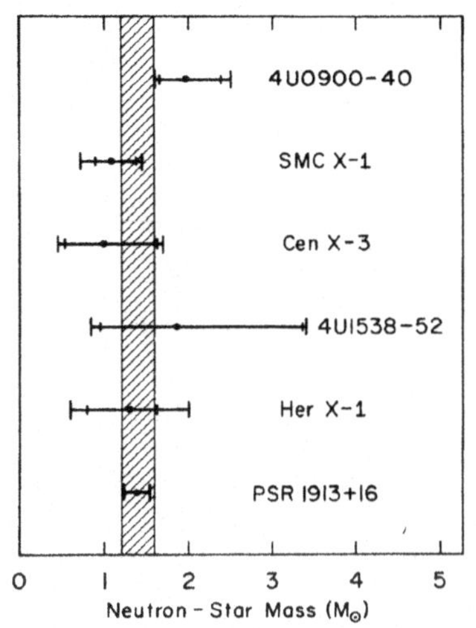

Figure 5: Empirical knowledge of neutron star masses (from Rappaport and Joss 1981). Five of the neutron-star masses are derived from observations of binary X-ray pulsars. PSR1913+16 is a binary radio pulsar (Taylor et al. 1979) and is added for completeness. The most probable value for the mass of each neutron star is indicated by the filled circle. For the X-ray binary systems, an inner set of error limits is also shown, corresponding to less conservative assumptions. The hatched region represents the range of neutron-star masses (1.2-1.6 $M_\odot$) that might be expected on the basis of current theoretical scenarios for neutron-star formation.

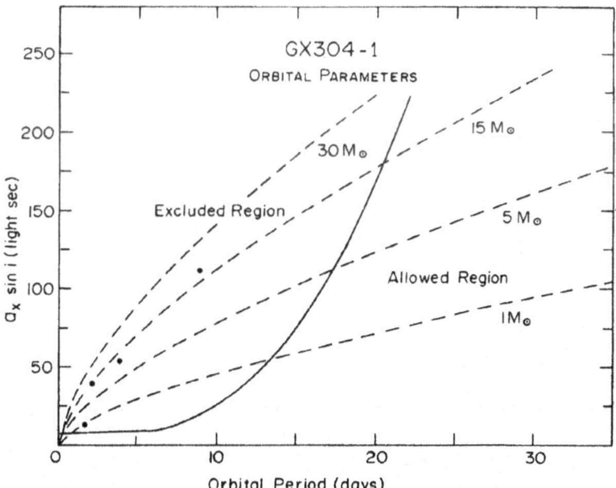

Figure 6: Allowed circular orbits for GX304-1 that are consistent with the SAS-3 timing data (Mc Clintock et al. 1977). *Heavy curve*, upper limits (95% confidence) on $a_x \sin i$ as a function of orbital period. *Dashed curves*, contours of constant mass function f(M). *Four dots*, the orbits of Her X-1 (P_{orb} = 1.7 days), Cen X-3 (P_{orb} = 2.08 days), SMC X-1 (P_{orb} = 3.89 days) and 4U0900-40 (P_{orb} = 8.96 days).

Table I
Properties of Be-star X-ray Binaries[a]

Source	Be-star counterpart	V	Spectral Type	Distance (kpc)	L_x[b] $(10^{36}$ergs/s)	Pulse Period (s)	$-\dot{P}/P$ (yr^{-1})	P_{orb} (days)	K_{opt} (km/s)	X-ray temporal variability
4U0352+30	X Per	6.3	O9.5(III–V)e	0.35	0.005	835	2×10^{-4}	580?	$\lesssim$50	steady
GX304–1 4U1258–61	MMV star	14.7	B2Vne	2.4	0.3	272	–	$\gtrsim$13	–	highly variable
A0535+26	HD245770	9.2	BOVe	1.5	5	104	–	$\gtrsim$18	$\lesssim$30	transient
A1118–61	Hen 3–640	12.1	O9.5(III–V)e	5	5	405	–	–	–	transient
4U0115+63	Johns star	15.5	Be	3	2	3.6	3×10^{-5}	24.3	–	transient, varies by $\gtrsim10^4$
4U1145–61	HD102567 Hen 715	9.0	B1Vne	1.5	0.03–0.3	292	$<10^{-4}$	$\gtrsim$35 187?	–	highly variable
A0538–66	Johnston star Q	15	Be	55 (LMC)	800	–	–	16.65?	–	transient flares periodically every 16.65d, varies by $>10^4$
2S0053+60	γ–Cas[c]	2.4	BO.5(II–V)e	0.30	0.003	–	–	–	$\lesssim$10	highly variable
2S0114+650	LSI+65°010	11.0	BOVe	1.4	0.01	–	–	–	–	variable
SMC X–2	Murdin star	16	B1Ve	65 (SMC)	100	–	–	–	–	transient
SMC X–3	Clark star 4	15	O9(III–V)e	65 (SMC)	70	–	–	–	–	transient
4U1735–28	Hen 3–1450?	11.2	Be	$\gtrsim$1	$\gtrsim$1	–	–	–	–	transient

a.Much of the data from this table are taken from Bradt, Doxsey and Jernigan (1979); most of the original references may be found therein. Additional references are cited in the text.
b.Average X-ray luminosity (2–10keV) when the source is bright.
c.Not yet established as a binary (see text).

similar to those of the other X-ray pulsars, already well established
as neutron stars, we conclude that the compact objects in these systems
are also neutron stars. Of the remaining 6 sources in Table I three have
peak luminosities greatly in excess of that which can plausibly be produced by accretion onto a white dwarf (Katz 1977; Kylafis and Lamb 1979);
therefore the compact objects in these systems could be either a neutron
star or perhaps a black hole, though there is no evidence for the latter
case.

Despite the fact that at least 11 of the 12 Be-star X-ray sources listed
in Table I are binaries, direct evidence for orbital motion is found in
only two of them. As discussed in § B the orbit of 4U0115+63 has been
directly measured (P_{orb} = 24 days). The source A0538-66 exhibits X-ray
flares at highly regular intervals of 16.65 days (Johnston, Griffiths
and Ward 1980; White and Carpenter 1978; Skinner et al. 1980). Such
regular outbursts of the X-ray source will probably ultimately be associated with motion of the compact star in an eccentric binary orbit with
P_{orb} = 16.65 days or, if the orbit is not coplanar with the equatorial
plane of the Be star, P_{orb} = 33.30 days, but to date this has not been
demonstrated conclusively. Similarly, seemingly regular X-ray outbursts
have been observed from 4U1145-61 with a recurrence time of ~ 187 days
(Watson, Warwick and Ricketts 1981), though the evidence for a strictly
periodic behavior is not as convincing as for A0538-66.

X-ray observations for 4 of the sources in Table 1 (GX304-1, A0535+26,
X Per and 4U1145-61) have allowed constraints to be set on the orbital
parameters of the binary system. Typically, the pulse arrival times
from a given source are closely monitored for an interval ranging from
a few days to a month. No compelling evidence for orbital motion was
detected in any of the above 4 sources. A limit to the size of the
semi-major axis of the orbit can then be set for any assumed orbital
period.

A sample of such orbital constraints for the source GX304-1 (McClintock
et al. 1977) is shown in Figure 6. Limits on $a_x \sin i$ vs. P_{orb}, as well as
contours of constant mass function in the $a_x \sin i / P_{orb}$ plane are shown.
The (X-ray determined) mass function in a Be star binary is

$$f(M) = \frac{4\pi^2 (a_x \sin i)^3}{G\ P_{orb}^2} = \frac{M_c \sin^3 i}{(1+q)^2} , \qquad (2)$$

where M_c is the mass of the Be star and $q \equiv M_x/M_c$. To a good approximation
$f(M) \approx M_c \sin^3 i$ because q is likely to be a small number (in § B we
showed that $M_x \approx 1.5\ M_\odot$ for several systems). Because the masses of
BO-2 IV, Ve stars lie in the range 10-20 $M_\odot$ we expect $f(M) \simeq (8-17)$
$\sin^3 i$. Therefore, if we adopt $f_0(M) = 1\ M_\odot$ as a reasonable lower limit
for a Be star X-ray binary, we can see from Figure 6 that P_{orb} for the

GX304-1 binary system should be greater than $\sim$ 13 days. In a similar manner it has been determined that P_{orb} in the A0535+26 (Li et al. 1979) and 4U1145-61 (White et al. 1978) systems should be greater than 18 days and 35 days, respectively. For X Per, a limit on $a_x \sin i$ of $\sim$ 15 lt-sec for orbital periods in the range 0.7 to 3 days has been obtained (Jernigan, Nugent and Rappaport 1981). This is only barely consistent with a neutron star in a short-period orbit of a few days around a Be star of radius $\lesssim$ 5 $R_\odot$, viewed with an inclination angle of $i \lesssim 60°$. Marginal evidence has been found for the 584-day orbital period (Hutchings et al. 1974) in X-ray data obtained over a 4-year interval (White, Mason and Sanford 1977).

We also note that no X-ray eclipses are seen from any of the sources in Table I, which provides additional evidence that these binary systems are not close (i.e. contact systems). In general, all of these systems are consistent with having binary periods greater than $\sim$ 15 days.

As a final point concerning the X-ray characteristics of the Be/X-ray systems, we note that all of them with the exception of X Per are highly variable or transient sources (see, e.g., Forman et al. 1978; Bradt, Doxsey and Jernigan 1979). Many of the sources are known to vary by a factor of 10 or 100; recent observations with the Einstein observatory indicate that at least at soft X-ray energies (0.1-3 keV) two of the sources 4U0115+63 (Kriss et al. 1980) and A0538-66 (Long, Helfand and Grabelsky 1981) have varied by more than a factor of 10^4.

D.   MODEL AND EVOLUTIONARY HISTORY

D.1 *Mass transfer mechanisms in Be X-ray binaries*

As argued above, at least 11 Be/X-ray systems are binaries. In the following important respects these binaries differ from the standard massive X-ray binaries such as Cen X-3 and SMC X-1 (see Figure 7). (1) The orbital periods of the Be/X-ray binaries are in excess of 15 days and probably range up to several years (see section C), while the standard massive X-ray binaries (with the sole exception of 4U 1223-62) have periods between 1.4 and 9 days. (2) The optical components of Be systems are generally unevolved stars of spectral types O9 Ve to B2 Ve. Such stars have relatively small radii, i.e. $\lesssim$ 5 - 10 $R_\odot$, absolute luminosities $\lesssim$ 3 x 10^4 $L_\odot$ and masses between $\sim$ 10 and 20 $M_\odot$. On the other hand, in the standard massive X-ray binaries the optical stars tend to be evolved (giant, Of or supergiant stars) with radii 10-30 $R_\odot$, $L_{opt} \gtrsim 10^5$ $L_\odot$ and initial masses (derived from their luminosities) in excess of $\sim$ 20 $M_\odot$). (3) Characteristics 1 and 2, together with the absence of X-ray eclipses and of periodic (ellipsoidal) light variations indicate that the Be stars in X-ray binaries reside deep inside their Roche lobes. This rules out Roche-lobe overflow as a possible source for the mass accretion. The standard massive systems, on the other hand, often eclipse and always show periodic ellipsoidal light variations, indicating that their optical components practically fill their Roche

lobes. The mass transfer in these systems is thought to be due to beginning Roche-lobe overflow or to enhanced winds that are expected from
stars that nearly fill their Roche lobes (cf. Savonije 1979,1980).
(4) The X-ray luminosities of the Be systems generally exhibit much
larger variability than those of the standard systems. In fact, with
the possible exception of the two nearby weak ($\sim 10^{33}$ ergs/s) sources
X Per and γ Cas, all known Be/X-ray sources are transients. This shows
that the mass transfer rate in the Be systems must be highly variable.

Since Roche-lobe overflow is ruled out, the mass transfer must be due
to the intrinsic mass-losing properties of the Be-component. Be stars
show two types of intrinsic mass loss: (i) by a stellar wind and (ii)
by mass ejection in their equatorial regions. This latter type of mass
loss produces the characteristic emission lines in their spectra. Since
Be stars are the most rapidly rotating B stars the equatorial ejection
of matter is presumably rotationally driven, though the physical processes involved are not yet well understood (Marlborough 1976; Slettebak
1979).

The stellar wind parameters of the Be stars X Per, γ Cas and HD 102567
(4U1145-61) (as well as of other early Be stars) derived from UV ob-
servations indicate typical wind mass loss rates $\dot{M}_c \simeq 10^{-9}$ to 10^{-7} $M_\odot$/yr,
and outflow velocities $v \sim 10^3$ km/s (Hammerschlag-Hensberge et al. 1980).
Using the standard stellar wind accretion theory (Davidson & Ostriker
1973) one finds for the accretion rate onto a compact companion with mass
M_x:

$$\dot{M}_x \approx 7\times10^{-6} \ \dot{M}_c \ (M_x/M_\odot)^2 \ \ (v/10^3 \text{ km s}^{-1})^{-4} \ \ (r_{orb}/5\times10^{12}\text{cm})^{-2} , \tag{3}$$

where r_{orb} is the orbital separation of the Be star and compact companion.
Equation (3) indicates that the wind capture accretion rates in Be
systems are typically expected to be 10^{-14} to 10^{-12} $M_\odot$/yr, corresponding
to X-ray luminosites of $\sim 10^{32}$ to 10^{34} ergs/s. It is very well possible,
therefore, that all Be/neutron star systems with $P_{orb} \lesssim 1$ year have steady
X-ray luminosities of the order of 10^{33} ergs/s during their low states.

These luminosities are very similar to those of the two nearby weak Be/
X-ray sources, X Per and γ Cas (see Table I). However, accretion from
a spherically symmetric wind falls short by 4 to 6 orders of magnitude
to explain the X-ray luminosities of the transient Be sources at X-ray
maximum. It seems most natural to ascribe these high luminosities to
the rotationally-induced mass ejection from the equatorial regions of
the Be stars since this type of mass ejection: (a) is actually observed,
and (b) is highly variable. The strength of the emission lines of rapidly
rotating Be stars is known to vary irregularly from complete absence to
large intensities, on timescales of months to years (Slettebak 1979).
Indeed, in at least one Be system (A0538-66) a clear correlation has been
established between X-ray high states and the presence of emission lines
in the spectrum (Pakull and Parmar 1981).

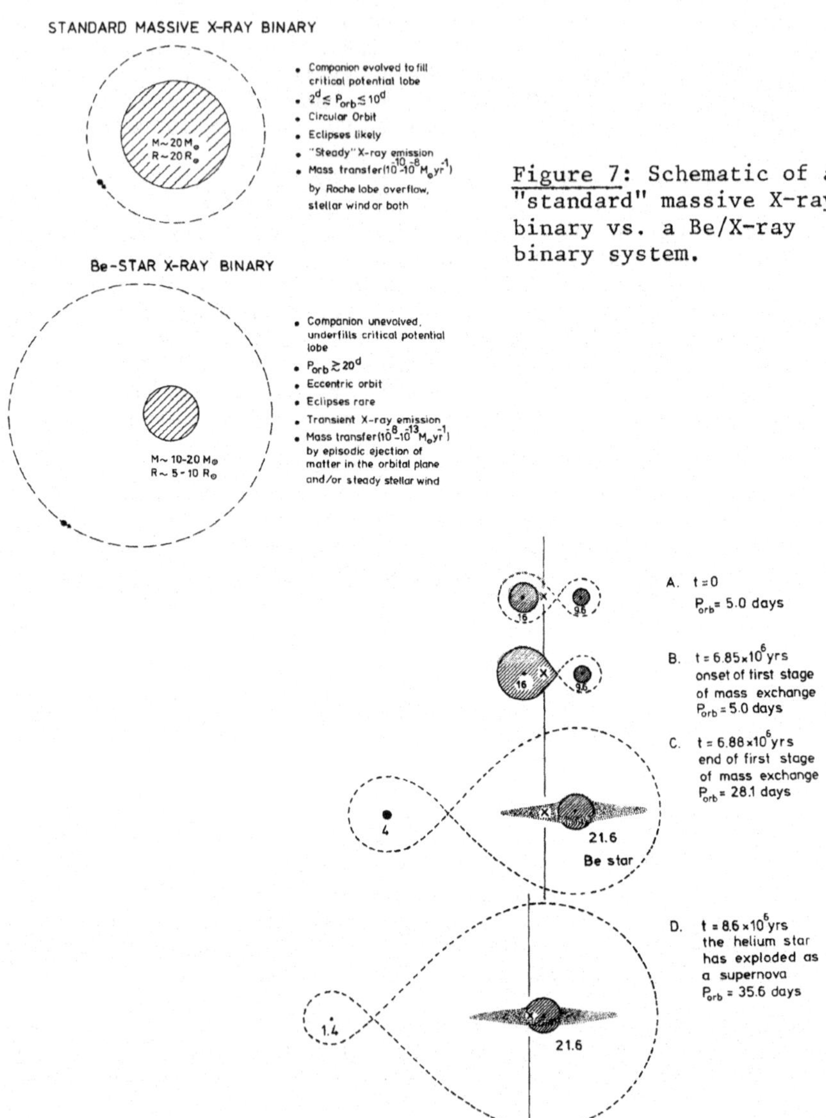

Figure 7: Schematic of a "standard" massive X-ray binary vs. a Be/X-ray binary system.

Figure 8: Evolution of a Be/neutron star binary from a close pair of early B stars. Conservative mass transfer is assumed. The mass of each component, in solar units, is indicated. At each epoch, the dashed curve represents the Roche surface. After the end of the first stage of mass exchange the Be star may have a circumstellar "disk or shell" of matter associated with its rapid rotation.

Although the precise causes and mechanism of the irregular mass outbursts
from Be stars are not known, the observed properties of the X-ray outbursts can be used to derive constraints on the gas density $\underline{n}$ in the
stellar envelope during an outburst, as follows. Equation (3) can be
rewritten as

$$\dot{M}_x \approx 6\times10^{-9}\ (M_x/M_\odot)^2\ (v_{rel}/10^3\ km\ s^{-1})^{-3}\ (n/10^{12}cm^{-3})\quad M_\odot/yr, \qquad (4)$$

where v_{rel} is the relative velocity between the outflowing matter and
the neutron star. Since during an outburst $\dot{M}_x$ can rise to $\sim10^{-8}$ $M_\odot/yr$,
one finds that temporarily: $(v_{rel}/10^3\ km\ s^{-1})^{-3}\ (n/10^{12}\ cm^{-3}) \approx 1.$
Hence, if the ejection velocity is of the same order as the stellar wind
velocity $(\sim10^3\ km/s)$, n must be $\sim10^{12}\ cm^{-3}$. On the other hand, if the
ejected cloud moves out with a somewhat lower velocity (e.g., ~200 km/s),
densities as low as $n\sim10^{10}\ cm^{-3}$ are sufficient to yield the observed X-ray
luminosities. These densities are in good agreement with the values of
$\sim10^{10}$-$10^{12}\ cm^{-3}$ inferred from the strengths of the Balmer emission lines
(see, e.g., Peters 1976; Slettebak 1979). For still smaller outflow velocities (i.e., less than or comparable to the orbital velocity of the neutron
star, as are often observed in Be shells; see Underhill 1966), the accretion rate given by equation (4) may no longer be appropriate. Instead,
the matter may form an accretion disk about the neutron star. This
could lead to even larger values of $\dot{M}_x$ than those given by equation (4).

D.2 *Galactic Number and Evolution*

The discovery of Be/neutron star systems is hampered by several observational selection effects. During low states their expected steady hard
X-ray fluxes are typically of the order of 10^{33} ergs/s, which implies
that beyond ~0.5 kpc such systems are not readily detectable. Beyond
~0.5 kpc the only systems that will have been discovered are those
that have exhibited large outbursts at times when X-ray satellites were
monitoring the appropriate parts of the celestial sphere. Since the sky
coverage has been far from complete, and since in most Be stars emission
or shell phases tend to occur not more than once per decade, one expects
that the actual number of Be/neutron star systems is perhaps an order
of magnitude larger than the detected number of bright Be X-ray transients.
With ~5 bright Be transients detected within a 2.5 kpc distance in 10 years
one thus expects the actual number of Be/neutron star binaries within
this distance to be at least $\gtrsim50$. (Such a number is consistent with an
expected number of ~100 within 2.5 kpc, derived by extrapolation from
the fact that two permanent weak sources - X Per and γ Cas - are found
within 0.35 kpc distance.)

With $\gtrsim50$ Be/X-ray systems within 2.5 kpc distance, against only 3
"standard" systems (4U0900-40, 4U1700-37 and Cyg X-1) the Be/X-ray
systems are by far the most abundant type of massive X-ray binary in
the galaxy.

If we adopt a radius of the galactic disk of 14 kpc and a uniform distribution of early type stars throughout this disk, we conclude that the

total number of Be/X-ray binaries must be $\geq 1.5 \times 10^3$, against a total number of "standard" massive X-ray binaries of ~ 50 (van den Heuvel 1978).

An important difference between the Be/X-ray systems and the "standard" massive X-ray binaries is in their expected lifetimes. The Be-systems have relatively unevolved (i.e., hydrogen-burning) optical components with an expected lifetime of order 5×10^6 yrs (i.e., they must correspond to the "quiet" stage in the standard evolutionary scenario for massive X-ray binaries; cf. van den Heuvel 1976, 1978). On the other hand, in the "standard" systems the optical component is an evolved star which is just beginning to overflow its Roche lobe (or is close to doing so); the lifetime in this stage is not expected to be longer than $\sim 10^5$ yrs (Savonije 1979, 1980). Since the Be/X-ray systems are expected to live some 50 times longer than the "standard" systems, and are some $\gtrsim$ 30 times more abundant in the galaxy, the galactic formation rates of both types of objects should be roughly similar.

The combination of systematically longer binary periods and lower optical companion masses for Be/X-ray systems as compared to the "standard" massive X-ray systems may find a qualitative explanation in an evolutionary scenario for the progenitor binary systems as follows. The most massive stars (M $\gtrsim 20$-30 $M_\odot$) suffer large mass loss by stellar winds during their evolution. [This mass loss is presumably generated by radiation pressure driven instabilities in the interior, connected with the increase in mean molecular weight, μ, during the evolution (see van den Heuvel 1979; Maeder 1980).]In a binary system, such a star may lose most of its hydrogen rich envelope by wind mass loss before it reaches its Roche lobe. Therefore, mass transfer by Roche-lobe overflow may play a relatively minor role in these systems (see also Vanbeveren and Conti 1980). Since in this case not much angular momentum transfer takes place between the components, the binary period need not change drastically during the evolution. This is evidenced by the fact that the Wolf-Rayet (WR) binaries, which are the products of this evolution, have about the same distribution of binary periods as their progenitors, the O-type spectroscopic binaries, although their mass ratios are very different (Massey 1981). Since the WR binaries tend to have short binary periods (mostly $\lesssim 10$ days) the same is expected for their descendants, the massive "standard" X-ray binaries (van den Heuvel 1973).

On the other hand, in stars with M $\lesssim 20$ $M_\odot$ stellar wind mass loss hardly affects the evolution (see Lamers 1981) and the primary star evolves to overflow its Roche lobe, which results in an extensive exchange of mass and (orbital) angular momentum with the companion. If this transfer takes place more or less conservatively (i.e., total mass and orbital angular momentum are approximately conserved), the orbital period will, in general increase by a considerable factor. This is due to the fact that unevolved close binaries tend to have mass ratios, q, fairly close to unity (i.e., in the range q $\simeq 0.5$ to 1.0, see Abt and Levy 1978; Lucy 1981), while after the transfer the mass ratio will be far from unity, since the primary star will have lost over 70% of its initial mass. Figure 8 shows

as an example the conservative evolution of a short-period system of
16 + 9.6 $M_\odot$ that produces a much wider Be/X-ray binary. In this example,
with an initial mass ratio of q = 0.6 and a final mass ratio of q = 5,
the binary period will increase by a factor of about 5 (see Paczynski
1971). Hence the difference between the mean binary periods of the
"standard" massive X-ray binaries and the Be/X-ray binaries would be
the result of *wind-mass-loss dominated evolution* in the very massive
close binaries vs. *mass transfer dominated evolution* among close binaries with primary mass $\lesssim$ 20 $M_\odot$.

If a number of the Be/X-ray binaries turn out to have orbital periods
greater than ~ 100 days, a complementary reason for the systematically
lower masses of the widest neutron-star binaries might be that the lower
mass limit for evolving to core collapse is lower for primaries in these
wide binaries than in short-period ones. This results from the fact that
in a long-period binary the core of the primary star still has time to
grow significantly during hydrogen shell burning, before the star over-
flows its Roche lobe and loses its envelope. Therefore, in binaries that
evolve according to Kippenhahn and Weigert's (1967) case C (i.e., the
primary does not overflow its Roche lobe before core-helium ignition)
the core mass that remains after Roche lobe overflow can be over 40%
larger than in the short-period systems that evolve according to the
cases A and B. Thus, for primaries in case C systems, the lower mass
limit for evolving to core collapse may be as low as 10 to 12.5 $M_\odot$,
whereas in case B and A systems it probably ranges from 15 to 20 $M_\odot$
(see van den Heuvel 1981). As a consequence, very wide neutron-star
binaries (products of case C) are expected to have, on the average,
lower masses than closer binaries (products of the cases A and B).

Natural direct progenitors of the Be/X-ray systems are binaries such as
AX Mon (B1e + K2II, P = 232 days), φ Per (B0e, P = 126 days), HR2142
(B1Ve, P = 81 days) (Peters 1976), and also the ζ Aurigae systems, which
consist of a late-type supergiant and an early Be star, and have binary
periods of over a year (Wilson 1960). The B-components in these systems
are always surrounded by a rapidly rotating ring or disk of accreted
matter. The accretion of mass with high specific angular momentum through
a disk will spin up the B-star and will cause it most probably to rotate
near break-up at the end of the accretion stage when the core of the
K-supergiant finally collapses. Hence, after the supernova, one expects
the remaining binary system to consist of a rapidly rotating Be star
and neutron star, in a relatively wide orbit.

E. CONCLUSIONS

(1) Be stars in binary systems with a neutron star companion are not
    uncommon.

(2) The Be star characteristics of the Be/X-ray systems are apparently
    indistinguishable from those of other Be stars. Whatever their past
    history, the presence of the neutron star does not seem to have any

influence on its companion. The neutron star acts solely as a probe
with which the intrinsic mass loss characteristics of the Be star
can be monitored. (A possible exception is A0538-66.)

(3) The Be characteristics need not arise from matter accreting onto
the Be star, as suggested by some authors.

(4) The rapid rotation of the Be stars in Be/X-ray binaries is most
probably due to their spin-up during a preceding mass transfer
phase. It may well be that many other Be stars derived their rapid
rotation from a similar evolutionary history with mass transfer
(cf. Kriz and Harmanec 1975).

(5) The transient X-ray behavior of most of the Be/X-ray systems indicates a high degree of variability in the flux of matter expelled
in the equatorial regions of the Be star. (A spherically symmetric
stellar wind is highly inadequate to power the flaring X-ray source.)
During periods of X-ray outburst, densities and outflow velocities
of the ejected matter of $10^9 - 10^{12}$ cm^{-3} and $10^2 - 10^3$ km/s,
respectively,are indicated.

(6) From evolutionary and statistical considerations thousands of Be/
neutron star systems are expected to exist in the galaxy. Therefore
many classically studied Be stars may well have neutron star com-
panions. In most cases, with $P \gtrsim 15$ days and $M_x \lesssim 1.5$ $M_\odot$ the radial
velocity variations in the optical companion will be $\lesssim 20$ km/s and
will be very hard to detect. Nonetheless, careful, precise studies
of a few selected objects (e.g., γ Cas and X Per) may prove very
fruitful in this regard.

ACKNOWLEDGEMENTS

We acknowledge stimulating discussions with R. Kelley of M.I.T. and
H. Henrichs of the University of Amsterdam.

This work was supported in part by the National Aeronautics and Space
Administration under contract NAS5-24441 and by the Netherlands
organization for pure research, ZWO, under contract Nr.B78-183.

REFERENCES

Abt, H.A., and Levy, S.G. 1978, Ap.J. Suppl., 36, 241.
Bahcall, J. 1978, Ann. Rev. Astr. Ap., 16, 241.
Bradt, H., Doxsey, R., and Jernigan, J. 1979, "X-Ray Astronomy", eds.
    W.A. Baity and L.E. Peterson (Oxford: Pergamon), p.3.
Cominsky, L., Clark, G.W., Li, F., Mayer, W., and Rappaport, S. 1978,
    Nature 273, 367.
Davidson, K., and Ostriker, J.P. 1973, Ap. J., 179, 585.

Forman, W., Jones, C., Cominsky, L., Julien, P., Murray, S., Peters, G.,
    Tananbaum, H., and Giacconi., R. 1978, Ap. J. Suppl., 38, 351.
Ghosh, P., and Lamb, F. K. 1979, Ap. J., 234, 296.
Gursky, H., and Schreier, E. 1975, "Neutron Stars, Black Holes and Bi-
    nary X-Ray Sources", eds. H. Gursky and R. Ruffini (Dordrecht: Rei-
    del), p. 175.
Hammerschlag-Hensberge, G., et al. 1980, Astr. Ap., 85, 119.
Hutchings, J.B., Cowley, A.P., Crampton, D., and Redman, R.O. 1974,
    Ap.J. (Letters), 191, L101.
Ives, J.C., Sanford, P.W., and Bell-Burnell, S.J. 1975, Nature, 254, 578.
Jernigan, J.G. 1976, IAU Circ., No. 2900.
Jernigan, J.G., Nugent, J., and Rappaport, S. 1981, unpublished SAS-3
    data.
Johnston, M., Griffiths, R., and Ward, M.J. 1980, Nature, 285, 26.
Joss, P.C., and Rappaport, S. 1976, Nature, 264, 219.
Katz, J. 1977, Ap.J., 215, 265.
Kippenhahn, R. and Weigert, A. 1967, Zs.f. Astrophysik, 65, 251.
Kriss, G.A., Cominsky, L., Remillard, R., and Rappaport, S. 1980, IAU
    Circ., No. 3543.
Kriz, S., and Harmanec, P. 1975, Bull. Astr. Inst. Czech, 26, 65.
Kylafis, N.D., and Lamb, D.Q. 1979, Ap. J. (Letters), 228, L105.
Lamb, F.K., Pethick, C.J., and Pines, D. 1973, Ap. J., 184, 271.
Lamers, H.J. 1981, Ap.J., 245, 593.
Li,F., Rappaport,S., Clark,G.W., and Jernigan,J.G.1979, Ap.J.,228,893.
Long, K., Helfand, D., and Grabelsky, D. 1981, preprint.
Lucy, L.B. 1981, "Fundamental Problems in Stellar Evolution", eds. D.
    Sugimoto et al. (Dordrecht: Reidel), p. 75.
Maeder, A. 1980, Astr. Ap., 92, 101.
Manchester, R.N., and Taylor, J.H. 1977, "Pulsars" (San Francisco:
    Freeman), p 18.
Maraschi, L., Treves, A., and van den Heuvel, E.P.J. 1976, Nature,
    259, 292.
Marlborough, J.M. 1976, "Be and Shell Stars", ed. A.Slettebak, IAU
    Symp. 70, 335.
Mason, K. 1977, M.N.R.A.S., 178, 81 P.
Massey, P. 1981, Ap. J., 246, 153.
Mc Clintock, J., Rappaport, S., Nugent, J. and Li, F. 1977, Ap. J.
    (Letters), 216, L15.
Paczyński, B. 1971, Ann.Rev. Astr. Ap., 9, 183.
Pakull, M., and Parmar, A. 1981, Astr. Ap., in press.
Pallavicini, R., Golub, L., Rosner, R., Vaiana, G., Ayres, T., and
    Linsky, J. 1981, CFA preprint No. 1446.
Peters, G.J. 1976, "Be and Shell Stars", ed. A.Slettebak, IAU Symp.
    70, 417.
Pringle, J.E., and Rees, M.J. 1972, Astr. Ap. 21, 1.
Rappaport, S., and Joss, P.C. 1977, Nature, 266, 683.
Rappaport, S., and Joss, P.C. 1981, "X-Ray Astronomy", ed. R.Giacconi,
    Proceedings of the HEAD-AAS Meeting, January 1980, Cambridge, MA
    (Dordrecht: Reidel), p. 123.
Rappaport, S., and Joss, P.C. 1982, "Accretion Driven Stellar X-Ray
    Sources", ed. W.Lewin and E.P.J. v.d.Heuvel, (Cambridge: Cambridge

Univ.Press),in preparation.
Rappaport, S., Clark, G.W., Cominsky, L., Joss, P.C., and Li, F.K. 1978,
    Ap. J. (Letters), 224, L1.
Rosenberg, F.D., Eyles, C.J., Skinner, G.K., and Willmore, A.P. 1975,
    Nature, 256, 628.
Rosner, R. 1981, private communication.
Savonije, G.J. 1979, Astr. Ap., 71, 352.
Savonije, G.J. 1980, Astr. Ap., 81, 25.
Schreier, E. 1977, Ann. N.Y. Acad. Sci., 302, 445.
Skinner, G.K., et al. 1930, Ap. J., 240, 619.
Slettebak, A. 1979, Sp. Sci. Rev., 23, 541.
Taylor, J., Fowler, L.A., and Mc Culloch, P.M. 1979, Nature, 277, 437.
Underhill, A.B. 1966, "The Early-Type Stars", (Dordrecht: Reidel).
Vanbeveren, D. and Conti, P.S. 1980, Astr. Ap., 88, 230.
van den Heuvel, E.P.J. 1973, Nature Phys. Sci., 242, 71.
van den Heuvel, E.P.J. 1976, "Structure and Evolution of Close Binary
    Systems", eds. P.Eggleton et al., (Dordrecht: Reidel), p. 35.
van den Heuvel, E.P.J. 1978, "Physics and Astrophysics of Neutron Stars
    and Black Holes", eds. R.Giacconi and R.Ruffini, (Amsterdam: North
    Holland), p. 828.
van den Heuvel, E.P.J. 1979, "Mass Loss and Evolution of O-Type Stars",
    eds. P.Conti and C.de Loore, (Dordrecht: Reidel), p. 491.
van den Heuvel, E.P.J. 1981, "Fundamental Problems in Stellar Evolution"
    eds. D.Sugimoto et al., (Dordrecht: Reidel), p. 155.
Watson, M.G., Warwick, R.S., and Ricketts, M.J. 1981, M.N.R.A.S.,
    195, 197.
White, N.E., Mason, K.O., and Sanford, P.W. 1977, Nature 267, 229.
White, N.E., Parkes, G.E., Sanford, P.W., Mason, K.O., and Murdin, P.G.
    1978, Nature, 274, 664.
White, N.E., and Carpenter, G.F. 1978, M.N.R.A.S., 183, 11 P.
White, N., Pravdo, S., Becker, R., Boldt, E., Holt, S. and Serlemitsos,
    P. 1980, Ap. J., 239, 655.
Wilson, O.C. 1960, "Stars and Stellar Systems, Vol. VI", ed. J.L.
    Greenstein, (Chicago: Univ. of Chicago Press), p. 436.

DISCUSSION

<u>Harmanec</u>: Is there some evidence from RV that the H emission is really connected with the optical components of x-ray/Be sources?

<u>Rappaport</u>: No optical Doppler velocity curves have yet been reliably determined for any of the x-ray/Be binaries.

<u>Stalio</u>: I am impressed by the <u>very large range of variability</u> you mentioned: for one source (4U0115+63) the x-ray luminosity varies by a factor of at least 3×10^4 and large variations are also observed for a number of others (SMCX-2, SMCX-3, A0538-66 etc.). Since these variations have been detected by different instruments, are you sure that this is not an instrumental problem?

<u>Rappaport</u>: The observed x-ray variability is certainly real.

<u>Persi</u>: What do you think about the transient x-ray phenomenon of A0595+262? Could it be due to sudden variations of the stellar wind of the Be star? Our IR observations taken during a quiescent x-ray phase very close to a flar-up show no significant variations of the characteristics of the envelope. X Per shows a long-term x-ray variability. Could this be correlated with an observed long-term IR variability?

<u>Rappaport</u>: The transient behaviour of A0535+26 can be explained by episodic mass ejection in the equatorial plane of the companion star HD245770. It is not obvious that a disc of matter, with density and flow velocity in the vicinity of the orbiting neutron star sufficient to power the x-ray emission, would necessarily be detectable in the IR. The source 4U0352+30, on the other hand, has a relatively more steady x-ray luminosity, which is sufficiently low that it may, perhaps, be accounted for in terms of the stellar wind on X Per. Studies of the long term IR and x-ray variability might prove to be a very useful diagnostic probe of this system.

<u>Giovannelli</u>: 1. Would you like to explain better the arguments you use to support your opinion that the orbital period of A0535+26 is ∼40 days, since we have found a weak evidence of ∼ 30, 40, 63 and 77 days orbital periods using our photometric measurements?
2. Do you believe it is possible that the short time variations are due to the orbital motion or to changes in the intrinsic rotation rate of the neutron star?

<u>Rappaport</u>: 1. Recent pulse timing data on A0535+26 obtained with the Hakucho's satellite by M. Oda and coworkers can be interpreted completely in terms of a spin-up of the neutron star. Even if the observed pulse arrival times represent orbital motion of the neutron star, it is obvious from an inspection of the raw data that the orbital period must be at least twice the duration of the observation i.e., $\geqslant$ 40 days.

2. It is difficult at this time to say, whether the observed short-term variations in pulse period are due to orbital motion or intrinsic changes; however, I tend to favor the latter hypothesis.

<u>Snow</u>: You have not mentioned soft x-rays at all. Can you clarify whether Be stars are known to be soft x-ray emitters at a level comparable to the coronal x-ray emission from O stars? This could be very important in assessing the relationship of Be phenomena to coronal activity, and their relationship (if any) to O stars.

<u>Rappaport</u>: Such results will hopefully be available shortly from the Einstein x-ray survey of Be stars.

<u>Pakull</u>: So far the optical identifications with Be stars have been made on positional coincidence alone I would like to comment that recently correlations between x-ray intensity and envelope characteristics have been established for 1145-61 and A0538-66 in the sense that the x-ray were turned off when the counterpart apparently lost part of its envelope being a more or less normal B V-IV star.

Be COMPONENTS IN X-RAY BINARIES

C. de Loore[1], M. Burger[1], E.L. van Dessel[2], M. Mouchet[1]
[1] Astrophysical Institute, Vrije Universiteit Brussel
[2] Royal Belgian Observatory

GENERAL CHARACTERISTICS

The Be X-ray binaries show rather weak and variable X-rays. They can be
divided into two types, the transient sources and the permanent sources.
Two ranges for the X-ray luminosity L_X can be discerned: a) $L_X \sim 10^{34}$erg
s^{-1} (X Per, γ Cas, 2SO114+65, all permanent sources); b) $L_X \sim 10^{36}$erg s^{-1}.
They have long periods, hence wide orbits, they ar not eclipsing and
their mass loss rates are low. The optical spectra are generally very
variable and irregular, masking periodic changes. No optical orbits exist
and only for 4U0115+63 an X-ray orbit is known (Rappaport et al.1978).
An overview of the Be X-ray binaries with some of their characteristics
is given in Table 1.

Object	V	Spectral type	d (kpc)	P (days)	P_{pulse} (s)	L_X/L_{opt}	M/M_o	R/R_o
2SO050-727T	15	O9(III-V)e	63	–	–	0.5	24	8
2SO052-739T	16.0	B1Ve	63	–	–	3	14	6
4U0053+604	2.3	BO.5(II-V)e	0.30	>1000?	–	6E-6	25	15
2SO114+650	11.0	BO.5IIIe	–	–	–	1.5E-4	–	–
4U0115+634T	15.6	Be	<7	24.3	3.6	2	–	–
4U0352+306	6.0-6.7	O9.5(III-V)e	0.35	581?	835	1E-4	–	–
AO535+262T	9.1	O9.7IIIe	1.8	28?77?	104	0.08	27	12
4U0538-669T	13.0-15.7	Be?	–	16.7	–	–	–	–
A1118-615	12.1	O9.5(III-V)e	5	–	405	2	33	15
4U1145-619	9	BO.5(III-V)e	1.5	187.5	291	0.2	15	6
4U1258-613T	14.7	B2Vne	2.4	>20	272	0.3	14	11

Table 1. Be X-ray binaries with some of their parameters (T=transient).

VARIABILITY AND THE BINARY HYPOTHESIS

Be primaries in X-ray binaries as well as classical Be stars present
variability with all sorts of timescales (minutes,days,years). For example
the rapid variations of the Balmer lines in γ Cas, X Per (Hutchings,1976)
and Hen 715 (Hammerschlag-Hensberge et al.1980)(see Fig.1) are similar to
these observed in o And and χ Oph (Doazan,1976). Changes in luminosity and
photometric irregular behaviour in the three Be X-ray sources mentioned
above are also found in most of the classical Be stars (Hutchings,1976).

347

due to heating by the X-ray flux) or in an accretion disk around the compact source. In some Be X-ray stars HeII λ4686 is also seen weakly in emission. The mechanism of formation of this line is not as well understood as for the close X-ray binaries. The behaviour of the line seems to be correlated with the ratio of the X-ray to the optical luminosity. Emission is observed in systems with a large ratio, absorption in systems with a small ratio.

X Per	1953-63	Hα , Hβ , Hγ emission
(Hubert-Delplace and Hubert,1979)		Hγ very weak emission on diffuse absorption FeII, SiII emission (FeII maximum in 1961) HeI filled
	1963-74	FeII lines disappear,other emission lines diminish
	after 1974	Hβ weak emission, Hγ large diffuse absorption
A1118-61	1975-77	Hα , Hγ emission
(Janot-Pacheco et al.1981)		Hδ partially filled in with emission HeI lines filled; FeII,|FeII| many emission lines
Hen 715	1978-80	Hβ emission Hγ absorption - central emission FeII,|FeII| detected

Few FeII lines observed in A1118-61 are present in Hen 715.

Table 2. The most important spectral lines in 3 Be X-ray sources.

EVOLUTIONARY SCENARIO FOR BE X-RAY BINARIES

The masses of the Be components in X-ray binaries can be derived from comparison of the position in the HRD (T_{eff},M_{bol}) with calculated evolutionary tracks (Table 1). The temperatures T_{eff} were taken from Underhill et al. (1979) and the bolometric corrections from Code et al.(1976). The Be components of the Be X-ray binaries are all situated near the ZAMS, hence they are not evolved stars (Figure 2). The masses range between 15 and 30 $M_\odot$. As a typical scenario for such systems the evolution of a 15$M_\odot$+8$M_\odot$ (De Grève and de Loore (1977)) can be used (Table 3). The primary evolves up to its supernova explosion and becomes a compact companion to the optical secondary. According to Packet (1981) matter accreted by the secondary in the case of Roche lobe overflow will spin up the star to rotational velocities near the brake-up velocity. Hence most of the matter expelled by the overflowing primary has to leave the system. Probably this overflowing material is stored in a ring around the secondary and then leaves the system in a radial symmetric outflow pattern, leading to a considerable widening of the system and consequently large orbital periods. In the case of a 15$M_\odot$+8$M_\odot$ system with 50% of the matter leaving the system the period is increased with a factor of 6 to 7. This is in contrast with the bulk of massive X-ray binaries where short periods are found, a consequence of substantial mass and angular momentum losses. The systematic longer periods and fast rotational velocities of the Be systems would point to a larger accretion by the secondary. A number of classical Be binaries contain red components; such systems could represent intermediate stages between massive main sequence systems and Be X-ray systems, either before or after mass transfer (17 Lep:B9Ve+M2III, P=276d - AX Mon:B0.5e+ K2II, P=232.5d - HD 218393:B3e+K1III, P=38d).

All these phenomena are obviously related to variations in the shell or
envelope structure. Probably the variations in the X-ray flux observed
in the Be X-ray sources are also due to changes in their envelopes.
4U1145-61(Hen 715) is a good illustration of the different types of variability that may occur in a Be X-ray source. A dramatic change in the behaviour of the Hβ line was observed in April 78. Normally Hβ is seen in
emission, sometimes superimposed on a broad absorption feature, but on
28 April 1978 Hβ appeared purely in absorption. Also Hγ and Hδ were
strongly in absorption, the HeI lines were enhanced and the λλ4640-4650
blend was weakened. Figure 1 illustrates this strong variation in Hβ and
changes in the emission profile on short and long timescales.

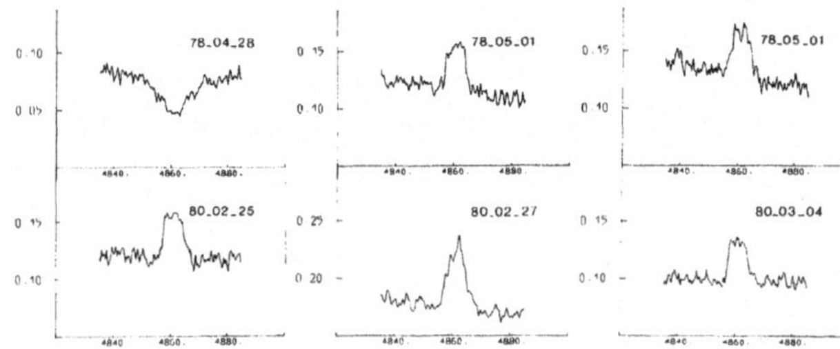

Figure 1. The variations of Hβ at different timescales. The intensities
are relative to the clear plate.

At the epoch when Hβ was in absorption no X-ray flux was detected. On
1 May 1978 Hβ appeared in strong emission (Hammerschlag-Hensberge et al.
1980). On 7 May 1978 the X-ray source was active and very bright (Jernigan
et al.1978). From 30 December 1979 to April 1980 Hen 715 became fainter by
0.25 magnitude. Simultaneously the IR excess disappeared, suggesting the
loss of the circumstellar envelope (Pakull et al.1980). Between 25 February and 8 March 1980 Hβ was in emission as well as the FeII lines.

It is necessary to study the variability over the whole spectral range
(from X-rays to infrared) to determine whether the Be X-ray sources are
binaries or not. Although for none of these sources an optical orbit has
been derived some arguments point to their binary character :
   a) X-ray pulsation which is the signature of a rotating compact object
(modulated by orbital motion in the case of 4U0115+63)
   b) several classical Be stars are known to be binaries (17 Lep, AX Mon,
HR 894). Such systems could be precursors of Be X-ray systems.

THE OPTICAL SPECTRA OF BE PRIMARIES IN X-RAY SOURCES

In comparing the optical spectra of some Be primaries in X-ray sources we
concentrate on the emission lines. Table 2 lists the behaviour of the
strongest lines of three Be X-ray sources : X Per, A1118-61, Hen 715.
We conclude that Hen 715 is a Be star intermediate between A1118-61
(strong emission lines) and X Per (1976-77). A1118-61 looks like X Per
during the beginning of the sixties.
For massive X-ray binaries with short orbital periods the HeII λ4686 line
is often observed in emission indicating that this line is formed in the
atmosphere of the primary (either by the process occurring in Of stars or

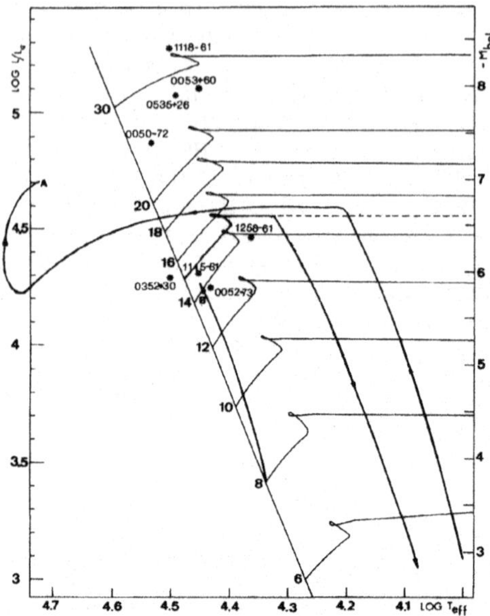

Figure 2. Evolutionary tracks and the positions of the Be components of Table 1. Note that X Per is found below the ZAMS (cf. Persi et al.1977). To reach the ZAMS a distance of 430 pc would be necessary.

	$T/10^6$yrs	$M/M_\odot$	$R/R_\odot$	log L	log T_{eff}	P(d)
Start ZAMS	O	15	5	4.26	4.48	5
Start mass transfer	8.4028	15	15	4.59	4.32	5
Min luminosity	8.4086	8.53	11.1	2.10	3.76	4.6
He ignition	8.4299	3.48	26	4.55	4.19	23
End mass transfer	8.4384	3.3	28.6	4.6	4.19	31
C ignition	9.446	3.3	1.8	4.64	4.79	31

Table 3. Evolution of a binary system with initial masses of $15M_\odot$ + $8M_\odot$.

REFERENCES

Code,A.D.,Davis,J.,Bless,R.C.,Hanbury Brown,R.,1976,Astrophys.J.203,417
De Grève,J.P.,de Loore,C.,1977,Astrophys.Space Sci.50,75
Doazan,V.,1976, in IAU Symp.N°70 "Be and Shell Stars", ed.A.Slettebak
    (Reidel,Dordrecht),p.37
Hammerschlag-Hensberge,G.,van den Heuvel,E.P.J.,Lamers,H.J.G.L.M.,Burger,
M.,de Loore,C.,Glencross,W.,Howarth,I.,Willis,A.J.,Wilson,R.,Menzies,J.,
Whitelock,P.A.,van Dessel,E.L.,Sanford,P.,1980,Astron.Astrophys.85,119
Hubert-Delplace,A.M.,Hubert,H.,1979, "Un Atlas des Etoiles Be", Observa-
    toire de Paris-Meudon.
Hutchings,J.B.,1976, in IAU Symp.N°70 "Be and Shell Stars", ed.A.Sletteba
    (Reidel,Dordrecht),p.13
Janot-Pacheco,E.,Ilovaisky,S.,Chevalier,C.,1981,Astron.Astrophys.,in pres
Jernigan,J.,Bradt,H.,van Paradijs,J.,Rappaport,S.,1978,IAU Circ.N°3225
Packet,W.,1981,preprint

Pakull,M.,Motch,C.,Lub,J.,1980,IAU Circ.N°3476
Persi,P.,Viotti,R,Ferrari-Toniolo,M.,1977,M.N.R.A.S.181,685
Rappaport,S.,Clark,G.,Cominsky,L.,Joss,P.,Li,F.,1978,Astrophys.J.224,L1
Underhill,A.,Divan,L.,Prevot-Burnichon,M-L.,Doazan,V.,1979, MNRAS 189,601.

DISCUSSION

<u>Viotti</u>: Since you mentioned the presence of Fe II emission lines in
the optical spectra of some x-ray/Be stars, I would like to stress the
importance of the "Fe II problem" in the Be phenomenon. These lines,
which have a large range of excitation potentials, oscillator strengths
and wavelengths might give important information on the outer envelopes
or rings or so of Be stars.
I also noted the large range of ionization energies (from Si II to
N V) you quoted. If I am correct this large range is mostly concerned
with resonance lines, while excited lines (N III, O IV) belong to a
smaller ionization energy range.

<u>Giovannelli</u>: About Hen 715: Have you searched the correlation between
the variations of the doubling in $H\beta$ and the orbital period?

<u>de Loore</u>: The analysis of the spectra of Hen 715 is in progress, but
what you mention has not been performed.

<u>Pakull</u>: There seems to be a correlation between the He II 4686
emission and L_x/L_{opt} in the close roche-lobe filling steady x-ray
binaries. In the case of the Be star x-ray systems, which are mostly
transients, however, optical observations have been carried out mostly
some time after the x-ray outbursts. Accordingly one would not expect
such a correlation.

<u>de Loore</u>: I agree with this remark. The He II 4686 A intensity is
in massive x-ray systems connected with the L_x/L_{opt} ratio (Hutchings,
1980). Probably this is related with the geometry of these systems,
presence of disks, etc., and in long period systems, as the Be x-ray
binaries, the conductors for the existence of the He II emission are
not there.
However, the correlation seems to be present, although I see no
explanation, since, as you mention, the x-ray value is unknown at the
moment of the observations in the visual for variable transient
sources.

ARE CLASSICAL Be STARS SOURCES OF HARD X-RAYS?

Geraldine J. Peters
Department of Astronomy
University of Southern California
Los Angeles, CA   90007   USA

ABSTRACT

A point summation technique was used to search the UHURU data base for X-ray emission from classical Be stars. Of the thirty-two stars considered, only three (γ Cas, HR 4009, and HD 187399) were detected at the 3.3σ level or higher. For reasons discussed in this paper, HD 45314, π Aqr, κ Dra, and 48 Per are considered to be <u>possible</u> detections. The X-ray emission from γ Cas from late 1970 through early 1973 is discussed.

1. INTRODUCTION

The optical counterparts of intrinsically strong galactic X-ray sources are usually early type emission line stars. But, in general, these objects are not the so-called "classical" Be stars (B stars of luminosity classes III – V whose spectra appear more or less normal except for the presence of Balmer and, perhaps, Fe II emission and/or shell features). The classical Be stars γ Cas and X Per are known to be the optical counterparts of hard X-ray sources but these sources are intrinsically weak ($<5\times10^{33}$ erg s^{-1}).

In order to determine whether weak X-ray emission is commonplace among classical Be stars a point summation technique (Ulmer and Murray 1976) was used to search the UHURU data base for evidence of 2-6 keV X-rays from these objects. The input data were the UHURU superposition data sets and, unless otherwise indicated, all data were from the narrow ($\frac{1}{2}°\mathrm{x}5°$) collimator. The individual superposition plots were visually inspected and data sets were eliminated if there was excessively high noise, an irregular or undulating background, or a source nearby which contaminated either the candidate object or the background. Thirty-two objects were considered which include Be stars which have recently been active in the visible spectral region, Be mass transfer binary systems including the peculiar binaries β Lyr and HD 187399, Be stars classified as B0ne (like γ Cas and X Per), and all Be stars within 100 pc of the sun.

M. Jaschek and H.-G. Groth (eds.), Be Stars, 353–357.
Copyright © 1982 by the IAU.

2.   GENERAL RESULTS

If the point summation produced a signal of 3σ or higher and only
data from the high resolution collimator were included, the object is
considered to be a <u>detected</u> source. If 2.5 < σ < 3.0, or if σ > 3.0
but data from the wide collimator were used, the star is considered to
be a <u>possible</u> source. The results appear in Table 1. Intrinsic fluxes
in the interval 2-6 keV are listed for all detected and possible
sources; otherwise, 3σ upper limits are quoted. Most distances used
for the computations are from Schmidt-Kaler (1964).

Only seven of the program stars are considered to be detected/
possible X-ray sources. In general, the upper limits for the nearest
objects are less than 10^{32} erg s^{-1}. Upper limits for the BOne stars
are comparable with the X-ray flux observed from X Per. Classical Be
stars do <u>not</u> appear to be sources of hard X-rays.

3.   INDIVIDUAL RESULTS

a. Detected Sources

<u>γ Cas</u> (MX0053+60) was discovered to be a hard X-ray source (in
SAS-3 data) by Jernigan (1976). The fact that γ Cas was not included
in the third <u>UHURU Catalog of X-ray Sources</u> has led to the popular
belief that the X-ray source was off during the early 1970's. Careful
examination of the UHURU data base has shown that the source, although
variable, persisted throughout the lifetime of the satellite. On the
average, the source was 4.1 ± 0.3 cts s^{-1} (from the point summation).
It was observed on the first useful UHURU superposition plot as well as
on one of the last. The flux results from 36 good data sets are presented in Figure 1. The source was detected in 20 of the observation
sets; 3σ upper limits were computed for the others. χ^2 tests performed
on the data (both summed data for each superposition set and the pass
by pass counting rates) show that the source is quite variable. In
fact, it was not uncommon to observe variations of a factor of three
between successive passes. Although the daily average UHURU flux was
always below the 15 cts s^{-1} first recorded by SAS-3, individual obser-

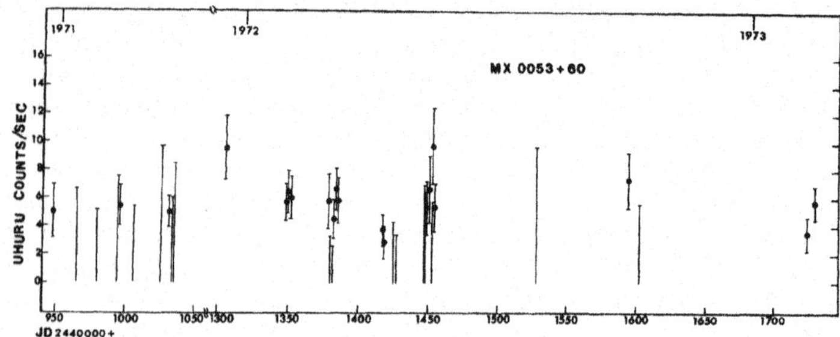

Fig. 1 - UHURU observations of MX0053±60 (γ Cas) from 1970 December
28 to 1973 February 16. 3σ upper limits are also indicated.

Table 1 - X-ray flux from Be stars

Star	Sp Type	d (pc)	F_{2-6} erg s^{-1}
γ Cas	BOIVne	220	4.0 x 10^{32}
HD 45314	BOIVne	550	6.5 x 10^{32}
HD 53367	BOIVne	670	<8.1 x 10^{32}
HD 153261	BOIVne	800	<2.2 x 10^{33}
HD 161306	BOIVne	910	<2.4 x 10^{33}
HD 203374	BOIVne	700	<1.2 x 10^{33}
HD 204116	BOIVne	950	<1.6 x 10^{33}
HD 206773	BOIVne	900	<1.8 x 10^{33}
HR 2855	BOIVpne	760	<1.2 x 10^{33}
AX Mon	BOpe	560	<5.7 x 10^{32}
ϕ Per	B0.5IV-Vnne	200	<1.2 x 10^{32}
π Aqr	B0.5Vne	325	5.0 x 10^{32}
HR 2142	B1IV-Vnne	400	<4.5 x 10^{32}
HD 173219	B1pe	900	<3.1 x 10^{33}
υ Cyg	B1.5IV-Ve	290	<1.5 x 10^{32}
HR 4009	B1.5IVe	715	9.8 x 10^{32}
HD 7636	B2Ve	305	<1.8 x 10^{32}
μ Cen	B2IVe	180	<6.9 x 10^{31}
η Cen	B2IVe	100	<3.4 x 10^{31}
66 Oph	B2Ve	280	<4.2 x 10^{32}
48 Per	B3IVe	175	2.2 x 10^{32}
HD 218393	B3IVne	550	<6.8 x 10^{32}
HD 51480	B3pe	500	<4.6 x 10^{32}
α Eri	B5IV(e)	39	<3.0 x 10^{30}
ψ Per	B5pne	160	<5.7 x 10^{31}
o And	B6pn(e)	100	<1.8 x 10^{31}
κ Dra	B7IVe	100	1.5 x 10^{31}
28 Tau	B8pne	130	<7.4 x 10^{31}
β Lyr	B8IIe	200	<1.4 x 10^{32}
HD 187399	B8IIIe	450	3.7 x 10^{32}
17 Lep	B9pe	100	<2.5 x 10^{31}
3 Pup	A2Ibpe	600	<6.1 x 10^{32}

vations often exceeded the latter value. The longest segment of semi-continuous data (1972 May 8-16) were analyzed for X-ray pulsations but no period was found. Further details of the UHURU observations of γ Cas will be published separately.

According to Polidan (Polidan, Locke, and Parmar 1981), ten Copernicus X-ray observations of γ Cas from 1973 November 4 to 1979 September 25 have revealed a behavior similar to that seen during the lifetime of UHURU. On two occasions, the source was definitely off. Quite noteworthy is an X-ray flare event which coincided with the UV event reported by Slettebak and Snow (1978). Apparently, during the course of the observation, the flux discontinuously increased from zero to about 10 UHURU cts s^{-1}.

HR 4009 (HD 88661) was detected at the 3.8σ level and, thus, is a
good candidate for a hard X-ray source. Thirty-four data sets were included in the superposition yielding a flux of 0.9±0.2 cts s^{-1}.
HR 4009 is in a crowded portion of the sky, however. An error box of
0.3 deg^2 generated from four lines of position in the UHURU files contains nine other SAO stars (all much fainter), one of which is close
enough to have contributed to the summed intensity. Recent IUE observations of HR 4009 reveal the presence of highly violet shifted
(≃600 km s^{-1}) C IV and, perhaps, Si IV lines.

HD 187399 is a spectroscopic binary with a large mass function and,
according to a suggestion by Hutchings and Redman (1973), may harbor a
black hole secondary. Although this star is in Cygnus, 28 data sets
were found suitable to be included in the point summation. If one
adopts a distance of 450 pc (Hutchings and Laskarides 1972), the summed
UHURU flux of 0.9 ± 0.3 cts s^{-1} suggests an intrinsic flux of 4x10^{32}
erg s^{-1}, 10^4 times lower than one would expect if the secondary is a
collapsed object. Hutchings (1981) failed to find soft X-ray emission
from HD 187399 in a single IPC observation with the Einstein satellite.
It is, of course, reasonable to assume that the source is highly variable

b. Possible Sources

HD 45314 displays a slightly variable, low contrast ground-based
spectrum. 16 good UHURU data sets suggested a flux of 1.1±0.4 cts s^{-1}
(2.6σ). The intrinsic flux is comparable with the one obtained for γ Cas.

π Aqr and κ Dra are also being considered as possible X-ray sources.
Since both are in regions of the sky relatively free from UHURU sources,
data from both narrow and wide collimators were used. When all data
were included, both showed 3σ results.

48 Per (MX Per) was suggested as a possible optical identification
for 4U0404+47. It is the only SAO or known variable star in the error
box. The point summation, including 44 data sets, gave only a 2.1σ
result (0.5±0.2 cts s^{-1}). Considering all data, a χ^2 test failed to suggest
variability. However, four daily averages gave fluxes of 3-8 cts s^{-1}.
If 48 Per is indeed associated with 4U0404+47, the X-rays are transient.

REFERENCES
Hutchings, J.B.: 1981, Publ. Astron.Soc.Pacific 93, pp.55-59.
Hutchings, J.B., and Laskarides, P.G.: 1972, Mon.Not.Roy.Astron.Soc.
      155, pp. 357-371.
Hutchings, J.B., and Redman, R.O.: 1973, Mon.Not.Roy.Astron.Soc.
      163, pp. 209-217.
Jernigan, J.G.: 1976, IAU Circ. No. 2900.
Polidan, R.S., Locke, M., and Parmar, A.N.: 1981, in preparation.
Schmidt-Kaler, T.: 1964, Bonn Veroffentl. 70.
Slettebak, A., and Snow, T.P.: 1978, Astrophys.J. 224, pp. L127-L131.
Ulmer, M.P., and Murray, S.S.: 1976, Astrophys.J. 207, pp. 364-366.

DISCUSSION

<u>Endal</u>: Can we assume that the Be stars in x-ray binaries are normal (typical) Be stars?

<u>Peters</u>: Not "classical" Be stars. With the exception of HD187399, all objects which showed up at the 3 σ level or higher are "normal" Be stars.

<u>Henrichs</u>:(additional answer): It is likely that in the case of (wide) x-ray binaries with Be companions, the Be star does not "know" that there is an orbiting compact object around. In other words: in these binaries the Be phenomenon is <u>not caused</u> by the presence of a companion <u>now</u>. It is, however, conceivable that in this case the Be phenomenon is caused by the nearby presence of the companion <u>in the past</u>, when the system was not yet evolved. This might be true for other Be binaries as well.

VII. UV OBSERVATIONS AND MASS LOSS

ULTRAVIOLET OBSERVATIONS, STELLAR WINDS, AND MASS LOSS FOR Be STARS

J.M. Marlborough
Astronomy Centre, The University of Sussex, Falmer, U.K.
and
Department of Astronomy, The University of Western Ontario,
London, Ontario, Canada.

1. INTRODUCTION

Although the first ultraviolet (UV) observation of an astronomical
source was obtained in 1946, the first UV observations of Be stars
were not obtained until 1964. In this review of UV data covering the
period since 1964, the term Be star will be assumed to include Oe
stars as well (Frost and Conti, 1976). An earlier review of this
subject is by Heap (1976).

Ultraviolet astronomy refers generally to the wavelength interval
$912 < \lambda < 3000$ A, these limits arising respectively from the absorbing
properties of the interstellar gas and the earth's atmosphere. This
wavelength range is sometimes further divided into the near UV,
$2000 < \lambda < 3000$ A, and the far UV, $912 < \lambda < 2000$ A. In the near UV balloons
can transport detectors to sufficiently high altitude at which a
significant residual signal remains. For the far UV however rockets
and satellites are the only practical vehicules. Observations of Be
stars have been obtained in a variety of space experiments. Some of
the characteristics of these are given in Table 1.

The UV spectra of Be stars resemble closely those of B stars,
although there are important quantitative differences. Qualitatively
the two obvious differences between the optical and UV for Be stars
are the lack of emission lines and the high density of absorption
lines. Although infrequently present, emission lines have been noted
or suspected at N V $\lambda 1240$ (Morton, 1976; Marlborough and Snow, 1980),
C IV $\lambda 1550$ (Bohlin, 1970), Mg II $\lambda 2800$ (Kondo et al., 1975) and
Si IV $\lambda 1400$ (Marlborough et al., 1978). There appears to be no
correlation between the presence of UV emission lines and the strength
of Balmer emission lines, even though at optical wavelengths when the
latter are strong, emission lines of Fe II are present (McLaughlin,
1961). Absorption lines can be either photospheric, circumstellar, or
interstellar. Separation of photospheric and interstellar lines is
straightforward, especially if v sin i is moderate or large. Distinction between circumstellar and interstellar lines is also possible,

M. Jaschek and H.-G. Groth (eds.), Be Stars, 361–376.
Copyright © 1982 by the IAU.

TABLE 1

Space Experiments in which Be Stars Have Been Observed

Experiment	Spacecraft	Launch Date	Wavelength range or λ_{eff} (A)	Resolution (A)	Reference
A. Filter Photometry					
——————	1964 83C	1964	1376		Smith(1967)
Celescope	OAO-2	1968	4 filters (110-3000)		Davis et al.(1972)
U. Wisconsin	OAO-2	1968	12 filters (1330-4250)		Code et al.(1970)
——————	rocket	1971	950		Troy et al.(1975)
B. Low Resolution Spectrophotometry					
U. Wisconsin	OAO-2	1968	1800-3800	20 or 200	Code et al.(1970)
			1050-2000	10 or 100	
UVS	Mariner 9	1971	1100-1900	7.5	Lillie et al.(1972)
			1500-3400	15	
S2/68	TD-1	1972	1350-2550	35-40	Boksenberg et al.(1973)
S-169	Apollo 17	1972	1180-1680	11	Henry et al.(1975)
S-019	Skylab	1973	1300-5000	variable	O'Callaghan et al.(1977)
Orion-2	Soyez 13	1973	2000-5000	variable	Gurzadyan(1975)
UV Exp	ANS	1974	1500-3300	150	Aalders et al.(1975)
——————	rocket	1977	912-3100	15	Brune et al.(1979)
C. High Resolution Spectroscopy					
——————	rocket	1968	1060-2130	2.0	Bohlin(1970)
——————	rocket	≤1970	~1200-~1600	0.5	Smith and Stecher(1971)
——————	rocket	1970	~1100-~1700	0.5	Morton et al.(1972)
S59	TD-1	1972	2060-2160	1.9	de Jager et al.(1974)
			2495-2595	2.3	
			2775-2875	1.8	
Princeton U.	Copernicus	1972	~750-1500	0.2(U2),0.05(U1)	Rogerson et al.(1973)
			~1500-3200	0.4(V2),0.10(V1)	
——————	rocket	1972	1100-2050	0.1	Heap(1975)
BUSS	balloon	1976	2200-3400	0.1	Kondo et al.(1979)
——————	IUE	1978	1150-1950	0.08-0.2,≤7	Boggess et al.(1978)
			1900-3200	0.2,≲8	

although more difficult, because circumstellar lines may be narrower
than photospheric ones. The standard approach for study of optical
spectra can be used to distinguish between photospheric and circumstellar lines.

As expected the UV spectra of B stars were found to contain a very
large number of absorption lines (Burger and van der Hucht, 1976;
Rogerson and Upson, 1977). This high line density leads to a number
of problems. Identification of spectral features on low resolution
spectra is complicated because all features are blends; assignment of
a single contributor to a particular feature may be very misleading if
not totally incorrect. In principle this difficulty can be reduced by
obtaining spectra of higher resolution. However as Kurucz (1974) has
demonstrated rapid rotation leads to an intrinsic blending which
cannot be overcome by higher instrumental resolution. Finally both
Kurucz (1974) and Peytremann (1975) have emphasized that the large
number of lines in the UV together with limited resolution, either
instrumental and/or intrinsic, make the location of the continuum
very uncertain.

2. THE CONTINUOUS ENERGY DISTRIBUTION

Observations of the continuous energy distribution in the UV for Be
and B stars are very important for meaningful tests of the predictions
of model atmospheres and of the effects of rotation on energy distributions, and also for better estimates of T_{eff} and bolometric corrections.
Furthermore the amount and distribution of radiation in the Lyman
continuum is an important input quantity in constructing models of the
circumstellar envelope (CE) of Be stars.

Beeckmans and Hubert-Delplace (1980) have determined the ratio of
the strength of the UV continuum (1650<λ<2500 A) to that of the
Paschen continuum in 63 stars using S2/68 spectra. For each Be star
they compared this ratio with the expected value for a normal star of
the same spectral type, the latter being obtained from analysis of
spectra of 200 normal B stars. They conclude that: Be stars earlier
than B5 have less UV flux at λ2100, relative to the optical V band,
than do normal B stars; Be stars later than B5 may also have this flux
deficit but the conclusion is less certain; and the magnitude of this
flux deficiency is correlated with optical emission line strength,
infrared excess and the presence of a shell. An earlier study by
Briot (1976) found essentially no difference in UV flux of Be stars
relative to B stars; her conclusions however are suspect because of the
method used to correct for interstellar extinction.

At the present time this UV flux deficiency may be explained by the
combined effects of line blocking together with hydrogen bound-free
absorption, especially in stars with conspicuous shells. Poeckert and
Marlborough (1978b) have shown that the predicted strength of both
Balmer emission lines and the infrared continuum for ad hoc models of

the CE increase with increasing envelope density. Snow et al. (1979)
discovered UV Fe III absorption lines in a number of Be stars whose
optical spectra showed no evidence for shell lines. Marlborough and
Snow (1980) noted that significant absorption by UV Fe III was
evident in 59 Cyg at a time when the optical spectrum was expected to
have strong Balmer emission but at most only weak shell lines. Finally
Beeckmans (1976) showed that the UV continuum of 59 Cyg was much fainter,
when the optical spectrum displayed strong shell lines, than at times
when no optical shell lines were present.

Various groups have obtained absolute UV spectrophotometry of B
and Be stars. In the range $912<\lambda<3100$ A, Brune et al. (1979) observed
5 stars including the Be star α Eri. Although α Eri has a stellar wind
(Marlborough and Snow, 1976) its spectrum shows Balmer emission only
occasionally (Andrews and Breger, 1966). Thus its UV flux deficiency
is expected to be small, since Beeckmans and Hubert-Delplace (1980)
noted that stars with weak Balmer emission had little, if any, UV flux
deficit. The absolutely calibrated flux of α Eri agrees well with
models for $\lambda> 1200$ A and T_{eff} obtained by Brune et al. is the same as
that of Code et al. (1976). If α Eri is representative of Be stars
with no significant UV flux deficit, then Brune et al.'s analysis,
together with the conclusions of Nandy and Schmidt (1975) for normal
stars of types B0-A2, suggest that theoretical models can reproduce
satisfactorily the observed UV fluxes for $\lambda> 1200$ A. Unfortunately
this is not so for $\lambda< 1200$ A. The best available models do not predict
the observed fluxes for $\lambda< 1200$ A. Even for a star as cool as α Eri
the model fluxes are 15% too large for $912<\lambda<1200$ A. This discrepancy
increases for hotter stars and appears now to be a serious problem for
O stars (Massa and Conti, 1980) and probably for Oe stars as well.
Since models overestimate the flux for $912<\lambda<1200$ A by at least 15%,
the Lyman continuum fluxes of models are probably too large as well.
Thus quantitative predictions based upon models of the circumstellar
matter around Be stars,e.g.Poeckert and Marlborough (1978a), are
perhaps even more uncertain than previously thought.

Collins and Sonneborn (1977) have demonstrated how the flux from a
star rotating at breakup speed and viewed equator on differs from that
of a non-rotator; the rotator can be up to 4^{m} fainter in the UV. This
is an extreme example however because there is no evidence that Be stars
rotate at breakup speed and because the flux difference increases
nonlinearly with rotation rate. Nevertheless differences between
rotating and non-rotating stars are expected; it should be recalled
however that the quantitative predictions are model dependent. If Be
stars are to be used to test these predictions corrections for line
blocking in the CE and for interstellar extinction must be applied.
Heap (1975,1976) compared observations of ζ Tau to the predictions of
Collins (1974) and found agreement could be obtained only if ζ Tau
rotated at breakup speed. There are, however, several problems with
her analysis. Because of an erroneous stellar temperature Heap compared the observations to a model of a B1 V star instead of a cooler
one. She estimated $E(B-V) \simeq 0.01$; Beeckmans and Hubert-Delplace (1980)

suggest instead 0.05 to 0.06. If these latter values are more
reliable Heap has underestimated considerably the correction for
extinction. Both effects act to increase the difference between the
model predictions and the corrected observations of ζ Tau. Because
of problems inherent in correction for line blocking in the CE plus
interstellar extinction, it may be prudent to use Be stars like α Eri,
which have little UV flux deficiency, or better still rapidly rotating
B stars which have never demonstrated Be characteristics.

3.   SPECTRAL CLASSIFICATION AND ANALYSIS OF ULTRAVIOLET SPECTRA

Numerous investigations support the contention that the physical
processes which control the formation of the optical and UV spectra of
both Be and B stars of the same spectral type are the same. If however
this proves not to be so, then one can assert that the effects of the
different processes on optical and UV spectra of Be stars are below the
present level of detectability. Cucchiaro et al. (1979 and references
therein) have examined many S2/68 spectra of B and A stars and have
shown that a self-consistent classification scheme can be produced
which correlates well with optical criteria. Peters (1979) has shown
that the same physical parameters determine the optical and UV spectra
of two Be stars. Brune et al. (1979) have concluded that the UV energy
distribution of α Eri for $\lambda > 1200$ A can be well represented by a model
whose T_{eff} is consistent with the optical spectral type. There is no
evidence to suggest that the optical spectra of Be stars, which have
lost their emission lines, differ in any way from those of rapidly
rotating B stars of the same spectral type. Line ratios used to assign
optical spectral types for Be stars do not seem to be affected greatly,
or perhaps at all, by the CE. Therefore if differences are found
between the UV spectra of Be and B stars of the same type, the spectral
features which differ are likely to have a circumstellar origin.

Panek and Savage (1976) examined 118 OAO-2 spectra of O,B stars.
For dwarfs they discovered that $\lambda 1400$ (Si IV) is strongest for types
BO-B1 and disappears by B3, while $\lambda 1550$ (C IV) decreases in strength
from O to B stars and disappears between B2 and B3. For a group of Be
stars mostly earlier than B5 they concluded that $\lambda 1400$ was normal in
strength but $\lambda 1550$ was stronger than expected in some cases. Note that
Barbier and Swings (1979 and references therein) contend that in low
resolution spectra cooler than about B1, the major contributor to
$\lambda 1550$ is Fe III. Henize et al. (1976) have found $\lambda 1400$ (Si IV)/$\lambda 1550$
(C IV) to be smaller in Be stars (BO-B2) relative to non-emission B
stars; Jaschek (1979) however did not note this on S2/68 spectra. Lamers
et al. (1980a) state that $\lambda 2070$, primarily Fe III, is weaker in Be stars
compared to normal B stars and suggest the weakness is due to line
emission in the CE. However the lower UV continuum in Be stars with
moderate optical emission strength may complicate their interpretation.

These investigations and others considered later demonstrate
conclusively that some features in the UV spectra of Be stars differ
significantly from normal B stars of the same spectral type. Some of

these differences are due to the cool circumstellar envelope (CCE),
cool in the sense that its kinetic temperature, T_{kin}, is the same order
as T_{eff} for the star. Others, like absorption lines of O VI $\lambda 1035$,
N V $\lambda 1240$, C IV $\lambda 1550$ and Si IV $\lambda 1400$, indicate directly the
existence of additional sources of ionization and indirectly the
presence of hot regions in the CE. Quantitative attempts to interpret
these latter as photospheric lines lead to inconsistencies with optical
results and as such support one of Jaschek's (1977) arguments for the
necessity of an UV classification scheme.

Quantitative investigations of high resolution spectra of a few Be
stars have been undertaken. Peters (1979) studied both Copernicus U2
and ground based spectra of υ Cyg and μ Cen and concluded that the
observed profiles of UV lines could be fitted with the same model atmosphere, normal abundances and v sin i as determined from the optical
spectra. Heap (1975,1976) analysed a high resolution UV spectrum of
ζ Tau and obtained a temperature, significantly higher than that expected
from the spectral type, from the Si III/Si IV and C III/C IV ionization
balance. The best fits to the observed profiles of Si and C lines
necessitated abundances 1/5 of solar values and the widths of C IV lines
required v sin i < 200 km s^{-1} , considerably smaller than that from
optical spectra. In her analysis Heap assumed that all strong resonance
lines in the UV spectrum were photospheric and that the CE would not
contribute to lines of high stages of ionization. With regard to C IV
both assumptions are incorrect.

Serious contamination of the $\lambda 1550$ feature by photospheric C IV
and Fe III or by Fe III from the CE can be eliminated. Photospheric
C IV is not expected in a star as cool as ζ Tau and photospheric Fe III
is not expected to be strong (Panek and Savage, 1976; Barbier and
Swings, 1979). The $\lambda 1930$ Fe III feature, which shows a marked increase
in strength in supergiants, shell stars and Be stars of large v sin i
(Beeckmans, 1975), is observed in ζ Tau; Heap (1977) has concluded that
the individual Fe III lines at $\lambda 1930$ are formed in the CE. But, as
Beeckmans (1975) has emphasized, the lower levels of these Fe III lines
are all metastable or quasi-metastable, whereas those Fe III lines
contributing to $\lambda 1550$ have non-metastable lower levels. Thus the
relative contribution of Fe III to $\lambda 1930$ and $\lambda 1550$ is expected to be
similar to that shown by permitted and forbidden lines in the optical
spectra of shell stars (Struve and Wurm, 1938). The lines at $\lambda 1550$
in ζ Tau are therefore most likely C IV, but contrary to Heap they have
a circumstellar origin. Although a re-analysis of Heap's data has not
been performed it seems safe to conclude that abundances closer to solar
values and a temperature more consistent with the optical spectrum would
be obtained.

The low value of v sin i obtained from UV lines (Heap 1975,1976)
is a common phenomenon in Be stars of large v sin i. In a study of
ζ Oph Morton et al. (1972) noted that photospheric lines, such as C III
$\lambda 1247$, were very narrow and yielded much lower values of v sin i than
expected. As one possible explanation they proposed that the narrow

lines were formed in the more slowly rotating polar regions. In providing
the first quantitative explanation Hutchings (1976a) pointed out that
gravity darkening due to rapid rotation would produce different continuum
brightness distributions in the optical and UV so that the ratio of
polar to equatorial flux would increase with decreasing wavelength.
For the simple case of a rigidly rotating, undistorted star, which
radiates like a black body but suffers gravity darkening according to
von Zeipel's law, Hutchings demonstrated that, if each line has a small
thermal width relative to rotational broadening, a ratio of the width
at half maximum of an idealized line at $\lambda 4500$ to that of one at $\lambda 1000$
as large as 2 could occur for hot stars rotating near breakup speed and
viewed at an angle to the rotation axis exceeding 60°. Ratios this
large agree with the data of Heap and Morton et al. At viewing angles
less than 60°, no significant difference is predicted, just as Peters
(1979) observed.

Hutchings et al. (1979 and references therein) have applied this
approach to determine the inclination angle, i, to the rotation axis
and ω/ω_c, the ratio of angular velocity to the critical value for breakup,
for a number of O,B stars including many Be stars. According to their
results, Be stars have a random distribution of angles i and a broad
distribution of ω/ω_c with a maximum near 0.8; significantly Be stars
do not as a group rotate at breakup speed.

Because of the importance of these results one must subject the
analysis to close scrutiny. Although the conclusions are definitely
model dependent the crucial questions to be answered are how sensitive
are the predictions to variations of the assumptions used and how
important are effects which were not included. Hutchings (1976a) argues
that his conclusions are not greatly sensitive to moderate departures
from any of his assumptions. On the other hand Sonneborn and Collins
(1977) performed similar calculations but used a rotationally distorted
star and included the temperature and gravity dependence of actual lines.
Quantatively their results resemble Hutchings's but quantitatively
their range of half width variation is smaller than both the observations
of Heap (1975,1976) and the predictions of Hutchings; they also note a
strong dependence on the choice of basic stellar parameters.

Many other factors may influence Hutchings's conclusions. These
include the problem of differential versus rigid rotation and errors in
the estimate of the breakup speed from uncertainties in the equatorial
radius of the distorted star and in the mass,if estimated from the
spectral type (Slettebak et al., 1980). Finally evidence exists to
suggest that some B and Be stars have dynamic and/or unstable photospheres. This evidence includes: stellar winds in Be stars (discussed
below); radial motions in the photosphere of τ Sco (Smith and Karp,
1979); and β-Cephei-type variations in the Be stars ω CMa (Baade,1979)
and 19 Mon (Balona and Engelbrecht, 1979). If such phenomena are common
the line of sight velocity component of a photospheric mass element may
contain macroscopic velocity components in addition to rotational ones.
Recently Mihalas (1979) has shown that atmospheric velocity fields can

produce significant centre to limb variations of profiles in expanding
atmospheres. The assumption of a unique intrinsic profile with which
to deduce v sin i from observed line widths is thus highly questionable.
For ζ Oph, Hutchings and Stoeckley (1977) derive i = 55°, rotation
speed = 475 km s^{-1} giving v sin i = 390 km s^{-1}, and ω/ω_c= 0.85 for a
breakup speed of 560 km s^{-1} . However Walker et al. (1979) observed
the variation of distinctive features of the He I λ6678 line, which
they presumed to arise from a non uniform surface brightness distribution, and derived either v sin i = 560 km s^{-1} , if these non
uniformities are equatorial, or v sin i = 508 km s^{-1} using i=55° from
Hutchings and Stoeckley. In either case their estimate of v sin i
differs greatly from Hutchings and Stoeckley.

Therefore there are numerous effects which may be important but
were not included by Hutchings (1976a). Whether inclusion of any of
these will lead to different predictions is unknown. This problem has
many similarities to the LTE, non-LTE debate of a decade or more ago.
Perhaps the only way to convince oneself that Hutchings's approach is
satisfactory is to redo the analysis including as many of the above
effects as possible. Until this is done all one can do is to emphasize
again the warning expressed by both Mihalas and Sonneborn and Collins
that properties of rotating stars determined from observational data
are model dependent.

4.   THE COOL CIRCUMSTELLAR ENVELOPE

In the past the CCE has been investigated primarily at optical and
infrared wavelengths. Although reliable estimates of its T_{kin} are
difficult to obtain, $T_{kin} \sim 10^4$ K is consistent with photoionization by
Lyman continuum radiation, with infrared excesses due to free-bound and
free-free emission (Gehrz et al., 1974) and with the presence of singly
and doubly ionized ions in shell star spectra. Various arguments suggest
the CCE is more disk-like than spherically symmetric (Marlborough, 1976).
Disk-like however does not mean a disk of constant thickness since the
scale height of the CCE perpendicular to the equatorial plane is expected
to increase with distance from the rotation axis.

Because the UV flux deficiency seems to increase with increasing
envelope density and because one of the contributors to absorption at
λ2100 are Fe III transitions whose lower levels are metastable (Beeckmans, 1975), a Be star with a small but significant UV deficit could
be called a shell star even though its optical spectrum shows no shell
lines. Examples of such cases, where individual Fe III resonance lines
are present in Copernicus spectra, were noted by Snow et al. (1979).
Perhaps then, as Hutchings (1976b) suggested, shell stars are simply
Be stars in which the CCE has a larger average density and/or is much
more extensive than average.

If all Be stars are rapid rotators and if the CCE is concentrated
to the equatorial plane one might expect to discover a dependence of

Fe III line strength and UV flux deficiency on v sin i. However other factors such as envelope density, ionization structure and extent of the CE perpendicular to the equatorial plane will also influence Fe III line strength, so it is perhaps not surprising that Snow et al. (1979) found only a weak dependence on v sin i. Beeckmans and Hubert-Delplace (1980) found no pole-on Be stars to have a statistically significant UV deficit; nor does χ Oph even though it has very strong Balmer emission and an infrared excess. Since these stars have smaller than average v sin i the lack of a UV deficiency may support weakly a non-spherical CCE.

The radial velocities of lines of various stages of ionization led Snow et al. (1979) to conclude that both ionization and expansion velocity of the CCE increase outward. Both Marlborough (1977) and Snow et al. have suggested that all Be stars have stellar winds which have low expansion velocity near the star and degree of asymmetry determined by factors peculiar to each star. The CCE is then the deeper, denser and more slowly moving region of such a wind.

5.   ANOMALOUS STAGES OF IONIZATION AND THE HOT CIRCUMSTELLAR ENVELOPE

One of the interesting and important results of UV observations of O,B and Be stars was the discovery of lines from stages of ionization, especially O VI and N V but also C IV and Si IV in B stars, whose photospheric number densities had been expected to be insignificant were photospheric radiation the only ionizing source. These ions must have an origin in the CE and their presence requires additional sources of ionization (Conti and de Loore, 1979). In main sequence stars and Be stars the resonance lines from these stages of ionization are usually absorption lines.

Lamers and Snow (1978) considered the ionization conditions in the CE of O,B stars, assuming the ionization source was dilute photospheric radiation, and concluded that in main sequence stars one should not see O VI at all, N V in stars with $T_{eff} \lesssim 40,000$ K (~O7), and Si IV in stars with $T_{eff} \lesssim 22,000$ K(~B3). By interpolation using ionization potentials C IV would not be expected in dwarfs with $T_{eff} \lesssim 30,000$ K(~B1). Kamp's (1978) results for the Si spectra in B stars agree with Lamers and Snow. Studies of high and low resolution UV spectra of normal dwarfs by Panek and Savage (1976), Barbier and Swings (1979) and Upson and Rogerson (1980) confirm respectively the absence of photospheric C IV in stars cooler than about B1 and photospheric Si IV cooler than about B3. Snow and Morton (1976) and Rogerson and Upson (1977) note that O VI and N V are present in dwarfs hotter than B1; in normal dwarfs cooler than B1 there is no evidence for either ion. Therefore normal main sequence stars cooler than B2-B3 do not show any evidence in their UV spectra for anomalous stages of ionization.

For Be stars, O VI is definitely present at O9 (Morton, 1979) but

not confirmed in any cooler Be star. Lines of N V have been reported
in Be stars hotter than about B1 (Morton, 1976; Marlborough, 1977),
but not detected at B2 (Peters, 1979). Lines of Si IV occur in stars
as cool as B5 (Lamers and Snow, 1978). Recent results indicate that
both C IV and Si IV are present at B5 and Si IV probably in stars as
cool as B8 (Marlborough and Peters, 1982). These results are summarized
in Figure 1.

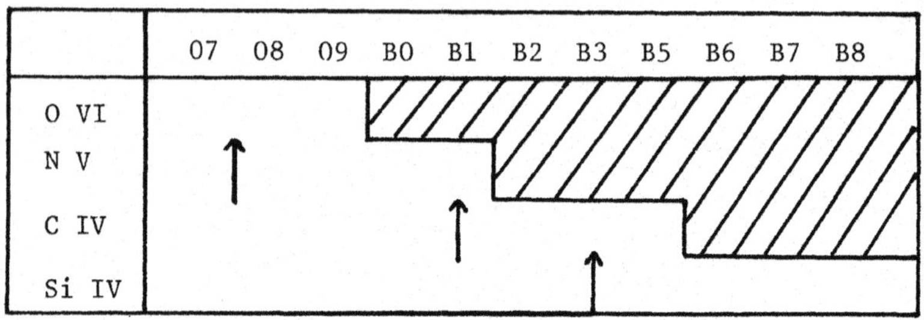

Figure 1.   Distribution of anomalous stages of ionization in Be stars.
The hatched region shows the range in which the ion is not detected in
Be stars; the vertical arrow, the coolest spectral type for the ion
to be present in normal dwarfs.

     Therefore, UV spectra covering the entire range of Be stars show
the presence of ions whose existence requires ionization sources in
addition to photospheric radiation. If this additional source is
collisional ionization, the CE must contain regions in which T_{kin} varies
from ~3 x 10^5K in late O and early B stars showing O VI and N V, to
~7 x 10^4K in late B stars showing C IV and Si IV (Lamers and Snow, 1978).
If the ionization is radiative some part of the CE must have $T_{kin} \gtrsim 10^6$K
to produce sufficient soft x-rays by bremsstrahlung (Cassinelli and Olson,
1979). In either case the CE contains hot regions (HCE) in addition to
cool ones.

6.   VARIABILITY OF ULTRAVIOLET SPECTRA

     Although few systematic searches for variations have been attempted
numerous reports document changes on time scales from hours to years.
Changes in the strength of Si III $\lambda\lambda$1388 and 1500 were detected by
Hammerschlag-Hensberge et al. (1980) in γ Cas over about 4 weeks. These
are probably real photospheric variations because, although λ1388 could
be affected by narrow, highly shifted components of Si IV λ1394
(Henrichs et al., 1980), there is no obvious contaminant for λ1500
and because the excitation energy of the common lower level of both
transitions is 17.7 ev.

Variability of spectral features ascribed to the CCE has been reported. Dramatic changes in the UV spectrum of 59 Cyg, concurrent with the formation and dissipation of optical shells, were discussed by Beeckmans (1976) and Marlborough and Snow (1980). Marlborough et al. (1978) noted changes over two years in the emission components of Mg II $\lambda 2800$ in the γ Cas; corresponding changes in Si IV core emission also seemed to occur (Slettebak and Snow, 1978). During simultaneous UV and optical observations of γ Cas, Slettebak and Snow (1978) reported that a sudden change in H α in a few hours seemed to be accompanied by an increase in Mg II emission and perhaps also in Si IV core emission.

Much data exist demonstrating variations in the HCE for the hotter Be stars. Large changes have occurred in the region around $\lambda 1240$: Marlborough and Snow (1980) and Doazan et al. (1980) discuss changes in N V in 59 Cyg over a period of years; Marlborough et al. (1978) show evidence for dramatic variations in γ Cas at $\lambda 1240$ in two hours. Modest C IV emission in γ Cas, noted by Bohlin (1970), was not detected later by Panek and Savage (1976) nor by Marlborough and Peters (1982). Panek and Savage did find significant variations in C IV and Si IV absorption in γ Cas over one year. Narrow absorption components in each of N V, C IV and Si IV in γ Cas, which appear in a time as short as a week and disappear within 2-3 months, were discovered by Henrichs et al. (1980). Features of this kind occur commonly in O,B star winds (Lamers et al., 1980b).

The UV variations that have been observed in the CCE seem to be analogous to the well-known optical variations, especially those of longer time scale. The observed changes in the HCE may be more closely related to those in the winds of O,B stars (Snow, 1979). Clearly systematic studies need to be undertaken to determine whether variations are common to Be stars of all spectral types and to discover the true time scales of variability.

7. STELLAR WINDS AND MASS LOSS

Radial velocities exceeding the escape speed in the region where the lines arise are generally considered to be direct evidence for mass loss. Mass loss and associated stellar winds have long been suspected in Be stars but never demonstrated conclusively until the advent of UV obser-vations. UV data which have been used to infer or demonstrate mass loss are summarised in Table 2. Column 3 contains the quoted radial velocity or the most negative value listed if several stars were dis-cussed. For main sequence stars of types O5,B0,B5 and A0, upper limits to the escape speed from the surface are respectively 920,960,800 and 700 km s^{-1} (Allen, 1973). Clearly the absolute values of many of the entries exceed these and thus provide evidence for mass loss. Note however that radial velocities exceeding the escape speed are obtained primarily from constituents of the HCE.

TABLE 2

Ultraviolet Features Indicating Winds and Mass Loss in Be Stars.

Star or Type of Star	UV Features Observed	RV (km s^{-1})	REFERENCE
γ Cas	C IV	<-450	Bohlin(1970)
ζ Oph	C IV	-1300	Smith and Stecher(1971)
ζ Oph	C IV,core	-900	Morton et al.(1972)
ζ Tau	UV resonance lines	-120	Heap(1975)
B0-B5,large v sin i	Si IV, short λ edge	~-1000	Snow and Marlborough(1976) Lamers and Snow(1978)
η Cen	Si IV, short λ edge	-250	Burton and Evans(1976)
B0-B3,large v sin i	resonance line cores	-240	Marlborough(1977)
γ Cas	Mg II, core	-215	Morgan et al.(1977)
ϕ Per	Fe III, λ 2062,2068,2079	-350	Bruhweiler et al.(1977)
59 Cyg	N V, short λ edge	<-830	Marlborough and Snow(1980)
59 Cyg	N V,C IV, cores	-750	Doazan et al.(1980)
γ Cas	N V,C IV,Si IV components	-1500	Henrichs et al.(1980)
γ Cas,X Per HD102567	Si IV, short λ edge	-800 to -1000	Hammerschlag-Hensberge et al.(1980)

Details of the determination of mass loss rates together with recent results are given by Snow (1982).

ACKNOWLEDGEMENT

I wish to thank Professor R.J. Tayler and the staff of the Astronomy Centre, University of Sussex, for their hospitality while this review was prepared. I have benefited greatly from discussions with friends and colleagues, especially P.K. Barker, G.W. Collins, D.G. Hummer, L. Mestel, G.J. Peters, R. Poeckert, R.C. Smith and T.P. Snow. I also wish to thank M. Rasche for her assistance. This work was supported by the Science Research Council (UK) and the Natural Science and Engineering Research Council (Canada).

REFERENCES

Aalders, J.W.G., van Duinen, R.J., Luinge, W., and Wildemann, K.J.:
     1975, Space Sci. Instrumentation 1, pp. 343-350.
Abbott, D.C., Bieging, J.H., Churchwell, E., and Cassinelli, J.P.: 1980,
     Astrophys. J. 238,pp. 196-202.
Allen, C.W.: 1973, 'Astrophysical Quantities', Athlone Press, London.
Andrews, P.J.,and Breger, M.: 1966, Observatory 86,pp.108-109.
Baade,D.: 1979, ESO Messenger, December, pp.4-6.
Balona, L.A., and Engelbrecht, C.: 1979, Monthly Notices Roy.Astron.Soc.
     189,pp.171-174.
Barbier,R.,and Swings, J.P.: 1979, Astron. Astrophys. 72,pp.374-375.
Beeckmans, F.: 1975, Astron.Astrophys. 45,pp.177-183.
Beeckmans, F.: 1976, Astron.Astrophys. 52,pp.465-466.
Beeckmans, F., and Hubert-Delplace, A.M.: 1980, Astron.Astrophys. 86,
     pp.72-86.
Boggess, A., Carr, F.A., Evans, D.C., Fischel, D., Freeman, H.R., Fuechsel,
     C.F., Klinglesmith, D.A.,Krueger, V.L., Longanecker, G.W., Moore,
     J.V., Pyle,E.J., Rebar, F., Sizemore, K.O., Sparks, W., Underhill,
     A.B., Vitagliano, H.D., West, D.K., Macchetto, F., Fitton, B., Barker,
     P.J., Dunford, E., Gondhalekar, P.M., Hall, J.E., Harrison, V.A.W.,
     Oliver, M.B., Sandford, M.C.W., Vaughan, P.A., Ward, A.K., Anderson,
     B.E., Boksenberg, A., Coleman, C.I., Snijders, M.A.J., and Wilson,
     R.: 1978, Nature 275, pp.2-7.
Bohlin, R.C.: 1970, Astrophys. J. 162,pp.571-587.
Boksenberg, A., Evans, R.G., Fowler, R.G., Gardner, I.S.K., Houziaux, L.,
     Humphries, C.M., Jamar, C., Macau,D., Macau, J.P., Malaise,D.,
     Monfils,A., Nandy, K., Thompson, G.I., Wilson R., and Wroe, H.:
     1973, Monthly Notices Roy.Astron.Soc. 163,pp.291-322.
Briot, D.: 1978, Astron.Astrophys. 66,pp.197-203.
Bruhweiler, F.C., Morgan, T.H., and van der Hucht, K.A.: 1978,
     Astrophys.J.Letters 225,pp.L71-L74.
Brune, W.H., Mount, G.H., and Feldman, P.D.: 1979, Astrophys.J. 227,
     pp.884-899.
Burger, M., and van der Hucht, K.A.: 1976, Astron.Astrophys. 48,pp.173-
     185.
Burton, W.M., and Evans, R.G.: 1976, in A.Slettebak (ed.), 'Be and
     Shell Stars, IAU Symp. 70',pp.199-207.
Cassinelli, J.P., and Olson, G.L.: 1979, Astrophys.J. 229,pp.304-317.
Code, A., Houck, T.E., McNall, J.F., Bless, R.C., and Lillie, C.F.:
     1970, Astrophys.J. 161,pp. 377-388.
Code, A., Davis, J., Bless, R.C., and Hanbury Brown, R.: 1976, Astrophys.
     J. 203,pp.417-434,
Collins, G.W.: 1974, Astrophys.J. 191,pp.157-164,
Collins, G.W., and Sonneborn, G.: 1977, Astrophys.J.Suppl. 34, pp.41-94.
Conti, P.S., and de Loore, C.B. (eds).: 1979, 'Mass Loss and Evolution
     of O-Type Stars, IAU Symp. 83', pp.169-234.
Cucchiaro, A., Macau-Hercot, D., Jaschek, M., and Jaschek,C.: 1979,
     Astron.Astrophys.Suppl. 35,pp.75-82.
Davis, R.J., Deutschman, W.A., Lundquist, C.A.,Nozawa, Y., and Bass,
     S.D.: 1972, in A.Code (ed,) 'The Scientific Results from the
     Orbiting Astronomical Observatory', NASA SP-310, Washington,pp.1-22.

Doazan, V., Kuhi, L.V., and Thomas, R.N.: 1980, Astrophys,J.Letters
      235, pp. L17-L20.
Frost, S.A., and Conti, P.S.: 1976, in A. Slettebak (ed.), 'Be and Shell
      Stars, IAU Symp. 70',pp.139-147.
Gehrz, R.D., Hackwell, J.A., and Jones, T.W.: 1974, Astrophys,J. 191,
      pp. 675 - 684.
Gurzadyan, G.A.: 1975, Space Sci.Rev. 18,pp.95-139.
Hammerschlag-Hensberge, G., van den Heuvel, E.P.J., Lamers, H.J.G.L.M.,
      Burger,M., de Loore,C., Glencross,W., Howarth,I., Willis,A.J.,
      Wilson,R., Menzies,J., Whitelock,P.A., van Dessel,E.L., and Sand-
      ford,P.: 1980,Astron.Astrophys. 85,pp.119-127.
Heap, S.R.: 1975, Phil.Trans.Roy.Soc.London A 279,pp.371-377.
Heap, S.R.: 1976, in A.Slettebak (ed.) 'Be and Shell Stars, IAU
      Symp. 70', pp. 165-178.
Heap, S.R.: 1977, Astrophys.J. 217,pp.90-94.
Henize, K.G., Wray,J.D., Parsons, S.B., and Benedict,G.F.: 1976, in A
      Slettebak (ed.) 'Be and Shell Stars, IAU Symp. 70',pp.191-195.
Henrichs,H.F., Hammerschlag-Hensberge,G., and Lamers,H.J.G.L.M.: 1980,
      paper presented at Second European IUE Conference.
Henry,R.C., Weinstein,A., Feldman,P.D., Fastie,W.G., and Moos,H.W.: 1975
      Astrophys.J. 201,pp.613-623.
Hutchings,J.B.: 1976a, Pub.Astron.Soc. Pacific 88, pp.5-7.
Hutchings, J.B.: 1976b, in A.Slettebak(ed.),'Be and Shell Stars, IAU
      Symp. 70',pp.13-27.
Hutchings, J.B., and Stoeckley,T.R.: 1977, Pub.Astron.Soc. Pacific
      89,pp.19-22.
Hutchings, J.B., Nemec,J.M., and Cassidy. J.: 1979, Pub. Astron.Soc.
      Pacific 91,pp. 313-318.
de Jager, C., Hoekstra,R., van der Hucht,K.A., Kamperman,T.M., Lamers,
      H.J., Hammerschlag,A., Werner,W., and Emming,J.G.: 1974, Astrophys.
      Space.Sci. 26,pp.207-262.
Jaschek,C.: 1977, in E. Muller (ed.), 'Highlights of Astronomy', 4 part
      II , pp.283-288.
Jaschek,C.: 1979, quoted by A.Slettebak: 1979,Space Sci.Rev. 23,pp.541
      -580.
Kamp,L.: 1978,Astrophys.J.Suppl. 36,pp.143-171.
Kondo,Y., Modisette,J.L., and Wolf,G.W.: 1975, Astrophys.J. 199,
      pp.110-119.
Kondo,Y., de Jager,C., Hoekstra,R., van der Hucht,K,A., Kamperman,T.M.,
      Lamers, H.J.G.L.M., Modisette,J.L., and Morgan, T.H.: 1979,
      Astrophys.J. 230,pp.526-533.
Kurucz,R.L.: 1974, Astrophys.J. Letters 188, pp.L21-L22.
Lamers, H.J.G.L.M., and Rogerson,J.B.: 1978,Astron.Astrophys. 66,
      pp.417-430.
Lamers,H.J.G.L.M., and Snow,T.P.: 1978, Astrophys.J. 219,pp.504-514.
Lamers,H.J.G.L.M., Faraggiana,R., and Burger,M.: 1980a,Astron.
      Astrophys. 82. pp.48-52.
Lamers,H.J.G.L.M., Gathier,R., and Snow,T.P.: 1980b, preprint.
Lillie,C.F., Bohlin,R.C., Molnar,M.R., Barth,C.A., and Lane,A.L.: 1972,
      Science 175, pp.321-322.
McLaughlin,D.B.: 1961, J.Roy.Ast.Soc.Can. 55,pp. 13-22 and pp.73-85.

Marlborough, J.M.: 1976, in A.Slettebak (ed.), 'Be and Shell Stars,
    IAU Symp. 70', pp. 335-370.
Marlborough, J.M.: 1977, Astrophys.J. 216, pp.446-456.
Marlborough, J.M., and Snow,T.P.: 1976, in A.Slettebak (ed.), 'Be and
    Shell Stars, IAU Symp. 70', pp. 179-189.
Marlborough, J.M., Snow,T.P., and Slettebak, A.: 1978, Astrophys.J.
    224, pp.157-166.
Marlborough, J.M., and Snow,T.P.: 1980, Astrophys.J. 235,pp. 85-96.
Marlborough, J.M. and Peters, G.J.: 1982 in M. Jaschek and H.G. Groth
    (eds.), 'Be Stars', IAU Symp. 98, pp. 387-390.
Massa,D., and Conti, P.S.: 1980, preprint.
Mihalas,D.: 1979, Monthly Notices Roy.Astron.Soc. 189,pp.671-699.
Morgan, T.H., Kondo ,Y., and Modisette , J.L.: 1977, Astrophys.J.
    216, pp.457-461.
Morton, D.C.: 1976, Astrophys.J. 203,pp.386-398.
Morton, D.C.: 1979, Monthly Notices Roy.Astron.Soc. 189,pp. 57-68.
Morton, D.C., Jenkins, E.B., Matilsky, T.A., and York, D.G.: 1972,
    Astrophys.J. 177, pp.219-234.
Nandy, K., and Schmidt,E.G.: 1975, Astrophys.J. 198, pp.119-125.
O'Callaghan, F.G., Henize, K.G., and Wray, J.D.: 1977, Applied Optics
    16, pp.973-977.
Panek, R.J., and Savage, B.D.: 1976, Astrophys.J. 206,pp. 167-181.
Peters, G.J.: 1979, Astrophys.J.Suppl. 39, pp.175-193.
Peytremann, E.: 1975, Astron.Astrophys. 39,pp. 393-403.
Poeckert,R., and Marlborough,J.M.: 1978a Astrophys.J. 220,pp.940-961.
Poeckert, R.,and Marlborough, J.M.: 1978b, Astrophys.J.Suppl.
    38, pp. 229-252.
Rogerson, J.B., Spitzer, L., Drake, J.F., Dressler, K., Jenkins, E.B.,
    Morton,D.C., and York,D.G.: 1973, Astrophys.J.Letters 181,
    pp.L97-L102.
Rogerson, J.B., and Upson, W.L.: 1977, Astrophys.J.Suppl. 35,pp.37-110.
Slettebak, A., and Snow,T.P.: 1978, Astrophys.J.Letters 224,pp.L127-L131.
Slettebak, A., Kuzma, T.J., and Collins, G.W.: 1980,Astrophys.J. 242,
    pp.171-187.
Smith, A.: 1976, Astrophys.J. 147, pp.158-171.
Smith, A., and Stecher, T.P.: 1971, Astrophys.J.Letters 164,pp.L43-L47.
Smith, M.A., and Karp, A.H.: 1979,Astrophys.J. 230,pp.156-161.
Snow, T.P.: 1979, in P.Conti and C.de Loore (eds.) 'Mass Loss and
    Evolution of O-Type Stars, IAU Symp. 83' pp.65-80.
Snow, T.P.: 1982, in M. Jaschek and H.G. Groth (eds.), 'Be Stars', IAU
    Symp. 98, pp. 377-385.
Snow, T.P., and Morton, D.C.: 1976, Astrophys.J.Suppl. 32,pp.429-465.
Snow, T.P., and Marlborough, J.M.: 1976, Astrophys.J.Letters 203,
    pp.L87-L90.
Snow, T.P., Peters, G.J., and Mathieu, R.D.: 1979, Astrophys.J.Suppl.
    39,pp.359-376.
Sonneborn, G.H., and Collins, G.W.: 1977, Astrophys.J. 213,pp.787-790.
Struve,O., and Wurm,K.: 1938, Astrophys.J. 88, pp.84-109.
Troy,B.E., Johnson, C.Y., Young, J.M., and Holmes, J.C.: 1975,
    Astrophys.J. 195, pp. 643-648.
Upson, W.L., and Rogerson, J.B.: 1980, Astrophys.J.Suppl. 42, pp.175-220.

Walker, G.A.H., Yang, S. and Fahlman, G.G.: 1979, Astrophys. J., <u>233</u>,
    pp. 199-204.

DISCUSSION

<u>Doazan</u>: I would like to comment on the UV "deficiencies" found by
Beekmans and Hubert-Delplace that you report. To accomplish the comparison of UV fluxes of Be stars relative to normal B stars, one has
to solve two separate problems. First, one has to find the correlation
of the interstellar extinction, because E(B-V) is contaminated by
intrinsic reddening. Second, one has to place the Be and B stars at
the same distance. If one normalizes the fluxes of B and Be stars at
the same magnitude V, one will systematically <u>underestimate</u> the UV
flux of the Be stars relative to the B stars because, as is well
known, Be stars are more luminous than B stars in the visual. The
"deficiency" obtained by the authors points toward the direction of
the expected results. So, the only conclusion the authors can derive
is that the <u>colours</u> of Be stars are <u>redder</u> than those of normal B's.
In no way the method used in this study allows to conclude anything
about the "excesses" or "deficiencies" of the UV fluxes of Be stars
relative to B's.

<u>Paterson-Beekmans</u>: In the work referenced, we have never considered
"absolute flux" deficiencies, but always UV fluxes relative to the
visible fluxes (in the V band).

<u>Slettebak</u>: With regard to the spectral type of ζ Tau from optical
spectra, I believe that the value you quote is too large. Morgan
classified this star as B1IV (it appears in the Morgan - Abt -
Tapscott spectral atlas of stars earlier than the sun) and I confirm
this type in my recent classification of all the Be stars brighter
than magnitude 6.0.

<u>Sonneborn</u>: First, the point of studying the properties of normal B
stars is very important. We should first determine the differences
(observationally and theoretically) in the UV spectrum between rotating and nonrotating normal B stars in order to help sort out the
effects of a circumstellar envelope. With respect to the UV energy
distributions, the models published in 1977 did not include the
effects of metallic line blanketing. Our new grid of models do
include line blanketing and therefore provide more realistic models
of the UV energy distribution in rotating stars. These models will
be published soon.

<u>Poeckert</u>: I would like to comment that winds and highly ionized
species are not found in SMC OB stars, according to Hutchings (Ap.J.
<u>237</u>, 285, 1980; paper in preparation 1981). Thus it appears that the
presence of coronae may be related to metal abundances.

STELLAR WINDS AND MASS-LOSS RATES FROM Be STARS

Theodore P. Snow, Jr.
University of Colorado at Boulder

Resonance-line profiles of SiIII and SiIV lines in 22 B and Be stars have been analyzed in the derivation of mass-loss rates. Of the 19 known Be or shell stars in the sample group, all but one show evidence of winds. It is argued that for stars of spectral type B1.5 and later, SiIII and SiIV are the dominant stages of ionization, and this conclusion, together with theoretical fits to the line profiles, leads to mass-loss rates between 10^{-11} and 3×10^{-9} for the stars. The rate of mass loss does not correlate simply with stellar parameters, and probably is variable with time. The narrow FeIII shell lines often seen in the ultraviolet spectra of Be stars may arise at low levels in the wind, below the strong acceleration zone. The mass-loss rates from Be stars are apparently insufficient to affect stellar evolution.

1. INTRODUCTION AND PREVIOUS RESULTS

The Be stars show a wide variety of phenomena, many of them seemingly unrelated or even in conflict with each other. No two stars behave exactly the same, and many, if not all, are variable with time.

In the 1970's a new ingredient was added to the Be star stew: stellar winds were discovered, first by Bohlin (1970), who found P Cygni profiles in the ultraviolet spectrum of γ Cas, and later by Snow and Marlborough (1976) and Marlborough and Snow (1976), who found asymmetric ultraviolet profiles indicative of high-velocity winds in a number of these objects. Later, Lamers and Snow (1978) added to the list of Be stars known to have winds, and it became apparent that high-velocity outflow from the stars has to be considered along with all the other phenomena associated with Be and shell characteristics.

Until now, very few attempts had been made to quantitatively determine the rates of mass loss associated with the winds from Be stars. Snow and Marlborough (1976) crudely estimated that 59 Cyg is losing matter at a rate between 10^{-10} and 10^{-9} $M_\odot$ yr^{-1}, but the first real quantitative analyses were carried out by Bruhweiler, Morgan, and van

M. Jaschek and H.-G. Groth (eds.), Be Stars, 377–385.
Copyright © 1982 by the IAU.

der Hucht (1978), who used metastable FeIII lines in the near-ultraviolet spectrum of ϕ Per to estimate $\dot{M} = 5 \times 10^{-11}$ for that star; and by Hammerschlage-Hensberge et al. (1980), who analyzed ultraviolet resonance lines in the spectra of γ Cas and X Per to derive values of $\dot{M}$ equal to 7×10^{-9} $M_\odot$ yr^{-1} and 1×10^{-8} $M_\odot$ yr^{-1}, respectively.

In the present study, Copernicus ultraviolet spectra of 22 early-to-mid-B objects, 19 of which have been recognized as Be or shell stars, have been studied with the aim of deducing mass-loss rates and then determining, if possible, how these rates are related to other stellar and circumstellar properties of the stars.

2. THE OBSERVATIONAL DATA

The Copernicus data used in this analysis were obtained primarily at a resolution of 0.2 A, although in a few cases a higher resolution of 0.05 A was used. All photometric corrections for stray light and backgrounds were carried out using standard procedures, and together are thought to introduce errors in intensity of no more than 20% of the continuum level. The details of the data reduction are outlined in a separate paper (Snow, 1981).

The stars observed were not selected in any systematic way, but simply represent the sampling of Be stars that happened to be observed by Copernicus at the appropriate wavelengths. In some cases the data were originally obtained specifically for studies of either stellar winds or Be stars, and nearly all the stars have been discussed in other publications by this author (Snow and Marlborough, 1976; Marlborough and Snow, 1976; Lamers and Snow, 1978; Snow, Peters, and Mathieu, 1979) or by others (e.g. Marlborough, 1977; Peters, 1976, 1979; Heap, 1976; Burton and Evans, 1976).

Table 1 lists the stars, along with assorted basic data taken from the references just mentioned. The values of v sin i are from Slettebak et al. (1975), or were scaled to their system, using the standard correlation. The T_{eff} entries are based on the listed values of $R_\ast$ and M_{bol}, which in turn are derived from standard values given in Code et al. (1976) for $R_\ast$; and from the compilations of Lesh (1968,1972), along with bolometric corrections from Code et al. for M_{bol}.

A number of the stars in Table 1 are known to undergo time variability. Particularly well known to have variable winds are γ Cas (Henrichs 1980), 59 Cyg (Doazan, Kuhi, and Thomas, 1980; Marlborough and Snow, 1980; and Doazan et al., this volume), and δ Cen (Snow, Oegerle, and Polidan, 1980). This variability, which is probably present in many of the other stars as well, will be mentioned again.

TABLE 1. THE STARS AND THEIR PROPERTIES

Star	HD	Spect.	V	E(B-V)	M_{bol}*	R_*+ ($R_\odot$)	T_{eff} ($^\circ K$)	v sin i** (km s^{-1})	$\dot{M}$ ($M_\odot$ yr^{-1})
γ Cas	5394	B0.5 IVe	2.58	0.08	-6.85	8.5	28,500	230:	6.0×10^{-11}
δ Sco	143275	B0.5 IV	2.33	0.16	-7.25	8.5	31,300	150	3.0×10^{-11}
η Ori	35411	B0.5 Vnn	3.42	0.11	-6.5:	8.5	26,300	40	9.3×10^{-11}
25 Ori	35439	B1 Vn	4.94	0.05	-5.71	8.0	22,600	260	2.2×10^{-9}
π Aqr	212571	B1 Ve	4.68	0.29	-6.08	8.0	24,600	300:	2.6×10^{-9}
	28497	B1.5 Ve	5.60	0.02	-5.21	7.5	20,800	290	7.8×10^{-10}
η Cen	127972	B1.5 V	2.3:	0.05	-5.21	7.5	20,800	260:	2.9×10^{-10}
59 Cyg	200120	B1.5 Ve	4.79	0.18	-5.21	7.5	20,800	350:	1.6×10^{-10}
δ Cen	105435	B2 IVne	2.59	0.12	-5.32	7.0	22,100	220:	7.4×10^{-11}
δ Cru	106490	B2 IV	2.80	0.00	-5.32	7.0	22,100	140	4.1×10^{-11}
δ Lup	138690	B2 IV	2.78	0.03	-5.32	7.0	22,100	210:	9.7×10^{-11}
ω CMa	56139	B2 IV-Ve	3.90	0.08	-4.76	7.0	19,430	80	1.4×10^{-10}
φ Per	10516	B2 IVep	4.06	0.22	-4.52	7.0	18,400	400	2.7×10^{-10}
μ Cen	120324	B2 IV-Ve	3.20	0.10	-4.72:	7.0	19,250	155	$<8.5 \times 10^{-11}$
υ Cyg	202904	B2 Ve	4.28	0.16	-4.52	7.0	18,400	200	6.3×10^{-11}
ζ Cen	121263	B2.5 IV	2.54	-.02	-4.22	6.8	17,400	160	3.8×10^{-11}
α Eri	10144	B3 Vp	0.48	0.04	-3.10	5.9	14,440	225:	1.3×10^{-10}
48 Per	25940	B3 Ve	4.03	0.17	-3.10	6.0	14,320	200	1.7×10^{-10}
ζ Tau	37202	B4 IIIp	2.95	-.01	-5.9:	6.0	27,300	300	2.4×10^{-10}
48 Lib	142983	B5 IIIp	4.87	0.06	-3.21	5.2	15,780	400:	6.7×10^{-11}
Ψ Per	22192	B5 Ve	4.25	0.12	-3.61	5.2	17,300	350	7.6×10^{-11}
o And	217675	B6p	3.62	0.05	-1.6:	4.7	11,400	280	1.2×10^{-10}

* Values of M_{bol} were taken from previous studies of these stars, primarily Lamers and Snow (1978) and Snow, Peters, and Mathieu (1979).

+ Values of R_* were based on standard calibrations of R_* with spectral type, most notably that of Code et al. (1976). In a few cases, stars from this list were specifically included in Code et al.

** Values of v sin i are from Slettebak et al. (1975), except for those that are underlined, which were taken from earlier compilations but converted to the system of Slettebak et al. using the correlation presented in their paper.

3. THE MASS-LOSS RATE ANALYSIS

Theoretical wind profiles from the atlas of Castor and Lamers (1979) were fitted to smoothed observational profiles of the SiIII and SiIV resonance lines at 1206 and 1393 A, respectively. In determining the fits, it was necessary to take into account the photospheric contributions to these lines, again using procedures given by Castor and Lamers.

The result of the fitting procedure was the derivation of the wind velocity law (which was not well-determined, but which also has very little effect on the mass-loss rate), a parameter that characterizes the changing ionization with height, and a total wind optical depth. These could then be combined with equations of radiative transfer and mass conservation to yield values of $\dot{M}$, the mass-loss rate.

A number of important uncertainties have to be kept in mind. Most significant of these is the ionization balance, since the mass-loss rates were based on the assumption that all the silicon is in the forms of SiIII and SiIV. It is well known that for the O and the earliest B stars, SiV is the dominant stage of ionization (e.g., Lamers, Gathier, and Snow, 1980), but the present data indicate that, for stars of type B1.5 and later, SiV is not important. Thus, the mass-loss rates for the program stars hotter than B1.5 may be underestimated due to the neglect of SiV, but for the cooler stars, this is probably not a serious omission.

When only SiIII or SiIV was observed, the mean SiIII/SiIV ratio for the winds of the other stars had to be assumed, introducing further potential errors.

Another major uncertainty was the measured value of the terminal velocity, which affects both the derived wind parameters from the profile fitting, and the calculation of $\dot{M}$. Detailed analysis shows, however, that the two effects compensate each other, and that the errors introduced into $\dot{M}$ by uncertainties in the terminal velocity are of the order of 30% or less, even if the velocity is underestimated by a factor of more than 2.

Other uncertainties arise from possible errors in the value of R_* that are adopted (a relatively minor effect) and from the possibility that the winds are not spherically symmetric. Furenlid and Young (1980) argue on the basis of Hα profiles that the winds are confined to the equatorial plane. Since the low v sin i stars in this sample show evidence for winds in the ultraviolet, this conclusion cannot be strictly true, but substantial enhancement in the equatorial plane is still possible. An attempt was made, following the method outlined by Hutchings (1976) and by Sonneborn and Collins (1977) to determine the stellar inclination angles from the visible and ultraviolet line widths, so that possible latitude-dependence of the winds could be assessed, but the data proved inadequate for this. Therefore, no sensible manner of allowing for latitude-dependence was found, and the rates of mass loss were derived on the assumption of spherical symmetry.

4. THE RESULTS

The derived values of $\dot{M}$ are listed in Table 1, where it is seen that they tend to lie between 10^{-11} and 3×10^{-9} $M_\odot$ yr^{-1}. These values are consistent with those found by earlier authors, except possibly for γ Cas, for which Hammerschlage-Hensberge et al. (1980) found a value about a factor of 100 higher. This may indicate that his hot Be star does have dominantly SiV in its wind.

The values of $\dot{M}$ in the present study do not correlate well with v sin i, T_{eff}, or M_{bol}. The scatter in these correlations may be due to intrinsic star-to-star variations of some unidentified parameter that controls the winds, to latitudinal dependence of the winds, or to time variations. In the latter regard, it has been shown that in at least one case (δ Cen) the rate of mass loss may vary by as much as a factor of 10 (Snow, Oegerle, and Polidan, 1980), and in two other cases (γ Cas and 59 Cyg) spectacular variations in the ultraviolet CIV and NV profiles have been seen (Henrichs, 1980; Doazan, Kuhi, and Thomas, 1980; Marlborough and Snow, 1980; Doazan et al., this volume) which reflect either changes in $\dot{M}$ or in the wind ionization.

5. THE FeIII SHELL LINES

The narrow FeIII shell lines seen in the ultraviolet spectra of many of these stars (Snow, Peters, and Mathieu, 1979) present a dilemma, for they are usually at or near rest, with velocity widths of about 50 km s^{-1}, in stars that, at the same time, have winds with speeds of several hundred km s^{-1}. There are two possibilities: (1) the shell lines arise outside of the wind zone, in a region where deceleration has occurred; or (2) they form in the wind, at a low level where little or no outwards acceleration has yet occurred.

From the present data it is possible to assess these alternatives somewhat quantitatively. First, the post-deceleration hypothesis (advocated by Thomas, 1980) presumably involves a build-up of material, perhaps in a stationary shock, at some distance from the star, probably quite far out. Snow, Peters, and Mathieu (1979) showed that typical shell column densities are 10^{19}-10^{21} cm^{-2}; at the rates of mass loss found in this study, such column densities would build up in times of order 10 years. Thus, for this hypothesis to work, there must be a loss mechanism capable of dispersing or ionizing FeIII to some other stage, otherwise column densities many orders of magnitude greater than those observed would build-up over the lifetime of a Be star.

To check the alternative hypothesis, the wind column density near zero velocity was computed by integrating the wind optical depth over

the appropriate velocity interval (about 50 km s^{-1}), with the result
that a shell of order 10^{18} cm^{-2} column density is easily maintained
within such an interval of velocity. Thus, the sharp FeIII lines could
be produced in the low-velocity portion of the winds, given two assumptions: (1) the ionization of iron changes abruptly from FeIII to high
stages at a height where the wind velocity is $\sim$50-100 km s^{-1}; and (2)
the wind in this region is not co-rotating with the star. Both may be
reasonable, but both will require further study.

6. SUMMARY

The Be stars in this study have mass-loss rates between 10^{-11} and
3×10^{-9} $M_\odot$ yr^{-1}, if the assumptions regarding the ionization balance of
silicon are correct. These values of $\dot{M}$ are 3 to 5 orders of magnitude
below those of the most luminous OB supergiants, but otherwise the Be
star winds appear similar to those of the more luminous stars, and may
simply represent an extension of those radiatively-driven winds to lower
luminosities.

The lack of strong correlations with stellar parameters indicates,
however, that the Be star winds are dependent on more complex parameters
than those of the OB stars, and the complexity may be due to some influ-
ential parameter not yet considered, to latitudinal dependence (or other
asymmetries) of the winds, or to time variability.

The narrow FeIII shell lines observed in the ultraviolet spectra of
many of the Be stars may arise in the low-velocity portions of the
winds, below the strong acceleration zone.

Further data are needed, particularly on other ions such as CIV,
NV, and especially OVI in order to reduce the uncertainties in the mass-
loss rates. It would also be extremely helpful if line profile analysis
could be carried out to yield estimates of the stellar inclination
angles.

This work has been supported by NASA grants NSG-5355 and NAG5-15
with the University of Colorado. Helpful discussions with T. Kallman,
G. Olson, J. Castor, and J.M. Shull are gratefully acknowledged.

References

Bohlin, R.C.: 1970, Ap. J., 162, p. 571.
Bruhweiler, F.C., Morgan, T.H., and van der Hucht, K.A.: 1978, Ap. J.
     Letters, 225, p. L71.
Burton, W.M., and Evans, R.G.: 1976, "IAU Symp. 70, Be and Shell Stars",
     ed. A. Slettebak (Dordrecht:Reidel), p. 199.
Castor, J.I., and Lamers, H.J.G.L.M.: 1979, Ap. J. Suppl., 39, p. 481.
Code, A.D., Davis, J., Bless, R.C., and Hanbury Brown, R.: 1976, Ap. J.,
     203, p. 417.
Doazan, V., Kuhi, L.V., and Thomas, R.N.: 1980, Ap. J. Letters, 235,
     p. L17.
Doazan, V., et al., this volume, p. 415.
Furenlid, I., and Young, A.: 1980, Ap. J. Letters, 240, p. L59.
Hammerschlage-Hensberge, G., et al.: 1980, Astr. Ap., 85, p. 119.
Heap, S.R.: 1976, "IAU Symp. 70, Be and Shell Stars", ed. A. Slettebak
     (Dordrecht:Reidel), p. 315.
Henrichs, H.: 1980, private communication.
Hutchings, J.B.: 1976, Pub. A.S.P., 85, p. 5.
Lamers, H.J.G.L.M., and Snow, T.P.: 1978, Ap. J., 219, p. 504.
Lamers, H.J.G.L.M., Gathier, R., and Snow, T.P.: 1980, Ap. J. Letters,
     242, p. L33.
Lesh, J.R.: 1968, Ap. J. Suppl., 17, p. 371.
Lesh, J.R.: 1972, Astr. Ap. Suppl., 5, p. 129.
Marlborough, J.M., and Snow, T.P.: 1976, "IAU Symp. 70, Be and Shell
     Stars", ed. A. Slettebak (Dordrecht:Reidel), p. 179.
Marlborough, J.M.: 1977, Ap. J., 216, p. 446.
Marlborough, J.M., and Snow, T.P.: 1980, Ap. J., 235, p. 85.
Peters, G.J.: 1976, "IAU Symp. 70, Be and Shell Stars", ed. A. Slettebak
     (Dordrecht:Reidel), p. 209.
Peters, G.J.: 1979, Ap. J. Suppl., 39, p. 175.
Slettebak, A., Collins, G.W., Boyce, P.B., White, N.M., and Parkinson,
T.D.: 1975, Ap. J. Suppl., 29, p. 137.
Snow, T.P., and Marlborough, J.M.: 1976, Ap. J. Letters, 203, p. L87.
Snow, T.P., Peters, G.J., and Mathieu, R.D.: 1979, Ap. J. Suppl., 39,
     p. 359.
Snow, T.P., Oegerle, W.R., and Polidan, R.S.: 1980, Ap. J., 242, p. 1077.
Snow, T.P.: 1981, Ap. J., submitted.
Sonneborn, G.H., and Collins, G.W.: 1977, Ap. J., 213, p. 787.
Thomas, R.N.: 1980, private communication.

DISCUSSION

Thomas: 1. I make you a wager that the F_M you list, and the lower
limits given by Stalio et al., are low by a factor 100.
2. The Fe II data make me very sceptical that in e.g. γ Cas and
59 Cyg, you can produce it in an atmosphere which is coolest at low
radii and hot at large radii.

Snow: 1. I accept.- I'll bet you a Coke that the mass fluxes in the
line of sight are not underestimated by a factor of 100 (except for
the hottest stars, which may have dominately Si IV, as I have noted).
2. At the moment there is no problem with Fe II, because little or
none is seen in any of the stars in my list. I agree that this might
be a difficult thing to explain, if we find stars with Fe II shell
lines, particularly the visible-wavelengths ones (because they require
high densities), but even then it may be possible to have Fe II close
to the star. I remind you of my comment that charge-exchange reactions
can produce Fe II from Fe III very efficiently.

Doazan: In many shell spectra one observes in the UV, in addition to
Fe III lines shell lines of Fe II. These lines cannot be formed near
the photosphere. I see no other possibility than to form them far in
the outer atmosphere. In this case, the observed low velocity imposes a
velocity law which is decelerated in the outer regions and which is
completely different from the one you assumed.

Snow: Let me elaborate on my answer to a similar question from Dick
Thomas. It is possible to have some Fe II near the star, if even a
little He I is present to form it through charge- exchange with Fe III.
Even without this, a little Fe II can survive the ionizing flux of
the star, although I admit it would be very little for the hotter Be
stars. However, I must stress that none of the stars in my sample
have notable Fe II shell lines, so at the moment there is no incon-
sistency to worry about. I know that one of the stars you are referring to is 88 Her, which is cooler than my sample stars, so the
presence of Fe II close to the star, may be even less difficult to
explain.

Persi: I agree with Dr. Snow that to determine the mass-loss rate
from IR observations the terminal velocities derived from UV observations cannot be used, because IR and UV radiations originate
in different regions of the envelope. In any case the ratio M/v_∞
derived for γ Cas from our IR observations is greater than derived
from UV observations by a factor > 100.

Snow: This discrepancy bears on a central issue that has come up
many times during this symposium, and perhaps is most obvious in this
case. There must be regions with very different physical conditions
in the circumstellar environment, so we must be very careful when
comparing data obtained in different ways in different wavelength
regions.

<u>Stalio</u>: I am glad to see that you have used the same method to
determine the wind components as we did; our ways of computing mass-
loss rates are, however, different. Because of the many uncertainties
that arise and you have mentioned, especially on ionization equilibrium, we made the choice of computing lower limits. I have the
impression that you made the same at least for some stars that we
have in common.
Concerning the correlation with $\underline{v \sin i}$ it would be interesting to
put together our results.

VARIATION OF ANOMALOUS STAGES OF IONIZATION WITH SPECTRAL TYPE FOR Be
STARS

J.M. Marlborough
Astronomy Centre, The University of Sussex, and Department of
Astronomy, The University of Western Ontario
Geraldine J. Peters
Department of Astronomy, University of Southern California

One of the interesting results of ultraviolet (UV) astronomy
was the discovery of ions from unexpected stages of ionization in spectra
of O,B stars. The most common ions concerned are O VI and N V, but
also C IV and Si IV in B stars. The presence of these ions is anomalous because generally their abundance is expected to be negligible if
they are produced by photoionization by stellar radiation, either in
the photosphere or in a cool circumstellar envelope (CE). The same ions
are observed in the UV spectra of Be stars. Previous investigations,
largely with Copernicus spectra, have reported O VI and N V in late Oe
and early Be stars and Si IV in stars as cool as B5 (Marlborough, 1981
and references therein). In this paper we present the results of a
preliminary survey of IUE spectra of Be stars covering a wide range of
spectral type.

We have high resolution IUE spectra, $1150 < \lambda < 1950A$ of 23 Be
stars, spectral types O9–B8, luminosity classes III–V. The stars are
listed in Table 1. We have two observations separated by about 6 months

TABLE 1
Ultraviolet Spectra Surveyed for N V, C IV, and Si IV

Spectral Type	Stars	
	$v \sin i > 150$ km s^{-1}	$v \sin i < 150$ km s^{-1}
O9	ζ Oph	
B0	γ Cas	
B1	59 Cyg, HD28497, HR4009, π Aqr	
B2	υ Cyg, 66 Oph, 48 Per	ω CMa, 31 Peg, 11 Cam, χ Oph
B3	α Eri, 28 Cyg, ω Ori, 16 Peg	6 Cep, HD58343
B5	ψ Per, ε Cap	
B8	o Aqr, α Col	

M. Jaschek and H.-G. Groth (eds.), Be Stars, 387–390.

for both ψ Per and HD58343; hence our sample consists of 25 UV spectra.

Each spectrum was examined for the presence of the resonance lines, in each instance a doublet, of N V, C IV, and Si IV. Our procedure consisted in assigning a number to each spectrum to indicate the presence of the particular ion. We assigned a '1' if both components of the doublet of a particular ion were present and a '0' if there was no indication of the ion. For some spectra, a weak feature is located at or near the rest wavelength of the stronger component of the doublet. For such cases we assigned '0.5' to indicate the ion may be present. We have computed the weighted mean of these numbers for each spectral subdivision, the weighted mean thus reflecting the chance of detecting the ion in stars of the specific subtype. Our results are illustrated in Figure 1. The vertical dashed lines represent the coolest spectral types at which the appropriate ion can exist under radiative equilibrium conditions with a sufficient abundance to produce a detectable photospheric absorption line (Marlborough, 1982 and references therein). Therefore the presence of N V in early Be stars and of both C IV and Si IV in Be stars as cool as B8 demonstrates that at least some Be stars in almost all spectral subdivisions show anomalous stages of ionization

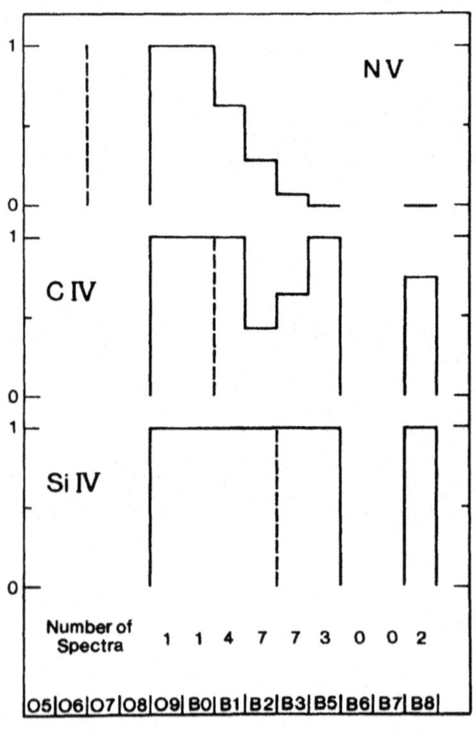

Figure 1. The chance of detecting N V, C IV, and Si IV in Be stars based upon 25 IUE spectra. The vertical dashed line represents the coolest spectral type at which the ion would be expected under radiative equilibrium conditions, either in the photosphere or in a cool circumstellar envelope.

in their UV spectra and thus require ionization sources in addition to the photospheric radiation field.

Marlborough (1982) has summarized the evidence to date regarding variability of N V, C IV, and Si IV in UV spectra. Although the strength, profile, and radial velocity of the lines vary there is no evidence presently to suggest that lines disappear totally. In each star cooler than about B3, Si IV lines were detected. Thus we suggest that the absence of N V in Be stars cooler than about B2 together with the presence of C IV and Si IV in later types is a real phenomenon and reflects some dependence of the additional ionization source on basic stellar parameters.

It is clear from Figure 1 that C IV lines seem to occur less frequently at B2 and B3 than at other subtypes. All stars in our sample, excluding types B2 and B3, have moderate to large v sin i. Only at B2 and B3, where we have a modest number of stars, is there a significant range of v sin i.

We have attempted to determine whether the behaviour at B2 and B3 might reflect some dependence on v sin i. Visual inspection of spectra reveals that the range of strengths of C IV lines depends on v sin i. While stars of large v sin i may have a range of C IV line strengths, stars of low v sin i seem to have only weak C IV lines. Furthermore there is some weak evidence, especially at B3, that the C IV line strength decreases with decreasing v sin i. The strength of Si IV lines at B3 seems to vary in a similar manner; however at B3 a photospheric contribution to Si IV may not be negligible. This apparent dependence of C IV line strength on v sin i is similar to the variation of Fe III line strength and linear polarization with v sin i. One possible interpretation is that the hot component of the CE is not distributed with spherical symmetry. More data however are needed to confirm this suggestion.

ACKNOWLEDGEMENT

We wish to thank C. Grady and M. Rashe for their invaluable assistance with data reduction. J.M. Marlborough wishes to acknowledge the hospitality of the Astronomy Centre, University of Sussex where his part of the work was carried out. This research was supported by the SRC (UK), NSERC (Canada), and NASA (USA)

REFERENCE

Marlborough, J.M.: 1982, in M. Jaschek and H.G. Groth (eds.), 'Be Stars, IAU Symp. 98', pp. 361.

DISCUSSION

<u>Viotti</u>: I presume that in your analysis you used spectral types
derived from optical spectra. Have you looked at the presence of
high ionization lines as a function of the strength of the UV excited
lines, which could be a better indication of the far_UV radiation
flux?

<u>Marlborough</u>: We have not done so yet.

<u>Thomas</u>: 1. You know as well as I the difficulties in detecting N V
in 59 Cyg on some of the plates; so I would be <u>very</u> cautious in your
conclusion.
2. I note that before IAU Montreal no N V had been observed in T Tau
stars; since then it has been seen in majority of those well-observed,
if my memory is correct.

<u>Marlborough</u>: 1. Our conclusions were based only upon the sample of
stars we observed. If the sample is not representative, the conclusions may not be correct.
2. Due to the low apparent brightness of T Tauri stars, there were
probably few, if any, ultraviolet observations of high resolution
obtained before the availability of IUE.

<u>Viotti</u>: I think that you have to include in your list of anomalous
stages of ionization Si II resonance lines which are present in
X Per and γ Cas and should not be present in their photospheres.

<u>Marlborough</u>: Conspicious changes in the Si II resonance lines occured
during the shell phase of 59 Cyg. Photospheric Si II would not be
strong in a star as hot as 59 Cyg either. Hence the Si II resonance
lines arise in the cool regions of the circumstellar envelope.

MASS LOSS FROM π AQUARII

J.A.de Freitas Pacheco
CNPq-Observatório Nacional
Rua General Bruce, 586 – São Cristovão
Rio de Janeiro – Brasil

SUMMARY: Fe III lines from metastable states have been detected in the
UV spectrum of π Aqr (HD 212571). The line profiles are asymmetric and
they are probably formed in a expanding envelope. From the observed line
widths and using the Sobolev approximation we have derived a mass loss
rate of about $2.5 \times 10^{-9} M_\odot$ y^{-1}.

Bruhweiler et al.(1978) have shown that the UV features appearing
at 2061, 2068 and 2079 Å in the spectrum of φ Per can be identified as
transitions of Fe III originating from a metastable state (a^5S) with an
energy of 5.01 eV above the ground level. The asymmetric profiles were
interpreted as being formed in a expanding envelope and from the
observed line widths they have derived a mass loss rate for φ Per of
about 5×10^{-11} $M_\odot$ y^{-1}. In their estimate Bruhweiler et al. have
considered a "static" envelope to establish the equivalent width
relationship with the column density and mass loss. However, if we
consider the presence of a velocity gradient, a large number of atoms
are necessary to produce the same absorption, resulting in a higher
mass loss rate as we shall see later.

We have made a search for Fe III lines in the UV spectrum of π Aqr
obtained on July 1979.High resolution IUE spectra in the short and long
wavelength range (images SWP 5886 and LWR 5141) were obtained at the
ESA tracking station at Villafranca del Castillo.

Besides the broad, rotationally enlarged, photospheric lines and
the asymmetric lines formed in a expanding envelope,interstellar (maybe
circumstellar) lines of Si II, NI, S II, CI, Cl I and Fe II are present
in the spectrum, having an average radial velocity equal to -15 km s^{-1}.

The resonance lines of Si IV and C IV present strong asymmetries and
striking differences when compared with the profiles obtained by
Ringuelet et al. (1981) six months before. The blue-edge of these lines
indicate velocities of -880 km s^{-1} and 1250 km s^{-1} respectively for the
Si IV and C IV ions.

M. Jaschek and H.-G. Groth (eds.), Be Stars, 391–392.

The Fe III and Al III lines are also asymmetric with blue edge velocities of -116 km s^{-1} and 210 km s^{-1} respectively. The ionization excitation increases towards the outer region of the envelope since the ions with higher ionization potential display higher velocities.

If the highest velocity value (-1250 km s^{-1}) indicates the wind terminal velocity, we may consider that Fe III ions are not present in the envelope for distances corresponding to velocities greater than 116 km s^{-1}. On the other hand, we assume that inside such a region iron is essentially twice ionized. Under this simplifying hypothesis and using the Sobolev approximation, we have derived equation (1), which relates the line equivalent width with the mass loss rate in the case of non saturated lines:

$$W \simeq 1.61 \times 10^{13} \; \gamma \; f_{ij} \; \lambda^2_{ij} \; f_B \; A_{el} \; \frac{\dot{M}}{RV_\infty} \; \overset{\circ}{A} \tag{1}$$

In the above equation f_{ij} is the oscillator strength of the ij transition, λ_{ij} is the wavelength (in Angstroms) of the considered line, f_B is the effective Boltzmann factor (equal to one if the ions are in the ground state), A_{el} is the abundance of the element, R is the stellar radius in solar units, V_∞ is the wind terminal velocity (in km s^{-1}) and $\dot{M}$ is the mass loss rate in solar masses per year. The factor γ is calculated numerically integrating the line profile on the blue side.

Using the data of the Fe III transitions $a^5S-z^5P^o$, from equation (1) we derived $\dot{M} \simeq 3.7 \times 10^{-9}$ M$_\odot$/year. Under the same assumptions, using the Al III resonance transition $3s^2S-3p^2P^o$, we have obtained a mass loss deduced of about 2.4×10^{-9} M$_\odot$/year. This value is comparable to that deduced for γ Cas (7×10^{-9} M$_\odot$ y^{-1}) by Hensberge et al. (1980) and to that deduced for the BOV star τ Sco by Lamers and Rogerson (1978). We have re-analysed the data of ϕ Per by Bruhweiler et al. (1978) using our theory and we have obtained a mass loss rate of about 6.5×10^{-10} M$_\odot$/year, assuming a terminal wind velocity equal to 1000 km s^{-1}.

All these results seem to indicate that, at least for the brighter Be stars, the mass loss rates are of the order of a few times 10^{-9} M$_\odot$y^{-1}, which are two orders of magnitude smaller than the previous results obtained from the optical recombination lines, suggesting that such a lines are formed probably in a denser disk around the equatorial region and not in the expanding envelope.

REFERENCES
- Bruhweiler, F.C., Morgan, T.H. and van der Hucht, K.: 1978, Astrophys. J., 255, L 71.
- Hensberge, H.G., van den Heuvel, E.P.J., Lamers, H.I.G., Burger, M., de Loore, C., Glencross, W., Howarth, I., Willis, A.J., Wilson, R. and Menzies, J.: 1980, Astron. Astrophys., 85, 119.
- Lamers, H.J.G. and Rogerson, J.: 1978, Astron. Astrophys. 66, 417.
- Ringuelet, A., Fontenla, J.M. and Rovira, M.: 1981 - preprint.

THE EXPANDING ATMOSPHERE OF HD 218393

F.Paterson-Beeckmans
ESA Space Science Dept.,Astronomy Div.,Noordwijk,Netherlands
A.M.Hubert-Delplace, H.Hubert, D.Ballereau
DEPEG, Observatoire de Paris, Meudon, France

Abstract. Visible and far ultraviolet high resolution spectra of the B2 binary star HD 218393 have been obtained at one day interval, shortly before the phase of maximum outwards velocity, adopting the period of 38.908 days. Both spectra show the existence of an extended atmosphere, accelerated outwards. The SiIV resonance lines show some indication of mass loss. With only one set of observations, the origin of the strong and wide (about 130 km/s FWHM) ultraviolet absorption lines of once ionized elements is difficult to determine.

1. INTRODUCTION

The star HD 218393 (KX And) is known since 1930 to show intensity and radial velocity (V_r) variations of the spectral lines (Struve, 1944, Merrill, 1949, Halliday, 1950). The radial velocity variations of the circumstellar lines have suggested a period of 35 to 40 days, but the amplitude and the period are not exactly reproducible. Doazan and Peton (1970) have interpreted the radial velocity variations as caused by an expanding envelope with variable acceleration. More recently, the star has been considered as an interacting binary with strong mass exchange (Kříž and Harmanec,1975). Polidan and Peters (1976) have discovered in the near infrared, lines of neutral elements which can be attributed to a KIIII companion. Photometric variations have also been observed, especially in the (U-B) colour, the star being reddest around maximum velocity (Harmanec *et al.*,1980). The binary period obtained by these authors is 38.908 days.

A visible spectrum has been obtained with the 152 cm telescope of the Haute Provence Observatory on 19th October 1979 (3500-5000 Å, 12.27 Å/mm dispersion). IUE observations of that star have been obtained one day later (1150-2100 Å, 0.15 Å resolution, and 1800-3200 Å, 8 Å resolution). These observations correspond to the phases 0.29 and 0.30 , considering the 38.908 days period, with phase 0.0 at the maximum of radial velocity. There is a good agreement between our values of V_r for the hydrogen lines and those obtained by Struve at the same phase.

M. Jaschek and H.-G. Groth (eds.), Be Stars, 393–397.
Copyright © 1982 by the IAU.

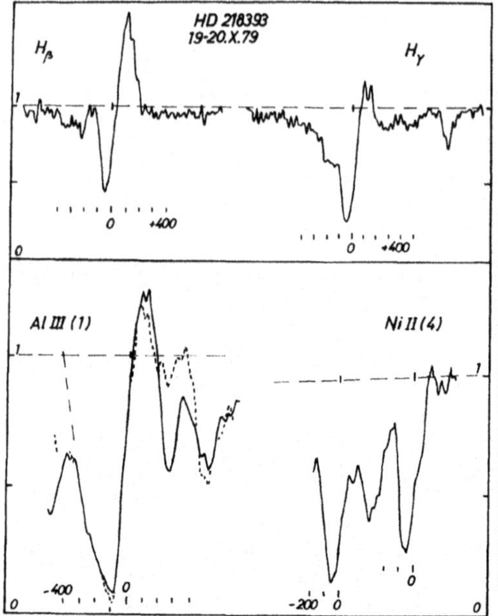

Fig.1 Characteristic profiles
of HD 218393; the continuum is
indicated by a dashed line; the
abscissa is in km/s.

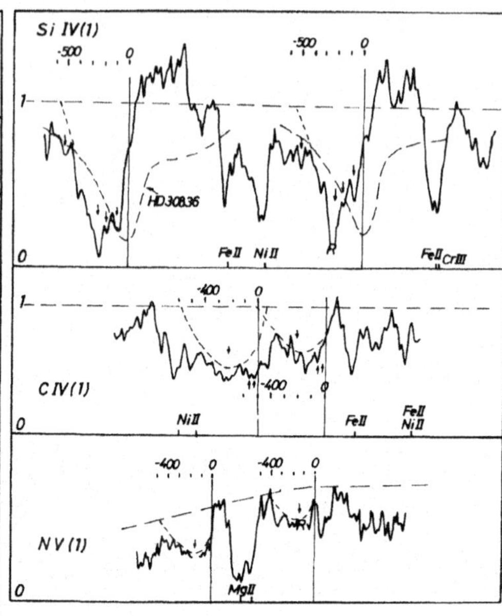

Fig.2 Same as in Fig.1; SiIV profiles of HD 30836(B2III) indicated
for comparison; for CIV and NV,
possible profiles are indicated
by dashed lines.

2. VISIBLE OBSERVATIONS

The good agreement between the Hγ and Hδ photospheric wings of HD 218393,
HD 37202 (B2III) and HD 33328 (B2IV) confirms the spectral type B2IV-III
already attributed to HD 218393 by Doazan and Peton (1970), from the ratio of the HeI λ4121 and 4144 lines. According to Divan (private comm.)
the BCD spectral type is approximately B2.

The visible spectrum has the same characteristics as those corresponding to the same phase (see above references). The first Balmer lines
show a P Cyg profile (Figure 1) and the red wing emission is visible to
H12; most of the Balmer lines have an extended violet absorption wing.
The circumstellar, narrow absorption lines are observed to H29, hence
Ne $\cong$ 2.10^{12}cm^{-3} in the inner part of the envelope. The CaII lines are
strong and asymmetric. The metallic absorption lines are faint, in agreement with Struve (1944) and Halliday (1959) observations : these lines
almost disappear at the minimum of V_r. The strongest FeII lines show emission in the wings.

Almost all the circumstellar absorption lines are blueshifted, as
indicated in Table 1. In addition, the Hγ and Hδ lines show a satellite

Table 1. Radial velocity of visible circumstellar lines.

H β	:	−50	km/s	MgII : −7.5 km/s
H γ, H δ	:	−42	"	FeII,MgI,NiII : −5 "
H12	:	−30	"	TiII 0.5 "
H22−H25	:	−15	"	SiII 1.5 "
CaII	:	−19	"	

violet absorption, shifted by 140 km/s with respect to the main component. The positive Balmer progression suggests the presence of an outwards accelerated envelope, the highest Balmer lines being formed closer
to the star than the first Balmer lines.

3. ULTRAVIOLET OBSERVATIONS

The ultraviolet spectrum of HD 218393, in opposition to the visible one,
is very rich in relatively broad absorption lines, most of which are
heavily blended, making the line identification very difficult. Wide
photospheric features (v sini= 250 km/s in the visible, Doazan and Peton,
1970) are observed for the strongest lines (CII,III,SiIII.IV,AiIII,FeIII).

By comparison with IUE spectra of other B2 stars (HD 36166(B2V),
HD 886(B2IV), and HD 30836(B2III), we notice a strengthening of mainly
lines of neutral and once ionized elements (OI,NI,CII,SiII.PII,AlII,NiII,

Table 2. Radial velocity of UV absorption lines; asymmetric
(edge velocity) and P Cyg profiles are indicated.

Ion	IP (eV)	V_r(max.abs.) EP=0	EP>0	Asym. Vedge	P Cyg
N V	77.5	−145?		−435?	?
C IV	47.9	−220?		−580?	?
SiIV	33.5	−245		−580	x
AlIII	18.8	−91		−365	x
NiIII	18.2		−28?	−	−
CrIII	16.5		−45	−	−
SiIII	16.3	−100		−200?	−
			−96	−	−
FeIII	16.2		UV34:		
			−100	−360	x
			UV48:		
			−90	−360	x
			−45	?	−
TiIII	13.6	?	−60	−	−
C II	11.3	−77		−360	x
P II	10.5	−43		−	−

Ion	IP (eV)	V_r(max.abs.) EP=0	EP>0	Asym. Vedge	P Cyg
S II	10.4	−53		−350	P
ZnII	9.4	−48		−220	−
SiII	8.2	−74		−300	P
			−35	−	−
FeII	7.9	−51		−	−
			−43	−	−
CuII	7.7	−40		−	−
MgII	7.6	−34		−	−
NiII	7.6	−45		?	−
MnII	7.4		−25	−	−
CrII	6.8	−40		−	−
AlII	6.0	−68		−360	x
			−50	−	−
O I	0	−44		−	−
N I	0	−30		−	−
			−30	−	−

FeII) which cannot be attributed to the faint companion and are relatively wide (FWHM = 100-130 km/s) as compared to classical shell lines (70 km/s in HD 37202). They are all blueshifted (see Table 2, velocity measured with respect to the IUE scale), and several resonance lines have extended violet wings, the edge velocity of which is indicated in Table 2.

Despite the uncertainty on the continuum level, several P Cyg profiles are also observed in the UV for strong resonance lines (Figures 1 and 2, Table 2) and the strongest transitions of FeIII (UV 34 and 48), the intensity of which is connected to that of the resonance lines (Bruhweiler *et al.*,1978). Broad absorption profiles have tentatively been identified with CIV and NV resonance lines (Figure 2); the CIV lines are weak in normal B2V-III stars; here, the features could result from a blend of other circumstellar lines. Hence, the degree of ionization of the envelope is uncertain.

Analysing Table 2 results, we distinguish a group of lines for which, as for other Be stars envelopes (48 Lib, φ Per,HR2142) the outflow velocity increases with the degree of ionization, being maximum for SiIV (column 3, left). The edge velocity of SiIV is equal to the escape velocity at 2 stellar radii. In the binary HR2142, the corresponding lines, formed in the envelope of the B2 primary, are strongly asymmetric but undisplaced (Paterson-Beeckmans,1980).

The lines of the second group (asymmetric or not) have a blueshift of about 45 km/s, as the first Balmer lines. They are too strong for normal B2 photospheric lines, and wider than classical shell lines. Are they formed at the bottom of the expanding envelope, or in a cool disk? Certainly more observations are needed before attempting to make a model.

The IUE data have been obtained at the European Ground Station in Villafranca. F.P.B. thanks the ESA Astronomy Div. where she is visitor.

REFERENCES

Bruhweiler,F.,Morgan,T.H. and van der Hucht,K.A.: 1978,Astrophys.J.225, pp.L71-L74.
Doazan,V. and Peton,A.: 1970,Astron.Astrophys.9,pp.245-251.
Halliday,J.: 1950.J.Roy.Astron.Soc.Canada 44,pp.149-157.
Harmanec,P.,Horn,J.,Koubský,P.,Ždárský,F.,Kříž,S. and Parlovski,K.: 1980,Bull.Astron.Inst.Czech.31,p.144.
Kříž,S. and Harmanec,P.: 1975,Bull.Astron.Inst.Czech.26,p.65.
Merrill,P.W.: 1949,Astrophys.J.110.p.420.
Paterson-Beecmans,F.: 1980,*Second European IUE Conference*,ESA SP-157, pp.51-54.
Polidan.R.S. and Peters,G.J.: 1976,IAU Symp.No 70,pp.59-65.
Struve,O.: 1944,Astrophys.J.99,p.75.

DISCUSSION

<u>Peters</u>: We have anlyzed <u>Copernicus</u> U1 observations of the gas stream lines of Fe III 1130 A in HR 2142 and find evidence of multiple components. I will discuss the importance of this analysis tomorrow.

<u>Paterson-Beekmans</u>: We also suspect some substructure in some resonance lines in HR 2142; in KX And, there may be some structure in strong resonance lines (Al III, Si IV, for ex.).

<u>Harmanec</u>: I want to mention preliminary results of our photometry of KX And:
1. There is no pronounced long-term variability.
2. Minimum light in the U colour coincides with maximum RV indicating that there is probably an eclipse of the star by a gas stream.
3. It is possible that the star will turn out to be an eclipsing binary with a partial eclipse.

THE PECULIAR Be STAR HD 87643

J.A.de Freitas Pacheco
CNPq-Observatório Nacional
Rua General Bruce, 586 - Rio de Janeiro - Brasil
D.Gilra
Kapteyn Laboratorium - University of Groningen
Postbus 800 - Groningen - Holland

SUMMARY: IUE observations as well as optical spectral of the peculiar star HD 87643 were obtained. The observed optical Fe II lines are probably formed in a expanding envelope and the rate of mass-loss derived from the analysis of these lines is about 7×10^{-7} $M_\odot$ y^{-1}.

HD 87643 is an emission line object which besides the H-Balmer lines present also Fe II and Ca II in emission.

The star is surrounded by a nebulosity, which direct narrow band plates taken recently indicate a reflective origin(Surdej et al.,1981). Besides that, this object seems to be part of a small group of early-type stars with the following characteristics: (a) spectra in the optical region without photospheric features; (b) Balmer lines with P-Cygni profiles;(c) optical Fe II lines in emission (sometimes presenting also P-Cygni profiles); (d) strong infra-red excess.

Our low resolution IUE spectra show a rather unusual appearence and they are unlike any other early-type star. In the long wavelength range we have identified the Mg II lines with P Cygni profile and absorption features due to Fe II resonance transitions and lines from low lying metastable levels. In the short wavelength range it was possible to identify the Si IV, Al III, Si III, C IV features attributable to the star. A high resolution spectrum in the short wavelength range convinced us that the spectrum is essentially an absorption spectrum. The most remarkable thing is that the major part of the spectrum is dominated by the envelope absorption lines. Fe II multiplets dominate the short wavelength spectrum also and some of the lines have only recently been identified in the laboratory.For example, the lower metastable level of some lines at 1400 Å is a^{4}G. From this level, forbidden emission lines are seen in the spectrum(multiplet 21F). Following our analysis, the IUE fluxes at the epoch of our observations (July 1979) were definitively higher than the ANS fluxes measured four years before. We atribute this difference probably to variable continuous opacity or to a weaker Balmer continuum. Recent infrared

399

M. Jaschek and H.-G. Groth (eds.), Be Stars, 399—400.
Copyright © 1982 by the IAU.

measurements (Epchtein and Lépine, 1980) are marginally consistent with a decrease in the circumstellar dust opacity. However this point still requires more observations to be cleared.

Our optical observations clearly display the P-Cygni characteristics of the Hβ line, showing a rather strong emission and a blue edge velocity exceding 1200 km s^{-1}. The optical Fe II emission lines are probably radiatively excited. Such a fluorescent mechanism is favoured by the strong UV Fe II observed absorption lines. The higher resolution spectra by Carlson and Henize (1979) would indicate that besides hydrogen, iron and calcium lines present also P-Cygni profiles. This lead us to the interpretation that all the lines of these ions are formed in the expanding envelope.

Using the Sobolev's approximation, the equivalent width of multiplet 42 is given by

$$W_{42} \simeq 8 \; (\frac{R}{D})^2 \; (\frac{V_\infty}{c}) \; \frac{F_{3UV}}{f_{42}} \; (\frac{\lambda_{3UV}}{\lambda_{42}}) \; \lambda_{3UV} \; G(\tau_0) \tag{1}$$

where R is the star radius, D is the distance to the star, V_∞ is the wind terminal velocity, c is the light velocity, F_{3UV} is the stellar photospheric flux at the weighted wavelength λ_{3UV} of multiplet 3UV, f_{42} is the observed (de-reddened) continuum flux at the weighted wavelength λ_{42} of multiplet 42 and $G(\tau_0)$ is a function of the effective (Sobolev) optical depth of the multiplet 3UV. From our data and our calculations using equation (1), we have obtained a rate of mass loss equal to 7.9×10^{-7} $M_\odot$ y^{-1}. A similar analysis of the Hβ line lead us to a mass loss rate of about 7×10^{-7} $M_\odot$ y^{-1}. In spite of the very simplified nature of our model, both mass-loss estimates are in very good agreement. The rarity of objects like HD 87643 means that this is a very short lived phase in stellar evolution, conclusion also supported by the fact that the estimated mass loss rate is about two orders of magnitude higher than the values derived for other Be stars through the analysis of the UV lines.

REFERENCES

Carlson, E. and Henize, K., 1979, Vistas in Astron., 23, 213.
Epchtein, N. and Lépine, J.R., 1980 - private communication.
Surdej, A., Surdej, J., Swings, J.P. and Wamsteker, W., 1981, Astron. Astrophys., 93, 285.

DISCUSSION

Thomas: You assume a mass-loss occuring in a cone-shaped sector, at and above a semi-quiet equatorial cylinder. How much would you increase your F_M if you assumed spherical symmetry?

de Freitas Pacheco: The equatorial disc occupies only a small fraction of the total solid angle viewed by the star. The assumption of sphe- rical symmetry introduces an error which is certainly smaller than the uncertainties in the mass-loss estimate that I have made.

EVIDENCE FOR MASS LOSS AT POLAR LATITUDES IN ω Ori AND 66 Oph

Geraldine J. Peters
Department of Astronomy
University of Southern California
Los Angeles, CA 90007   USA

ABSTRACT

IUE observations of the "pole-on" Be stars ω Ori and 66 Oph have revealed the unexpected presence of high velocity ($v \simeq -750$ km s^{-1}), relatively narrow ($\Delta\lambda \simeq 1A$) absorption components in the resonance lines of C IV, Si III, and Si IV. The C IV features show structure indicative of multiple shells or clouds. Similar high velocity lines were not observed in other pole-on Be stars considered in the program. The nature of these unusual features and the column densities and mass loss rates implied by them are discussed in this paper.

1. INTRODUCTION

Recently, the importance of rapid rotation in the establishment of the circumstellar envelopes in Be stars has been a subject for debate. Whereas it was previously assumed that all the matter in the envelope was ejected from the equatorial region of a rapidly rotating star, ultraviolet observations of variable spectral lines from super-ionized species (Doazan, et al. 1980, Doazan, et al. 1982) and the repeated failure of researchers to find observational evidence that Be stars rotate in excess of 0.85 V_{cr} (Hutchings, Nemec, and Cassidy 1979, Slettebak 1979) have inspired a search for other explanations for the Be phenomenon.

In order to establish the extent of mass loss at polar latitudes and gain some insight into the mechanism of mass loss in Be stars, twelve "pole-on" (vsin i $\leq$ 280 km s^{-1}) stars were observed at high resolution ($\Delta\lambda \simeq .15A$) with IUE. The UV spectra of two of the program stars, ω Ori and 66 Oph, were found to contain highly violet shifted, shell-type features of C IV, Si III, and Si IV. These observations are discussed in this paper.

The visible spectra of ω Ori and 66 Oph have a history of variability (Hubert-Delplace and Hubert 1979). ω Ori recently displayed unusual variable polarization (Hayes 1980) while 66 Oph was reported

M. Jaschek and H.-G. Groth (eds.), Be Stars, 401–404.

to have shown optical flares (Page and Page 1970). Values of vsin i
are 160 and 280 km s^{-1}, respectively, for ω Ori and 66 Oph (Slettebak
1976); if $V_{eq} \simeq 400$ km s^{-1}, then inclinations of 24° and 44°, respectively, are implied.

2. NATURE OF THE IUE OBSERVATIONS

The C IV and Si IV resonance features observed in ω Ori and 66 Oph
are shown in Figures 1 - 4. Superimposed on the C IV profiles (dotted
lines) is the C IV feature observed in υ Cyg, a Be star of comparable
spectral type and rotational velocity (B2IVe, vsin i $\simeq$ 175 km s^{-1}).
It can be said with confidence that the C IV lines in ω Ori and 66 Oph
are almost entirely circumstellar. Multiple components are clearly
seen in C IV. In ω Ori, the "clouds" responsible for the C IV absorp-

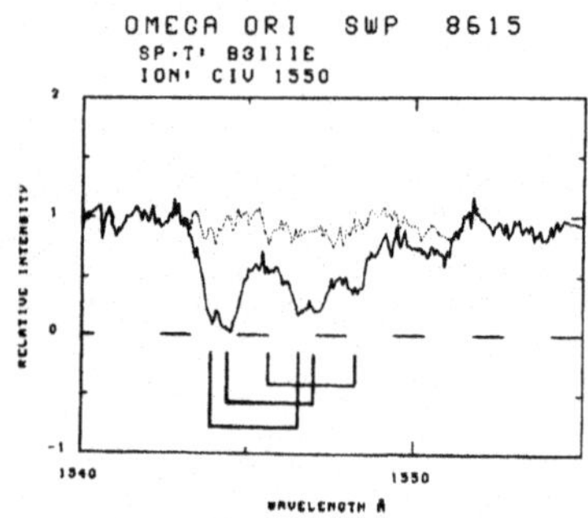

Fig. 1 -- The C IV doublet in
  ω Ori. Multiple components
are indicated. Rest wavelengths are 1548.2A and
1550.8A.

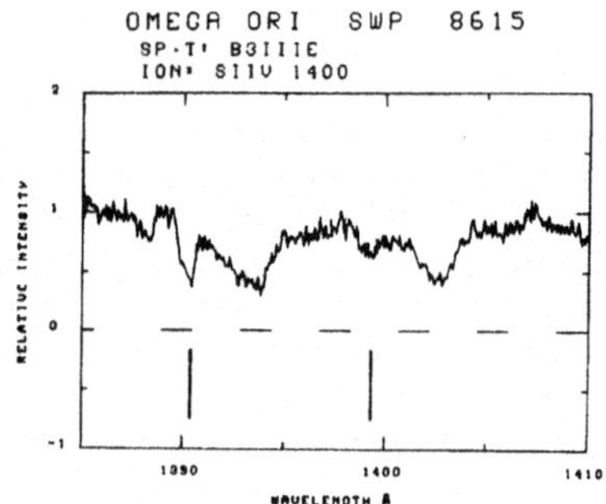

Fig. 2 -- The Si IV doublet in
  ω Ori. Vertical lines point
to the violet shifted shell
components. The broader
features are photospheric.

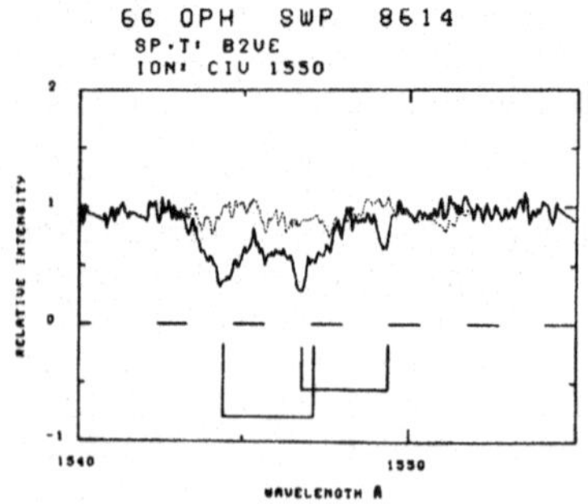

Fig. 3 -- Same as Fig. 1 for
  66 Oph.

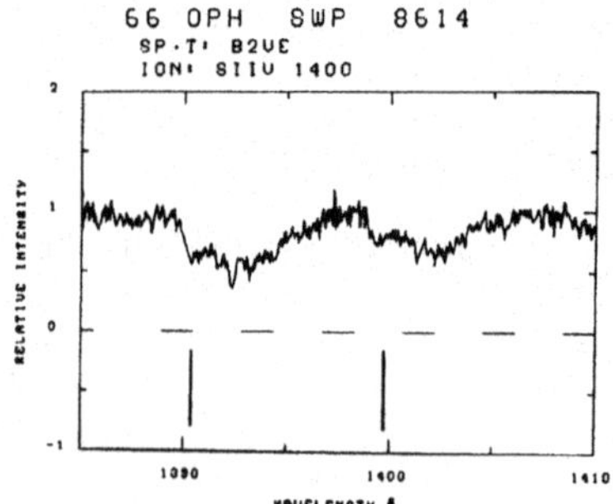

Fig. 4 -- Same as Fig. 2 for
  66 Oph.

tion have outflow velocities (relative to the photosphere) of 850, 750, and 550 km s^{-1}. Similarly, in 66 Oph, the features are violet shifted by 700 and 250 km s^{-1}. In Figures 2 and 4, one can easily see the shift of the narrower shell component of Si IV relative to the photospheric feature. This shift is -800 km s^{-1} in ω Ori and -670 km s^{-1} in 66 Oph. Violet shifted components are also seen in the Si III resonance line at 1206.5 A. In ω Ori the outflow velocity is -795 km s^{-1}, while in 66 Oph one measures a velocity of -685 km s^{-1}. The Si IV and Si III velocities in ω Ori agree with the mean of the two more negatively shifted components in C IV and, thus, imply the presence of unresolved structure in the former features. In general, the velocity data for both stars strongly suggests that C IV, Si IV, and Si III are formed in the same region. Furthermore, the fact that the shell features are narrow ($\Delta\lambda \simeq 1A = 200$ km s^{-1}) indicates that the mass is being lost in discrete "clouds" or "shells" instead of via a conventional wind.

3. ANALYSIS

The interpretation of the high velocity features is complicated by the fact that the lines are on the flat portion of the curve of growth. Although fine structure cannot be seen in the individual components, their strengths imply that each is composed of numerous closely spaced saturated lines. The <u>observed</u> half-width of the features suggest b values of 80 and 128 km s^{-1}, respectively, for ω Ori and 66 Oph. Using these values, we obtain the following column densities in C IV, Si IV, and Si III, respectively: 1) 4×10^{14}, 9×10^{13}, and 4×10^{13}cm^{-2} for ω Ori, and 2) 4×10^{14}, 7×10^{13}, and 2×10^{13} cm^{-2} for 66 Oph. These values should be regarded as lower limits. If unresolved fine structure exists, then the actual column densities could be as much as two orders of magnitude higher.

A non-LTE situation most certainly exists in the line formation region. In LTE, the observed Si III/Si IV implies that T≃15000K. At such a temperature, one would not expect to observe C IV. C IV/Si IV ≃ 5 in both stars. Bruhweiler, Kondo, and McCluskey (1980) observe a similar ratio in some parts of the interstellar medium and suggest that a temperature of 40,000K prevails in the line forming region. However, the presence of the Si III line in the spectra of the Be stars suggests a lower temperature for these objects.

If we assume that C IV is the dominant ion of carbon in the "high velocity" region, we obtain a minimum hydrogen column density of 3×10^{18} cm^{-2}. If $N_p = N_e \geq 10^9$cm^{-3}, small path lengths (<1R$_\odot$) are implied (even if fine structure prevails). The rate of mass loss is uncertain because we lack information on the density in the line formation region. However, if $N_e \simeq 10^9$cm^{-3}, then $\dot{M} \simeq 10^{-9}$ M$_\odot$ yr^{-1}.

4. CONCLUDING REMARKS

At the onset of this investigation, the existence of such high velocity features, some of which greatly exceed the escape velocity of the star, was not predicted. Their presence is not compatible with models of Be envelopes based upon rapid stellar rotation or binary mass

exchange. Evidence that matter does exist at polar latitudes in some
Be stars has already been presented (Peters 1976, Snow, Peters, and
Mathieu 1979) but, until now, only low outflow velocities were observed
in the "pole-on" stars. High velocity, "narrow" features have been
seen in 59 Cyg (Doazan, Kuhi, and Thomas 1980) and γ Cas (Henrichs 1980)
but these stars have higher values of vsin i.

If ω Ori and 66 Oph are indeed viewed "pole-on", then the observations presented in this paper demonstrate that mass loss in Be stars
is not restricted to the star's equatorial regions and, perhaps, rotation does not play an important role in the establishment of the circumstellar envelope. But if this is the case, then ω Ori and 66 Oph
may not be rapidly rotating after all and could instead be viewed
equator-on! The large polarization sometimes observed in ω Ori (Hayes
1980) seems to be compatible with the latter suggestion.

Future IUE observations are planned to search for similar high
velocity features in other "pole-on" Be stars and to assess the sta-
bility of the line formation region. At a velocity of 700 km s^{-1}, the
absorbing "cloud" will travel in excess of $10R_*$ in one day. One should,
therefore, be able to observe profile variations between observations
spaced a few days apart. An interpretation of this phenomenon will be
deferred until the forthcoming observations have been completed. At
this time, one cannot rule out that we are observing a "chromospheric"
phenomenon similar to the one suggested for 59 Cyg (Doazan, Kuhi, and
Thomas 1980). Although ω Ori and 66 Oph were only observed once, N V,
a signature of such "chromospheric" activity, was not detected. There
may be several explanations for what we call the Be phenomenon. Continued observations of the strong UV resonance lines in Be stars should
allow us to determine which of these are valid.

This project is being supported, in part, by NASA grant NSG 5422.

REFERENCES

Bruhweiler, F. C., Kondo, Y., and McCluskey, G. E.: 1980,
       Astrophys. J. 237, pp.19-25.
Doazan, V., Kuhi, L.V., Thomas, R.N.: 1980, Astrophys.J. 235,pp.L17-L20.
Doazan, V., Grady, C., Kuhi, L. V., Marlborough, J. M., Snow, T. P.,
       and Thomas, R. N.: 1982, this volume, pp.415-418.
Hayes, D. P.: 1980, Publ. Astron. Soc. Pacific 92, pp.661-665.
Henrichs, H. F.: 1982, this volume, pp.431-435.
Hubert-Delplace, A. M. and Hubert, H.: 1979, An Atlas of Be Stars,
       Observatoire de Meudon, France.
Hutchings, J.B., Nemec, J. M., and Cassidy, J.: 1979, Publ. Astron.
       Soc. Pacific 91, pp. 313-318.
Page, A. A. and Page, B.: 1970, Proc. Astron. Soc. Austral. 1,pp.324-325.
Peters, G. J.: 1976, in A. Slettebak (ed.), 'Be and Shell Stars'
       IAU Symp. 70, pp.209-217.
Slettebak, A.: 1976, in A. Slettebak (ed.), 'Be and Shell Stars'
       IAU Symp. 70, pp.123-135.
Slettebak, A.: 1979, Space Science Reviews 23, pp.541-580.
Snow, T. P., Peters, G. J., and Mathieu, R. D.: 1979, Astrophys. J.
       Suppl. 39, pp.359-376.

ULTRAVIOLET OBSERVATIONS OF INTERACTING BINARY Be STARS

Geraldine J. Peters
Department of Astronomy, Univ. of Southern California
Los Angeles, CA 90007   USA

Ronald S. Polidan
Princeton University Observatory
Princeton, NJ 08540   USA

ABSTRACT

Initial results from the analysis of a series of timed, high resolution IUE observations of HR 2142, ϕ Per, CX Dra, KX And, AU Mon, and
TT Hya are presented. The data base for HR 2142 also includes Copernicus U1 and U2 observations. Variable absorption lines, indicative of
mass flow in the system, are observed in all objects except ϕ Per. We
also, in general, find evidence of mass outflow in the form of winds
and/or discrete components. We observe variable N V absorption in
CX Dra and AU Mon and emission features in KX And and ϕ Per (C IV only).
U1 data reveals the presence of complex structure in the gas stream in
HR 2142. These observations are compared with those of Be stars which
are not thought to be interacting binaries.

1. INTRODUCTION

The ultraviolet spectra of interacting binary Be stars contain a
wealth of data on the gas streams, general matter flow, and disks in
these systems. In this paper, we present the initial results from a
detailed study of the UV spectra of some of the brighter systems. The
properties of the program stars and the phases at which the observations
were obtained are given in Table 1. All IUE observations were made
from 1979 October 23 to 1981 February 25. Copernicus observations of
HR 2142 were from 1976 December to 1979 December.

2. RESULTS

a. HR 2142

Evidence that HR 2142 is an interacting binary and the behavior of
the visible "gas stream" lines throughout the observed two-component
shell phase are given in Peters (1976) and the references quoted therein. Orbital parameters for the system (which lend support to the proposed model) are included in Peters (1982). Striking phase dependent
spectral variations have also been observed in the far ultraviolet
(Peters 1981, Paterson-Beeckmans 1980, Polidan and Peters 1980). Some
UV gas stream lines include the resonance lines of C II, Mg II, Al II,

405

M. Jaschek and H.-G. Groth (eds.), Be Stars, 405–409.

Table 1 - The Program Stars

Stars	Sp Types	Period	Phase coverage
HR 2142	B1IV-Ve+?	$80\overset{d}{.}860$	IUE ($0.8 < \phi < 0.1$: 11 obs. + $\phi = 0.17, 0.42, 0.45, 0.47$) <u>Copernicus</u> ($0.8 < \phi < 0.1$: 14 obs. + $\phi = 0.30, 0.44, 0.49, 0.70$)
ϕ Per	B0IV-Ve+?	$126\overset{d}{.}696$	0.00, 0.22, 0.25, 0.46, 0.46, 0.66, 0.85
CX Dra (HR 7084)	B3Ve+?	$6\overset{d}{.}697$	0.40, 0.56, 0.57, 0.60, 0.83, 0.86, 0.90
KX And (HD 218393)	B3IVe+K1III	$38\overset{d}{.}9$	4 obs., 3 close in phase, one $0\overset{p}{.}6$ later (ϕ arbitrary)
AU Mon (HD 50846)	B5Ve+F	$11\overset{d}{.}113$	0.30, 0.48, 0.85, 0.93
TT Hya (HD 97528)	B9Ve+G5	$6\overset{d}{.}953$	0.40, 0.90

Al III, Si II, III, IV, S II, III, Ti III, and Fe III and numerous subordinate lines of Fe III and Si III. In general, the variations in the
strengths and velocities of the UV features parallel those observed in
the visible. However, there are some important differences. The UV
"shell phase" or <u>gas stream phase</u> persists from 0.80<ϕ<0.10. Gas stream
features are observed prior to the visible <u>primary shell phase</u>, at conjunction, and after the short-termed visible <u>secondary shell phase</u>. IUE
observations reveal that the absorbing stream and/or disk cuts into the
line of sight abruptly between 0.83<ϕ<0.86. During this interval of
time, many of the stronger resonance lines saturate. The strengths of
the gas stream lines remain fairly constant between 0.90<ϕ<0.95. Then,
from 0.95<ϕ<0.97, there is an additional strengthening of the UV gas
stream lines which coincides in time with the appearance of the conspicuous Balmer shell spectrum ($\Phi_s \simeq 0.0$). The features subsequently weaken;
at conjunction the column densities are about 100 - 150 times lower than
the values observed from 0.91<ϕ<0.96. We further observe a strengthening
of the gas stream lines after conjunction (coincident with the visible
secondary shell phase) and the column densities from these features are
comparable with those obtained from the lines observed during the primary shell phase. The UV gas stream lines disappear by $\phi = 0.10$.

A detailed line profile analysis of <u>Copernicus</u> U1 observations of
the Fe III 1130 A resonance line has clearly shown that the gas stream
features are complex. Each line is composed of at least 2 - 3 components, one of which yields a column density 100 times greater than the
others. The denser component (at $\phi = 0.91$, 0.96, and 0.03), which appears
to be fixed in velocity prior to conjunction ($V - V_{phot} = 30$ km^{-1}), suggests a column density (in 1130 only) of about 10^{18} cm^{-2}. If the path
length is about $10R_\odot$, then the latter value implies a particle density
in excess of 10^{12} cm^{-3}. The velocity shifts of the <u>weaker</u> 1130 components
(at $\phi = 0.91, 0.96$) are more in accordance with those from the Balmer lines.

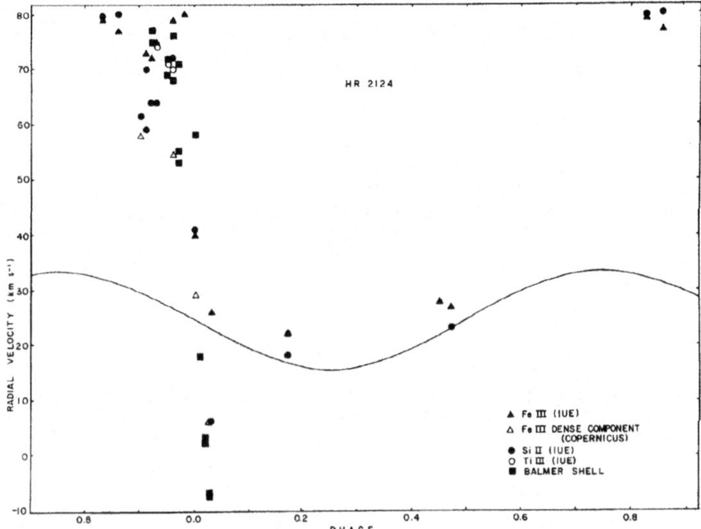

Fig. 1 - Radial velocities from selected gas stream lines versus phase. The solid curve shows the motion of the primary.

The radial velocity data from various gas stream features is summarized in Figure 1. Note that without exception the motion of the absorbing matter is toward the photosphere before conjunction; the opposite is observed after conjunction. However, if the lines are indeed formed in a conventional gas stream and counter stream, the velocities are significantly lower than theory predicts. Perhaps, the streams have large inclinations with respect to the line of centers or we are simply viewing through a density enhancement in the disk which results from a stream/ disk interaction. Many resonance lines show evidence of outflow in discrete components near $\phi = 0.5$ (Paterson-Beeckmans 1980). Our observations suggest that this <u>outflow phase</u> peaks at $\phi \simeq 0.45$ (before L_3). Subordinate lines do not show the outflow (cf. Fig. 1). We also observe an overall stellar wind (in C IV, Si III, IV) which appears to enhanced at conjunction.

b. ϕ Persei

An orbit and model for this enigmatic binary have recently been published by Poeckert (1981). Whereas striking phase dependent spectral variations are observed in the visible, our IUE observations show this star to be relatively invariant in the UV (unlike HR 2142). With one exception, the UV spectrum of ϕ Per is typical for a Be-shell star. The C IV resonance doublet stands as unique, however. As shown in Fig. 2, the C IV lines have P Cygni profiles with a peak emission line strength of twice the continuum value. Of the twenty-five classical Be stars observed to date with IUE, only ϕ Per displays C IV emission. Perhaps even more novel is that we do <u>not</u> observe emission at N V or Si IV. Our IUE observations were obtained over four orbital cycles and, in general, no statistically significant profile variations were observed.

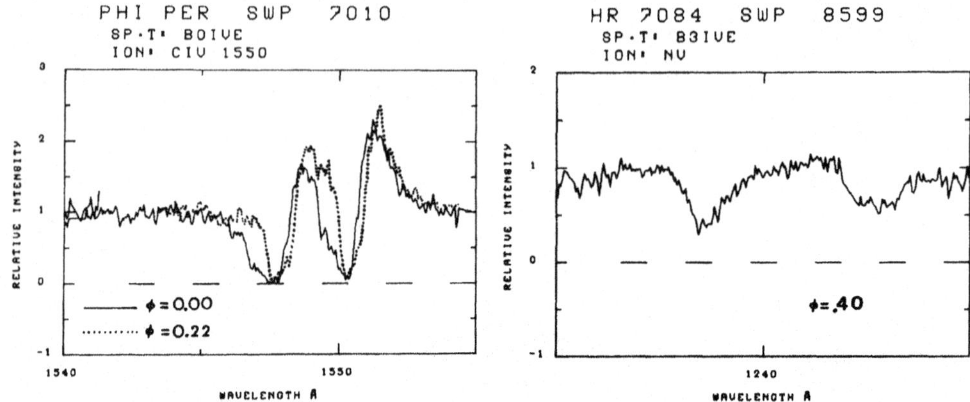

Fig. 2 – The resonance lines of   Fig. 3 – The N V resonance lines
C IV in φ Per at two phases.   in CX Dra on 1980 March 30.

In order to determine whether the C IV feature is associated with
the proposed high temperature secondary (Poeckert 1981), radial velocities were measured for the C IV doublet using neighboring interstellar
lines for wavelength calibration. Reference points on the profile included the emission line peak and wings and the cores of the lines. The
data clearly show that the feature is associated with the primary. Within the uncertainties of the measurements, the motion of the C IV region
follows the radial velocity curve of the primary. Motion with respect
to the secondary would be quite conspicuous since $K_s = 100$ km s^{-1}. Possibly the only observed UV features associated with He II object (secondary) are redshifted cores to Si III (UV 4) observed at $\phi = 0.66$.

c. CX Draconis (HR 7084)

Koubsky (1978) proposed that this object is a mass transfer binary
system. IUE observations reveal redward displaced absorption components,
indicative of a gas stream, in the resonance lines of C II, Si II,III,IV,
and Al III and the subordinate lines of Fe III at $\phi = 0.83, 0.86$, and 0.90.
These features are broader than their counterparts in HR 2142 and show
a larger velocity shift from the photospheric lines ($V \approx 200-250$ km s^{-1}).
We also observe violet asymmetries in the above mentioned lines at
$\phi = 0.56, 0.57$, and 0.60 which suggest the presence of mass loss.

Quite unexpected was the discovery of strong N V features in the
spectrum of this "B3e" star (Fig. 3). Under normal circumstances, a
temperature of at least 40000K is required to form N V. The strength of
the N V doublet is quite variable ($0.25 < r_v < 0.95$) but does not appear to
be phase dependent. Since the radial velocities from the N V lines
apparently follow Koubsky's published radial velocity curve, we conclude
that the N V is formed close to the stellar photosphere (perhaps the result of an impacting gas stream). Plavec (personal communication) recently has found CX Dra to be a soft X-ray source. We suggest that the X-
rays and N V both arise in an accretion heated region near the primary.

d. KX Andromedae (HD 218393)

Observations of this system at essentially two different phases ($\Delta\phi = 0.6$) reveal the presence of broad emission at the resonance lines of C II, Si IV, Al III, and Mg II. Numerous shell lines which vary in strength and width, pervade the UV spectrum. The IUE data support the spectral classification of B2 (Doazan and Peton 1970).

e. AU Monocerotis (HD 50846)

If AU Mon were not an eclipsing binary, it would be considered an average B5e star. IUE observations show this system to be quite similar to CX Dra. A similar gas stream is seen ($\phi = 0.85, 0.93$) and equally strong, variable N V (and C IV) resonance lines are found.

f. TT Hydrae (HD 97528)

TT Hya is a well-known Algol-type eclipsing binary. The prominent Hα emission feature observed in this system displays phase dependent V/R variations (Peters 1980). A rich shell spectrum, which does not appear to be highly variable, is seen in the UV. The gas stream was detected in the resonance lines of C II and Si II.

3. DO THE UV SPECTRA OF INTERACTING BINARY Be STARS DIFFER FROM THOSE OF "SINGLE" Be STARS?

IUE observations of six Be stars which are known to be mass transfer binaries have, unfortunately, not revealed a single spectral feature, set of features, or spectroscopic behavior which alone can be considered a signature of an interacting system. However, there are a number of criteria which can be employed to find <u>possible</u> binaries. These include the presence of 1) numerous shell lines, especially those which arise from low lying and/or metastable levels, 2) redward asymmetries in the resonance features, and 3) variable N V absorption. Normally, N V features are not observed in Be stars later than B1 (Marlborough and Peters 1982) but, among our program binaries, we have observed strong N V as late as B5!

We gratefully acknowledge support from NASA grants NSG 5422, NSG 5356, NAG-5-126 (GJP), and NAS 5-23576 (RSP).

REFERENCES

Doazan, V., and Peton, A.: 1970, Astron. Astrophys. 9, pp.245-251.
Koubsky, P.: 1978, Bull.Astron.Inst.Czech. 29, pp.288-298.
Marlborough, J.M., and Peters, G.J.: 1982, this volume, pp.387-390.
Paterson-Beeckmans, F.: 1980, Proc.Sec.European IUE Confer., pp.51-54.
Peters, G.J.: 1976, in 'Be and Shell Stars',IAU Symposium 70,pp.417-428.
Peters, G.J.: 1980, in 'Close Binary Stars:Observation and Interpretation', IAU Symposium 88, pp. 287-292.
Peters, G.J.: 1981, in 'The Universe at Ultraviolet Wavelengths:The First Two Years of IUE;, NASA publ., in press.
Peters, G.J.: 1982, this volume, pp.311-314, 353-357, 401-404, 411-414.
Poeckert, R.: 1981, Publ.Astron.Soc.Pacific, 93, in press.
Polidan, R.S., and Peters, G.J.: 1980, in 'Close Binary Stars: Observations and Interpretation', IAU Symposium 88, pp. 293-297.

RECENT CHANGES IN THE ULTRAVIOLET SPECTRUM OF THE Be STAR HR 2855

Geraldine J. Peters
Department of Astronomy
University of Southern California
Los Angeles, CA 90007   USA

High resolution IUE observations of the Be star HR 2855 (HD 58978) spaced five months apart have revealed striking variations in the resonance lines of C IV and N V. The nature of these variations can be seen in Figures 1 and 2.

When HR 2855 was first observed on 1979 October 23, each member of the C IV resonance doublet appeared to have three distinct absorption components. A similar pattern was seen in the N V resonance doublet. The centroid of each complex feature is violet shifted by 150 km s^{-1} relative to the photospheric lines. The C IV and N V features which were observed on 1980 March 31, did not display the complex structure observed five months earlier. Even though the profiles varied considerably, the velocity shift of the features remained the same. The

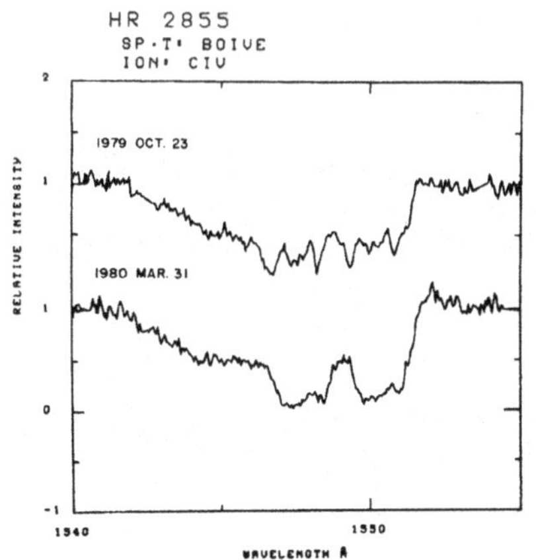

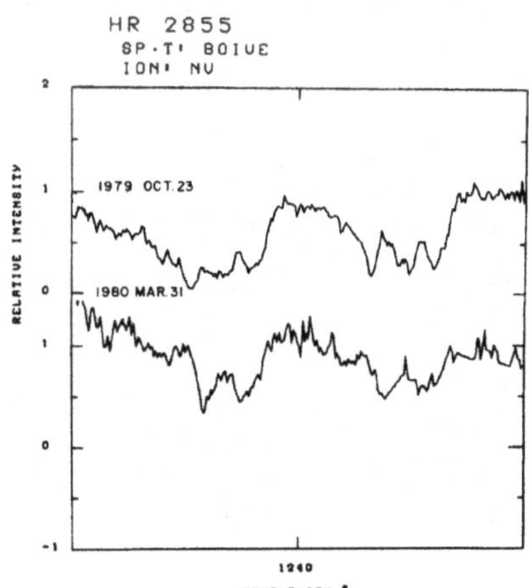

Fig. 1 - Conspicuous differences are seen between these normalized C IV profiles observed five months apart.

Fig. 2 - Same as Fig. 1 for N V.

411

M. Jaschek and H.-G. Groth (eds.), Be Stars, 411–414.
Copyright © 1982 by the IAU.

observations suggest that, in late October of 1979, the broader absorption features contained centrally reversed emission components. A weak emission component persisted in N V during the 1980 March
observation.

In contrast to C IV and N V, the Si
IV resonance lines remained invariant (Fig.
3). Although the presence of a wind is
evident from the shallow slopes of the
violet wings of the Si IV features, there
is no evidence of mass outflow in discrete components. Radial velocities from
the Si IV cores ($\simeq 30$ km s^{-1}) agree very
well with those from the photospheric features. It should be noted that the de-
pressed violet wing observed in C IV is
simply a blend of photospheric features
which are strong in B0 stars; it is not
the result of a wind.

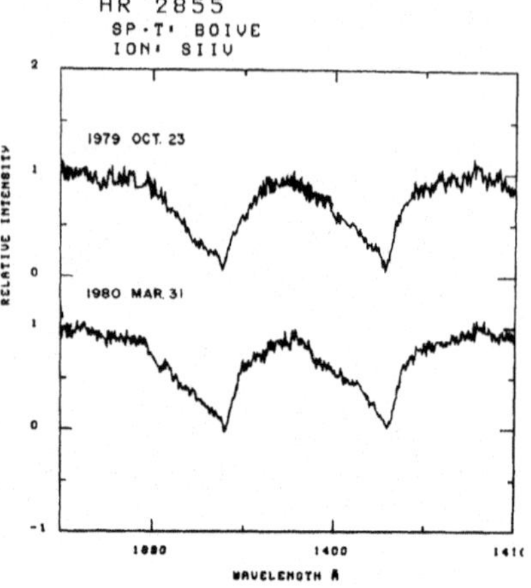

Fig. 3 — The Si IV doublet
observed simultaneously
with C IV and N V.

The spectral variations reported in this paper bear some resem-
blance to those recently observed in the active Be star 59 Cyg (Doazan,
Kuhi, and Thomas 1980, Doazan, et al.1982). In both stars, it is the
superionized species which undergo the changes; Si IV shows no statistically significant variation. In both cases, the active regions appear to be separate from the stellar photosphere.

During the past decade, the behavior of the Balmer lines in HR 2855
seems to have been slightly different from that reported for 59 Cyg
(Doazan, Kuhi, and Thomas 1980, Hubert-Delplace and Hubert 1979).
Spectrograms of HR 2855 obtained at Lick Observatory, Kitt Peak National
Observatory, and UCLA's Ojai Observatory from 1969-1981 have shown the
Balmer emission to be persistent (though relatively weak). V/R variations and transient shell features have frequently been observed. No
period has been discovered but our data show that the lifetime of the
shell is of the order of a few months or less.

In conclusion, HR 2855 should be included in a growing class of
Be stars which display a "chromospheric" type activity. Future UV observations which are planned will allow us to determine the time scale
for the variations and whether the activity parallels that seen in the
extensively observed Be star 59 Cyg.

This project is being supported, in part, by NASA grant NSG 5422.

REFERENCES

Doazan, V., Kuhi, L.V., Thomas, R.N.: 1980, Astrophys. J. 235,pp.L17-L20
Doazan, V., Grady, C., Kuhi, L.V., Marlborough, J.M., Snow, T.P., and
     Thomas, R.N.: 1982, this volume, pp. 415-418.
Hubert-Delplace, A.M. and Hubert, H.: 1979, An Atlas of Be Stars,
     Observatoire de Meudon, France.

DISCUSSION OF PAPERS BY G.J. PETERS

a. ω Ori, 66 Oph

Harmanec: May I ask you how safe your identification of different UV lines is? Some of your identifications appear to be done somewhat arbitrary in a flat portion of broad blends. I have no personal experience with the UV spectra and I feel a bit worried after seeing the theoretical UV spectra, full of lines, shown yesterday by Dr. Marlborough, and after hearing that many unidentified lines were found in the UV spectra.
2. How do you explain the two other components of the UV lines found in ω Ori?

Peters: 1. The lines are strong in ω Ori. The Si IV features in 66 Oph are only about 4σ, However, in view of the agreement between the Si IV velocities and those of the stronger C IV and weak Si III, I believe the identifications of the weaker lines. The v sin i's for these stars are high enough to completey rotationally broaden the numerous photometric lines which exist in the vicinity of C IV and Si IV. Therefore the sharp components are most assurely formed in the circumstellar matter.
2. I interpret the presence of the multiple component as evidence for the existence of multiple discrete "clouds" or "shells". Further observations should be able to tell us more about this interesting pattern of mass loss.

Henrichs: The high velocity features in ω Ori and 66 Oph are exactly like those observed in γ Cas (see page 431). If you plan new observations you might consider our experience with γ Cas that these high velocity absorptions are observable only during a few weeks.
Their column densities decay according to t^{-2} (like we had predicted, 2nd IUE Conf. Tübingen), which is very rapid.

Slettebak: With regard to the spectral type of ω Ori, B2 III is a better type than B3 III, in my opinion. Also, I now believe that v sin i for 66 Oph is closer to 240 than 280 km/s.

b. Interacting binaries

de Groot: I have just heard that Sahade has found a change in behaviour of the mass outflow in AU Mon, but I don't have further details. Do you know anything about this?

Peters: In the UV, N V (and C IV) are quite variable. H_α observations reveal that the disc is extended (comparable in size of the Roche surface of the primary) and that there is enhanced mass-loss near $\phi = .5$.

Henrichs: The variable high excitation lines you described remind me of the x-ray binary 4U900-40 where in the optical counterpart a

nice orbital phase dependence was found.
This was predicted by MacCray '75, as being a direct consequence of
the wind—ionization dependence of the nearby x-ray companion. Therefore, x-ray observations, if they are feasible, might help towards
an explanation.

<u>Peters</u>: I agree, but to the best of my knowledge only one of the
program stars, HR7084, is a x-ray source (Plavec, personal communication).

c. HR 2855

<u>Doazan</u>: I would like to emphasize the importance of G. Peters
observations which give direct evidence of 1. mass ejection in pole-
on stars, i.e. mass ejection in polar regions if one accepts the
rotational model, 2. variability of the superionized lines, which
seem now to be a general phenomenon. You will see that in 59 Cyg
this variability is observed in a striking manner.

<u>Snow</u>: We know from ample evidence presented here that the Be stars
themselves have winds, almost always highly variable in velocity and
strengths of multiple components. In view of this, I just don't see
how you can even hope to find phase effects at the relatively low
velocity amplitudes of these binary systems. The mass-loss components
are broad and variable. I am very skeptical that binary phase effects
can be seen in the ultraviolet wind profiles, especially with IUE data.

<u>Peters</u>: Generally speaking, we have <u>not</u> observed the winds in our
program binaries to be highly variable. If variations exist at all,
they are slight (more like those seen in δ Cen as opposed to the
activity observed in 66 Oph, ω Ori, 59 Cyg, etc.). We have found IUE
data to be quite adequate for detecting the major phase dependent
changes that exist such as the doubling of the equivalent width of
certain spectral features around $\phi \approx 0.9$. Indeed, using IUE, one can
easily detect ($\gg 3\sigma$) redward asymmetries in numerous resonance lines
due to the mass transfer. This absorption prevails throughout the
phase interval predicted by theory. We have observed most of the stars
over several cycles at the same phase points to check for transient
activity and have reported suspected instances in this paper (N V in
AU Mon and HR 7084). Evidence for the phase dependence of the wind in
HR 2142 is found in both IUE and <u>Copernicus</u> data.

THE ACTIVE UV PHASE OF 59 Cyg

V. Doazan[1], C. Grady[2], L.V. Kuhi[3], J.M. Marlborough[4], T.P.
Snow [2], R.N. Thomas[5].
Observatoire de Paris[1], Univ. Colorado[2], Univ. California[3],
Univ. western Ontario[4], Inst. d'Astrophysique[5], Paris.

Abstract : Coordinated UV and visual observations of 59 Cyg in 1978-81
show strong mass ejection activity and strong variability in displacements and profiles of superionized lines during the new Be phase, starting from a "quasi normal B" phase in 1977, and increasing irregularly
through 1981 to a low and then moderate Hα emission. These data show that
visual data alone cannot describe the activity of the star.

I. Introduction

Mc Laughlin (1948) characterized 59 Cygni's behavior by "long periods of quiescence and short periods of activity" which, if understood
as "Balmer emission line" activity has been largely confirmed by subsequent visual observations. However, UV observations have shown that "mass
ejection" activity of the star cannot be infered from only the strengths
of the Balmer emission line. A calm, inactive mass ejection phase can be
assigned to the star on the basis of visual observations, at an epoch
where indeed UV observations show a highly active one. These conclusions
come from a comparison of UV and visual observations of 59 Cyg, in the
interval 1972-present. It was shown that : (i) at "almost normal B phase", near minimum Hα emission, line displacements exceeding escape velocities are observed in superionized species (Doazan et al 1980a). By
contrast, much smaller displacements in these lines were observed during
the last episode of spectacular variations in the visual (Snow and
Marlborough 1980). (ii) Near that minimum Hα emission phase, and during
the subsequent new Be phase of low level Hα emission, a strong variability in displacements and profiles of the superionized lines is observed (Doazan et al 1980 b). So, the period of highest activity thus
far observed in the UV corresponds to one which is considered uninteresting in the visual spectrum, and which was thought to characterize
"a calm and quiescent" period in that star. This demontrates that any
description of the mass-flux activity in Be stars, from visual data
alone, can be misleading, and that an interpretation of the Be phenomena must necessarily combine visual and UV observations. To illustrate
the activity of 59 Cyg in 1980, where Hα emission has reached a moderate level, we present an abstract of UV data from a coordinated program

415

M. Jaschek and H.-G. Groth (eds.), Be Stars, 415–418.
Copyright © 1982 by the IAU.

of UV and visual observations, which began in 1978 and which continues
through 1981 in an enlarged collaboration.

  II. <u>The observations</u>.
   From 1978 to 1981, 59 Cyg has been regularly observed in the UV,
almost monthly, at high dispersion with IUE, and with Copernicus (in
1979). The IUE data were reduced using software written in the Interactive Data Language (IDL), at the Laboratory for Astrophysics and
Space Sciences (Univ. of Colorado). Fig. 1. shows selected profiles of
CIV resonance lines observed in 1980. One sees clearly the presence of

Fig. 1. Representative profiles of CIV resonance lines of
        59 Cyg in 1980 ; the arrows show laboratory wavelengths.

multiple components, whose positions and strengths vary strongly. The
majority of the spectra obtained so far show 2 such components : one of
low expansion velocity (< 100 km/s), one of high (in the range 600 to
1000 km/s) if one measures the displacement of the deepest part of the
absorption profiles. The position of the blue edge varies from -900 to
-1200 km/s. The most frequent configuration is the one where the high
velocity component is the stronger, but the reverse is also sometimes
seen. At other times, a broad profile is observed, where one can identify the different components from similar features corresponding to the
difference in velocity of the doublet. Large variations are observed on
time scales of a month, but some observations taken 24 or 48 hours apart,
show smaller variations which suggest a progressive change in the profile. Regular observations made at interval of days, over a period of
 one or two months, will help to understand the pattern of variability.

By contrast to the superionized resonance lines, the Si IV resonance lines showed almost no variations. In 1978-79, their profiles are
quasi-symmetric, quasi-undisplaced and quasi-nonvariable. In 1980, some
asymmetry begins to appear with a larger extension of the violet wings.
A detailed discussion of all the combined UV and visual observations
from 1978 to 1981 will be given in a more extensive paper.

III. Discussion

Our data lead to the 2 important conclusions :
(i) The activity of a Be star consists of a whole cycle : from the
beginning of the recharging of the extended atmosphere to its complete
dispersal, wherein the Hα minimum emission phases are very important.
Thus, one cannot infer from visual observations only, the details of
the mass ejection activity of the star. The prespatial description of
the activity of the star, related to strong Hα emission only, where phases of minimum Hα emission, or quasi-normal B phases, were neglected
and considered "uninteresting", can give a wholly erroneous picture of
the Be phenomena. This implies that the history of the star over the
whole cycle be considered and not only some selected spectacular phases
observed in the visual (Doazan 1981).
(ii) From the above, to base a model on only the visual spectrum and
only on observations at one or a few epochs, will be wholly misleading.
From the visual spectrum alone, one will model for the Be star an outer
"cool" atmosphere producing Hα emission. From UV data alone, one will
model an outer superionized atmosphere. An empirical model of the real
Be star must necessarily combine both UV and visual data. The above
underlined importance of the history of the star throughout a whole
cycle, from "shell build up" phase to "shell dispersal" phase, requires
the mass flux, thus the model to be time-dependant.

REFERENCES
Doazan, V., Kuhi, L.V., Thomas, R.N., 1980 a, Ap. J. 235, L17
Doazan, V., Kuhi, L.V., Marlborough, J.M., Snow, T.P., Thomas, R.N.,
        1980 b, The Second IUE Conference, ESA-SP-157, p151
Doazan, V., 1981, A. A. submitted
Mc Laughlin, D., 1948, as quoted in Merrill P.W. and Burwell, C.G.,
        1948, Ap. J. 98, 163
Snow, T.P., Marlborough, J.M., 1980, Ap. J. 235, 85

DISCUSSION

<u>Dachs</u>: Do your UV observations of 59 Cyg permit the following
conclusion: During phases of weak or no H_α emission, the velocity
and strength of the stellar wind in this star are maximum, and the
stellar wind blows away the H_α emitting envelope. When the strength
of the wind declines, the envelope will recover. Is this a possible
model of the activity cycle of a Be star?

<u>Doazan</u>: The observations seem to suggest the following:
1. For some reason the H_α emission shell disappears and the star
becomes a "quasi-normal B star".
2. Then, about this "minimum Be phase", the mass flux "activity" is
large and continues so, for at least two years, up to the present.
It is during this 2-3 years that the "extended atmosphere" where H_α
emission is formed builds up, i.e. the F_M "fills the balloon".
3. The important question is what fixes the outer boundary of the
atmosphere; the observations seem to suggest that it is the "enhanced"
F_M running into a previous low F_M, not the ISM, which will build the
extended atmosphere.
4. Once the extended atmosphere is built up, this situation seem to
constitute the quasi-state Be star phase.

<u>Viotti</u>: I agree with you that one has to go down to UV to make a
good modelling. My question is the following: Did you look for excited
lines (photospheric) that would tell you more about the temperature
deep in the atmosphere, and what about the continuum energy distribution in UV, is it variable?

<u>Doazan</u>: From a rough examination of the spectra one would conclude
that the photospheric lines do not show conspicious variations. These
results and those for the continuum will be published in a seperate
paper.

FAR-ULTRAVIOLET COLORS OF B STARS WITH AND WITHOUT EMISSION LINES

J. ZOREC[1], D. BRIOT[2], L. DIVAN[3]
[1]Laboratoire d'Astrophysique Théorique, Collège de France ;
[2]Observatoire de Paris, France ; [3]Institut d'Astrophysique ,
Paris, France

ABSTRACT. A method free of interstellar reddening of comparing the
energy distributions in the far-UV of Be/Shell stars to those of normal
B stars is presented. The deviations of Be/Shell stars from the sequences of normal stars are correlated with other physical parameters
observed in these stars. The largest UV color differences to the sequences of normal stars are found for those Be/Shell stars having the
largest IR color excesses.

I. INTRODUCTION

A big problem to be solved in analysing the energy distribution
of Be/Shell stars in any wavelength is the separation of the effects
due to the ISM and the circumstellar envelope. In the far-UV, this is
illustrated by the disagreement between the results obtained by Briot
(1978) and Beeckmans and Hubert-Delplace (1980) using different methods of interstellar dereddening.

When the UV fluxes of Be/Shell stars are compared to those of
normal B stars, they have to be normalized to a similar photospheric
flux. However as probably no spectral region in Be/Shell stars is produced only by a photosphere, it does not seem possible to determine if
a flux excess or deficiency exists as compared to normal B stars. We
are then limited to only compare slopes of flux distributions or
colors.

Due to the irregular variations of the radiation of Be/shell
stars it is necessary to define colors only from simultaneously observed fluxes. With this in mind, the method proposed here avoids : a)
the determination of the ISM extinction and b) the scatter due to the
non-simultaneity of the observations in the different wavelength
ranges.

II. THE METHOD

We have defined a color index which is independent of a mean ISM
extinction and strongly sensitive to colour differences due to the
spectral type in the UV spectral range covered by the low resolution
S2/68 observations made with the TD-1 satellite (Jamar et al.,1976 ;

419

Macau-Hercot et al., 1978). This index is in fact independent of all extinctions with the same absorption law than the mean one for the ISM adopted here (Savage and Mathis, 1979). The color index which characterize the UV behavior is defined as follows :

$G = \Delta_{13} - K.\Delta_{12}$

where $\Delta_{13} = m_{1460} - m_{2740}$; $\Delta_{12} = m_{1460} - m_{2350}$ and K is the ratio of the colour excesses $E(\Delta_{13})/E(\Delta_{12}) = -11.21$. The m_λ values are magnitudes and the wavelengths are chosen to be as free as possible from strong line absorption, as seen from the high-dispersion IUE spectra.

It was noted by Divan (1979) that neither the value of the first Balmer discontinuity D_0 attributed to the stellar photosphere nor the value of the parameter λ_1 characterizing the luminosity class change during the variations of Be/Shell stars. These parameters are then used to classify the Be/Shell stars (BCD system of classification), which is in good agreement with the MK spectral classification. Chalonge and Divan (1977) have shown that the type given by (λ_1, D_0) may be also characterized by a single parameter S_{70} free of the luminosity class.

Using MK standard stars with BCD classification, we have obtained two well defined sequences (G vs.S_{70}), one for the luminosity classes IV and V and the other for the luminosity class III. As the dispersion of the G values around each mean sequence is not higher than $\sigma = 0.6$, this procedure may be used to classify normal B stars from the continuous energy distribution in the far-UV.

It is interesting to note that the ßCMa stars are placed in this diagram between both sequences (they are normally considered to be of luminosity class IV and are known to have the UV colours of normal stars of the same spectral type (Beeckmans and Burger, 1977), but σSco which has a different ISM absorption law (Snow and York, 1976), has such a different value of G that it does not fit into our scheme. So our method immediately indicates special absorption laws.

Using the parameter S_{70}, which represents a non-variable spectral type of the Be/Shell stars (or just their MK spectral type when BCD classification were not available),we have compared the G values of these stars to the sequence of normal B stars of the same luminosity class. The differences $\Delta G = G(Be/Shell) - G(B)$ were then correlated to other physical quantities.

III. <u>RESULTS AND DISCUSSION</u>

The Be/Shell stars plotted in the (G, S_{70}) diagram have a higher dispersion than the dispersion due to the method itself around the main sequence of normal stars of the same luminosity class. The majority of the Be stars show redder colors in the far-UV than the normal B stars, i.e. $\Delta G > 0$, while a smaller number of them show bluer colours, i.e. $\Delta G < 0$. These results are compatible with those obtained by Beeckmans and Hubert-Delplace (1980, fig.7) for the ultraviolet colors of Be and Shell stars. The mean ΔG for each spectral type seems to decrease towards later types. On the other hand, among the Be and Shell stars with $\Delta G < 0$, some like XOph have even bluer colors than those of the bluest normal stars (around spectral type B0). Such a very blue

color may be due to a different interstellar absorption law absorbing
less in the short wavelengths than the mean law used here, as for
σSco. But this cannot explain cases like 59 Cyg varying from $\Delta G=+1.3$,
during a Be phase (1972), to $\Delta G=-5.5$ during a strong shell event
(1973) (these two points are connected in fig.1).The UV observations
are taken from Beeckmans (1976).This UV color change has been noticed
in Beeckmans and Hubert-Delplace (1980). Very blue colours, like those
of some well developed shell stars (e.g. 48 Lib) might be explained by
a strong circumstellar absorption with a λ^3-like law. But this is not
the case for ΧOph for which an explanation might be given assuming a
different interstellar law. In any case,we can note that as the number
of stars with $\Delta G<0$ is rather low, it is possible to think that the
time the stars spend during such events (when they have them) must
generally be short.

   We have studied the UV color index G of Be/Shell stars according
to their IR color exceses using Allen (1973) and Jaschek et al.
(1980). The Be/Shell stars classified F in Allen (1973) (stars with an
infrared color excess attributed to free-free radiation) correspond

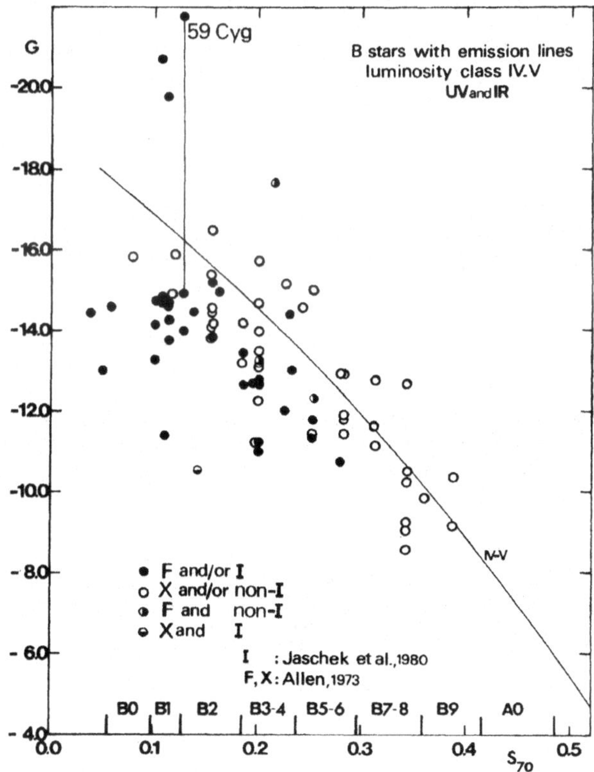

Fig. 1.

almost entirely to the stars classified I in Jaschek et al. (1980) while the stars classified X in Allen (1973) (stars with little or no IR color excess) correspond to the stars not classified I in Jaschek et al. (1980) with a few exceptions. In figure 1 the Be/Shell stars of luminosity class IV and V as classified above and the mean sequence of normal stars of these luminosity classes are plotted. Even though the observations in the IR are not strictly simultaneous with the UV, we can see that there exists a quite different distribution of stars according to their IR color excesses : the stars classified X and/or non-I have UV colors close to those of normal B stars (small values of $|\Delta G|$)and those classified F and/or I have very different UV colors compared to normal B stars (large values of $|\Delta G|$). This indicates that some connection may exist among the processes' or the regions which produce such IR and UV color behaviors.

The differences ΔG were also compared to the strength of emission lines using the photometric index in the Balmer lines observed during February 1972 (quasi-simultaneously with the UV observations) by Feinstein (1974). A general tendency was found of high values of ΔG corresponding to strong line emission. However a complete lack of correlation is seen between ΔG and vsini, indicating that there is little probability of detecting some rotational darkening in the far-UV.

It appears that with the very simple method described here, it is possible to clearly relate the UV colors of Be/Shell stars to their physical properties in other wavelength ranges.

REFERENCES

Allen, D.A.:1973,Monthly Notices Roy. Astron. Soc.161,145
Beeckmans, F.: 1976, Astron. Astrophys. 52,465
Beeckmans, F., Burger,M., 1977, Astron. Astrophys.61,815
Beeckmans, F., Hubert-Delplace, A.M.: 1980, Astron. Astrophys. 86,72
Briot, D.: 1978, Astron. Astrophys. 66, 197
Chalonge, D., Divan, L.: 1977, Astron. Astrophys. 55, 121
Divan, L.: 1979, IAU Coll. 47, Ed. G.V. Coyne, M.F. McCarthy, A.G.D.
     Philip, p.247
Feinstein, A.: 1974, Monthly Notices Roy. Astron. Soc. 169, 171
Jamar, C., Macau-Hercot, D., Monfils, A., Thompson, G.I, Houziaux, L.,
     Wilson, R.: 1976, Ultraviolet Bright-Star Spectrophotometric
     Catalogue, ESA SR-27
Jaschek, M., Hubert-Delplace, A.M., Hubert, H., Jaschek, C. : 1980,
     Astron. Astrophys. Suppl. Ser., 42, 103
Macau-Hercot, D., Jamar, A., Monfils, A., Thompson,G.I, Houziaux, L.,
     Wilson, R. : 1976, Ultraviolet Bright-Star Spectrophotometric
     Catalogue, ESA SR-28
Savage, B.D, Mathis, J.S.: 1979, Ann. Rev. Astron. Astrophys. 17,73
Snow, T.P., York, D.G: 1976, Astrophys. Space Sci.55,19

THE PROBLEM OF X PERSEI[+]

R. Viotti, M. Ferrari-Toniolo, A. Giangrande, P. Persi
Istituto Astrofisica Spaziale, CNR, Frascati, Italy

G. B. Baratta
Osservatorio Astronomico, Roma, Italy

X Per is a variable emission line star which shows among other peculiarities a weak X-ray emission (4U 0352+30) and a strongly variable IR excess (Ferrari-Toniolo et al. 1978, Viotti et al. 1980). In the past decade the star has undergone three phases of enhanched "activity" (1972–73, 1978 and 1980) characterized by brighter visual luminosity, excess in the Balmer continuum and in the IR, stronger X-ray emission, with intermediate periods of minimum activity (1974-77, 1979) when the optical-infrared energy distribution was closer to that of a normally reddened early type star (figure 2). But during most of its history the energy distribution largely deviated from that of a non-emission line early type star, and the first problem is to determine the interstellar extinction, disregarding any "local" effect. The strength of the 2200 A band in the UV spectrum of X Per is consistent with $E(B-V)=0.35$, a value close to the extinction towards other Per II stars: ζ Per (0.34), o Per (0.31), ξ Per (0.32, Viotti & Lamers 1975). The i.s. Ly_α line observed in the high resolution IUE spectrum of X Per obtained on 1979, December 23, has a FWHM of 11.0 A corresponding to $N(HI)=4.9\ 10^{20}cm^{-2}$. The Copernicus observation of H_2 lines(of not good quality) gives $N(H_2)\simeq1.1\ 10^{21}$(Mason et al. 1976). A much lower value of $3-5\ 10^{20}$was derived by Snow (1976, 1977) for o and ζ Per. Taking for X Per $N(H_2)\simeq5\ 10^{20}$we have $N(H\ total)\simeq1.5\ 10^{21}$, yielding to $N(H)/E(B-V)=4.3\ 10^{21}cm^{-2}/mag$ in agreement with Bohlin law (1975).

At minimum luminosity the IR excess vanishes or is much smaller (figure 2) and the Balmer discontinuity is closer to that of a BOV star (de Loore et al. 1979). We then derive from the colours at min.$(U-B=-0.78, B-V=0.08;\ V=6.72)\ E(B-V)=0.40$ and $(B-V)_o=-0.31$.

+ Based on observations with the International Ultraviolet Explorer (IUE) collected at the Villafranca Satellite Tracking Station of ESA.

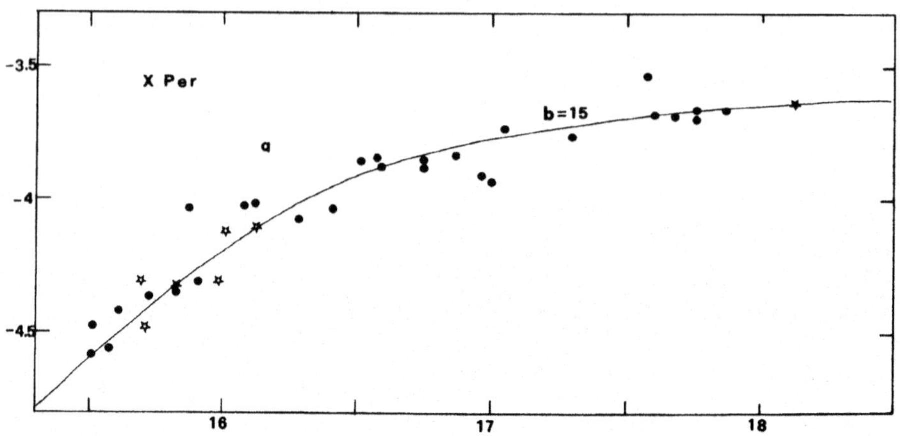

Figure 1   Curve of growth for the UV interstellar lines of ionized species and of NI, OI. Stars: CIV, SiIII,IV, AlIII. The solid line is the
theoretical curve for b=15 km/s from Spitzer (1978).

    We have identified in the IUE spectrum of X Per 63 i.s. lines of
neutral and ionized elements (H,C,N,O,Mg,Al,Si,S,Mn,Fe,Ni,Cu,Zn) and of
CO. Their intensity is systematically stronger than those of o and ζ Per
measured by Snow. This should be ascribed to a large velocity dispersion
of unresolved absorption components, implying that in front of and close
to X Per there are additional low and high temperature i.s. media.CO is
stronger than expected from E(B-V) but we note that a strong CO emission
towards X Per was found by Knapp & Jura (1976). A preliminary curve of
growth analysis following the method described by Spitzer (1978) gives a
good fit for a velocity dispersion of 15 km/s for all the ionized lines
and for NI, OI (figure 1), and of 5 km/s for the other neutral elements.
Taking logN(H)=21.18 as above, a large element depletion is derived for
all the elements with respect to the Sun, in particular for Al,Mg,C,Si
and Fe. The high ionization i.s. lines of CIV, SiIII,IV, AlIII are much
stronger than in the nearby stars implying the presence of a hot local
medium, probably resulting from photoionization by a hot radiation field
rather than from interaction of the stellar wind with the i.s. medium,
also because there is no sistematic velocity difference with the other
i.s. lines. The UV spectrum of X Per shows broad asymmetric resonance
lines belonging to a large ionization range (from SiII to CIV and probably NV, Viotti et al. 1980) formed in the outer layers of the expanding
envelope where large temperature differences seem to be present. The

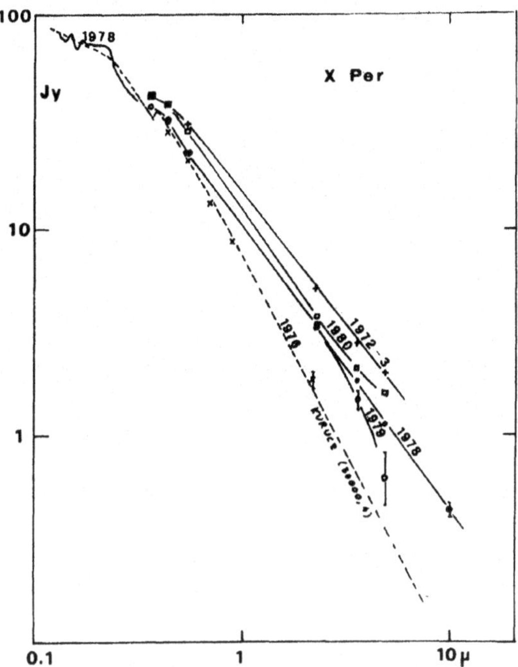

Figure 2 : The energy distribution of X Per in different epochs.
Crosses : 1972-73, x : November 1976, dots : Nov. 1978, circles : Dec.
1979, squares : Aug.-Sept. 1980. The IUE spectrum of Oct. 1978 and the
Kurucz (30000,4) model atmosphere are shown. See Ferrari-Toniolo et al.
1978. Persi et al. 1977 and this Symposium, Viotti et al. 1982.

excited lines of HeII, NIII, OIV that are formed in the deepest parts of
the atmosphere (and might be considered as "photospheric"), are broad
and symmetrical indicating a large rotational velocity of about 200 km/s
of the "star". We conclude that a large acceleration of the stellar wind
should take place in the upper atmosphere, up to about 650 km/s (Bianchi
& Bernacca 1981). No significantly large variation of the line intensity
and profile, in particular of the resonance lines was found from a preli
minary analysis of the UV spectrum of X Per during 1978-79 notwithstand-
ing the large luminosity changes observed during that period, supporting
the idea that the observed optical and IR variations probably took place
in the higher atmospheric layers. Figure 2 shows the energy distribution
of X Per during different activity phases. It is evident that the whole

optical-IR spectrum is varying with a larger amplitude in the IR. This suggests that it could be the result of two contributors, one constant and close to the spectrum of an early type dwarf with $V_o \simeq 5.62$, $(B-V)_o = -0.31$, and the other one strongly variable which could be associated to emission from the variable outer layers. However difficult is to say if the observed variations are related to strong changes of the mass loss rate, or it is largely due to structural changes of the expanding envelope. We have estimated the rate of mass loss from the IR excess of $4\ 10^{-7}$ $M_\odot$/yr (Persi et al. this Symposium) which is probably variable in time and largely depends on the adopted model, but that is definitely much larger than the rate of 10^{-8}–10^{-9} estimated from the asymmetric profiles of the UV resonance lines (Hammershlag-Hensberge et al. 1980, Bernacca & Bianchi 1981). We hope that a more detailed analysis of the UV line profiles will overcome this discrepancy.

Finally, variations of the X-ray flux was found that seem to be associated with the activity of X Per (e.g. de Loore et al. 1979, White et al. 1976), indicating that it should be originated in a region, or compact star, not too far from X Per. However the presently published data on the X-ray emission are not sufficent to give a clearer picture.

REFERENCES

Bernacca, P.L., Bianchi, L. 1981, Astron. Astrophys. 94, 345

Bohlin, R.C. 1975, Astrophys.J. 200, 402

de Loore, C., Altamore, A., Baratta, G.B., Bunner, A.N., Divan, L., Doazan, V., H. Hensberge, G., Sterken, C., Viotti, R. 1979, Astron. Astroph. 78, 287

Ferrari-Toniolo, M., Persi, P., Viotti, R. 1978, Mon.Not.R.A.S. 185, 841

Hammerschlag-Hensberge, G. et al. 1980, Astron. Astrophys. 85, 119

Knapp, G.R., Jura, M. 1976, Astrophys. J. 209, 782

Mason, K.O. et al. 1976, Mon. Not. R.A.S. 176, 193

Snow, T.P. 1976, Astrophys. J. 204, 759

Snow, T.P. 1977, Astrophys. J. 216, 724

Spitzer, L. 1978, Physical Processes in the Interstellar Medium, J. Wiley and Sons, New York

Viotti, R., Lamers, H.J.G.L.M. 1975, Astron. Astrophys. 39, 465

Viotti, R., Ferrari-Toniolo, M., Giangrande, A., Persi, P., Bianchi, L. Grasdalen, G., Kalv, P., Stalio, R. 1980, Second European IUE Conference ESA SP-157, 165

White, N.E., Mason, K.O., Sanford, P.W., Murdin, P. 1976, Mon. Not. R.A. S. 176, 201

Persi, P., Viotti, R., Ferrari-Tóniolo, M., 1977, Mon.Not.R.A.S. 181, 685

THE SPECTRUM OF HD 51585 IN THE BLUE AND IN THE
ULTRAVIOLET (1) (2)

L. Houziaux, Y. Andrillat, A. Heck and K. Nandy
Institut d'Astrophysique , Université de Liège,
Avenue de Cointe, 5,
B - 4200 Liège (Cointe-Ougree), Belgium

Observatoire de Haute-Provence, C.N.R.S.
F - 04870 St. Michel l'Observatoire, France

Astronomy Division, E.S.T.E.C.,
European Space Agency

Royal Observatory, Edinburgh, U.K.

ABSTRACT

New visible and ultraviolet spectra of the peculiar
emission line star HD 51585 are described. Interstellar
lines and the λ 2200 feature are rather weak. A colour
excess E(B-V) = 0.33 is derived. The extinction curve resembles the one obtained from LMC stars.

HD 51585, mentioned by Merrill and Burwell (1933) as
a P Cygni star has been studied spectroscopically by Andrillat and Houziaux (1969, 1973), and at a higher dispersion by Klutz and Swings (1977).

---

(1) Les observations utilisées dans le présent travail ont
été effectuées partiellement à l'Observatoire de Haute
Provence (C.N.R.S.).

(2) Based on observations by the I.U.E. collected at the
Villafranca Satellite Tracking Station of the European
Space Agency.

A new spectrum has been obtained on Jan. 2, 1981, using an RCA 3 stages image-tube spectrograph attached at the Cassegrain focus of the 193 cm telescope at the Observatoire de Haute-Provence. The region from Hβ to about 5000 A is covered with a reciprocal dispersion of 80 A.mm^{-1}. The spectrum does not differ from the description given by two of us in 1973. Nebular lines due to [O III] and [S II] appear in the observed region where all the lines are seen in emission and due to Fe II (multiplets 42, 43, 37, 38 and 27), to [Fe II] (multiplets 4F, 6F, 7F, 11F, 20F, 21F, 23F), N II (multiplets 11, 55, 61), He I (multiplets 12, 14, 16, 18, 51, 53). Hydrogen lines Hβ to Hϵ are seen as strong single emissions.

I.U.E. spectra in the low resolution mode have been obtained in March 1979 and May 1980. Like in the blue, the spectrum is dominated by strong emissions, several of which are also seen in planetary nebulae. O I λ 1302 is a strong emission, as it is also the case for Mg II around 2800 A. Stellar emission features include [N IV] λ 1483, C IV λ 1548 which shows a double emission with a moderate absorption, He II λ 1640 (?), [O III] at λ 1666, [N III] at λ 1760, an emission at λ 1888, which might be due to Fe II, the [C III] line at λ 1909, a strong unidentified emission appears at 2191 A. There might also be weak emissions between λ 2320 and λ 3000 due to Fe II, but identifications are uncertain. S II is present at 2849 A, while two features at $\lambda\lambda$ 3174 and 3187 may be attributed respectively to Mg II and He I.

Weak but definite interstellar features appear in the spectrum, notably S II (1235 A), Si II (1262 A), C II (1335 A), Si II (1527 A) and Al II (1668 A). The λ 2200 feature is weak but present. This may seem abnormal for a 11.5 magnitude star but this region in Monoceros has been found relatively free of interstellar absorption. From UBV measurements, two of us had deduced a color excess E(B-V) = 0.68. Continuum colors lead to a spectral type around B1. From measurements on the S2/68 spectra, we know that

$$\frac{E(2200-2740)}{E(B-V)} = 3$$

From the intrinsic color for a B1 star, we then deduce a value of 0.33 for E(B-V), showing that the visible colors are contaminated reemission in the Paschen continuum. The object is also known for exhibiting a strong infrared excess (Allen, 1973) which affects the V_o magnitude of the star. The continuum energy distribution has been dereddened

according to the $A_\lambda/E(B-V)$ curve by Nandy et al.(1976),
with $E(B-V) = 0.33$. Then $(m_\lambda-V)_0$ has been plotted over λ.
It is found that no model atmosphere fits
the energy distribution curve over the entire range. A
$25,000°K$, log g = 4 cgs model gives nevertheless a good
agreement in the $\lambda\lambda\ 2000-2500$ region. If this model is
forced to represent the continuous energy distribution of
the star in the region $\lambda\ 1200 - \lambda\ 2000$, we have to adopt
another reddening curve with a larger absorption below
2000 A. Such a curve is not unlike the one found for
several LMC stars observed with I.U.E. (Nandy et al., 1980).

REFERENCES

Allen, D.A. : 1973, Monthly Notices R. Astron. Soc., 161,
        p. 145.
Andrillat, Y. et Houziaux, L. : 1969, Les transitions
        interdites dans les spectres des astres, Mém. Soc. R.
        Sc. Liège, sér. V, XVII, p. 343.
Andrillat, Y. and Houziaux, L. : 1973, Les nébuleuses pla-
        nétaires, Mém. Soc. R. Sc. Liège, sér. VI, V, p. 377.
Klutz, M. and Swings, J.P. : 1977, Astron. and Astrophys.,
        56, p. 143.
Merrill, P.W. and Burwell, C.G. : 1933, Ap. J., 78, 87.
Nandy, K., Thompson, G.I., Jamar, C., Monfils, A. Wilson,
        R. : 1976, Astron. and Astrophys., 44, p. 195.
Nandy, K., Morgan, D.H., Willis, A.J., Wilson, R.,
        Gondhalekhar, P.M. and Houziaux, L. : 1980, Nature,
        283, p. 725.

DISCUSSION FOLLOWING ZOREC

<u>Paterson-Beekmans</u>: In our paper (Beekmans and Hubert-Delplace, A&A, 1980) we have also investigated the behaviour of Be and shell stars concerning the slope of their UV energy distribution and found essentially the same results as you. The stars having a bluer energy distribution as normal stars have a strong metallic shell in the visible region at the time of the observations.

<u>Sonneborn</u>: Among the normal B stars you have considered, are there many with large v sin i's?

<u>Zorec</u>: Not higher than 200 km/s, if I remember well.

DISCUSSION FOLLOWING VIOTTI

<u>Thomas</u>: I support your structure of atmospheric regions, and your idea to classify data according to which region they give information on, but please add a "region of deceleration" just near or before the "accretion" region.

DISCUSSION FOLLOWING HOUZIAUX

<u>Selvelli</u>: 1. The unidentified 2191 emission is a <u>constant</u> feature present in all the IUE low resolution LWR spectra. It is a spurious "emission" feature due to the constant presence of a saturated pixel. 2. The emission at $\approx$1890 A is unlikely to be Fe II because the other strong Fe II emissions are not present. It is more likely to be [Si III] 1890 that is usually present together with [C III] 1908 in many nebular spectra.

UV OBSERVATIONS OF γ Cas: INTERMITTENT MASS-LOSS ENHANCEMENT

H.F. HENRICHS
Astronomical Institute, University of Amsterdam
Roetersstraat 15, 1018 WB Amsterdam.

Introduction

Twenty one high resolution IUE observations made over a two year interval revealed remarkable changes in the profiles of the resonance lines
of C IV, N V and Si IV in the B 0.5 IVe star γ Cas. Narrow (FWHM < 100
km s^{-1}), blue shifted (v < 1500 km s^{-1}), absorption components of varying strenghts are observed in addition to the broad, asymmetric, low-
velocity absorption lines which seem to be steady.

An earlier proposed model to explain this phenomenon (the "UV-shell
model" by Henrichs et al. 1980, Paper 1) has now been tested and probably confirmed by new observations which were taken with high time resolution, as was suggested in Paper 1.

We arrive at the following conclusions:

1) Intermittent enhancement (factor >> 2) of the stellar wind mass-loss
   rate (duration < 1 day) gives rise to a rapidly expanding (thin),
   high-density shell. A higher density might be correlated with a
   lower terminal velocity of the shell, which is less than the terminal velocity of the wind.

2) The decay of the column density of these shell lines is expected to
   be proportional to (time)$^{-2}$ (Paper 1). This means that they are observable only during a few weeks.

3) The "UV-shell" phases of the star occur very frequently, typically
   on timescales of a week to a month. The strength and/or duration is
   variable. No periodicity has been found in γ Cas.

4) It is suggested that these UV-shell lines are not necessarily associated with the Be nature of γ Cas, but rather they may be common
   property of all early-type stars. This conclusion is based additionally on a statistical study of 26 OB stars with Copernicus (Lamers
   et al. 1981) where the majority of this sample shows more or less
   similar UV-shell lines.

5) A spin-off conclusion of the present work is that in OB binary X-ray
   sources the often observed irregular X-ray variability might well be
   explained by the UV-shell model: intermittent enhancement of the X-
   ray flux will be observed if the high-density shell passes through
   the orbit of the compact object. Simultaneous UV and X-ray observations of γ Cas might reveal its binary nature (see Paper 1).

M. Jaschek and H.-G. Groth (eds.), Be Stars, 431–435.
Copyright © 1982 by the IAU.

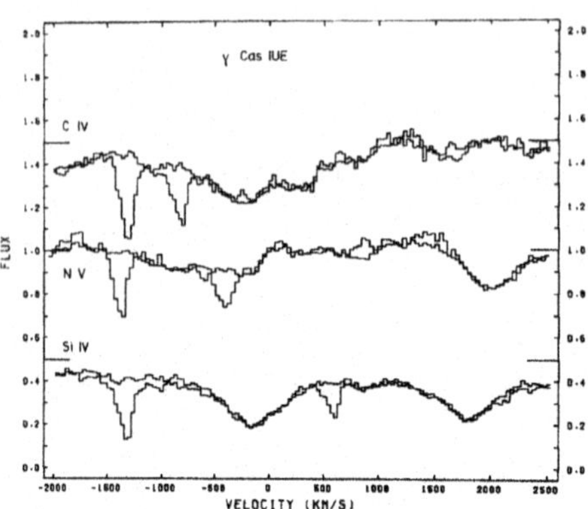

Fig.1. Spectra with and without
"UV-shell" lines

A detailed paper containing
much of the observational material
and the line profile analysis will
appear elsewhere (Henrichs et al.
1981). Here we describe only a few
important observations leading to
some of the conclusions mentioned
above.

Line profile analysis

Figure 1 shows the strongest
UV-shell episode ever recorded
from γ Cas (26 may '80) superposed
on a "quiet" spectrum near the
resonance lines of C IV, N V and
Si IV. The horizontal velocity
scale is relative to the principal
line of each doublet. Fluxes are
normalized to the stellar continuum (indicated by horizontal lines).
The "quiet" spectrum is a mean of
4 spectra which were taken between
different UV-shell episodes in
1978 and 1979. A mass-loss rate of
about 10^{-8} $M_\odot$ year^{-1} was derived
from this spectrum (Paper 1). Note
that the centers of the broad absorptions are all shifted by -200
km s^{-1}, indicating mass outflow.

In fig. 2 a time sequence of
19 days duration is shown. The

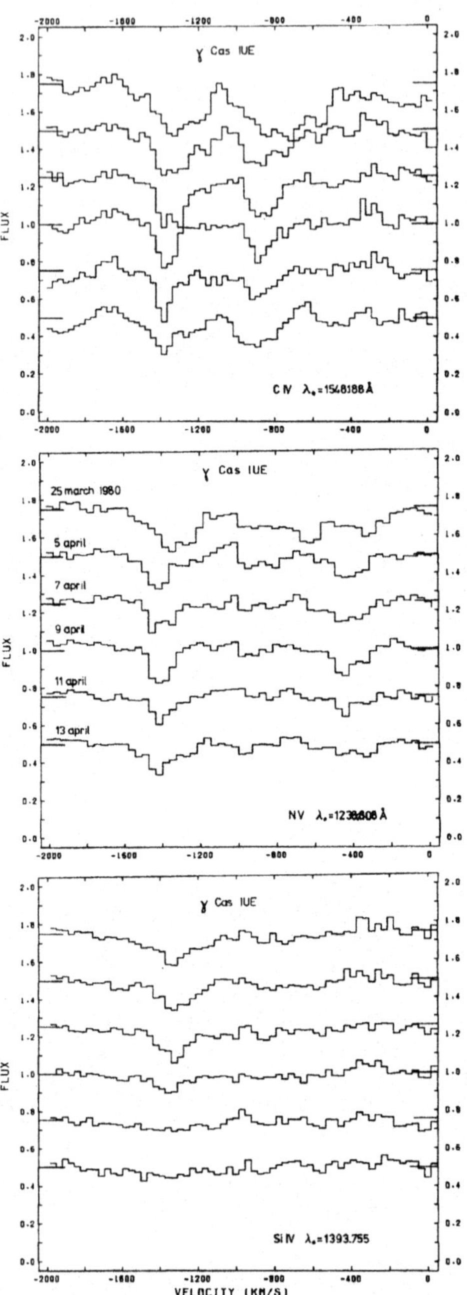

Fig.2. Normalized shell spectra of
the april 1980 UV-shell episode.

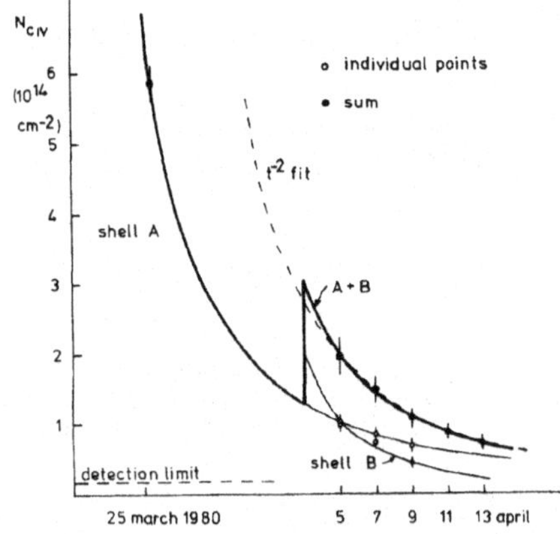

Fig.3. C IV column density decay
and possible t^{-2} behaviour.

scales are those of fig.1. Here the spectra are normalized to the local underlying stellar + line flux. These UV-shell spectra are therefore ready for direct comparison with theoretical predictions. It is clear from fig.2 that all three ions do indeed show equal (sometimes several) shell velocities. In complicated cases the identification is aided by the presence of both doublet lines from all three ions.

Spectral line fits (corrected for the IUE instrument) were made to derive the column density according to profiles of the type

$$F(v) = \exp(-\tau_o \exp(-((v-v_c)/v_t)^2))$$

where F is the flux, τ_o the central optical depth of the line, v_c the central velocity and v_t the "turbulent" width, respectively. If necessary, more lines were included in the fit.

The resultant C IV column densities (not corrected for abundance) are displayed in fig.3. The dashed curve is a t^{-2} fit through the last 5 datapoints, suggesting that at 25 March we are dealing with not the same shell. For the first four datapoints the sum of more components was used. Therefore a more appropriate description would be with two sets of shell lines, which are observed at v_A = -1400 kms^{-1} and at v_B = -1300 kms^{-1}. A two shell fit based on 4 free parameters and 9 points (solid line in fig. 3) gives a very good agreement with the observed behaviour, implying that the second UV-shell episode should have started a few days before 5 April.

Shell parameters

From the observed column density we may estimate the mass M in a shell using M = 4 π r_s^2 m_H N_H. Here r_s is the inner radius of the shell, which is estimated from its velocity and the length of time it has been visible. We assumed that 10% of Si is in Si IV. This prescription yields ~ 3.10^{24} g for shell A and $\gtrsim$ 3.10^{23} g for shell B. If we express the formation time τ of the shell in days and the mass-loss enhancement by p (Paper 1) we obtain τp ≃ 50 and 5 for shell A and B respectively. No attempt has yet been made to solve the ionization balance to obtain the density in the shell (giving the thickness and formation time). In any case the high values of τp indicate a very strong mass-loss enhancement.

References

Lamers, H., Gathier, R., & Snow, T., 1981, in preparation.
Henrichs, H., Hammerschlag-Hensberge,G.,Howarth,I.,Barr,P., in preparation.
Henrichs, H., Hammerschlag-Hensberge, G., & Lamers,H.,Proc. 2nd European
        IUE Conference,Tübingen,Germany, ESA Sp-157, 147,April 1980(Paper 1)

DISCUSSION

<u>Stalio</u>: For some of the O stars you mentioned as having shells we
started a program of monitoring their shell variability with IUE.
From the observations already made we found several other O stars
having shells. The most striking evidence of changes is found in
HD 175754 (O8 III) (Carasco, Costero, Stalio, 1981, A&A in press).

<u>Snow</u>: Many Be star observers are not familiar with the observations
of O star winds. The more we learn about winds in Be stars, the
greater the similarity that can be seen between O and B stars.
Whether or not the winds are the underlying-cause of the Be phenomenon,
they are clearly of great importance, and the comparison with the O
stars is constructive.

<u>Doazan</u>: I want to emphasize 3 points:
1. Your term of "shell" lines to designate the narrow absorption
superionized lines is confusing because this term designates the
<u>low</u> excitation lines observed in shell spectra in the visible. To avoid
needless confusion it would be better to find an other designation.
2. The variability exhibited at the present epoche in the UV in γ Cas,
the high velocity component remains at almost the same velocity, the
variability affects principally the strength of the line. In 59 Cyg
we observe large velocity changes simultaneously with profile changes.
3. Your model tries to explain <u>only</u> the superionized region surrounding
the star and I see no way of explaining the low velocities observed
in the H_α emission with the velocity law you have adopted. The Be
phenomenon is defined by the presence of this cool and low velocity
atmosphere and UV observations show that we have in addition a high
velocity expanding region. Any realistic model should explain both
regions.

<u>Henrichs</u>: 1. I propose: ultraviolet, high-velocity, high-excitation,
narrow absorptions as sometimes has been observed in γ Cas, in order
to avoid <u>any</u> possible confusion... Perhaps we might abreviate it to
<u>high-velocity</u> (shell) lines because they only appear at high velocity.
2. If you look closer to the γ Cas data you would find in several
cases (for instance in the 25 March and 13 April spectra) absorptions
at velocities between 600 and 800 km/s, just what you observed in
59 Cyg. I argued that during one "hight-velocity shell" episode the
probability is rather low to observe the absorptions at low velocity,
simply because the acceleration goes very fast (paper 1). So in 59
Cyg apparently the episodes occur very frequently. They are observed
not to have such a high strength as in γ Cas. The latter point is in
favour of this interpretation: the ratio formation time/ decay time
is higher for 59 Cyg than for γ Cas, which is expected for "weak" and
"strong" episodes, respectively.
3. Olson (1980, priv. communication) calculated the H_α emitted by the
"high velocity shells" of the type I described. The result was that
the effect on H_α is completely neglegible. So you may form your H_α
whenever you like: this "high velocity shells" are not going to

effect this. In other words: I have no model for the H_α emission, the
only thing I know is that it does not come from these high velocity
shells. But I agree definitely with you that ultimately it is still
the same which displays so many phenomena which has to fit together
in one model.

Thomas: We shall emphasize that γ Cas is in the "quasi-steady" bright
Be phase; while 59 Cyg is in the minimum, but presently low-level
Be phase. So, that one observes always the same 1400 $\pm$ 200 km/s for
γ Cas can equally be interpreted as a stationary velocity pattern,
with density rising and falling, and not an outward moving shell.
If it is an outward moving shell, then it must interact with the
"storage bubble", which produces the H_α emission. Further, if the
shell is continously accelerated, why is then not a similar acceleration acting on this "ambient H_α medium" through which the shell moves?

Henrichs: I agree completely on the first point; the Be phase might
be very important. For your second point I refer to my answer to the
question (2) of Dr. Doazan. I think that it is really difficult to
construct a temperature distribution from which you would observe
lines of ions with 1. a wide spread in ionization potential at exactly
the same velocity, independent of the density, and 2. multiple
structure, i.e. two or more sets of different (high) velocities. Both
of these properties have frequently be seen. The last point you
mentioned is true. It would probably be observable, but don't forget
the short time scales (hours). How strong this effect will be is not
calculated: it would be extremely interesting to know the magnitude
of this effect.

Endal: Can you estimate from the velocity dispersion within a blob,
the time duration of an enhanced mass loss event?

Henrichs: The problem is that a velocity dispersion cannot be dis-
entangled from a density gradient. Otherwise this would give indeed
an estimate for the thickness of these shells.
Assuming a Lamers and Morton velocity law we arrived at a lower limit:
$p \gtrsim 2$ (see paper 1), based on the assumption that there is no density
gradient. I am afraid that we really have to solve the ionization
balance to obtain the density. This is however a difficult task.

" IUE OBSERVATIONS OF 17 Lep"[*][o]

P. Molaro – Astronomical Observatory of Trieste;
P.L. Selvelli and R. Stalio – Astronomical Observatory of
Trieste – International School for Advanced Studies, SISSA,
Trieste.

SUMMARY

The spectrum of 17 Lep is dominated by numerous and strong shell
lines of once-ionized metals. The shell shows a composite radial velocity structure which indicates the presence of multiple components with
a velocity range between −50 and −250 km s^{-1}. Unlike in the visible,
where these multiple components are present only during outburst, in the
ultraviolet such components seem to be a permanent feature.

1. INTRODUCTION

The visual spectrum of 17 Lep has been the object of many detailed
analyses, starting with Struve in the thirties (1932) and followed by
Slettebak (1950), Wright (1957), Widing (1965) and Cowley (1967).

17 Lep is a binary system composed of a late B primary and an M1
giant filling its Roche lobe. The orbit is quite eccentric with e=0.132.
The orbital period is about 260^d, and the inclination of the system is
estimated to be around 20°÷30°.

The spectrum of the primary is composite, showing the presence of
broad and diffuse absorptions of photospheric origin, corresponding to
a late B spectral type, along with sharp absorptions of once-ionized
metals which originate in a shell expanding at v$\simeq$ −55 km s^{-1}. These absorptions correspond to an early A spectral type.

The most spectacular feature of 17 Lep is the occurence of outbursts
in the shell surrounding the primary. The outbursts, which occurr at
intervals centered around 150^d, and last $\simeq$ 20 days, are manifested by
the presence of additional violet-displaced components in the shell

* Based on IUE observations obtained at the ESA Satellite Tracking
  Station of VILSPA (Spain)
o Details of this analysis will be published in Astron. Astrophys.

lines. No related variation in the luminosity is observed. A correlation has been found (Cowley, 1967) between the occurence of outbursts and the passage at the periastron, suggesting a possible mass-exchange effect.

2. THE SPECTRUM.

2 IUE high resolution images (SWP 4790 and LWR 4144) covering the range 1175 - 3200 Å have been obtained at VILSPA in March 1979. Moreover, 2 LWR high resolution images (LWR 5104 and LWR 6808) taken by other observers some months later and available from the VILSPA data-bank have been used to investigate possible spectral variations.

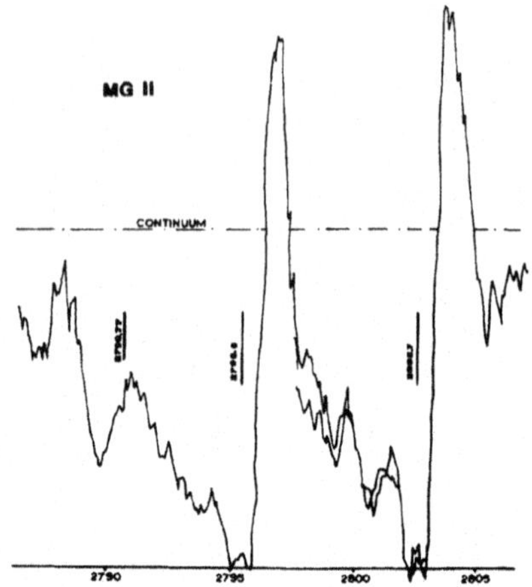

Figure 1. - The MgII resonance doublet.

Besides the strong emissions in the P Cyg profiles of the MgII resonance doublet (Fig. 1), the UV spectrum is characterized by a very large number of strong "shell" absorption lines shortward displaced with respect to the nominal wavelength. Almost the totality of the absorptions show a typically quasi-rectangular profile with FWHM $\geqslant$ 1 Å and central intensity close to zero (Fig. 2). A composite structure is clearly evident in the shell lines. They show a well marked absorption peak in the least displaced component (v $\simeq$ -55 km s^{-1}) and additional less marked

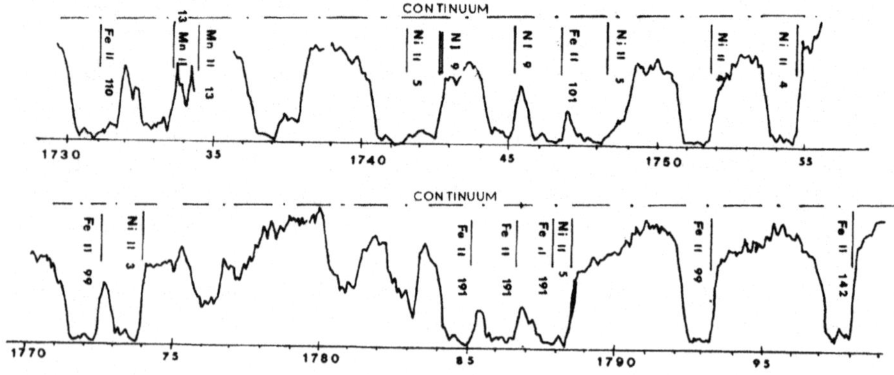

Figure 2. - A portion of the spectrum of 17 Lep.

peaks shortward displaced up to -250 km s^{-1}. It is notable that this structure is present in the visible only during outbursts.

About 75% of the strongest spectral features have been identified; lines of CI, CII, NI, MgII, SIII, CrII, MnII, FeII, CoII and NiII are definitely present. About 95% of the identified features belong to the once-ionized metals, and principally to FeII, which is the main contributor to the line opacity. It is remarkable, however, that in the far UV range there are various regions, mainly below λ 1500, showing strong unidentified absorption blends which reach zero intensity. There is no positive evidence of the presence either of doubly ionized species such as SiIII, TiIII, CrIII, FeIII etc, which are common in the spectra of B stars, or of the "superionized" species such as SiIV, CIV and NV which are found in the B supergiants. No photospheric line, i.e. centered around the nominal wavelength and with a rotationally broadened profile, have been definitely detected. The sole candidates are the two resonance doublets of CII λ 1335 and SiII λ 1530 which fall in regions affected by severe blends.

No significant changes are present in the two additional LWR spectra. The overall multiple-shell structure of the absorption profiles has remained essentially the same. The chance that all the three LWR spectra have been taken during the outbursts phase is very low ($\simeq 10^{-3}$). This means, therefore, that the UV spectrum, unlike the visible, shows the constant presence of multiple-shell structure. However, the possibility remains that 17 Lep has recently undergone a permanent outburst phase also in the visible, where no recent spectra are available.

3. THE RADIAL VELOCITIES.

RV's have been measured for all reasonably unblended lines. As reported above, the red component of the absorption profile falls, on an average, around -55 km s^{-1}, a value which is very close to that found in the visible for the sharp absorption component formed in the expanding shell. There are, however, small but definite deviations for different elements around the mean value. CrII and TiII fall around -42 km s^{-1}, NiII around -66 km s^{-1}, while the MgII doublet falls around -80 km s^{-1}. For the line-rich FeII spectrum we have also tried to

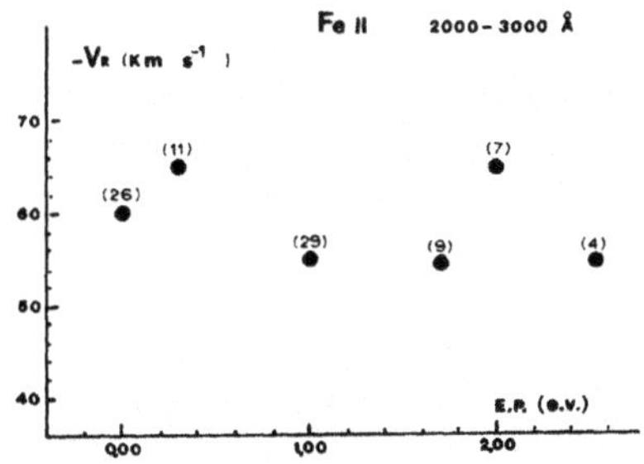

Figure 3. – The relation between the RV of the main shell component and the excitation potential; in brackets, the number of lines used.

ascertain whether there is any dependence of the RV's on the excitation potential. We have found no such dependence (Fig. 3).

The fairly good but not ideal resolution of the IUE spectrograph and the scarcity of lines definitely free from blending in the UV range set serious limits to accurate RV analysis of the secondary components of the multiple shell structure of 17 Lep. Nonetheless, using the most reasonably unblended lines, we have found a clustering of the secondary absorption peaks around −180 and −230 km s^{-1}, with other less important contributions up to $\simeq$ −300 km s^{-1}.

An attempt has also been made to measure the "edge velocity" for all the unblended lines. A value of around −300 km s^{-1} has been found for most of the lines, except for the MgII doublet which reaches −700 km s^{-1}. There is a definite relation between v_{edge} and the excitation potential, with the resonance lines having the highest v_{edge} (Fig. 4).

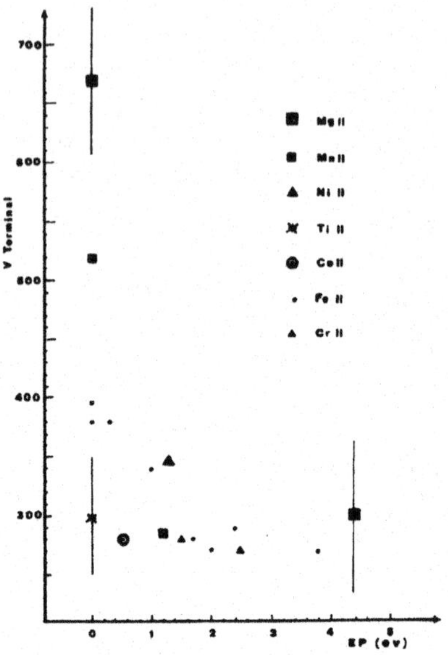

Figure 4. – The relation between v_{edge} and the excitation potential.

4. CONTINUUM−TEMPERATURE−COLUMN DENSITIES.

The very severe line blocking makes the tracing of the continuum very uncertain. Using the few "windows" which are present in the spectrum, a "pseudo-continuum" has been derived and compared with the Kurucz models, after having determined the flux using the method described by Cassatella et al. (1981). The best fit with the models gives a value of $\simeq$ 8000°K for the temperature. This value is lower than that of a B9 star (visible) and could be explained either by the uncertainties in the continuum determination, or, from a quite different viewpoint, by assuming that the observed UV continuum is not formed in the photosphere of the primary but in the optically thick shell that surrounds it.

The curve of growth method applied to the FeII lines yields a value of $\simeq$ 7700°K for the excitation temperature.

Column densities have been calculated for the once-ionized metals on the assumption that all the atoms of these elements appear in the first ionization state. The results are reported in Table 1.

5. THE MgII RESONANCE DOUBLET.

These lines are the only undoubted emissions in the UV spectrum of 17 Lep. The profile is a typical P Cyg, with the emission intensity reaching a peak height of about 60% of the continuum intensity.

The absorption have a minimum around $v = -215$ km s^{-1}. The emission component, although strong, is weaker than the absorption component, as in common P Cyg profiles produced by resonant scattering. This indicates the absence of additional excitation mechanisms in the shell itself, such

TABLE 1.

ION	lg N
FeII	16.8±0.5
NiII	16.0±0.5
MnII	16.3±0.5
CrII	15.4±0.5

as collisions, and confirms the low value found for the temperature. The emission width obeys the relation found by Kondo et al.(1976) for the supergiants. The value derived (M_V = 2.1) is in satisfactorely agreement with that derived by the trigonometric parallax (1.7±1.2).

CONCLUSIONS.

a). The UV line spectrum of 17 Lep is entirely produced in the optically thick shell fully covering the primary.

b). The once-ionized metals are the main contributors to the shell spectrum. There is no evidence either of "photospheric" lines or of lines of higly-ionized species.

c). The shell shows a constant presence of multiple components (with velocity range between -40 and -200 km s^{-1}), which in the visible are present only during the outburst phase.

d). The presence of P Cyg profiles for the MgII resonance doublet, and the strength of its emission component are quite uncommon in Be and B-shell stars (Dachs 1980), where the emissions are very weak and the absorption sharp.

BIBLIOGRAPHY.

Cassatella, A., Ponz D., Selvelli, P.L.: 1981, Communic. to the IUE Users
        ESA   IUE Newsletter No. 10.
Cowley, A.: 1967, Astrophys. J. 147, 609.
Dachs, J.: 1980, Proceeding of the 2nd European IUE Conference held at
        Tubingen, Germany. ESA SP-157.
Kondo, Y., Morgan, T.H., Modisette, J.L.: 1976, Astrophys. J. 209, 489.
Slettebak, A.: 1950, Astrophys. J., 112, 559.
Struve, O.: 1932, Astrophys. J. 76, 85.
Widing, K.G. 1965, Astrophys. J. 143, 121.
Wright, K.O.: 1957, Publ. Astron. Soc. Pacific 69, 552.

DISCUSSION

<u>Paterson-Beekmans</u>: There are in the UV and the visible Fe II lines
arising from the same unstable level (vis. 42, UV 191, 192, 193).
In KX And, for ex., all these lines are strong. Do you see a different
behaviour of these lines in 17 Lep?

<u>Selvelli</u>: We don't have visual tracings for comparing the behaviour
of visual and UV Fe II lines. The lines of m = 191, 192, 193 are very
strong in 17 Lep.

<u>Koubsky</u>: What is the phase coverage of your UV observations?

<u>Selvelli</u>: We would like to know it, but unfortunately there are no
simultaneous visible and UV observations. Our three LWR observations
are seperated by 120 and 180 days, while the frequency of the outburst
in the visible is near 150 days, so we can exclude that all 3 spectra
correspond to an abnormal phase.

<u>Andrillat</u>: Have you compared the P Cyg H_α profile with the P Cyg Mg II
profile? In April 1980 we have observed the P Cyg H_α profile with a
good resolution.

<u>Selvelli</u>: No, we do not have visual informations.

<u>Harmanec</u>:Is there some similarity between the UV spectrum of 17 Lep
and the UV spectrum of some symbiotic stars?

<u>Selvelli</u>: No.

<u>de Groot</u>: You said the spectral "outbursts" are not reflected in
increases in the stars luminosity. How well, then, is the visual light
curve known?

<u>Selvelli</u>: UBV photometric observations of Widing (1967) exclude
variations greater than .1 mag. in coincidence with visible outbursts.

SIMULTANEOUS IUE GROUND-BASED SPECTROSCOPIC OBSER-
VATIONS OF THE VARIABLE LMC STAR R 71

B. WOLF, I. APPENZELLER, O. STAHL
Landessternwarte, Königstuhl
D-6900 Heidelberg 1

ABSTRACT:

Using the IUE satellite we obtained high resolution UV spectrograms (1200 < λ < 3200 Å) of the S Dor type variable R 71
in the LMC. The IUE observations were supplemented by coordinated groundbased high dispersion spectroscopy and by
photometric observations. From these observations we derive
for the minimum state of R 71 the following stellar para-
meters: $L = 2.0 \times 10^5 L_\odot$, $R \approx 81 R_\odot$, $T_{eff} \approx 13\,600$ K. For
the expanding envelope we find a surprisingly low tempera-
ture of only about 6000 K and an apparently decelerated velocity field with a maximum outflow velocity of ~ 127 km s^{-1}.
The minimum state mass loss rate is in the order of 3×10^{-7}
$M_\odot$ y^{-1}. Our results support the suggestion that the visual
light variations of the S Dor type variables are produced
by strong density variations of the expanding envelopes of
these objects. A detailed study is forthcoming in Astronomy & Astrophysics.

M. Jaschek and H.-G. Groth (eds.), Be Stars, 443.

SPECTRAL ENERGY DISTRIBUTION (119 - 685 nm) IN 16 SHELL STARS AND A
TENTATIVE MODEL FOR ACCRETING Be STARS

Mirek J. Plavec, Jan J. Dobias, Janet L. Weiland
Department of Astronomy, University of California, Los Angeles

Remington P.S. Stone
Lick Observatory, University of California

ABSTRACT   I.U.E. low-dispersion spectra and spectral scans made with
the Lick Observatory IDS scanners have been combined for 16
shell stars.   Eleven objects can be represented by Kurucz model atmospheres, although some of them display strong shell-type line spectra.
Five among them are known binaries. The six remaining objects (all
interacting binaries) display complex spectra. A model involving continuum and line radiation from a hydrogen cloud surrounding the accreting component is proposed. A generalization of this model with optically
thick segments of the cloud promises to explain even more exotic objects
such as β Lyrae, W Serpentis and possibly ε Aurigae.

This paper will ultimately contain the claim that the exotic
eclipsing binary system W Serpentis contains a Be star, and should
therefore be represented at the Be Stars Symposium. The chain of arguments leading to this conclusion will start with a survey of 16 shell
stars. Spectral energy distributions of these shell stars were obtained
by combining low-dispersion I.U.E. spectra with scans made with the
Image Dissector Scanners attached to the 120-inch and 24-inch telescopes
of the Lick Observatory. With the gratings used, the resolving power of
the scanners has been very nearly the same as that of the I.U.E. spectra.
In favorable cases, the entire spectral interval 119 through 685 nm was
covered. More often, there exists a gap between about 320 and 365 nm,
since the I.U.E. calibration is not reliable beyond 320 nm and the red-
sensitive tube of the 24-inch scanner does not give good response shortward of 365 nm. A complete spectral scan consists of two I.U.E. spectra
(short camera, SWP, between 119 - 196 nm, and long camera, LWR, between
190 - 320 nm), and of two grating settings of the Lick Scanner.
The complete spectral scans were compared with Kurucz (1979)
model atmospheres with normal chemical composition. (We tested Kurucz
models against normal stars and found very good overall agreement.
Several local discrepancies will be discussed elsewhere.) Monochromatic
fluxes were extracted from the observed continuous energy distributions
at up to 191 wavelengths corresponding to those used by Kurucz. A least-
squares fit was then made in an effort to solve simultaneously for the

445

M. Jaschek and H.-G. Groth (eds.), Be Stars, 445–450.
Copyright © 1982 by the IAU.

color excess E(B-V), the effective temperature of the star, T_{eff}, and
its surface gravity (log g). Using two quite different acquisition
systems may lead to systematic displacements on both sides of the Balmer
jump, the more so because it was impossible to obtain I.U.E. and optical
data simultaneously. For that reason, separate solutions were first
made for the two parts of the spectrum. Surprisingly, for all the
"simple" shell stars, the mutual agreement was perfect. The only devi-
ating case, HD 206773, required only a vertical displacement of the
optical spectrum, most likely because of an incorrectly recorded scanning time.

SIMPLE SHELL STARS

        Eleven shell stars could be fitted by the Kurucz atmospheres
over the whole scanned spectral interval (in two the coverage is incom-
plete, though). No additional source of radiation is required, except
that in five of them, a later-type companion is to be expected, since
they are known binary systems. In such cases, we at first fitted only
the blue optical scans (shortward of about 500 nm) and included longer
wavelengths only if they were consistent. In practice, only in 17
Leporis is the companion clearly present in our scans. A list of the
most important parameters for the 11 "simple" shell stars is in Table 1.
Please note that the spectral types and luminosity classes were assigned
from the directly determined values of T_{eff} and log g. This inverse
calibration is based on the recent tables by Flower (1977) and Hayes
(1978), while the luminosity classes were derived from the log g calibration given by Schmidt-Kaler (1965).
        Our values agree in general very well with spectral types
derived by various authors from the photospheric line spectra. It is
rather surprising to see that the circumstellar shells seem to have
little influence on the continuous energy distribution, although they
affect the line spectra quite conspicuously.
        Perhaps the most surprising result is the low effective tem-
perature found for 17 Leporis. Cowley (1967) and others saw or suspected
spectral features characteristic of a B9 spectrum. Yet, the I.U.E.
spectra of 17 Lep led quite unmistakably to the low temperature listed,
and no other source of radiation appears to be present -- at least not
at the observed phase.

COMPLEX SHELL STARS

        Five objects cannot be fitted by normal Kurucz atmospheres;
they require inclusion of an additional source of radiation, different
from the companion which is either clearly visible in our scans (as in
AX Mon and HD 51480) or known to exist from various other observations.
Surprisingly, the simplest of these complex systems appears to be ϕ
Persei. The optically thin free-free radiation of circumstellar hydrogen observed, e.g., by Gehrz, Hackwell and Jones (1974), combined with
bound-free radiation, can be traced on both sides of the Balmer jump,

but contributes relatively little, so that the whole spectrum can be
fitted by a Kurucz model with T_{eff} = 20750°K, log g = 3.75 with very
good accuracy, except longwards of about 600 nm where the hydrogen radiation becomes more prominent. Also, the surprisingly low color excess
E(B-V) = 0.04 appears to be well-established.

The systems KX And, AX Mon, HD 51480 and HD 72754 seem to be
rather similar to each other, and all of them bear a good deal of resem-
blance to the Serpentids as defined and described by Plavec (1980). The
first three stars have been more completely observed (the fourth one is
invisible from Lick) and have these peculiar properties: i) The optical
continuous spectra resemble those of A stars, in agreement with the
shell lines, but in sharp disagreement with the presence of the He I
lines. For example, KX And can be formally fitted by a T_{eff} = 9000°K
atmosphere in the optical region, although its photospheric line spectrum has been classified as B3; ii) The far ultraviolet spectrum dis-
agrees with the nominal A-type optical spectrum, and is much more indicative of the B-type spectrum suggested by the optical photospheric lines.
However, it shows unusually deep absorption lines, or rather clusters of
lines, not observed in any of our comparison standard B spectra obtained
with I.U.E. by Plavec and Koch; iii) There tends to be an excess flux,
over even the B-type spectrum, between about 240 - 320 nm (and possibly
further to the Balmer jump). Another such excess flux is observed
between about 190 - 210 nm, except that HD 72754 shows anomalously low
flux there.

These characteristics can be in principle explained by two
alternative models: A) The stars are indeed A stars as their optical,
continuous and shell spectra indicate. Since they are the accreting
components of interacting binaries, one can anticipate the existence of
an accretion disk. The hottest part of the disk would be the transition
layer between the disk and the star (Pringle 1977). It would produce a
B-type spectrum in the FUV, but owing to its small surface area, it cannot dominate in the optical region. This model was discussed by Plavec
(1980); B) The stars are B stars as their FUV spectra and optical photospheric lines suggest. The B star is heavily obscured by a disk. Outside the star, the disk itself radiates, usually as an optically thin
hydrogen cloud. This continuous radiation of the disk becomes relatively
stronger as we proceed toward longer wavelengths in the optical region,
and the combined continuous spectrum appears to have a much less steep
slope of the Paschen continuum, thereby simulating a cooler star.

We are more and more convinced that model B) represents the
typical case. This is confirmed by our observations of the primary component of β Lyrae. This star is usually classified as B8 II. Actually,
its spectral energy distribution resembles that of a B8 II star only in
the optical region, between about 380 - 600 nm. Continuous radiation
of hydrogen is quite prominent longward of 600 nm and between about the
Balmer jump down to about 240 nm. The above-mentioned excess flux near
190 - 210 nm is also present in β Lyrae, and is probably due to a superposition of numerous emissions of Fe III, as already suggested by several
authors before. In HD 72754, Fe III lines are in equally conspicuous
absorption.

The disk we envisage here is not the optically thick but geometrically quite thin (with respect to a non-degenerate star) disk obtained by simply scaling up the classical alpha-disks used to interpret the cataclysmic variables. Rather, we anticipate a disk that is quite thick geometrically, even perpendicular to the orbital plane, but often optically thin for continuous radiation in the spectral region considered here. Such disks have been advocated by Hall (1969), Wilson (1974), Polidan (thesis, 1979) and Kříž (preprint, 1980). Rapid variability seems to be their common property and we detected such variability, e.g., in KX And.

EVEN WILDER MODELS FOR THE WEIRDEST SYSTEMS

W Serpentis, although classified as F5 II (with probably a cooler invisible companion) displays emission lines of hydrogen in the optical region, and we (Plavec, Polidan and Peters) discovered even emission lines of He I. The far UV spectrum resembles that of a B star, and we claim that the accreting component most likely indeed is a B star, but largely obscured by that portion of the circumstellar disk that is projected against it (Plavec and Sakimoto 1978).

How do we explain the observed F-type continuum in the optical region? Assume that the density of absorbing hydrogen atoms decreases with some power of distance from the accreting star. If so, then as our line of sight approaches the star, the optical thickness of the disk must increase steeply and will at some point make the disk optically thick. We are viewing this eclipsing system essentially edge-on, so that the edge of the disk will appear to us as a spurious photosphere, largely hiding the B star inside. What kind of radiation will be emitted by the edge of the disk? Stellar radiation will hardly reach it; most of it will be either absorbed inside the disk or scattered perpendicularly to the orbital plane. The source of the energy of the disk edge will more likely be the dissipated kinetic energy of an impacting stream. Madej and Paczynski (1977) suggested for a similar case of U Gem that the edge will mimic a photosphere of approximately solar type.

In the 1940's, Otto Struve led us to realize that some absorption lines in interacting binaries need not come from stellar surfaces. Today, we have to take one step further and realize that continuous and whole absorption spectra do not have to come from stellar photospheres. If we accept this principle, we may hope to resolve other puzzling systems. The Serpentids (Plavec 1980) are among them; similarly, RZ Ophiuchi may actually very well be an earlier-type star inside a thick disk. Perhaps the most audacious idea is to re-examine ϵ Aurigae once more. The puzzling quasi-total eclipse does not seem to be fully explained yet, in spite of several ingenious theories (Huang 1965; Hack 1962; Hack and Selvelli 1978). It would be explained rather naturally as an eclipse of a large disk (seen edge-on) by a smaller, cooler object. However, the large luminosity of the alleged F2 Ia supergiant is not easy to explain by this model, and there are other difficulties. Nevertheless, further explorations of the complex nature of accreting stars are promising.

Table 1

Parameters of Simple Shell Stars

Star	HD	T_{eff}	log g	Spectral type		E(B-V)	Date I.U.E.	Observed Lick
Cas	698	12,000	3.0	B8.5	III	0.25	5 Feb 79	
Cas	220300	16,000	4.0	B4.5	V	0.15	10 Feb 79	
Cep	206773	30,000	4.0	B0	V	0.50	18 Aug 79	18 Jul 79
CX Dra	174237	19,000	4.0	B3	V	0.05	6 Feb 79	13 Sep 79
4 Her	142926	12,000	4.0	B8	V	0.03	6 Feb 79	28 Apr 79
88 Her	162732	13,000	4.0	B7	V	0.03	6 Feb 79	28 Apr 79
EW Lac	217050	19,500	4.0	B2.5	V	0.12	6 Feb 79	13 Sep 79
17 Lep	41511	8,000	3.0	A6	III	0.05	5 Feb 79	13 Sep 79
FX Lib	142983	15,250	4.0	B5	V	0.06	8 Feb 79	28 Apr 79
HR 716	15253	9,400	3.5	A1	III	0.04	8 Feb 79	12 Sep 79
28 Tau	23862	10,750	3.0	B9	III	0.07	9 Feb 79	12 Sep 79

Remarks: HD 698: alternative fit: T_{eff} = 14,000°K, log g = 3.0,
SP B6 III, E(B-V) = 0.33. HD 220300 and 28 Tau: LWR
spectrum only.

Table 2

Tentative Stellar Parameters of Complex Shell Stars

Star	HD	T_{eff}	log g	Spectral type		E(B-V)	Date I.U.E.	Observed Lick
KX And	218393	13,000	3.0	B7	III	0.15	14 Nov 78	Many
AX Mon	45910	17,000	4.0	B4	V	0.18	9 Feb 79	12 Sep 79
Mon	51480	14,000	4.0	B6.5	V	0.15	9 Feb 79	12 Sep 79
ϕ Per	10516	20,750	3.7	B1.5	III	0.04	8 Feb 79	12 Sep 79
FY Vel	72754	22,500	3.0	B1	III	0.45	14 Nov 78	

We have passed rather smoothly from the "classical" shell stars to interacting binaries. It is hard to draw a line between them, even for shell stars listed as "simple" in our Table 1.

In dealing with these complex and often variable objects, one realizes that coordinated, nearly simultaneous I.U.E. optical and infrared (spectroscopic as well as photometric) observations are essential. We wish therefore to emphasize the importance of the coordinated campaigns initiated by Harmanec and Kříž from the Ondřejov Observatory.

A more complete article on these observations and problems will be published elsewhere. Our thanks are due to the I.U.E. Observatory staff, to N.S.F. and NASA for supporting grants, to C.D. Keyes, M.P. Lesser and Z. Plavcová for considerable programming assistance, and to R.L. O'Daniel for typing the manuscript.

REFERENCES

Cowley, A.P.: 1967, Astrophys. J., 147, 609.
Flower, P.J.: 1977, Astron. Astrophys., 54, 31.
Gehrz, R.D., Hackwell, J.A. and Jones, T.W.: 1974, Astrophys. J., 191, 675.
Hack, M.: 1962, Mem. Astron. Soc. Ital., 32, 3.
Hack, M. and Selvelli, P.L.: 1978, Nature, 276, 376.
Hall, D.S.: 1969, Bull. Amer. Astron. Soc., 1, 345.
Hayes, D.S.: 1978, in "The HR Diagram" (ed. A.G.D. Phillips and D.S. Hayes): Reidel, p. 65.
Huang, S.S.: 1965, Astrophys. J., 141, 976.
Kurucz, R.L.: 1979, Astrophys. J. Suppl., 40, 1.
Madej, J. and Paczynski, B.: 1977, Veröff. Bamberg 11 #121, 313.
Plavec, M.J.: 1980, in "Close Binary Stars: Observation and Interpretation" (eds. M.J. Plavec, D.M. Popper and R.K. Ulrich): Reidel, p. 251.
Plavec, M.J. and Sakimoto, P.J.: 1978, Bull. Amer. Astron. Soc., 10, 609.
Pringle, J.E.: 1977, Mon. Not. R.A.S., 178, 195.
Schmidt-Kaler, Th.: 1965, in Landolt-Börnstein (ed. H.H. Voigt): Springer, Bd. I, p. 309.
Wilson, R.E.: 1974, Astrophys. J., 189, 319.

DISCUSSION

Slettebak: With regard to the spectral type of 17 Lep, my recent spectrograms show no trace of helium. I would classify the star as A-type rather than late B.

VIII. ATMOSPHERIC MODELS

MODEL ATMOSPHERES OF Be STARS

Roland Poeckert
Dominion Astrophysical Observatory
Herzberg Institute of Astrophysics
Victoria, B.C.

1. INTRODUCTION

The basic concept of what a Be star is we owe to Struve (1931).
His proposal that these stars are rapidly rotating and have an
extended, disc-like, circumstellar envelope is still the basis for all
models of Be stars. Unfortunately, progress in understanding the
dynamics of Be stars has been painfully slow, a consequence of the very
complex nature of the problem. As yet there does not exist a self-
consistent unique model of any Be star.

Marlborough (1976) reviewed the state of Be star models at the
last IAU Symposium (No. 70) on Be and Shell Stars. The emphasis of
this review will be on the advancements in the modelling art made in
the five years since that symposium. Then, as now, all models which
attempted to relate directly to observed properties of Be stars were,
to a greater or lesser extent, <u>ad hoc</u>.

2. GENERAL THEORY

The equations governing mass flow, the radiative transfer
problems, and the role that turbulence and magnetic fields might play
have been outlined by Marlborough (1976). Generally, the mass flow
problem and the radiative transfer problems are dealt with
separately, simultaneous solutions being very complicated.

2.1 Mass Flow

Much of what we think we know about mass loss in early type stars
comes from models of stellar winds in O type stars. These models
usually assume spherical symmetry so the problems are somewhat simpler
than in the case of Be stars. It is useful to consider the O star wind
models because much of what is learned there can be applied to Be
stars.

M. Jaschek and H.-G. Groth (eds.), Be Stars, 453–483.
Copyright © 1982 by the IAU.

The radiatively-driven-wind models of Cassinelli and Castor (1973) and Castor, Abbott and Klein (1975) have, in recent times, had to be modified to incorporate a high temperature $(T > T_{PHOTOSPHERE})$ region. The need for such a modification arises out of the observations of lines from highly ionized species (e.g. O VI) in the spectra of O stars. A summary of the various wind models is given by Cassinelli, Castor and Lamers (1978) and Cassinelli (1979). Three models are proposed, a modified cool-wind model, a warm-wind model, and a corona-plus-cold-wind model. All of these models require some non-radiative heating, but all assume that it is solely the radiation pressure that drives the wind. Also, these models constrain the wind to accelerate smoothly through the transonic region. The radiatively-driven-wind models are successful in matching terminal velocities. Moreover, low metal abundance O stars in the LMC and SMC show little or no sign of stellar winds, which is probably linked to the lack of lines by which the wind is accelerated (Hutchings 1980; Prevot et al. 1980; Hutchings, private communication).

Cannon and Thomas (1977) argue that the non-radiative heating governs the flow and it is an artificial constraint to require a smooth transition through the sonic point. If the flow through the transonic region is not smooth, i.e. a shockwave exists in the flow, the flow may be unstable. This is consistent with the observed variability of the flow in O star winds. A hybrid model has been proposed by Mazurek (1980) in which the mass loss rate and the flow to the sonic point are controlled by the non-radiative energy input at the base of the wind. Once the flow becomes supersonic the wind cools rapidly and it is accelerated by radiation pressure.

Non-radiative heating is a common feature of all the models. The most commonly invoked sources of non-thermal energy involve either acoustic or magnetic waves which become shock waves when entering the low density regions above the photosphere. The dissipation of these shock waves results in the heating of the wind. Meridonal circulation currents may give rise to turbelence in the photosphere of a rotating star (Smith and Roxburgh 1977; Kodaira 1980). There is also the possibility of generating acoustic waves in the wind directly by radiative mechanisms (Martens 1979) and of shock waves in the flow as discussed by Cannon and Thomas (1977). Non-thermal structure in stellar atmospheres is the subject of a series of papers introduced by Cram (1980).

Be stars, unlike O star wind models, are probably not spherically symmetric. A key point here is that in spherically symmetric models one can only separate wind regions radially. Therefore the flux from the star must pass through all regions. This is not the case in Be stars. Very few of the O star wind models incorporate stellar rotation in their formulation, but the suggestion of enhanced equatorial mass loss has come up in the case of WR stars (Rempl 1980). Marlborough and Zamir (1975) extended the solution of Cassinelli and Castor (1973) to include the effects of rapid rotation. They found that a wind could exist in the equatorial plane of a rapidly rotating star, while at the

same time the polar regions remained hydrostatic. Nerney (1980) points
out that rotation can generate magnetic fields which could easily
initiate mass loss. The field strengths required are small ($\sim$ 1 gauss
at the stellar surface) and below the level of detection at the present
time (Landstreet 1980). A search for magnetic fields in A type shell
stars by Clayton and Marlborough (1980) proved negative at the 300
gauss level.

2.2 Radiation Transfer

The envelopes of Be stars appear to be rotating and moving
radially with velocities much greater than the mean thermal velocity.
The densities within these envelopes are such that one cannot assume
LTE, nor can one ignore collisional transitions in determining level
populations. Thus we have the worst possible case when trying to solve
the radiation transfer problem.

A popular method of treating the line radiation transfer problem
is the "mean escape probability" method (Sobolev 1960). Rybicki and
Hummer (1978) give a generalized formulation of the method, and Surdej
(1979) shows the results of the application of the method to a variety
of radially accelerating and decelerating envelopes. The method is
very useful when velocity gradients are large and line opacity small,
but its usefulness at large line opacities has been questioned (Bernat
and Robbins 1974; Hamman 1981).

In addition to the transfer problems one must also consider the
non-LTE nature of the atomic level populations. Drake and Ulrich
(1980) have calculated the spectrum of a slab of hydrogen at moderate
densities and temperatures between 5000 K and 40000 K. Their calculations are based on a 20 level hydrogen atom, complete with radiative
and collisional transitions, and an escape probability approach to the
transfer problem. Drake and Ulrich assumed complete frequency redistribution in the line. It should be noted that if the line opacity is
very large this assumption may not be valid (Chugai 1980). Figure 1
gives some examples of the computed spectra.

2.3 Rotation

A general discussion of stellar rotation is beyond the scope of
this review, and the reader is directed to an excellent comprehensive
look at stellar rotation given by Tassoul (1978). However, some
aspects of rotation that bear directly on the models to be discussed in
the remaining sections deserve mention.

Magnetic fields, apparently inevitable in rotating stars, have
been incorporated in some models (Barker 1979, Nerney 1980, Barker
et al. 1981). The consensus is that relatively weak fields, $\sim$ tens of
gauss, may dramatically affect the dynamics of circumstellar envelopes.

The effects of rapid rotation on line profiles have been investi-

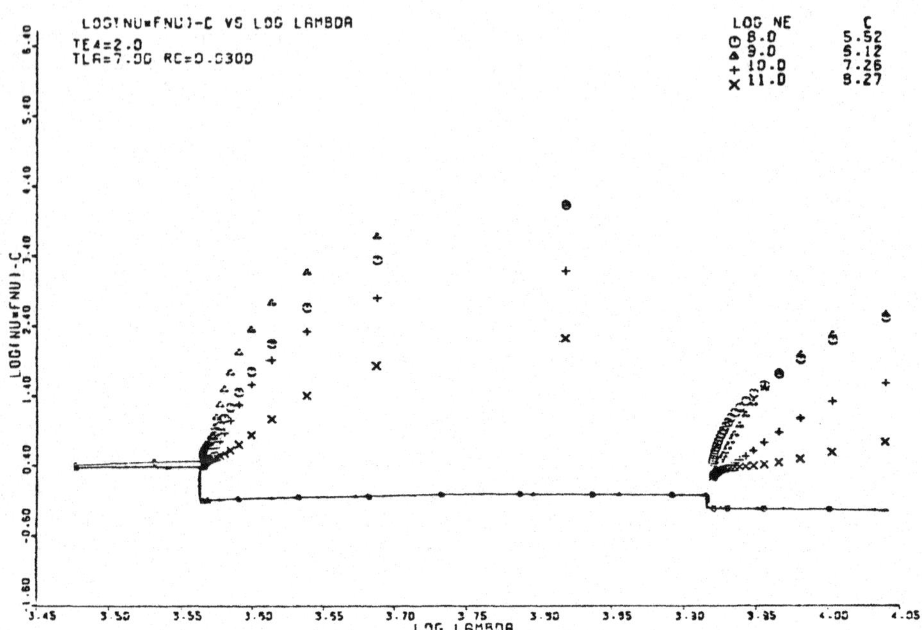

Figure 1. Energy distribution and line strengths for a slab of hydrogen at 20000 K and a large Lyman α optical depth. Various symbols
indicate different densities (Drake 1980).

gated by Slettebak, Kuzma and Collins (1980). It appears that rapid
rotators seen equator on will be classified later, by about 1.5
subclasses, and somewhat higher in luminosity (based on Hγ equivalent
width) than an equivalent non-rotating star. Another aspect of rapid
rotation relates to the fact that the measured V sin i is dependent on
the spectral line chosen for analysis. The effect is particularly
noticeable when one compares ultraviolet and visible lines (Heap
1976). Hutchings (1976) and Hutchings, Nemec and Cassidy (1979) have
used the difference to determine the inclination and equatorial of
several stars. Sonneborn and Collins (1977) and Collins and Sonneborn
(1977) give the results of detailed calculations comparing the Mg II λ
4481 and Si III λ 1113 line profiles in rapid rotators seen at various
inclinations. It is heartening to find that the Von Zeipel gravity
darkening law, used almost universally, is valid in early-type stars
(Smith and Roburgh 1977).

The problem here, as far as modelling is concerned, is deciding on
what value one chooses for V_{equ}, sin i and T_{eff} to represent the star.

3.   Ad Hoc Models

The raison d'etre of models is to give us an indication of the dynamics and physical conditions in and around stars. In principle, if one completely understood the dynamics (and could solve the relevant equations) the physical conditions would be known by default. We are far from this ideal situation. Observations reveal to us information on the physical conditions, not the dynamics. The role of the ad hoc model is to determine the physical conditions, from which we might infer the dynamics.

This is a highly interactive (often intuitive) process because ad hoc models are rarely unique. To decide which model is closer to the truth one must consider the number of observations a model fits, the quality of the fit and the dynamical picture inferred from the model. Ad hoc models are quite successful at matching observations, in part because they have the interesting property of gaining free parameters at will; few if any are dynamically sound.

3.1. The Elliptical Ring Model

Struve (1931) first proposed and Huang (1972, 1973) and Albert and Huang (1974) gave a quantitative framework to the elliptical ring model. The circumstellar material is thought to be contained in a narrow elliptical ring. Long term, apparently periodic, V/R variations are explained as uniform apsidal motion of the ring. Detailed line profiles have not been computed for the ring models, but a schematic picture is given by Huang (1975).

One of the difficulties faced by the elliptic ring model is that there are times when V/R variations are cyclic, separated by abrupt changes in V/R. In addition there is the case where the period and amplitude of the variations slowly changes over several cycles. Huang (1976, 1977) has proposed that sudden variations in V/R are due to gaseous outbursts from the star which disrupts the existing ring and forms a new, different ring. Huang (1978) has also investigated the effect that a continuous weak flow of gas would have on the elliptic ring. He finds that both apsidal motion and changes in the eccentricity and semi major axis can result from an interaction between the wind and the ring.

In order for the apsidal motion concept to be valid the ring should be thin, both radially and perpendicular to the ring plane. Marlborough (1976) outlined the problem of a small perpendicular extent. In essence it is difficult to form deep shell absorption lines with thin rings.

Kriz (1976, 1979a,b) has calculated line profiles (shown in Figure 2) for elliptical rings. He finds that such rings must be optically thick in Hα and the ring must have a large radial extent, ~ 5 R$*$, in order to match observed emission line profiles. Thus we are really

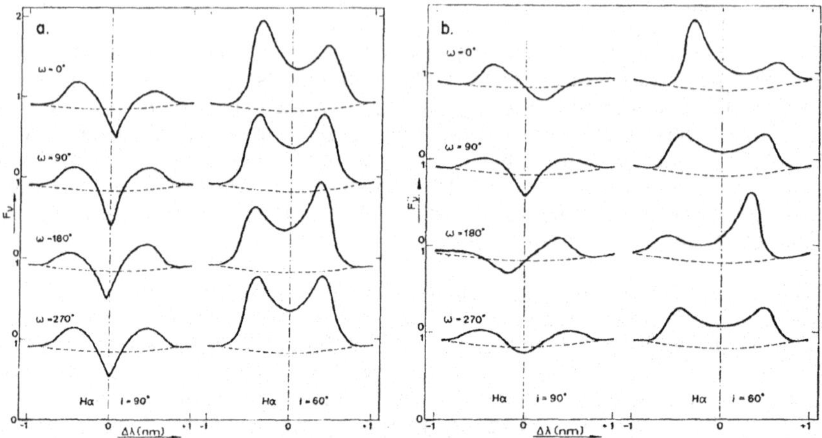

Figure 2. Line profiles for elliptic disks. Eccentricity of the disk is a) 0.1 and b) 0.3. The profiles are shown at two inclinations and four azimuthal angles.

talking about elliptic disks, not rings, and most of the dynamical arguments in favor of the ring model fail.

The problem with the elliptic disk model is that it is difficult to understand why the disk should rotate uniformly. Apsidal motion of an orbiting ring is no longer an attractive theory because a whole family of rings, all with the same eccentricity, must precess in unison. A hybrid model, in which only the outer part of the envelope is elliptic has not been considered in detail as yet. Such a model bears some semblance to the distorted disk model proposed by Marlborough, Snow and Slettebak (1978) for γ Cas.

The case for elliptical, or a least oval, structure in circumstellar envelopes is more convincing when the Be star is a member of a binary system.

3.2. Binary Star Models

Binary Be stars are the subject of another review paper (by Harmanec) in this volume, and will not be discussed at length here. Several Be stars are members of mass exchanging systems, e.g. HR 2142 (Peters 1972, 1976). In this case the secondary is believed to be losing mass through Roche lobe overflow.

A second scenario, one which does not explicitly require any mass exchange, has been proposed by Suzuki (1979). His model consists of a ring of gas in a stable periodic orbit. Quasi-circular, non intersecting stable periodic orbits exist for a range of binary configurations (Hernon and Gyot 1970). Suzuki applied his interpretation to the binary Be star ϕ Persei and predicted a mass ratio and separation for the system. Poeckert (1981) has confirmed Suzuki's parameters from independent observations and using a standard spectroscopic orbit analysis. One of the exciting results of this model is that the disk radius which best explains the ϕ Persei data is also close to the maximum radius for stable orbits.

Marlborough, Snow and Slettebak (1978) proposed that γ Cas has a neutron star companion. This could explain the observed X-ray flux from γ Cas and the V/R variations. The X-rays are produced when some of the stellar wind emanating from γ Cas is accreted onto the neutron star. The V/R variations arise because the outer envelope is tidally distorted. A similar model has been proposed for GX304-1 (Parkes, Murdin and Mason 1980) and X Persei (Persei, Viotti and Ferrari-Toniolo 1977). The secondary in the ϕ Persei system is also peculiar (Poeckert 1979).

3.3 Disk-like Models

Disk models are usually used in interpreting a restricted set of observations. The models tend to be highly simplified generally, but can be quite complex in the treatment of a specific problem.

3.3.1 Line Profiles

Kogure (1975), Hirata and Kogure (1977, 1978) and Kogure, Hirata and Asada (1978) analysed the residual flux in the Balmer shell lines of several Be stars. They derive the Hα optical depth, the fraction of the star that is occulted, and the outer radius of the envelope. A schematic representation of their model is shown in Figure 3, and an example of how the model fits the data is shown in Figure 4. Higurashi and Hirata (1978) applied a similar analyses to the metallic shell lines in the spectrum of the Be star HD 23862. They found that the fractional occultation increases with ionization potential, suggesting that the cooler part of the envelope is surrounded by the hotter region.

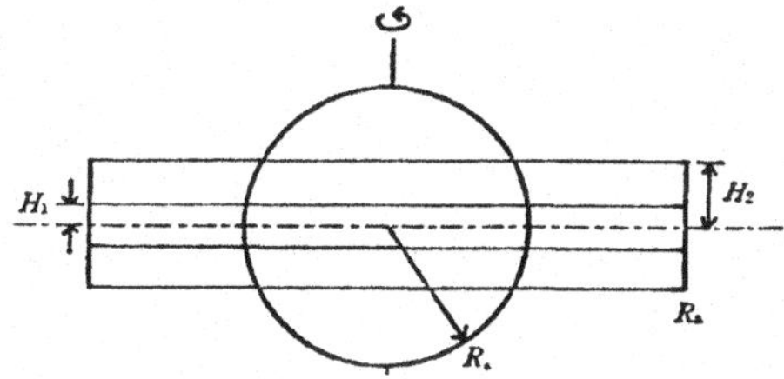

Figure 3.   An equatorial view of the two disk model as proposed by Hirata and Kogure (1977).

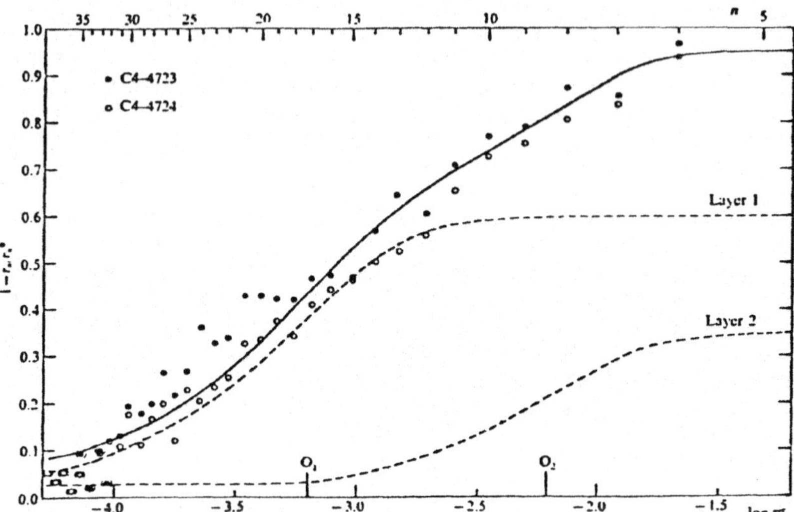

Figure 4.   The fit of the two disk model to the observed residual flux in the Balmer shell lines of Pleione.

Van Blerkom (1978) and Kunasz and Van Blerkom (1978) have developed a model for P Cygni based on the Balmer line profiles.   Ordinarily P Cygni is not included among classical Be stars, but their model has an expansion velocity which increases gradually and linearly with radius resulting in an inner, moderate-density envelope, much like Be star envelopes.   Since rotation does not appear to be a factor in P Cygni the model is spherically symmetric.

3.3.2 Continuum Energy Distribution

Modelling the energy distribution of Be stars is usually approached as a problem concerning flux excesses.   The comparisons made are with "normal" stars or "normal" model atmospheres.   There is some degree of danger in this approach in that we already know that the stars are not "normal".   One is apt to think of Be stars as "normal", but with an external, detached peculiarity.   Cram (1980) has commented on the inconsistencies of this kind of modelling.   On the other hand, Be stars that observationally lose their Be characteristics appear to be "normal" B-type stars, albeit rapidly rotating, and pole-on Be stars appear to be "normal" (Peters 1979) so comparison with "normal" stars may, after all, be justified.

Another problem in modelling the continuum energy distribution is correcting for interstellar extinction.   There is substantial evidence for intrinsic reddening in Be stars (Schild 1976) which immediately calls into question the use of E_{B-V} in correcting for interstellar extinction.

The infrared excess observed in many Be stars is usually ascribed to free-free emission from the circumstellar envelope (Gehrz, Hackwell and Jones 1974), but in some "peculiar" Be stars, e.g. HD 45677, hot dust may also be present (Coyne and Vrba 1976).

Ferrari-Toniolo et al. (1978) have compared their infrared observations with models for the Be stars γ Cas, X Per, φ Per and ζ Tau. Their models consist of isothermal, constant density, spherical envelopes. They calculate the emission spectrum of such an envelope including free-free and bound-free processes. Scargle et al. (1978) also observed and modelled the infrared flux of γ Cas. They considered both disk and spherical models and found that either geometry would give satisfactory results. Figure 5 compares the γ Cas observations

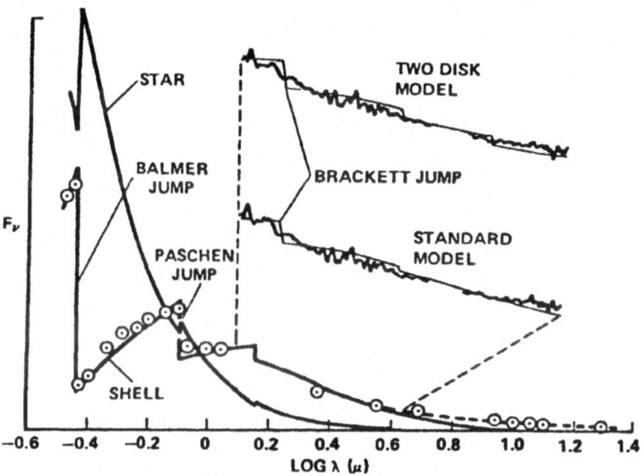

Figure 5. The flux excess (circles) and the model results for γ Cas. At 3751A the shell contribution is 11% of the total flux.

and best disk model from Scargle et al. Note that there is a significant flux contribution by the envelope in the Paschen continuum. This explains why Be stars appear to have intrinsic reddening as found by Schild (1976), and why interstellar reddening corrections based on E_{B-V} may be incorrect.

Haisch and Cassinelli (1976) and Hartmann (1978) consider spherically symmetric, scaled-down Wolf-Rayet star models, which they truncate into disks to produce linear polarization (see also section 3.3.3). These models are fundamentally different from those discussed so far in that they treat the envelope like an extended photosphere, rather than a separate disk or sphere surrounding a "normal" star.

3.3.3. Polarization

It is now generally accepted that the intrinsic linear polarization of Be stars is due to electron scattering in circumstellar disks. There are two problems which must be addressed, the transfer of radiation, which will determine the wavelength dependence of polarization, and geometry, which will determine the overall degree of polarization (although not exclusively).

Haisch and Cassinelli (1976) considered the polarization produced by a rotationally distorted star and disc-like envelope. The rotationally distorted (Roche) models produce only small, < 0.2%, net polarization. This is consistent with the fact that rapidly rotating stars are not observed to have large intrinsic polarizations. Peraiah (1976), modelling early-type rotating supergiants, also looked into rotationally distorted models, but these had much more extended atmospheres than the models of Haisch and Cassinelli. Peraiah was able to produce polarizations up to 25% in envelopes with large scattering optical depths, $\tau_{sc} \sim 5$.

The disk models considered by Haisch and Cassinelli (1976) are based on the WR star models of Cassinelli and Hartmann (1975). A spherical model is truncated to a disk to produce a net polarization. Jones (1977, 1979) also considered a disc model. His approach was to divide the envelope into many ($\sim$ hundreds) cells and calculate the contribution of each cell to the total flux. The models incorporated free-free and free-bound processes, single and double scattering of stellar photons, and single scattering of photons originating within the envelope. Figure 6 shows the results of various models and the effects of various processes. It is clear that absorptive opacity is very important in determining the wavelength dependence of polarization. Jones' models are somewhat more successful than those of Haisch and Cassinelli in that they fit the observations of the Balmer jump polarization better and the overall polarization is higher. Observed Be star polarizations go as high as 2%, a value easily obtained by Jones' models, while the models of Haisch and Cassinelli were limited to net polarizations under 1.2%.

Another approach to the disc model was made by Johns (1975). He determined the net polarization due to a simple disk and star by using a Monte Carlo technique to solve the radiation transfer problem. Johns concludes that the wavelength dependence of polarization is the result of competing opacities, bound-free and free-free absorption and electron scattering. Free-free and free-bound emission, "diluting" the polarization, plays only a minor role. Johns adds that the disc which best reproduces the observed wavelength dependence and magnitude of polarization in Be stars is optically thick to electron scattering and is relatively thin compared to the stellar radius. These conclusions are different from other studies which usually find $\tau_{sc} < 1$, and a disk thickness of ~ 2 R*.

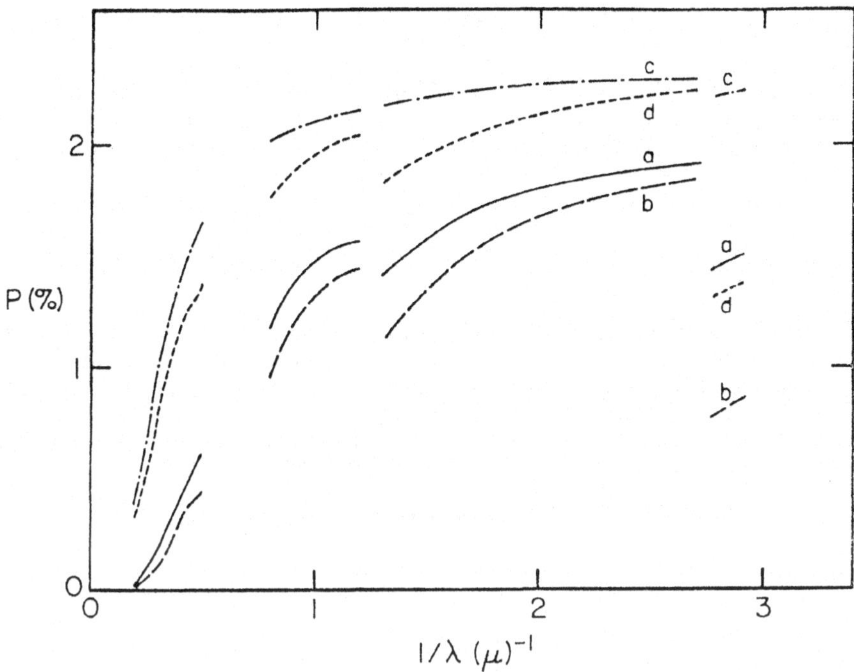

Figure 6.   Polarization from a disk at 10000 K, 1 R_* thick and 5 R_* in radius.   The curves represent a) only free-free opacity included, b) free-free and bound-free opacity included c) scattering of envelope photons included, but no bound-free opacity, and d) everything included (Jones, 1979).

Brown and McLean (1977) outline the basic geometry problem in electron scattering from circumstellar envelopes.   The simplifying assumptions that are made in order to give an analytic solution are that the star is a point source, the density distribution can be described by a series of uniaxial ellilpsoids or cylinders, and the envelope is optically thin.   The result is that the polarization of the scattered radiation is given by

$$P(\beta,\underline{i}) \;=\; \sin^2\underline{i}/(2\alpha + \sin^2\underline{i}),$$

where

$$\alpha \;=\; (1+\beta)/1-3\beta),$$

$\underline{i}$ is the inclination and β is a "shape factor" describing the density distribution.   The importance of such an analytic solution cannot be overemphasized because it can give far greater insight into a problem than the more involved numerical analyses.

Brown, McLean and Emslie (1978) extended the results of Brown and McLean to binary stars (two illuminating sources) and McLean (1979) applied their results to Be star envelopes. McLean discusses the variation of polarization across emission lines in Be stars. The decrease in polarization within an emission line (as observed by Clarke and McLean 1974 and Poeckert 1975) is due to an increase in absorptive opacity and "dilution" by the line emission. The variations in position angle of polarization as observed in γ Cas by Poeckert and Marlborough (1977) are due to the doppler shifted absorptive opacity. McLean found that the amplitude of these position angle variations is expected to vary as cos $\underline{i}$.

Powerful as the analytic analyses are one must keep in mind the assumptions which make them possible. The assumption of a point source star is invalid if the envelope is close to the star. Rudy (1978) has looked into the effects of a finite sized illuminating source and finds that the analytic solutions are valid if the envelope is symmetric about the equatorial plane. If the envelope is comparable in size with the illuminating source there will be occultation effects. For example, Piirola (1980) has observed eclipse effects in U Cephei.

The assumption that envelopes are optically thin must also be questioned. The models of Johns (1975), Haisch and Cassinelli (1976) and Jones (1977) all have substantial scattering optical depths, $\tau_{sc} > 0.5$. Absorptive opacity is usually comparable to or greater than the scattering opacity. All these points must be kept in mind when applying the analytic results to the interpretation of polarization data.

Daniel (1980) compared the analytic results of Brown, McLean and Emslie (1978) to his own calculations, which involved a Monte Carlo approach. Daniel determined the polarization of ellipsoidal envelopes in which the scattering optical depth was large. He found that the analytic results were valid only for very thin envelopes, $\tau_{sc} < 0.1$. The problems of interpretation of data using the analytic results is demonstrated by Daniel (1981) in the case of Cyg X-1.

Electron scattering is not the only mechanism for producing intrinsic polarization. Coyne and Vrba (1976) measured the polarization of HD 45677. They did not find the canonical Be star wavelength dependence of polarization. They argue that the polarization and peculiar infrared excess are consistent with a dust ring 45 AU from the star, $\backsim$ 15 AU thick and clumpy.

3.3.4 Time Variations

Temporal variations in flux and polarization are a well established facet of the Be problem. The variations occur over long time periods, e.g. the $\backsim$ 1000 day V/R variations (see section 3.1), and short time periods, $\backsim$ minutes. Purely radial axisymmetric variations should have periods > 1 day (Morgan 1975).

The long-term variations (P > week) can be understood in terms of binaries (section 3.2), elliptic disks (section 3.1) or a variable mass loss rate (e.g. Limber's (1969) model for Pleione). Short-term (< 1 day) variations in the continuum flux have been attributed to pulsation (cf. Percy, Jakate and Matthews 1981) and temperature changes in the atmosphere of the central star (Lester 1975). In the case of EW Lac Lester argues that the lack of variations in the equivalent width of Hα rules out changes in envelope emission or absorption causing the continuum variations. However, Lester assumes that the envelope is optically thin, which is certainly not the case in Hα in EW Lac.

Short-term variations in polarization in the star γ Cas have been reported by Poeckert and Marlborough (1977), Rodriquez (1979) and Piirola (1979). The variations occur on a time scale of hours, but are not large with respect to the total polarization. Hayes (1980) looked for polarization variations in pole-on Be stars. If azimuthal variations occur in the envelope one can expect to see them in pole-on stars as well as the more active high V sin i stars. Hayes found no short-term (days) variations in the pole on stars suggesting that the envelopes are axisymmetric, at least in the regions near the star.

3.3.5. Stellar Wind Models

Poeckert and Marlborough (1977, 1978a, 1978b) reinvestigated the stellar wind models first presented by Marlborough (1969). A typical density distribution is shown in Figure 7. Also shown in Figure 7 is the ionization structure in such an envelope (from Poeckert and Marlborough 1981).

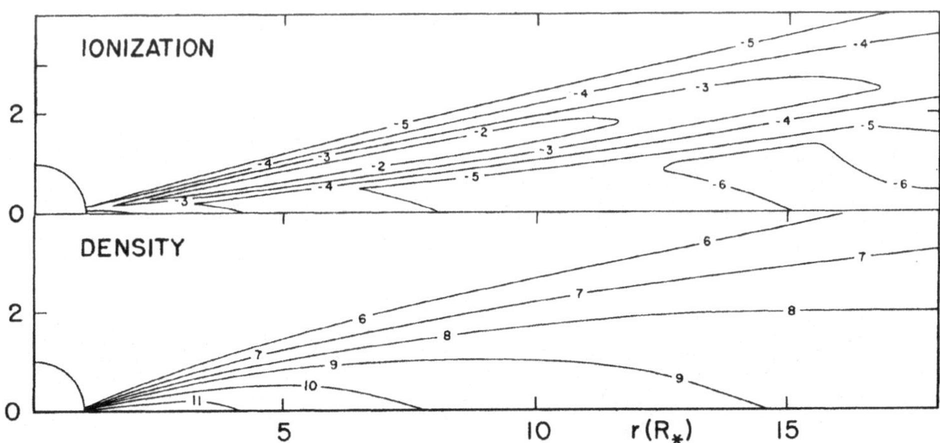

Figure 7. Density distribution and fractional ionization for a model envelope (T_{eff} = 15000 K, T_e = 10500 K). Contours are labelled in a) log N_H/N, b) log N. The solid circle represents the sun to scale.

The emergent flux is calculated at 200 frequencies in the Balmer lines and 33 frequencies in the continuum, between 1000 A and 11 cm. Polarized radiation (singly scattered stellar photons) is incorporated in all the calculations. The central star is spherical and its flux is assumed to be given by a single normal model atmosphere. Gravity darkening is neglected. The finite size of the star is taken into account explicitly.

The model for γ Cas (Poeckert and Marlborough 1977, 1978a) matches the observed Hα profile and polarization (Figure 8), and the continuum polarization. Note the position angle variations across the Hα line. The model was also used to predict the continuum energy distribution. Figure 9 shows the various components of the continuum energy distribution between 1000 A and 20μ. At 3.7 and 11 cm the flux is predicted to be 4.2 and 0.5 mJy, respectively, which is at the threshold for detection (Purton 1976).

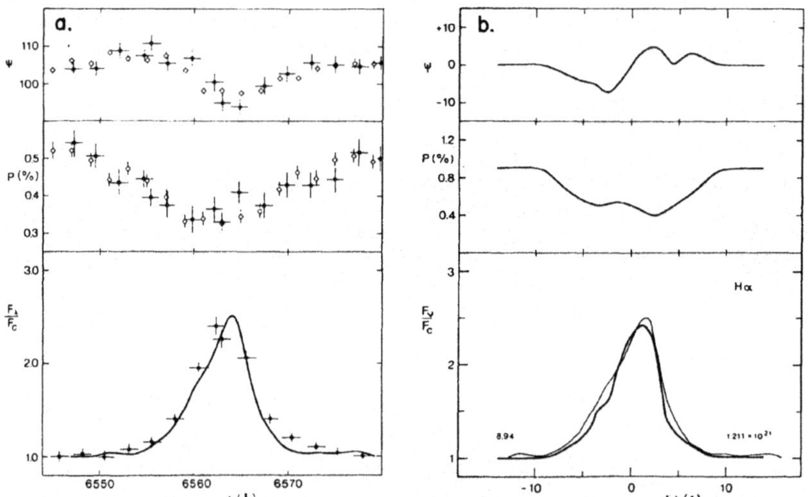

Figure 8.   a)   Observed Hα profile and polarization for γ Cas.
b) Computed profile and polarization.   (Position angle (ψ) is given in degrees).

The effects of inclination and envelope density were investigated by Poeckert and Marlborough (1978b).   Figure 10 shows the effect of inclination on the Hα line profile and polarization. Note the increase in position angle variations at smaller inclinations.   McLean (1979) determined that the amplitude of such variations is proportional to cos i.   The Hα profile for the equator-on view is a typical type III P Cygni profile (Beals 1951).   This profile is <u>not</u> typical of Be stars, but is seen in some other early-type stars.   (cf. Mihalas and Conti 1980).

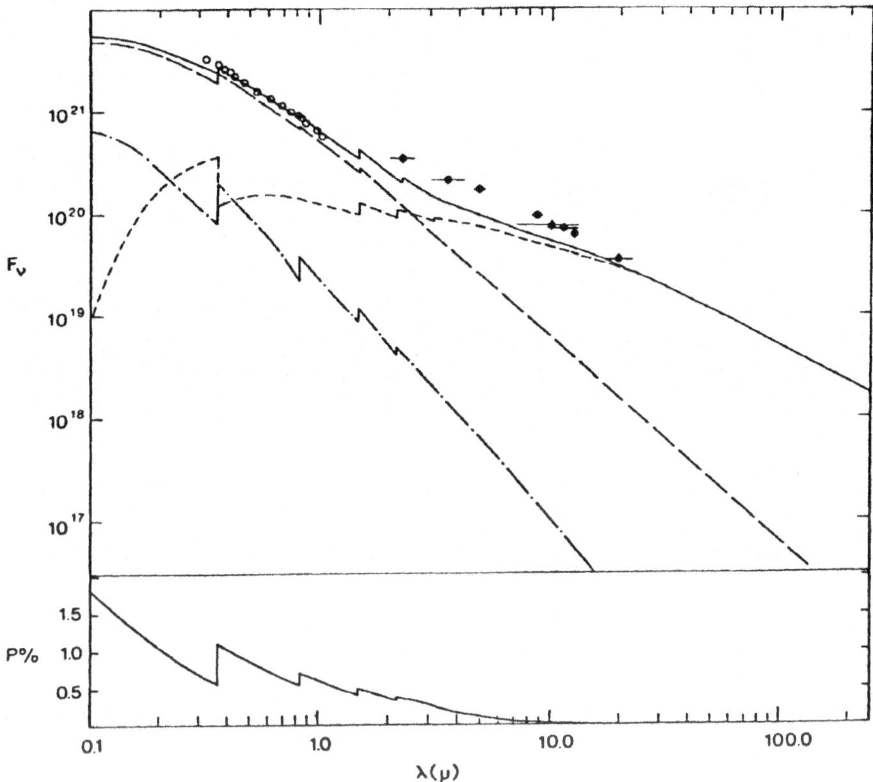

Figure 9.   Continuum energy distribution and polarization for γ Cas
model.

Figure 11 shows the effects of inclination on magnitude excess and
colours. Pole-on stars are brighter and redder (in B-V) than equator-
on stars.   Figure 12 shows the effects of envelope density on the
continuum energy distribution.

Figure 13 shows the effect of envelope density on polarization.
Note that at large densities polarization shortward of the Balmer jump
decreases, while polarization in the Paschen continuum increases, with
increasing density.   Several Be stars show this effect (Poeckert,
Bastien and Landstreet 1979).

The effect of electron scattering-on the Balmer line profiles was
investigated by Poeckert and Marlborough (1979). Figure 14 shows the
Hα and Hβ profiles for a dense envelope. The calculations confirm the

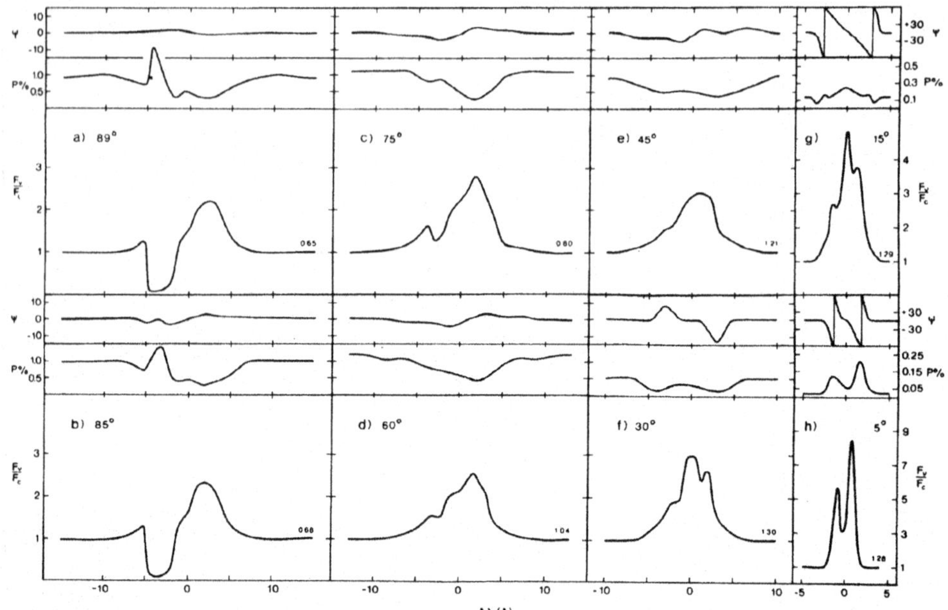

Figure 10. Hα profile and polarization as a function of inclination. Note the scale change in panels g and h. The continuum flux levels are given in the lower right of each panel.

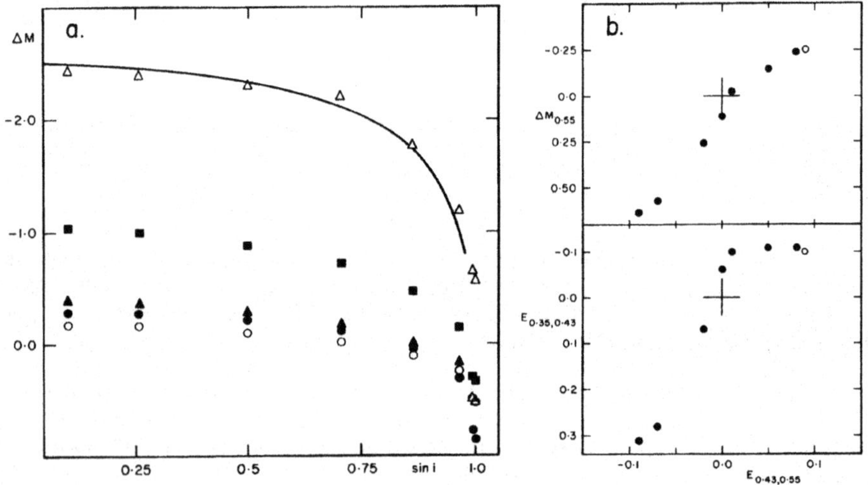

Figure 11. a) Magnitude excess as a function of sin i and wavelength (open triangles), 10μm; filled circles, 0.35μm; open circles, 0.43 μm). b) ΔV and E_{U-B} as a function of inclination (from 5° (open circle) to 15°, 30°, 45°, 60°, 75°, 85° and 90°).

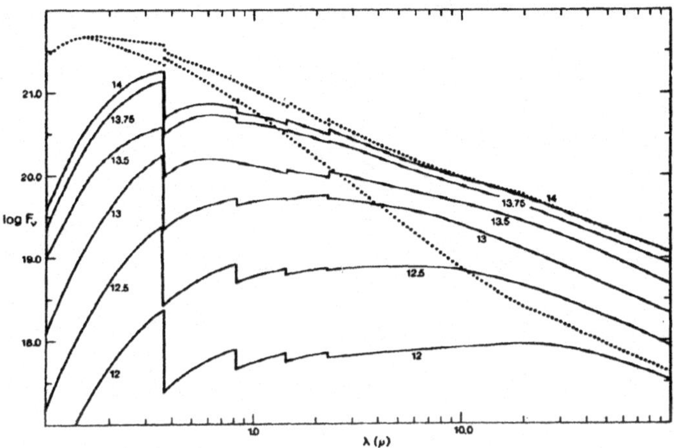

Figure 12. Continuum energy distribution for envelopes of various density (at 45° inclination). Solid lines represent envelope emission (labelled in log density). Dotted lines represent the extremes for the total flux.

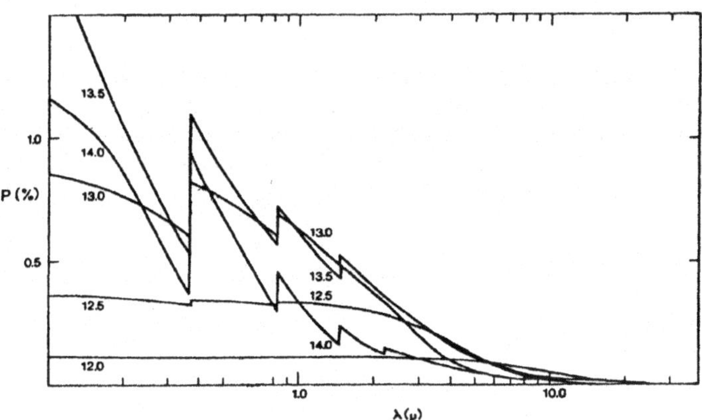

Figure 13. Continuum polarization for envelopes of various density (at 45° inclination).

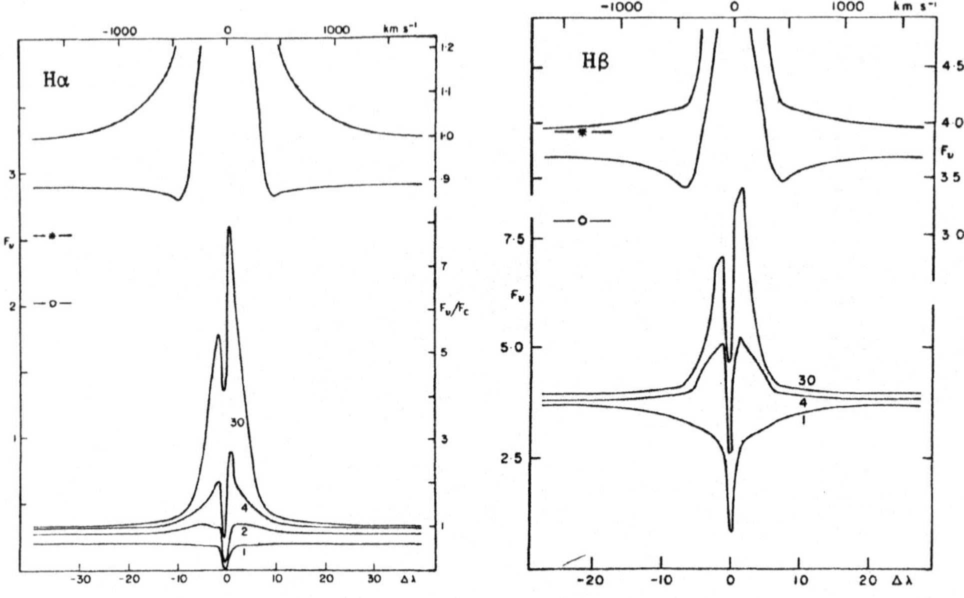

Figure 14.   Hα and Hβ line profiles showing the effects of electron scattering.   The upper curves show the details of the line wings with and without the effect of electron scattering.   The lower curves show the line profiles broken down by impact parameter (in R_*).

suggestion made by Marlborough (1969) that the broad wings seen at Hα in some Be stars are due to electron scattering.   Bernat and Lambert (1978) invoked electron scattering as the cause of the broad Hα wings in P Cygni.

Poeckert, Gulliver and Marlborough (1981) have computed a stellar wind model for the Be star o And.   They interpret the waning of the most recent shell episode in that star as a decrease in envelope density propagating outward.   Figure 15 summarizes the observations while figure 16 presents the results for the model.   The mass loss rate is assumed to decrease by a factor of 20 over 300 days, and the density decrease propagates outward at 12 km s^{-1}.   It takes several hundred days for the entire envelope to respond to the decreasing mass loss rate.

3.3.6 Coronal Models

Many studies of the ultraviolet spectra of Be stars have found lines of highly ionized species, e.g. N V and O VI (cf. Snow and Marlborough 1976; Marlborough 1977a; Marlborough and Snow 1980; Doazan, Kuhi and Thomas 1980; Dachs 1980; Heinrichs, Hammerschlag-Hensberge and

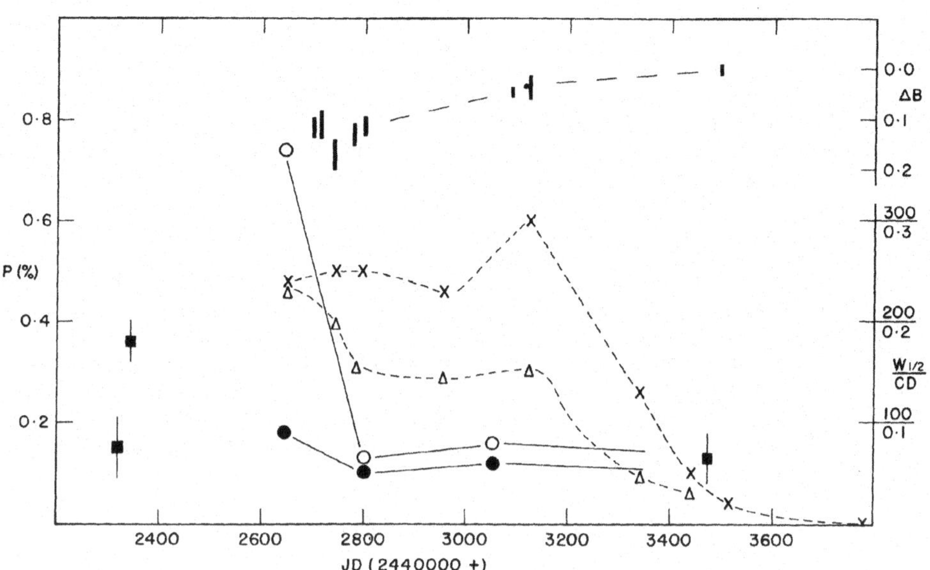

Figure 15.   Intrinsic polarization (open circles, 4050A; filled
circles, 3450A; filled squares, V), ΔB (solid bars; length of bar indicates diurnal variations), shell line depth (crosses) and shell line
half width (triangles) for o And between JD 2442200 and JD 2443800.

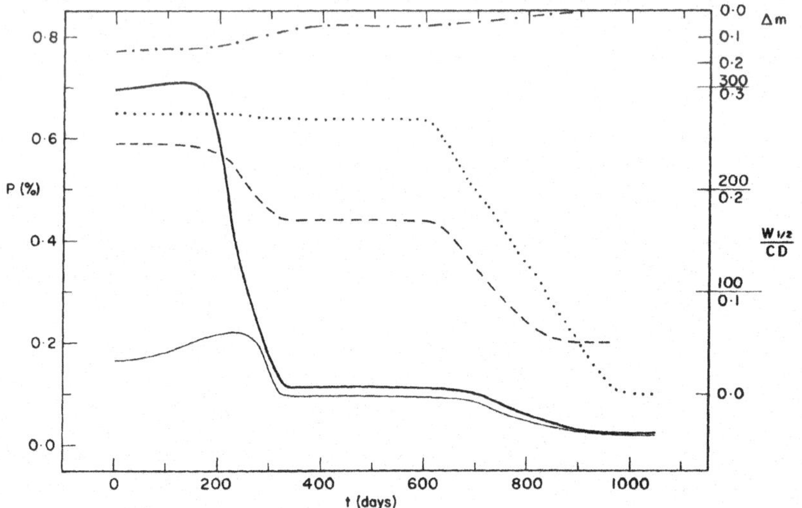

Figure 16. Polarization (light line, 3500A; heavy line 4000A), ΔB (dot
dash), Hγ shell line depth (dots) and half width (dashed line) for the
model as a function of time (days after start of density decrease).

Lamers 1980). In almost all cases such lines indicate suprathermal outflow velocities. The spectra of OB supergiants also show such lines and their presence is taken as a sign of coronal regions in the winds of these stars. Many OB supergiants, and some Be stars, also have unexpectedly large X-ray fluxes, another possible indicator of coronae. The models for OB supergiants generally assume spherical symmetry and the corona is placed just above the photosphere (see section 2.1). The situation in Be stars is somewhat more ambiguous.

Marlborough, Snow and Slettebak (1978) proposed a model for γ Cas in which the cool region of the envelope can be described by the stellar wind model of Poeckert and Marlborough (1978a), the coronal region lies above the higher density disk. It is suggested that the differential rotation in the cool disk leads to turbulence, and this turbulence forms shock waves when it enters the lower density region above the disk resulting in a corona. Icke (1976) proposed just such a scenario for accretion disks. He found that temperatures up to 10^6 K can be achieved, well above the temperature required to provide N V lines.

Marlborough (1977b) has shown that the X-ray emission from γ Cas requires a corona with $T_e \sim 2 \times 10^7$ K if the emission is due to bremsstrahlung. Marlborough, Snow and Slettebak suggest that the X-ray emission arises from an accretion disk around a neutron star companion. There is no direct evidence for such a companion, but the period (4 yrs) and the mass ratio (17:1) would make radial velocity variations undetectable. However, it is suggested that the outer region of the cool disk is tidally distorted, resulting in V/R variations in the Balmer emission lines. Such V/R variation should be synchronized with the binary period and their amplitude should increase with increasing Balmer emission.

Time variability in the lines of highly ionized species is common place in most early-type stars with winds. In OB supergiants it appears to be primarily a variation in optical depth not velocity, i.e. the terminal velocity is constant. The Be stars show variability in both optical depth and velocity. The latter can be quite dramatic, e.g. in 59 Cyg the velocity has changed from -50 km s^{-1} to -750 km s^{-1} (Doazan et al. 1980a).

Heinrichs, Hammerschlag-Hensberge and Lamers (1980) propose a model for γ Cas which attempts to explain the sudden appearance and disappearance of high velocity features in the ultraviolet lines. They suggest that there is a stellar wind which has a rapid transition from low to high velocity. A "puff" of material is blown off the star and it rapidly accelerates to the terminal velocity. The result is that lines are not seen at intermediate velocities, only at low velocities, because the density is high, and at high velocities, because the "puffs" reach the terminal velocity quickly.

A competing model has been proposed by Doazan et al. (1980b).

They suggest that the lack of intermediate velocity lines is due to the temperature structure in the corona. In this model the temperature rises to $\sim 10^6$ K and then falls again. Lines of N V, for example, are only formed at temperatures $\sim 10^5$ K, so we would expect two regions, one pre-coronal and one post-coronal, in which such lines can be formed. The post-coronal region has a large outflow velocity. It is also suggested that the highest temperature region is responsible for the X-ray flux.

Some correlation between the coronal line variations and Balmer emission line variations is apparently observed in 59 Cyg (Doazan et al. 1980a), but a model tying the coronal region to the cooler region has not been developed (with the exception of the model of Marlborough, Snow and Slettebak (1978) which did not address the problem of coronal line variations).

4.   Conclusions

A solution to the basic equations governing the structure and dynamics of Be star envelopes is as remote as it was five years ago. In fact, the advent of coronae in stellar wind models has made the entire process even more complex than it was. At present the _ad hoc_ models still provide our best picture of Be stars.

There are several aspects of the Be phenomenon which desperately need clarification. First, the location of the coronal region. Marlborough, Snow and Slettebak (1978) suggest such a region exists above the higher density cool disk. If such is the case one might expect to see substantial differences between pole-on and equator-on Be stars. At present the limited amount of data indicates no dependence on inclination of the coronal lines (Dachs 1980). The possibility of having a coronal region at the base of the disk is small since we see no high outflow velocity in the disk and a high temperature high density region is ruled out by the polarization observations (Marlborough and Poeckert, in preparation).

The V/R variations are also a puzzle. The elliptic ring model proposed by Huang (1975) has serious difficulty in matching all, but the V/R variations. The models of Kriz (1979) are more compatible with the data, but the dynamical arguments in favor of precessing elliptic rings are lost in the case of a disk. The fact that V/R variations might be due to a secondary star distorting the envelope has been proposed by Marlborough, Snow and Slettebak (1978), but the changes in period of V/R variations argues against such a model for all Be stars.

Several Be stars are X-ray sources and this has been interpreted as evidence for very hot coronae or degenerate secondaries. It is interesting that ϕ Persei, a star which shows He II λ 4686 in emission and appears to have a peculiar companion, is not an X-ray source (Hutchings 1981).

Finally, it is worthwhile repeating the admonition given by Marlborough (1976), do not believe everything you see in print!

I wish to thank R. Haapala and D. Duncan for help in preparing the manuscript, and all those, particularly J.M. Marlborough, with whom I have had discussions in the last few years concerning Be star models.

REFERENCES

Albert, E., and Huang, S.S.: 1974, Astrophys. J. **189**, 479.
Barker, P.K.: 1979, Ph.D. Thesis, Univ. of Colorado.
Barker, P.K., Landstreet, J.D., Marlborough, J.M., Thompson, I., and
     Muza, J.: 1981, preprint.
Beals, C.S.: 1951, Publ. Dom. Astrophys. Obs. **9**, 1.
Bernat, A.P., and Lambert, D.L.: 1978, Publ. Astron. Soc. Pacific
     90, 520.
Brown, J.C., and McLean, I.S.: 1977, Astron. Astrophys. **57**, 141.
Brown, J.C., McLean, I.S., and Emslie, A.G.: 1978, Astron. Astrophys.
     68, 415.
Cannon, C.J., and Thomas, R.N.: 1977, Astrophys. J. **211**, 910.
Cassinelli, J.P.: 1979, Ann. Rev. Astrophys. **17**, 275.
Cassinelli, J.P., and Castor, J.I.: 1973, Astrophys. J. **179**, 189.
Cassinelli, J.P., and Hartmann, L.: 1975, Astrophys. J. **202**, 718.
Cassinelli, J.P., Castor, J.I., and Lamers, H.J.G.L.M.: 1978,
     Publ. Astron. Soc. Pacific **90**, 496.
Castor, J.I., Abbott, D.C., and Klein, R.I.: 1975, Astrophys. J.
     195, 157.
Chugai, N.N.: 1980, Soviet Astron. Letters **6**, 91.
Clarke, D., and McLean, I.S.: 1974, Monthly Notices Roy. Astron. Soc.
     167, 27P.
Clayton, G.C., and Marlborough, J.M.: 1980, Astrophys. J. **242**, 165.
Collins, G.W., and Sonneborn, G.H.: 1977, Astrophys. J. Suppl. **34**, 41.
Coyne, G.V., and Vrba, F.J.: 1976, Astrophys. J. **207**, 790.
Cram, L.E.: 1980, Comments on Astrophys. **9**, 25.
Dachs, J.: 1980, Proc. Second European IUE Conf., p. 139.
Daniel, J.Y.: 1980, Astron. Astrophys. **86**, 198.
Daniel, J.Y.: 1981, Astron. Astrophys. **94**, 121.
Doazan, V., Kuhi, L.V., and Thomas, R.N.: 1980, Astrophys. J. **235**, L17.
Doazan, V., Kuhi, L.V., Marlborough, J.M., Snow, T.P., and Thomas, R.N.:
     1980a, Proc. Second European IUE Conf., p. 151.
Doazan, V., Selvelli, P., Stalio, R., and Thomas, R.N.: 1980b, Proc.
     Second European IUE Conf., p. 145.
Drake, S.A.: 1980, Ph.D. Thesis, Univ. California (Los Angeles).
Drake, S.A., and Ulrich, R.K.: 1980, Astrophys. J. Suppl., **42**, 351.
Ferrari-Toniolo, M., Leonetti, O., Persi, P., Spada, G., and Vioti, R.:
     1978, Memoirs Soc. Astron. Italiana, **49**, 223.
Gehrz, R.D., Hackwell, J.A., and Jones, T.W.: 1974, Astrophys. J.
     191, 675.
Gulliver, A.F., Bolton, C.T., and Poeckert, R.: 1980, Publ. Astron.
     Soc. Pacific, **92**, 774.

Haisch, B.M., and Cassinelli, J.P.: 1976, Astrophys. J. 208, 253.
Hammon, W.R.: 1981, Astron. Astrophys. 93, 353.
Hartmann, L.: 1978, Astrophys. J., 224, 520.
Hayes, D.P.: 1980, Publ. Astron. Soc. Pacific 92, 661.
Heap, S.R.: 1976, in A. Slettebak (ed.) "Be and Shell Stars",
     IAU Symposium, 70, 165.
Hennon, M., and Gyot, M.: 1970, in G.E.O. Giacaglia (ed.) "Periodic
     Orbits Stability and Resonances", Reidel, Dordrecht, 428.
Henrichs, H.F., Hammerschlag-Hensberge, G., and Lamers, H.J.G.L.M.:
     1980, Proc. Second European IUE Conf., 147.
Higurashi, T., and Hirata, R.: 1978, Publ. Astron. Soc. Japan 30, 615.
Hirata, R., and Kogure, T.: 1977, Publ. Astron. Soc. Japan 29, 477.
Hirata, R., and Kogure, T.: 1978, Publ. Astron. Soc. Japan 30, 601.
Huang, S.S.: 1972, Astrophys. J. 171, 549.
Huang, S.S.: 1973, Astrophys. J. 183, 541.
Huang, S.S.: 1975, Sky and Telescope 49, 359.
Huang, S.S.: 1976, Publ. Astron. Soc. Pacific 88, 448.
Huang, S.S.: 1977, Astrophys. J. 212, 123.
Huang, S.S.: 1978, Astrophys. J. 219, 956.
Hutchings, J.B.: 1976, Publ. Astron. Soc. Pacific 88, 5.
Hutchings, J.B.: 1980, Astrophys. J. 237, 285.
Hutchings, J.B.: 1981, Publ. Astron. Soc. Pacific 93, Feb.
Hutchings, J.B., Nemec, J.M., and Cassidy, J.: 1979, Publ. Astron. Soc.
     Pacific 91, 313.
Icke, V.: 1976, in P. Eggleton, S. Mitton and J. Wehlan (eds.)
     "Structure and Evolution of Close Binary Stars", IAU Symposium
     73, 267.
Johns, M.W.: 1975, Ph.D. Thesis, Dartmouth College.
Jones, T.J.: 1977, Ph.D. Thesis, Univ. of Hawaii.
Jones, T.J.: 1979, Astrophys. J. 228, 787.
Kodaira, K.: 1980, Publ. Astron. Soc. Japan 32, 435.
Kogure, T.: 1975, Publ. Astron. Soc. Japan 27, 165.
Kogure, T., Hirata, R., and Asada, Y.: 1978, Publ. Astron. Soc. Japan
     30, 385.
Kriz, S.: 1976, Bull. Astron. Inst. Czech. 27, 321.
Kriz, S.: 1979a, Bull. Astron. Inst. Czech. 30, 83.
Kriz, S.: 1979b, Bull. Astron. Inst. Czech. 30, 95.
Kunasz, P., and Van Blerkom, D.: 1978, Astrophys. J. 224, 193.
Landstreet, J.D.: 1980, Astron. J. 85, 611.
Lester, D.F.: 1975, Publ. Astron. Soc. Pacific 87, 177.
Limber, D.N.: 1969, Astrophys. J. 157, 785.
Marlborough, J.M.: 1969, Astrophys. J. 156, 135.
Marlborough, J.M.: 1976, in A. Slettebak (ed.) "Be and Shell Stars",
     IAU Symposium 70, 335.
Marlborough, J.M.: 1977a, Astrophys. J. 216, 446.
Marlborough, J.M.: 1977b, Publ. Astron. Soc. Pacific 89, 122.
Marlborough, J.M., and Snow, T.P.: 1980, Astrophys. J. 235, 85.
Marlborough, J.M., Snow, T.P., and Slettebak, A.: Astrophys. J.
     224, 157.
Marlborough, J.M., and Zamir, M.: 1975, Astrophys. J. 195, 145.
Martens, P.C.H.: 1979, Astron. Astrophys. 75, L7.

Massa, D.: 1975, Publ. Astron. Soc. Pacific **87**, 777.
Mazurek, T.J.: 1980, preprint.
McLean, I.S.: 1979, Monthly Notices Roy. Astron. Soc. 186, 265.
Mihalas, D., and Conti, P.S.: 1980, Astrophys. J. **235**, 515.
Morgan, T.H.: 1975, Astrophys. J. **195**, 391.
Nerney, S.: 1980, Astrophys. J. **242**, 723.
Parkes, G.E., Murdin, P.G., and Mason, K.O.: 1980, Monthly Notices Roy.
     Astron. Soc. **190**, 537.
Peraiah, A.: 1976, Astron. Astrophys. **46**, 237.
Percy, J.R., Jakate, S.M., and Matthews, J.M.: 1981, Astron. J. **86**, 53.
Persi, P., Viotti, R., and Ferrari-Toniolo, M.: 1977, Monthly Notices
     Roy. Astron. Soc. **181**, 685.
Peters, G.J.: 1972, Publ. Astron. Soc. Pacific **84**, 334.
Peters, G.J.: 1976, in A. Slettebak (ed.) "Be and Shell Stars",
     IAU Symposium **70**, 417.
Peters, G.J.: 1979, Astrophys. J. Suppl. **39**, 175.
Piirola, V.: 1979, Astron. Astrophys. Suppl. **38**, 193.
Piirola, V.: 1980, Astron. Astrophys. **90**, 48.
Poeckert, R.: 1975, Astrophys. J. **196**, 777.
Poeckert, R.: 1979, Astrophys. J. **233**, L73.
Poeckert, R.: 1981, Publ. Astron. Soc. Pacific 93, June.
Poeckert, R., Bastien, P., and Landstreet, J.D.: 1979, Astron. J.
     83, 812.
Poeckert, R., Gulliver, A.F., and Marlborough, J.M.: 1981, in
     preparation.
Poeckert, R., and Marlborough, J.M.: 1977, Astrophys. J. **218**, 220.
Poeckert, R., and Marlborough, J.M.: 1978a, Astrophys. J. **220**, 940.
Poeckert, R., and Marlborough, J.M.: 1978b, Astrophys. J. Suppl. **38**, 22
Poeckert, R., and Marlborough, J.M.: 1979, Astrophys. J. **233**, 259.
Poeckert, R., and Marlborough, J.M.: 1981, preprint.
Prevot, L., Laurent, C., Paul, J., Vidal-Madjar, A., Audouze, J.,
     Ferlet, R., Lequeux, J., Maucherat-Joubert, M., Prevot-
     Burnichon, M.L., and Rocca-Volmerange, B.: 1980, Astron. Astrophys.
     90, L13.
Purton, C.R.: 1976, in A. Slettebak (ed.) "Be and Shell Stars"
     IAU Symposium **70**, 157.
Rodriguez, M.H.: 1979, Astrometriya I Astrofisika 39, 3.
Rudy, R.J.: 1978, Publ. Astron. Soc. Pacific **90**, 688.
Rumpl, W.M.: 1980, Astrophys. J. **241** 1055.
Rybicki, G.B., and Hummer, D.G.: 1978, Astrophys. J. **219**, 654.
Scargle, J.D., Erikson, E.F., Witteborn, F.C., and Strecker, D.W.: 1978
     Astrophys. J. **224**, 527.
Schild, R.E.: 1976, in A. Slettebak (ed.) "Be and Shell Stars", IAU
     Symposium **70**, 107.
Slettebak, A., Kuzma, T.J., and Collins, G.W.: 1980, Astrophys. J.
     242, 171.
Smith, B.L., and Roxburgh, I.W.: 1977, Astron. Astrophys. **61**, 747.
Snow, T.P., and Marlborough, J.M.: 1976, Astrophys. J. **203**, L87.
Sobolev, V.V.: 1960, "Moving Envelopes of Stars" English Transl.,
     S. Gaposchkin, Harvard University Press, Cambridge.
Sonneborn, G.H., and Collins, G.W.: 1977, Astrophys. J. **213**, 787.

Struve, O.: 1931, Astrophys. J. **73** 94.
Surdej, J.: 1979, Astron. Astrophys. **73**, 1.
Suzuki, M.: 1979, Publ. Astron. Soc. Japan **32**, 331.
Tassoul, J.L.: 1978, "Theory of Rotating Stars", Princton Univ. Press,
     Princeton.
Van Blerkom, D.: 1978, Astrophys. J. **221**, 186.

DISCUSSION

<u>Traving</u>: The models which are actually computed should depend criti-
cally on the radiative transfer in Ly - α line. So the proper choice
of the redistribution function should matter.

<u>Poeckert</u>: Treating the Lyman line transfer problem correctly is very
difficult and is only handled in a very simplified form in our model.

<u>Endal</u>: Do your models predict the type of movement in the colour-
magnitude diagram discussed by Hirata for Be stars with disks of
time-varying density?

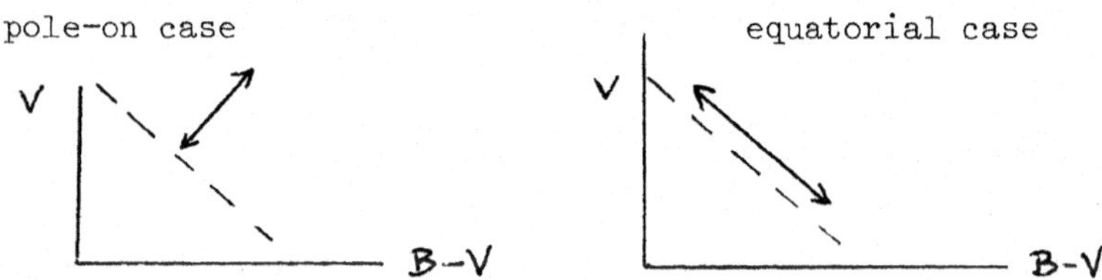

<u>Poeckert</u>: In the pole-on case, yes. However, this type of behaviour
is what you expect whenever you add a cool envelope, without obscuring
the star. In the equatorial case the models suggest an increase in V
is accompanied by a decrease in B-V; due to bound-free absorption.
I should stress that this only happens in cases when the envelope
density is high (resulting Hα equivalent widths in excess of 90 A),
and no large V or colour changes are predicted for less dense envelopes.

<u>Peters</u>: According to a recent paper by Hayes, the polarization
observed in ω Ori secularly decreased from .7% to .2% during a
7 months interval of time just prior to my IUE observation which
showed the presence of the high velocity shell lines. Would you care
to comment on this observation? What would the two observations taken
together imply about the causes (and/or nature) of the interesting
mass loss in this star?

<u>Poeckert</u>: Hayes also found that some pole-on stars are not variable
polarimetrically. I don't find it surprising that during an outburst
the polarization is variable since asymmetric (non-axissymmetric)
mass loss is certainly possible.

<u>Harmanec</u>: Have you some idea what could happen to 88 Her which - on
a time scale of years - became <u>brighter</u> and <u>bluer</u> and at the same
time almost lost the Hα emission?

<u>Poeckert</u>: The models presented by Marlborough and myself do not show
this kind of behaviour. It may be possible to get the kind of behaviour you describe in a dense envelope which partially obscurs the
stellar disk.

<u>Stalio</u>: 1. When you say that radiatively driven wind models predict
the terminal velocity of the wind, do you mean the relation $V_\alpha = 3 \times V_{esc}$?
If yes, I inform you that nobody considers it true anymore.
2. The word "puffs" has been coined by Lamers, Kondo and myself (ApJ
220, 1978) in order to explain the appearance and disappearance of MgII
shell lines in β Ori.

<u>Poeckert</u>: 1. As I am not an expert on OB wind theory I accept whatever
results appear in print. If that has changed, then I'm sorry that I'm
not as up-to-date as I should be.
2. The term "puffs" is not something I attribute to anyone, but myself,
as I did not see it in the literature. I apologize for not recognizing
your prior coining of the word.

<u>Hirata</u>: If my memory is correct, o And became fainter before the shell
appearance. If so it seems to me that this is an important information
on the shell activity mechanism.

<u>Poeckert</u>: As far as I am aware the actual onset of the shell was not
observed. The shell was well developed spectroscopically when it was
discovered. Hence, I am not sure the decrease in brightness preceeds
the shell phase.

<u>Snow</u>: In reviewing the model for O-star winds, you omitted one that
might be relevant to the Be stars. The x-ray observations show that
soft x-ray emission arises from high levels in the wind; that is, far
from the star. This suggests that heating occurs in the wind, not
just at the stellar surface. One proposed explanation is that the
star emits a steady stream of blobs or "bullets", which have a substantial velocity relative to the lower-density material, and this
causes shock-heating, producing x-rays. If this idea is realistic for
O-stars, it ought to be considered for Be stars as well, especially
in view of what we heard about "puffs" and blobs in the winds represented by the narrow, high-velocity components.

<u>Thomas</u>: 1. All the Be- and B star data I know are compatible with a
low-lying, coronal x-ray emission (cf τ Sco and ρ Oph). Same, for
T Tauri stars. I.e., the more extended the atmosphere, the "brighter"
Hα the less I(x-ray).
2. The "bullet" model you mention, by White and Lucy(?) at Columbia
is aerodynamic nonsense. They <u>assume</u> the "bullets" (which are gas
bubbles) are not distorted when moving at 1-3000 km/s. <u>Impossible</u>.

<u>Thomas</u>: The 26-level hydrogen calculation that ignores photoionisa-
tion for levels n≥ 2 is arithmetic only, not applicable to any Be-
star or even solar situation $(\tau(Lyc) \sim 10^7)$. So it is a good example
of what not waste your time with. Better refer to Avrett et.al. at
Harvard Smithsonian, who do 7-level atoms with all transitions,
radiative and collisional, included. Myself, for our present leveled
Be calculations, I think 3-levels with transfer in LyC and BaC and Hα
is quite sufficient.

<u>Poeckert</u>: Now I think it is probably worthwile to consider an over-
all scenario for Be stars. We have seen in the preceding four days
that the Be stars have cool and hot envelopes (or winds), are variable
on a variety of timescales, and rotate rapidly. We have heard each of
these aspects discussed individually, but there has been no attempt
to conceil any large fraction of the data in one model. I hesitate
to include what follows in my review paper because there is a great
danger that the wild speculations presented here will be taken too
literally by some and perhaps even be cited as a discussion of Be
star models in some future paper. This discussion is meant primarily
to present in a concrete manner what has been discussed informally
over the last week and what has been suggested in the literature in
order that we may have a frame work for future debate. I will address
myself to what I consider the dominant problem, the juxaposition of
the cool and hot envelopes. I will assume that any time variability
can be incorporated in some way in each of the scenarios discussed
below. A related problem to the positioning of the cool and hot
region is how and where the non-thermal energy input occurs, a region
which I will refer to as the transition zone.
The first scenario I present is one suggested several years ago by
J.M. Marlborough and formally put forward by Marlborough, Snow and
Slettebak (1978). The cool envelope is thought to be a disk à la
Poeckert and Marlborough (1978a). The inner part of the disk rotates
rapidly and differentially leading to turbulence. The turbulence
becomes super sonic along the upper and lower edge of the disk (the
density falls rapidly) resulting in heating (transition zone) of the
low density region above (and below) the disk. The expansion velocity
in the disk is small, but in the hot region it is large. This scenario
is nice in that it ties the cool and hot region together and also
suggests a possible source for the non-thermal energy (cf. Icke 1976).
One might expect to see differences in the hot wind depending on the
inclination. "Pole-on" stars would be expected to have somewhat different winds from "equator-on" stars. Also, a star whose cool region
decreases in density, so that the star is no longer considered to be
a Be star, should show a marked change in the hot wind.
The second scenario and the one I prefer at the moment also has a
cool disk, but in this case the transition zone is at the stellar
surface. The transition zone covers the entire star and under normal
circumstances would lead to a hot wind in all directions. However,
the density in the equatorial plane is sufficiently high that the
gas cools almost immediately resulting in the formation of a cool
disk. This scenario is different from the preceding one in that there
is no direct link between the cool and hot envelopes. Moreover this
scenario leads itself to explain hot winds in all early type stars.
The only distinction between "normal" OB stars and Be stars is that
the letter have dense, and therefore cool equatorial regions. This
is most likely a consequence of rapid rotation. The transition zone
must be small ($\leqslant .3R_*$) otherwise it will be detectable in the
visible polarization and Balmer emission line profiles. One would
also expect that a Be star that loses its Balmer emission would
continue to show the presence of the hot wind.

The third scenario places the transition region at some large ($\gtrsim 20$ R_*)
radial distance out in the disk. In this case the region above and
below the disk (and $r < 20$ R_*) does not contribute in any way to the
observed spectrum. Once the gas passes through the transition zone
it is very hot and it expands rapidly, both radially and latitudinally.
It must expand latitudinally otherwise we would not see hot winds in
"pole-on" stars.

The fourth, and final, scenario is one suggested by R.N. Thomas and a
discussion of the proposal follows this afternoon. In essence, the
transition zone is at the stellar surface and a hot wind flows away
from the star more or less spherically symmetric. At some large radial
distance ($r > 10$ R_*) the gas decelerates, passes through a shock wave
(cooling it) and forms a high density cool envelope. Rotation plays
no major role in this proposal; the envelope is essentially spherically
symmetric. The difficulty with this proposal is, that it does not
address itself to the vast amount of data which suggest a non-spherically symmetric envelope (eg. polarization) and the overwhelming
circumstantial evidence which points to v sin i (rotation) as a
major influence on the observed spectrum (e.g. shell lines are seen
predominatly in the rapid rotators).

Again I must emphasize that these scenarios are very speculative and
should not be considered too literally. Moreover, the actual situation
in the case of a Be star may not necessarily be any one of the scenarios discussed, but it may be a combination; or more likely, a scenario which has not even occured to me as yet.

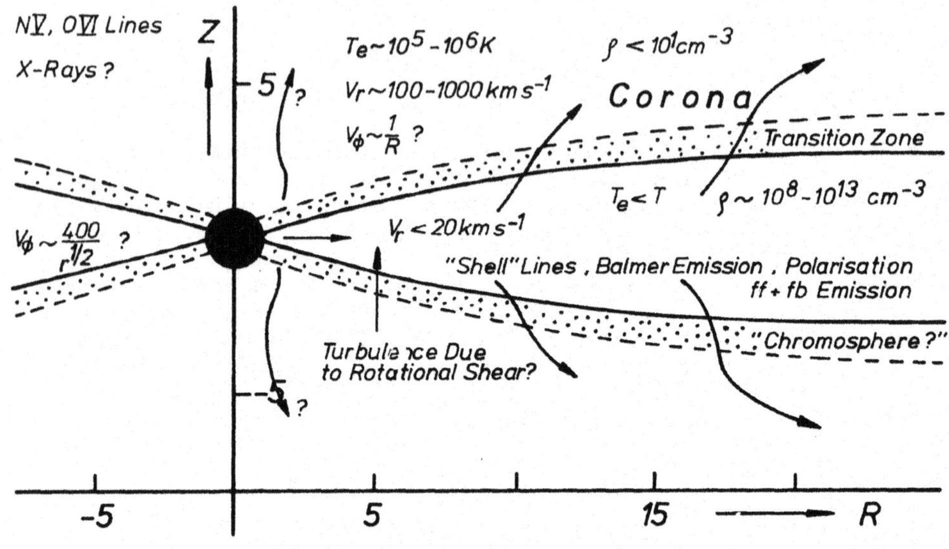

Scenario 1

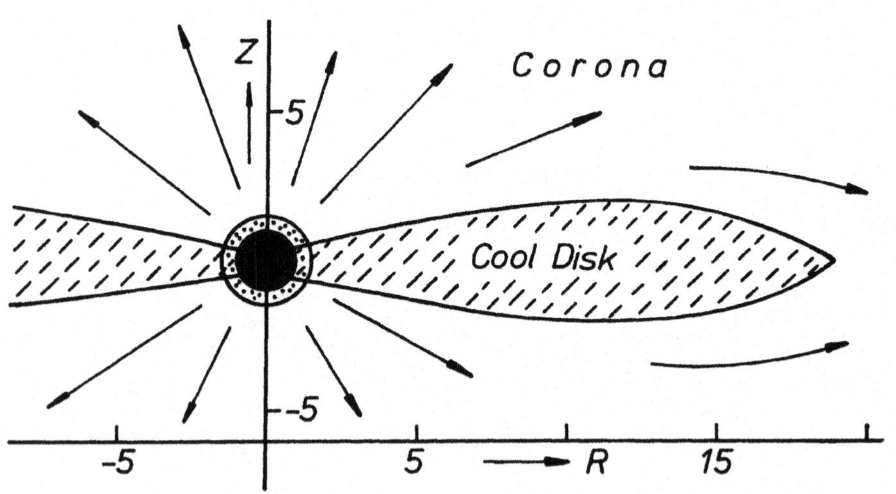

Scenario 2

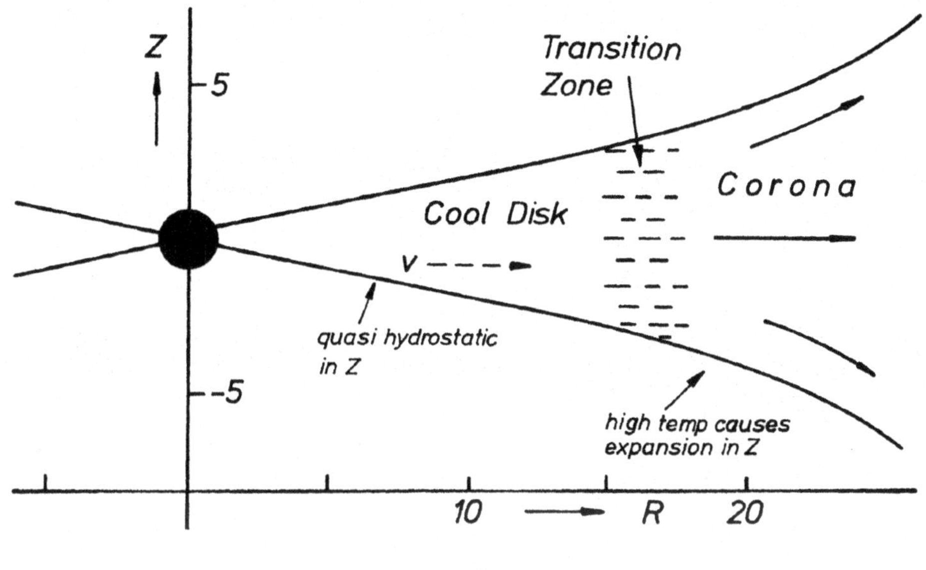

Scenario 3

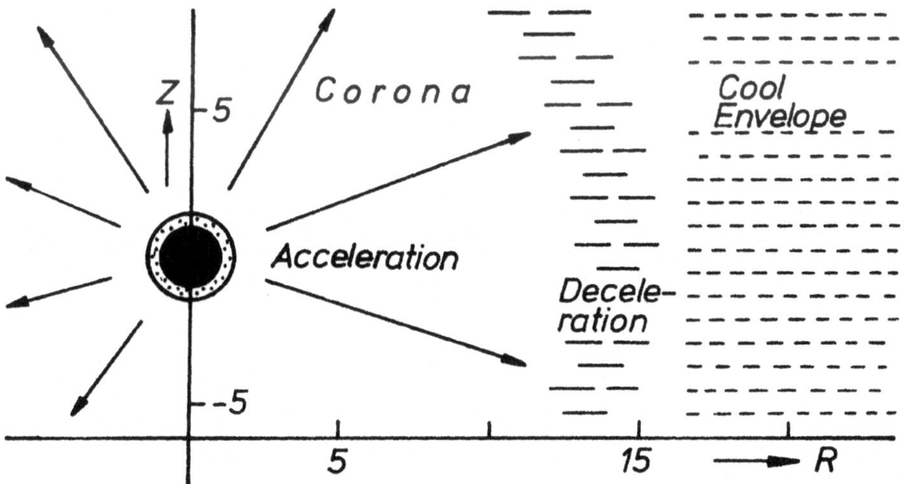

Scenario 4

HYDRODYNAMICAL MODELS OF ROTATING MAGNETIC WINDS

Paul K. Barker
The University of Western Ontario

ABSTRACT. Numerical investigations of rotating stellar winds driven by
a combination of magnetic and line radiation forces show a clear
dichotomy: shallow winds have their azimuthal velocity enhanced by
magnetic corotation, whereas steep winds may have a <u>diminished</u> circular
velocity.

Limber (1974) and Saito (1974) amongst others have considered the
nature of early-type stellar winds driven by magnetic corotation. These
models were constructed before the development of modern radiatively
driven wind theories, and also before the discovery of high ionization
species (important for line driven winds) in Be envelopes. This paper
reports some results of exploratory calculations which include the
combined effects of both line radiation and magnetic forces on rotating
stellar winds.

Attention is focussed on the equatorial plane of a steady-state
isothermal spherically symmetric expanding rotating wind of infinite
electrical conductivity and zero viscosity. It is supposed that
magnetic field lines originating at the stellar equator are confined to
the equatorial plane, so that a one-dimensional model may be constructed
in that plane.

The azimuthal momentum equation and the centrifugal and magnetic
terms in the radial momentum equation are treated according to the
formalism of Weber and Davis (1967). In the radial equation, the force
due to line radiation is represented through its dependence on the
hydrodynamical density and velocity variables, in the manner derived by
Castor, Abbott, and Klein (1975; CAK). The continuity and radial
momentum equations are solved simultaneously by a Henyey relaxation code
described in detail by Barker (1979), for a variety of combinations of
parameters such as mass loss rate, stellar angular velocity, and photospheric radial field strength. Two sets of numerical calculations have
been performed: one using an expression for the line radiation force
such that the parent non-rotating non-magnetic wind accelerates steeply

485

M. Jaschek and H.-G. Groth (eds.), Be Stars, 485–488.
Copyright © 1982 by the IAU.

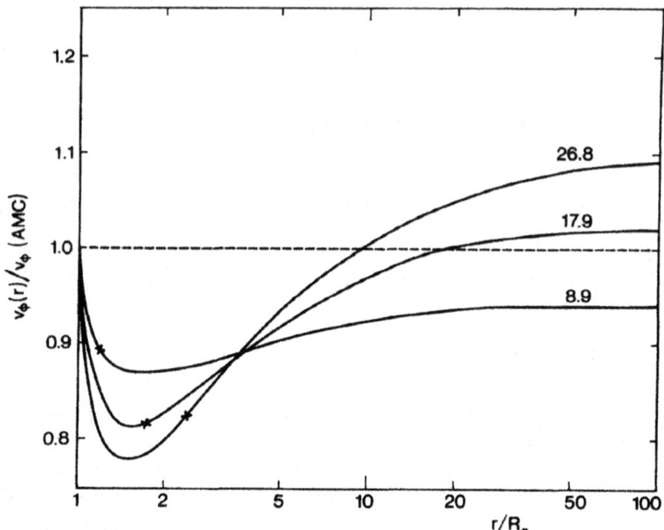

Figure 1. Azimuthal velocity v_ϕ in the equatorial plane of magnetic winds arising from a star of 13 M⊙ and 6 R⊙ rotating at 62% of the break-up velocity, with $\dot{M}$ = 5 x 10^{-9} M⊙ yr^{-1}; v_ϕ is shown relative to the value from a non-magnetic 1/r conservation law. These models result from a hard (steeply accelerating) line radiation force law; curves are labelled with the photospheric radial field strength, and crosses mark the Alfvén radius.

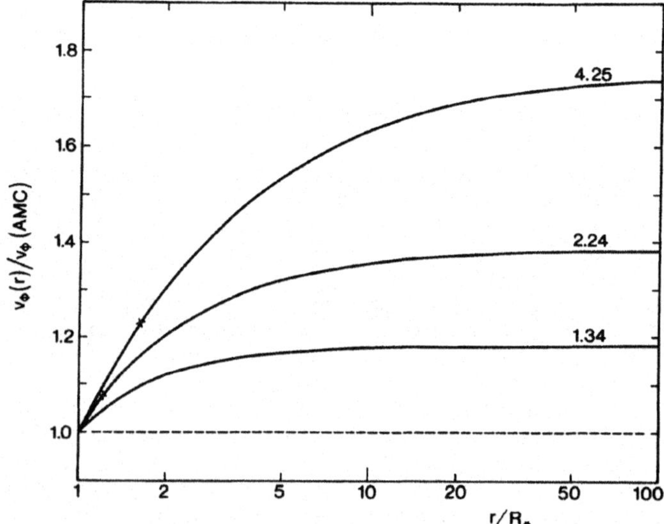

Figure 2. As in Figure 1, but for a semi-soft (gradually accelerating) line radiation force law.

from the photosphere (a hard wind) as described by CAK; the second
using a new expression for a gradually accelerating wind. The shallow
wind form approximates the soft wind law of Abbott (1977) and is
designated semi-soft as defined by Barker (1979).

The calculations were carried out in particular for a rapidly
rotating B1V star with a Be-like mass loss rate of 5 x 10^{-9} M⊙ yr^{-1},
and it was found that the hard and semi-soft winds respond in character-
istically different ways to the effects of magnetic rotation. Thus far
only extremely weak radial photospheric fields have been considered,
with strengths far below the current observational limits of 1-200 Gauss
(Landstreet, private communication) for Be stars.

In hard winds, the radial velocity law is not substantially
affected by global radial stellar fields below 30 Gauss, but the
azimuthal velocity is modified in a remarkable way. At small radii the
circular velocity is <u>reduced</u> below the 1/r value corresponding to
conservation of angular momentum in the fluid particles alone. Figure 1
shows the circular velocity in the hard magnetic winds--notice especially
the sharp drop in v_ϕ (relative to 1/r) near the photosphere. This
result is totally unexpected and is contrary to the prevalent intuitive
picture that (by analogy with the solar wind) the rotation velocity in
a stellar wind is necessarily enhanced by magnetic corotation.

Figure 2 shows the corresponding results for semi-soft winds.
When the line radiation force produces gradual radial acceleration,
the azimuthal velocity is indeed magnetically enhanced--by almost a
factor of two even for exceedingly weak stellar field strengths. The
radial velocity structure is also appreciably modified in these winds.

Numerical difficulties were formidable even for the current weak-
field models; development of the code is in progress to investigate
the cases of stronger magnetic fields and higher rotation rates.

This work was supported by the National Science and Engineering
Research Council of Canada through a grant to Dr. J.M. Marlborough and
by the National Science Foundation through grant NSF 78-20135 to
Dr. P.S. Conti.

REFERENCES
Abbott, D.C. 1977, Ph.D. Thesis, University of Colorado.
Barker, P.K. 1979, Ph.D. Thesis, University of Colorado.
Castor, J.I., Abbott, D.C., and Klein, R.I. 1975, Astrophys. J.,
          195, p.157.
Limber, D.N. 1974, Astrophys. J., 192, p.429.
Saito, M. 1974, Publ. Astron. Soc. Japan, 26, p.103.
Weber, E.J., and Davis, L. 1967, Astrophys. J., 148, p.217.

DISCUSSION

<u>Endal</u>: Is the effective moment-arm of your winds given by the radius
at the Alfven point, as in the Weber-Davis solar wind model?

<u>Barker</u>: The Weber and Davis theory is formally correct only in the
case of <u>zero</u> photospheric mass loss. These calculations were performed
using a corrected version of the theory, in which the solar-wind
result you quote is no longer valid.

<u>Harmanec</u>: Could you estimate how strong the magnetic field must be to
obtain detectible difference between magnetic and non-magnetic wind?

<u>Barker</u>: Even 5 Gauss fields have substantial effects on the wind
velocity structure; calculations are planned to investigate the resultant observable line profile changes.

GROSS STRUCTURAL PATTERN FOR THE ATMOSPHERES
OF Be, AND SOME CLOSELY RELATED, STARS

V. Doazan, R. Stalio, R.N. Thomas
Observatoire de Paris, Osservatorio di Trieste, Institut
d'Astrophysique, Paris

Summary : Abstracted comparison of Struve's 1942 quasi-empiri-
cal model of the shell-atmosphere of Be and similar stars, based on visual data and quasi-radiative equilibrium, to that demanded by current
visual+nonvisual data, requiring nonradiative+mass fluxes characterized
by variability and individuality.

In 1941-3, Edlen's identification of the superionized coronal lines,
and Redman's demonstration of the mass-dependence of chromospheric line-
widths, destroyed RE modeling for the observed extended-atmosphere of
the Sun. Just then, Struve (1942) abstracted existing visual-spectral observations of a variety of hot stars, mainly peculiar, with prototype the
Be, into a universal, ad hoc, cool-structured pattern of their inferred
extended-atmospheres, then called "shells". The observed visual subionization, relative to photosphere, implied the cool, hence quasi-RE but not
He, shell. The 3-phase behavior of Be stars suggested variability for
some shell properties. Observation of displaced, sometimes multiple, line-
components in a variety of stars suggested that the shell could have
multiple regions, some of which moved, differentially. Struve proposed
a two-region structure, by analogy with the pre-1942, RE, "solar-shell":
a lower, quasi-static region, the "chromosphere" ; an upper, differen-
tially-expanding region, the corona. Such shells differed from star to
star, even from sg to MS, not in their structural pattern, just in their
visual continuous opacities. The shell-regions merged continuously with
those above and below ; there were no nonradial asymmetries in opacity,
no opaque areas adjacent to one where the photosphere could be seen.
Struve conjectured that all hot stars, not only peculiar ones like Be
and Sun, had such extended-atmospheres at some phases ; thus that the
ability to produce such a pattern was a rigid test of the capacity of
modern atmospheric modeling. He further emphasized that it was premature,
at that epoch, to speculate on the cause of such extended-atmospheric
pattern until it was established, observationally, what modifications
must be made in the proposed pattern. While he suggested a rotationally-
unstable mass-ejection might, in some stars, be one possible cause, he
subordinated its general importance. Those rough estimates he made of

489

theoretical ejection-velocity focused on radiative-acceleration ;hence,
his spherically-symmetric model.

At this epoch, we are better-equiped than was Struve to ask for
modifications in his proposed pattern, from the farUV, x-ray, farIR,
radio and more high-resolution visual observations. Ours is also similar
to the 1942 epoch, in that now the hot-star shell, as then the solar-
shell, cannot be forced into an RE pattern ; and we try to decide between
nonRE but thermal extension, and nonHE nonthermal extension to represent
the atmospheric density gradient : empirically, not speculatively. So we
summarize the change in Struve's shell-pattern suggested by modern data,
especially nonvisual, but still emphasizing the equal importance of the
visual, subionization data to the farUV, superionization data in giving
the complete extended-atmospheric pattern. At Struve's epoch, the sub-
ionized data characterized the whole shell ; at our epoch, only the post-
coronal, cool decelerated outermost shell regions. Then, the only superionized data were on WR symbiotic,novae, etc "similar-to-Be" stars ;
now, they are universal ; and they refer to the inner shell regions. We
still focus on the "Be+similar" stars, and on the striking thermodynamic
change for these stars, as for the 1942-Sun, being the universal obser-
vational-existance of nonradiative-energy and mass fluxes instead of their
pre-1942 speculative-exclusion.

We summarize the changes in Struve's shell-structural pattern by
region, identifying the data suggesting such change. Details, and references, will appear in Underhill and Doazan (1981), Hack and Stalio
(1982), Thomas (1982).

Struve's solar-analogy chromosphere had 2 roles : (1) provide that
continuous, thermal (quasi-static) transition from the photosphere which
preserved high-enough shell-densities to produce the observed Balmer-line
emission : (2) to produce sufficient cooling from photospheric T_e to
permit FeII absorption lines of the observed strength, (1) and (2) being
capable of time-variation. Role (1) is now replaced by that of a hot
chromosphere, with RE failing at a maximum height, and minimum density,
fixed by the observed size of the mass-flux. For F_M of 10^{-5}-10^{-9} ; log=4;
$R=10R_\odot$; T=25,000K ; these minimal densities correspond to particle concentrations $3(10^{13}- 10^9)$. Given a concentration 4.10^{14} at $\tau_5\sim1$, nonradiative heating starts no more than 2-10 density scale-heights, or 0.0008-
0.004 radii, above $\tau_5\sim1$. Thus the post-1942, hot chromosphere,begins
low. The nonradiative energy flux,which is demanded by the observed superionization, can only decrease these beginning heights ; and it is its properties, unfortunately not known other than speculatively,which fix the
outward increase of Te. The only data give a lower limit on the T_e(max)
somewhere above these minimal heights : some $x.10^6$ (x>1) : from x-ray and
the superionization observations. Role (2) of Struve's 1942 chromosphere
is, in 1981, transferred to a post-1942 post-corona, where expansion velocities $\lesssim$100km/s and FeII exist.

Struve's solar-analogy corona had 2 roles : (1) to move,differential-
ly, at velocities up to $\gtrsim$ 150km/s, sufficient to produce either or both
of strongly-displaced absorption lines and P Cyg profiles ; (2) maintain
sufficiently-large densities that such profiles (1) can be observed ; (3)
have sufficient range of excitation/ionization to produce lines from
HeII to FeII but to remain "cool". Role (1) is now expanded to permit V

up to 2-3000km/s. The densities of Role (2) are maintained by a combination of thermal, HE suport as T_e increase from T_{eo} (RE, HE) up through $x.10^6$ (x>1) ; giving way to the nonthermal support corresponding to the velocity-distribution of (1) ; with the transition-height from thermal to nonthermal support being highly individualistic, fixed by a combination of $(T_e,g,\text{rotation},\text{mass-flux},\text{radiative-acceleration})$ effects. Role (3), expanded to include lines from FeII through solar-coronal ions and soft x-rays, unrestricted to being RE and cool, is now divided between 3 regions : the post-1942 hot corona, containing T_e (max), and two post-coronal regions of lesser T_e.

The post-Struve, post-coronal, modern Stromgren HII region has 2 functions : (1) to cool the coronal gas to values where the HII-HI transition occurs ; (2) to decelerate the gas to V $\stackrel{<}{\sim}$ 100km/s relative to the photosphere. To accomplish these functions, the gas must cool, radiatively outwards except for a rise and fall at the shock-front necessary to decelerate the wind to the V $\stackrel{<}{\sim}$ 100km/s observed in FeII and H. Note that a lower-limit on the pre-deceleration wind-velocity is the coronal thermal velocity. Note that the properties of this modern Stromgren HII region are not fixed by the RE photospheric radiation field, as was the classical one, but by the properties of the combined, nonradiatively-heated, hot chromosphere-corona, plus those of the wind. Further, note that the decelerating medium cannot be the normal ISM, which would decelerate in distances $\sim$ psc, because of the visual, farIR, and radio data which fix maximum radii of this HII region $\sim$100 photospheric radii. One recalls McLaughlin's characterization of Be stars as "little planetary nebulae". Then, note that the mass-flux variability shown by Be stars, and the idea that one requires a variable mass-flux to produce a planetary nebula (Kwok, et al, 1978) accord in this picture. For empirical modeling of these low-density regions, one must depend heavily on data from the Bep, and symbiotic stars which show both forbidden lines of varying ionization and the largest IR and radio excesses.

The post-Struve, post-coronal, HI region is subject to debate as to whether the gas alone ---from all the interior regions, hot and cool--- suffices to produce the IR and radio data, or whether dust regions are also necessary. Again, it is the Bep and symbiotic stars, among the "Be and similar" star group which provide the most information. At present, we do not have sufficient data to say much except that these regions, for at least some stars, extend out some 10^3 photospheric radii.

We present several slides and calculations illustrating characteristics of the proposed pattern, which there is not enough space to reproduce in this summary.

REFERENCES

Edlen, B., 1942, Zs. Ap. 22, 30
Kwok, S., Purton, C.R., Fitzgerald, P.M., 1978, Ap. J. 219, L125
Redman, R.O., 1942, MNRAS, 102, 134, 140
Struve, O., 1942, Ap. J. 95, 134
NASA-CNRS monograph series, Nonthermal Structure of Stellar Atmospheres: 1981 et seq.
Hack, M., Stalio, R., Cataclysmic Stars, 1982
Thomas, R.N., Stellar Atmospheres, 1982
Underhill, A.B., Doazan, V., B and Be Stars, 1981.

THEORETICAL SURFACE BRIGHTNESS DISTRIBUTIONS AND CONTINUUM POLARIZATION
OF RAPIDLY ROTATING B STARS

George Sonneborn
Department of Astronomy, The Ohio State University
Columbus, Ohio 43210 USA

In order that a rotating star be in hydrostatic and radiative
equilibrium, the effective temperature of the stellar atmosphere must
decrease between the star's poles and its equator. Wavelength differences in this "gravity darkening" have been suggested to account for
the observed differences in the rotational broadening of lines in the
ultraviolet and visual spectral regions (hutchings, 1976; Sonneborn
and Collins, 1977). This paper investigates the problem, using
detailed model atmospheres to examine the surface brightness distribution of a rapidly rotating star.

The intensity distribution on the projected disk of a rapidly rotating B2 V star was computed for models with w=0.9 and 0.95
(V_{eq}=340km/sec and 384km/sec, respectively; $w=\omega/\omega_{crit}$) at wavelengths of 1325Å and 4400Å. The specific intensities were calculated
for a grid of 1000 surface points from a model of the star's global
atmospheric structure. The non-LTE, line-blanketed model atmospheres
were computed with the ATLAS6 code (Kurucz, 1979). The models assume
rigid rotation and use a Roche model for the shape distortion.

The isophote contours on the rotationally distorted star are shown
in Figure 1. Important features of these results are: (1) The isophote
topology in the visual and UV is nearly identical (cf. 1b and 1e). The
severity of the UV "gravity darkening" compared with that in the visual
is clearly demonstrated, since the UV contour interval is twice that in
the visual. (2) The brightness distribution is a complex function of
inclination. (3) The magnitude of the "gravity darkening" is a steep
function of w (cf. 1b and 1d).

Since these brightness distributions are not spherically symmetric
one expects the continuum to exhibit a net linear polarization due to
electron scattering. The net polarization of rapidly rotating main-
sequence stars (B0-B5) was computed using the methods described by
Harrington and Collins (1968) and Collins (1972). The results are
shown in Figure 2 for models with w=0.95, where the net degree of polarization is defined as $p = (I_\ell - I_r)/(I_\ell + I_r)$. I_ℓ and I_r are parallel

493

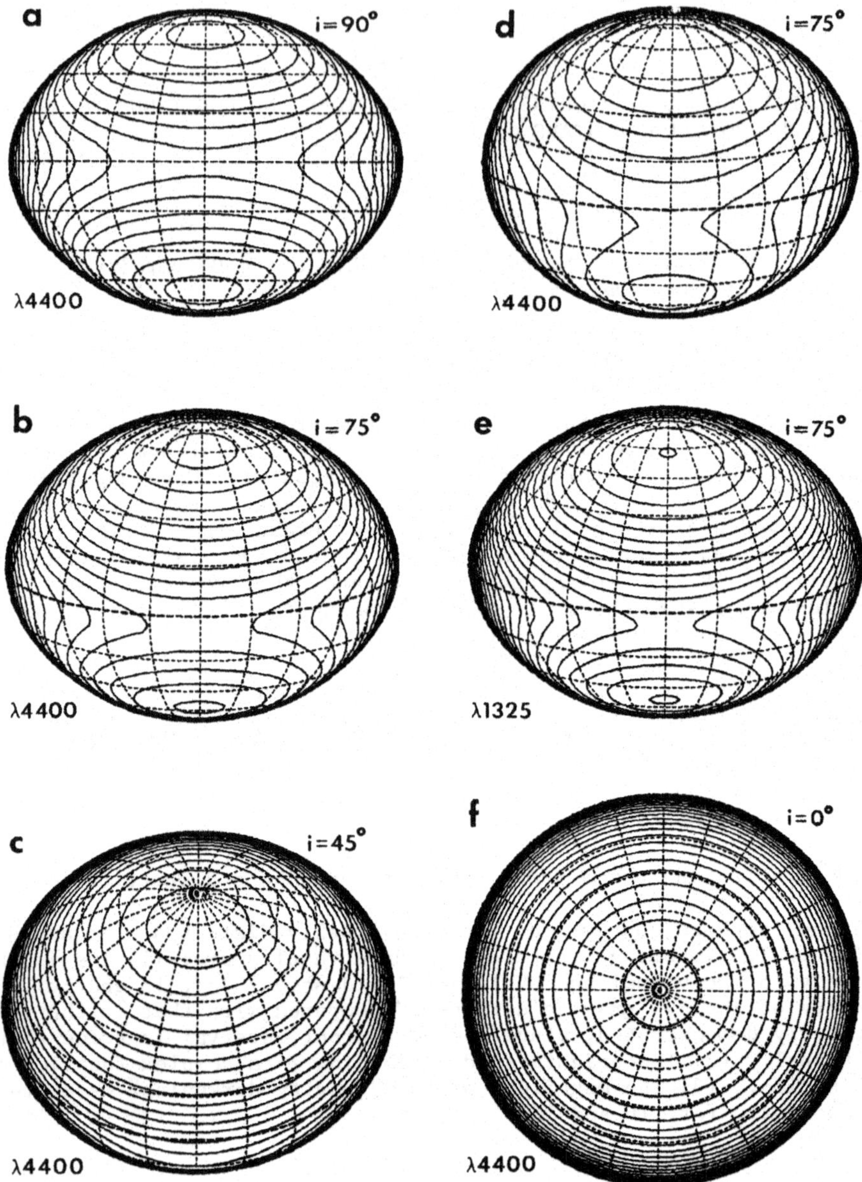

Figure 1. - Theoretical surface intensity distributions at 4400Å
(except 1e, where λ=1325Å) for a B2 V star with w=0.95 (except 1d,
where w=0.90) are shown for the indicated inclinations. The contour
interval is 0.05mag at 4400Å and 0.1mag at 1325Å. A latitude -
longitude grid with 15° intervals is also shown (dashed-lines).

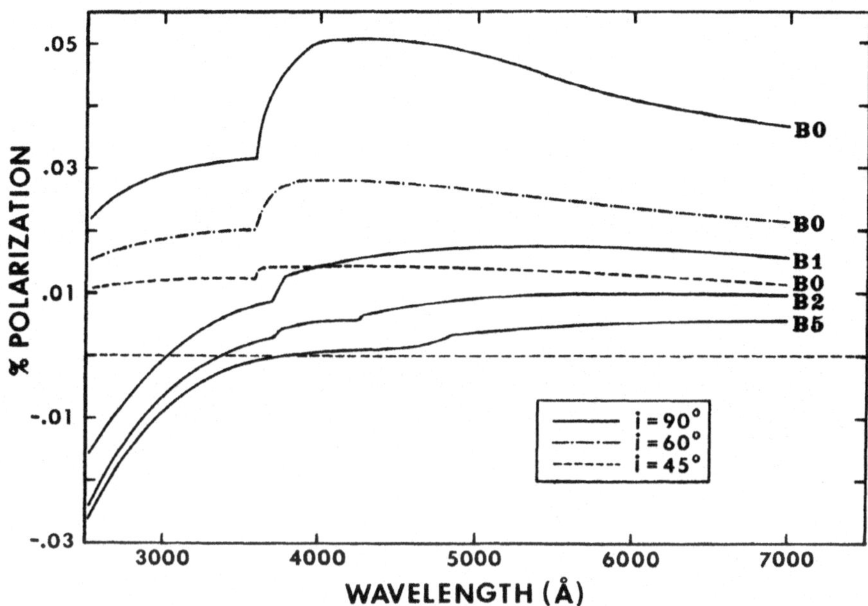

Figure 2. - The net linear polarization in the visual and near ultraviolet continua is shown as a function of wavelength for several rotation models, all with w=0.95. Zero polarization is shown by the thin
dashed-line. The Balmer discontinuity coicides with the discontinuities in the B0 data.

and perpendicular, respectively, to the rotation axis. Figure 2 shows
the continuum polarization to be up to a factor of 100 less than that
found by Harrington and Collins (1968) for a gray atmosphere. Furthermore, p decreases into the UV and becomes negative. It is found that
the UV continuum of rotation models viewed equator-on is polarized by
-0.7% (B0, w=0.95) to -2.5% (B5, w=0.95). The large UV polarization is
not surprising, in view of the extreme asymmetry of the UV surface
brightness distribution. The extremely small visual polarization found
for these models implies the observed continuum polarization in Be
stars is not produced in the stellar photosphere.

The author wishes to thank the American Astronomical Society and
the International Astronomical Union for financial assistance to
attend Symposium No. 98.

Collins, G.W.,II.: Astrophys. J. 175, pp.147-156.
Harrington,J.P., and Collins,G.W.,II.: Astrophys. J. 151, pp.1051-1056.
Hutchings,J.B.: Pub.Astron.Soc.Pacific 88, pp.5-7.
Kurucz,R.L.: Astrophys. J. (Suppl.) 40, pp.1-340.
Sonneborn,G. and Collins,G.W.,II.: Astrophys. J. 213, pp.787-790.

DISCUSSION

<u>Poeckert</u>: If you had .1 A spectral resolution, what magnitude of polarization would you predict one might observe in the line?

<u>Sonneborn</u>:Unfortunately we have not yet looked at polarization across lines.

<u>Endal</u>: Are the Sackmann-Arand models you use in your calculations ZAMS models?

<u>Sonneborn</u>:Yes.

<u>Metz</u>: 1. A pole-on star is brighter and cooler as well compared to a star observed edge-on. Is that reproduced by your model?
2. Normally Be stars do not rotate at break-up velocity. Therefore the difference of polar and equatorial flux (according to von Zeipel) is too low to produce a relevant net polarization. Isn't that in contradiction to observations?

<u>Sonneborn</u>: 1. The discussion earlier (after Dr. Poeckert's review paper) on this subject concerned the effects of rotation <u>and</u> an equatorial disc. For a rotating star viewed pole-on will be bluer than the same mass star which is not rotating.
2. This is one of the points of this paper. The intrinsic polarization of a rotating star is much less than that observed in Be stars, implying that the observed polarization is produced completely in the circumstellar envelope.

ON THE BALMER PROGRESSION PHENOMENA IN Be STARS

Ryuko Hirata
DEPEG, Observatoire de Paris, Meudon,
on leave from Department of Astronomy, University of Kyoto

Typical Balmer progression phenomena in Be stars are characterized by
$dv_n/dn<0$ when $v_n<0$(negative progression) and $dv_n/dn>0$ when $v_n>0$(positive
progression), where v_n is the observed radial velocity of the Hn line.
We examine the conditions, under which the observed progression occurs.
The shell absorption-line profile, $R_n(v)$, expressed in depth, is written
as
$$R_n(v)=\sigma(v)\{1-\exp[-\tau_n(v)]\},$$
where $\sigma(v)$ is the fractional area of the stellar disk which is screened
by the velocity zone of v, and $\tau_n(v)$ is the mean optical thickness of
the velocity zone for the Hn line(Kogure et al. 1978). We define v_σ and
v_τ as the line-of-sight velocities at which $\sigma(v)$ and $\tau_n(v)$ take respective maximum. Then, the negative and positive progressions could occur
for $v_\tau<v_\sigma<0$, and $v_\tau>v_\sigma>0$, respectively. We examine various velocity fields
and geometries from this point of view. Numerical simulations have been
also performed in order to check the validity of our simple consideration
in the three-dimensional envelope, and also to examine the emission effect.
     We obtained the following conclusions:
1) The cyclic(positive$\leftrightarrow$negative) progression phenomenon observed in 48
Lib and ζ Tau can be most easily explained by the precession of the elongated disk. Such a model can also account the characteristics of the observed variation of shell-line profiles, in addition to the V/R variation.
2) In the case of axially symmetric envelope, the observed negative(positive) progression could occur only in the decelerating outflow(accelerating inflow), only if the line optical thickness decreases outwards.
The latter condition requires special temperature structure. It is impossible to account for the negative progression observed in Pleione's shell
episode by Limber's(1969) model.
     The full paper will be published elsewhere.

REFERENCES

Kogure,T.,Hirata,R.,and Asada Y.: 1978,Publ.Astron.Soc.Japan,30,pp.385-407.
Limber,D.N.: 1969,Astrophys.J.157,pp.785-797.

ON THE SPECTRUM OF THE HERBIG Be STAR HD 200775

B. Baschek, M. Beltrametti, J. Köppen, G. Traving
Institut für Theoretische Astrophysik
Im Neuenheimer Feld 294, 6900 Heidelberg

ABSTRACT

On the basis of high-dispersion spectra (I. Appenzeller, C. Bertout)
and other observational material HD 200775 has been reanalyzed using
l Her as a comparison star. By comparing the depths in the wings of
$H\gamma$ and $H\delta$ and the equivalent widths of HeI-lines with predictions from
model atmospheres we find for both stars $T_{eff} = 17000 \pm 1000$ k (in the
scale of Kurucz' models, 1979) and $\log (g\ cm\ s^{-1}) = 3.6 \pm 0.3$
corresponding to a spectral type of B3V. With a colorexcess $E_{B-V} = 0.56$
and the standard interstellar reddening also the continuum energy
distribution of HD 200775 corresponds to the above photospheric parameters if one allows for some circumstellar emission. However, the
interstellar absorption features (4430 A and 2200 A) are of unusual
weakness. The strength of the $H\alpha$ emission requires either a nonthermal
source of energy or a substantial rate of photo ionization from the
level n = 2. We tend to conclude that for this second reason the
dilution factor in the HII-region cannot be less than 10^{-2}, a conclusion
which is in accord with the finding that the radio emission at $\lambda = 2$ cm
is less than 0.3 mJ which also indicates a very compact HII-region.

M. Jaschek and H.-G. Groth (eds.), Be Stars, 499.

SPECTROSCOPIC INVESTIGATIONS OF HERBIG-Ae-Be-STARS

Ulrich Finkenzeller
Landessternwarte
Heidelberg-Königstuhl

ABSTRACT

"Herbig-Ae-Be-Stars" are assumed to be pre-main sequence objects of moderate mass with line emitting envelopes
of an unknown nature. From our present theoretical knowledge
it is not clear whether the physical structure of these envelopes is dominated by mass accretion or mass loss induced
by a stellar wind or radiation pressure effects. Radial velocities and remarks on peculiarities are given for the star
HD 200 775, which seems to represent a typical Herbig-Ae-Be-
star fairly well. A catalogue of about 60 supposed Herbig-
Ae-Be-stars is presented and comments, in particular on the
brighter members, are invited.

I. INTRODUCTION

At Landessternwarte Heidelberg high resolution spectroscopic work on Herbig-Ae-Be-stars (HBe's) is in progress.
Spectra of about 35 stars are already available and are currently being reduced. Though these objects are not 'classical' Be-stars, it is intended to give Be-observers some in-
sights into the spectral appearance and invite critical
comments on a preliminary Herbig-Ae-Be-star catalogue since
that list may include related objects as well.

According to model computations of star formation, for
mass values of 3-4 M_O the time scale of the accretion phase
of the protostellar envelope is equal to the core age at
which hydrogen burning begins (Larson 1969). Present theoretical considerations cannot predict whether the circumstellar envelope is affected by mass accretion processes or by
mass loss incuced by a stellar wind or radiation pressure
effects. In 1960 G. Herbig made a first discussion of 'Spectra of Be- and Ae-type Stars associated with Nebulosity'
(Herbig 1960). In a more recent paper it is demanded that

M. Jaschek and H.-G. Groth (eds.), Be Stars, 501–507.

the following criteria should be fullfilled by HBe's (Strom
et al. 1972):
   - presence of emission line spectra
   - association with regions of heavy obscuration
   - illumination of fairly bright (reflection) nebulosity.

II. HD 200775: A TYPICAL HERBIG-AE-Be-STAR.

        Since HD 200775 seems to be representative of the bulk
of HBe's this star was chosen to illustrate the work on radial velocities and to show spectral peculiarities which
seem to be present in other HBe's as well (see also contribution No. 69). 9.5 Å/mm coudé plates, obtained by Appenzeller and Bertout in June 1977 at the Observatoire de Haute
Provence, France, and 42 Å/mm plates from Calar Alto Observatory, Spain, were considered. In the latter case three
tracings were superposed in the computer to improve the signal to noise ratio (see Fig. 1).

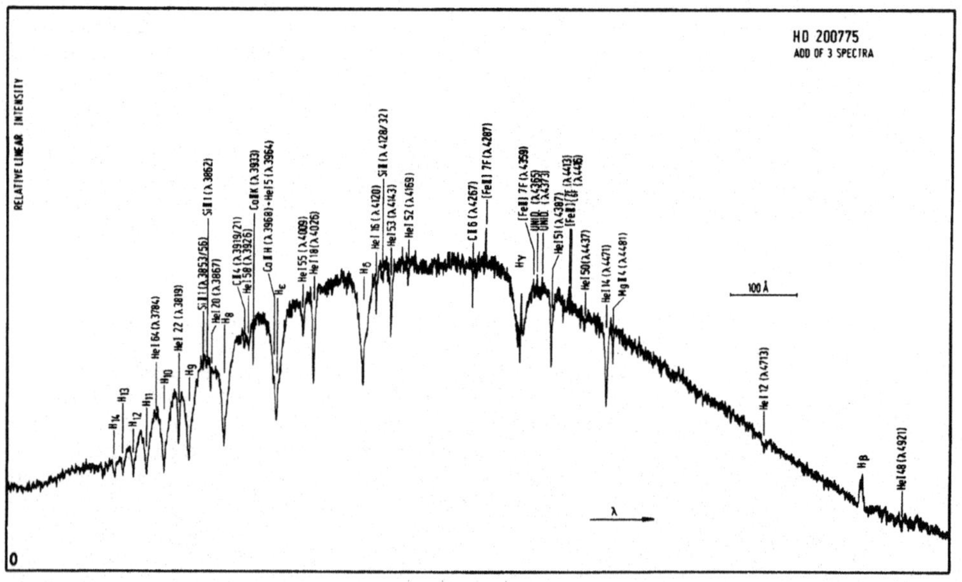

Figure 1. A spectrum of HD 200775, obtained from 3 42 Å/mm
spectra taken at Calar Alto Observatory, Spain, on June
30th, 1980. (Since only radial velocities and relative intensities are considered no corrections for the instrument
sensivity were applied).

TABLE I

Heliocentric Radial Velocities [km/sec] of HD 200775

Line	Plate W 6421 9.8 Å/mm		Plate N 740 42 Å/mm	
H_{12}	–		– 7 $\pm$ 30	
H_{11}	–		+ 2	
H_{10}	–		+14	
He I 3819	–		– 9	
H_9	–41 $\pm$ 7		–41	
He I 3867	–		–16	
H_8	(–66)		– 2	
Ca II K	–14	X	– 5	X
He I 3964	–		+26	X
H_ε	–10		+ 3	
He I 4026	–25	X	–10	X
H_δ	– 1		–16	
He I 4120	–16	X	+ 3.5	X
He I 4143	–		–10	
C II 4267	+ 2	X	+ 0	X
He I 4387	– 9	X	+13	X
He I 4471	+ 0	X	– 5	X
Mg II 4481	–10	X	– 8	X
He I 4713	–23		– 1	
Mean	–16.2 $\pm$ 6.0	–7.7 $\pm$ 4.5	–3.6 $\pm$ 3.1	+1.8 $\pm$ 4.3
[FeII] 4243.98	–11		–	
4276.83	– 4		–	
4287.40	– 9		+ 2	
4352.78	– 8		–	
4359.24	–		–	
4413.78	+ 0		+20	
4416.27	–14		– 3	
Mean	–6.3 $\pm$ 2.8		+6.3 $\pm$ 4.9	
H_γ Blue Peak	–119		–103	
Red Peak	+102		+120	
Central Abs.	0		+ 13	
H_β Blue Peak	118		– 60	
Red Peak	84		+111	
Central Abs.	– 8		+ 20	

The crosses indicate lines which are generally accepted for
system velocity determination (see for example A. Underhill,
The Early Type Stars, Dordrecht, 1966, p. 121). These values compare well with those of the forbidden lines.

The photospheric spectrum can be estimated as B3 IV - B3 V. The following points seem to be noteworthy:

- H_β and H_γ possess complex emission components with a central absorption. The emission on the red side is the stronger one. At H_δ and H_ϵ a disturbed red slope can be discerned. The Balmer series can be followed up to H_{14}. HD 200775 has extremely broad shallow absorption wings of about 80 Å total width at H_β.

- Most of the HeI lines show a disturbed blue wing. The perturbations are not restricted to triplet transitions but can be observed in singlet transitions as well.

- There are several lines of [Fe II] present, the strongest of which is [Fe II] 4287. Forbidden lines of other elements and ionization stages were not observed. Two sharp emission lines, probably originating in the envelope, could not be identified (4365.5 and 4373.1).

Since the structure of many lines is complex and asymmetric one should be cautious when assigning radial velocities. For that reason each line is listed separately. For the determination of the system velocity Balmer lines were not considered. The system velocity of -7.7 + 4.5 km/sec compares fairly well with literature values (Plaskett & Pierce 1931) who give the mean of -6.1 (1925), -1.8 (1927), +4.2 and +5.9 (1928) as +0.6 $\pm$ 1.8 km/sec.

Thus the visual and blue spectrum is dominated by features that appear rather static. Only the asymmetry of the He I lines may indicate mass loss, which has been determined by various authors and different techniques to be in the range of 10^{-7} $M_\odot$ yr^{-1} (Garrison 1978, Altamore 1980). Theoretical models should explain the strong, basically unshifted central absorption phenomenon of the Balmer lines. No prominent variations of the spectrum of HD 200775 have been found or are known in the literature.

III. OTHER HERBIG-Ae-Be-STARS

From the observational material available at present (35 objects out of about 65 are covered by spectra with $\approx$10 Å/mm dispersion) we feel that most of the HBe's are like HD 200775, i.e. they show in the lower Balmer lines a double peaked emission feature with a sharp central absorption. About 65% fall into that category. Another group ($\approx$ 15%) displays prominent P-Cygni profiles, e.g. BD +61°154 or HD 250550. A third group consists of stars which have an emission line rich spectrum very similar to T Tauri stars, e.g. T CrA or HK Ori.

TABLE II

LIST OF SUPPOSED HERBIG-Ae-Be-STARS

Star	α (1900)	δ (1900)	m_{pg}	References
Lk H$_\alpha$198	0^{h}06.1	+58°17'	15	(1)
BD +61°154	0 37.5	+61 22	10.6	(1)
BD +30°549	3 23.2	+31 05	9.5	(2)
XY Per A+B	3 42.9	+38 40	8.8-10.1	(3)
Elias 1	4 12.5	+28 04	15	(4)
AB Aur	4 49.4	+30 23	7.2-8.4	(1)
HK Ori	5 25.9	+12 05	4.4-12.5	(1)
BD +9°880	5 29.2	+ 9 58	9.3	(5)
T Ori	5 30.9	- 5 32	9.5-12.6	(1)
V380 Ori	5 31.6	- 6 47	9.7-10.3	(1)
BF Ori	5 32.4	- 6 39	9.8-13.4	(6)
RR Tau	5 33.3	+26 19	10.2-14.2	(1)
HD 37490	5 33.9	+ 4 04	4.37	(2)
HD 250550	5 56.2	+16 31	9.7	(1)
Lk H$_\alpha$208	6 02.1	+18 42	13.0	(1)
MWC 137	6 13.0	+15 19	11.2	(6)
Lk H$_\alpha$341	6 25.3	+10 37	13	(6)
HD 46060	6 26.0	- 9 36	8.6	(2)
Lk H$_\alpha$215	6 27.2	+10 14	10.7	(1)
HD 259431	6 27.6	+10 24	8.7	(1)
R Mon	6 33.7	+ 8 50	11.3-13.8	(1)
LH$_\alpha$25	6 35.2	+ 9 53	13.0	(1)
HD 52721	6 57.2	-11 09	6.40	(2)
Lk H$_\alpha$218	6 57.5	-11 17	12.2	(7)
ZCMa	6 59.0	-11 24	8.8-11.2	(1)
Lk H$_\alpha$220	6 59.6	-11 18	12.3	(7)
HD 53367	6 59.7	-10 18	7.0	(1)
CoD -44°3318	7 16.4	-44 24	10.4	(8)
HD 76534	8 51.5	-43 05	7.5	(2)
RCW 34	8 52.8	-42 43	11	(2)
RCW 36	close to RCW 34		11	(2)
Herbst 28	8 54.8	-43 03	11.5	(2)
HR Car	10 19.4	-59 07	8.2-9.6	(1)
Anon.	10 50.0	-59 55	10.5	(1)
GG Car	10 52.0	-59 52	9.1-9.5	(1)
AG Car	10 52.6	-59 55	7.1-9.0	(1)
HD 97048	11 05.3	-77 07	8.4	(6)
CED 110	11 03.4	-76 50	11.3	(9)
HD 97300	11 06.8	-76 04	9.2	(9)
Anon.	11 11.9	-77 14	11.5	(9)
HR 5999	16 01.9	-38 50	6.44	(10)
HD 150193	16 34.3	-23 42	8.6	(4)
Herbst	16 49.6	-40 06	10.5	(2)

TABLE II (continued)

CoD -42°11721	16 52.0	-42 33	11	(11)
KK Oph	17 03.9	-27 08	11.8-12.7	(6)
HD 163296	17 50.3	-21 56	6.8	(12)
Lk H$_\alpha$118	17 59.7	-24 16	12	(6)
Lk H$_\alpha$119	18 05.0	-24 17	13	(13)
MWC 297	18 22.4	- 3 55	11.0	(1)
VV Ser	18 23.7	+00 05	10 - 13	(6)
MWC 300	18 24.1	- 6 09	10 - 13	(6)
AS 310	18 28.0	- 5 03	10	(6)
R CrA	18 55.2	-37 06	10.0-13.6	(1)
TY CrA	18 54.9	-37 01	8.9-12.1	(6)
T CrA	18 55.0	-37 06	11.8-13.6	(1)
BD +40°4124	20 17.0	+41 03	10.6	(1)
BD +41°3731	20 20.8	+41 58	9.9	(1)
HD 200775	21 00.4	+67 47	7.4	(1)
V645 Cyg	21 36.4	+49 47	15.1	(14)
BD +65°1637	21 40.6	+65 39	11	(1)
Lk H$_\alpha$234	21 40.8	+65 39	11 - 13	(1)
BD +46°3471	21 48.7	+46 46	10.1	(1)
Sh -125	21 49.1	+46 27	10.5	(15)
Lk H$_\alpha$257	21 50.5	+46 43	13.5	(6)
Lk H$_\alpha$233	22 30.3	+40 08	14.5	(1)
HD 216629	22 49.4	+61 36	9.0	(16)
Sh -158	23 08.9	+61 06	10.8	(15)
MWC 1080	23 12.9	+60 18	13.0	(1)

(1) Herbig, G., 1960, Astrophys. J. Suppl. 4, 337
(2) Herbst, W., 1975, Astron. J. 80, 683
(3) Herbig, G., 1980, private communication
(4) Elias, J., 1978, Astrophys. J. 224, 857 & 453
(5) Penston, M. et al., 1976, Obs. 96, 22
(6) Herbig, G. and Rao, K., 1972, Astrophys. J. 174, 401
(7) Cohen, M., and Kuhi, L.V., 1979, Astrophys.J.Suppl. 41, 743
(8) Irvine, N., 1975, PASP 87, 87
(9) Rydgren, A., 1980, Astron. J. 85, 444
(10) Thé, P., and Djié, H. 1977, IAU Coll. 42, 137
(11) Glass, I., and Allen, D., 1975, Obs. 95, 27
(12) Polidan, R., IAU Symp. 70, 414
(13) Appenzeller, I., 1980 private communication
(14) Humphreys, R., Merrill, K., 1980, Astrophys. J. Letters, 237, L 17
(15) Carsenty, U., 1980, private communication
(16) Assousa et al., 1977, Astrophys. J. Letters 218, L 13

A catalogue of Balmer line profiles at ≈ 10 Å/mm for as
many HBe's as possible is in preparation and will allow discussions on a more quantitative and representative basis.

The accompanying list contains supposed HBe's and related
objects. The brightest members are HD 37490, HD 52721,
HD 53367, HD 76334 and HD 163296. It is hoped that future
research will take these objects into consideration.

References

Altamore, A., Baratta, G.B., Cassatella, A., Grasdalen,
    G.L., Persi, P., and Viotti, R.: 1980, Astron.Astrophys.
    90, 290

Garrison, Jr., L.M.: 1978, Astrophys.J. 224, 535

Herbig, G.H.: 1960, Astrophys.J.Suppl. 4, 337

Larson, R.B.: 1969, MNRAS 145, 271

Plaskett, J.S., and Pierce, J.R.: 1931, Publ. Dominion
    Astrophysical Observatory, V, 1

Strom, S.E., Strom, K.M., Yost, J., Carrasco, L., and
    Grasdalen, G.: 1972, Astrophys.J. 173, 353

DISCUSSION FOLLOWING THE PAPER PRESENTED BY U. FINKENZELLER

SWINGS: What is the origin of your catalogue, since
                 I recognize several B [e] stars, and at least
                 one symbiotic?

FINKENZELLER: Entries are from various sources. Among the
                 most important are Herbig's original work, the
                 Herbig-Rao Catalogue of Emission-Line stars
                 of the Orion Population, a paper of Herbst
                 from 1975 on the Vela R2 Association, and va-
                 rious private communications with G. Herbig
                 and I. Appenzeller.

SWINGS: The sample seems very heterogeneous: there are
                 objects at very different stages of evolution.

FINKENZELLER: I agree that, due to the selection criteria
                 applied, objects other than HBe's may be in-
                 cluded. However, it is hoped to cancel some
                 when all the spectroscopic material is avai-
                 lable and compared accurately.

THE CONTINUING SAGA OF THE Be STARS: A SUMMARY

Theodore P. Snow, Jr.
University of Colorado at Boulder

1. INTRODUCTION

This paper, written in summary of IAU Symposium 98 on the Be
Stars, contains an outline of the proceedings, organized according to
physical parameters of the stars, the circumstellar environment,
models explaining the data, and a summary of the outstanding problems.
Suggested future observations and needed theoretical work are
described.

At the end of a very full week, the participants in IAU Symposium 98 had heard some 70 contributed papers and 7 invited review
talks. All of the observed phenomena and a substantial variety of
theoretical interpretations were thoroughly discussed, and in the
process, many important points of agreement and of controversy were
emphasized.

A few important themes appeared repeatedly throughout the week.
One was that the Be stars, whatever they are, must not be studied in
a vacuum, without consideration at the same time of their astronomical
relatives, including especially the Oe stars and the A shell stars,
but also other classes of objects, such as O stars with stellar winds
and mass-exchange binaries, all of which may have important lessons
to teach us about the Be stars.

Another very important theme of the Symposium was that we all
must use caution in the interpretation of data; the Be star phenomena
are quite complex, and care must be taken not to confuse the issue
further by failure to fully recognize observational uncertainties.
Vorsicht!

While the Symposium was organized primarily by observing technique, this summary will follow a different course in reviewing the
proceedings. Hindsight permits the information to be rearranged
according to scientific aims, presenting a more cohesive and coherent

M. Jaschek and H.-G. Groth (eds.), Be Stars, 509–519.
Copyright © 1982 by the IAU.

view. Perhaps the outline of this summary could be taken as a suggested agenda for the next symposium on Be stars.

The review is organized as follows: it begins with a brief overview of the nomenclature and catalogs, followed by sections on properties of the stars; characteristics of the circumstellar envelopes; the
evidence for winds, coronae, and mass loss; the astounding variety of
information on time variability; models of the Be phenomenon; a discussion of related objects; a summary of the major remaining questions
about Be stars; and, finally, an outline of important future research.

The author apologizes in advance for any omission or oversight;
it was a very full Symposium, the subject is complex, and the time and
space for writing this review are short.

2. NOMENCLATURE AND CATALOGUES

Some discussion took place on the question of what we mean by the
term "Be star", and by other words used in describing related phenomena. While it proved difficult to agree on what a Be star is, a general consensus was reached that it is not an OB emission-line supergiant; nor a pre-main sequence Ae or Be star (the so-called Herbig Ae
and Be stars); nor a B[e] star, a peculiar Be star associate with nebulosity. At the same time, it was agreed that the Oe and A shell
stars probably represent straightforward extensions of the Be phenomena to higher and lower temperatures, and should be included.

Several associated terms, such as "shell star", "extreme Be star",
"pole-on", and others were discussed, and it was agreed that even
though we may not clearly understand the phenomena to which these terms
refer, we should at least agree on what we mean by them.

Substantial discussion on bibliographic organization led to the
adoption of an outline proposed by Koubsky, which will be followed by
Jaschek in the future in listing papers on Be stars in the Be Star
Newsletter. This outline will also be published in the Newsletter.

Active efforts on both the cataloging of Be stars and surveys
for new emission-line objects were described. Egret and Jaschek have
completed a new general catalog of some 1100 Be stars, which will be
available from the Stellar Data Center, and Ducati described a catalog
of the emission-line strengths that he has compiled. Survey work for
faint new Be stars was discussed by MacConnell and Cardon.

3. PROPERTIES OF THE STARS

Whatever causes the Be phenomenon, it must have something to do, at a very fundamental level, with the stars themselves. The circumstellar envelopes must some how arise from the stars, and it is appropriate that we begin by compiling our information on them. Quite possibly more time could have been devoted to this very important topic.

On the subject of spectral classification, Slettebak in his review outlined the general distribution of spectral types, and in later contributions, Egret and Jaschek, Hubert-Delplace, and others discussed statistical properties of the Be stars as a function of spectral type. An important question came up regarding the effect of rotation on classification, with Slettebak suggesting that high values of v sin i can result in assignment of a star to a later class, by as much as 2 or 3 subclasses. It is evident that spectral classification of Be stars can be very complicated, due not only to rotation, but also to the presence of optically thick circumstellar envelopes, which can completely hide the underlying star, as reported by Plavec et al. and others.

The fundamental properties of the stars, such as luminosities, masses, radii, and effective temperatures, were not discussed extensively, except for a number of papers on flux distributions, such as the reviews by Mendoza and Houziaux, and contributions by Divan et al.; by Zorec and co-workers; and by Plavec et al.; and reports on luminosity indicators by Kozok and Zeuge. A particularly important problem was stressed in the reviews by both Marlborough and Poeckert, who pointed out that recent far-ultraviolet data indicate serious errors in the assumed extreme ultraviolet flux distributions of early-type stars. This is significant because models of circumstellar envelope radiative transfer have had to rely on predicted stellar Lyman continuum fluxes, and these now appear to be in error.

Equatorial darkening due to rapid rotation, an idea whose possible importance was brought to the fore at the time of the previous IAU Symposium on Be Stars (in 1975), has now begun to receive the attention it deserves. If sufficiently complete and accurate models of line profiles as a function of rotation and stellar inclination angle can be developed, we will have in our hands the ability to untangle the confusion between projection and true rotation effects that pervades all Be star observations to date. Papers presented by Marlborough (review), by Ruusalepp, and by Sonneborn showed that progress is being made, although simplifying assumptions of the models leave us short of our ultimate goal.

The subject of pulsational instabilities in Be stars was an exciting new ingredient brought forth at the Symposium. While it has been recognized for some time that many of the Be stars lie near the

β Cephei instability region in the H-R diagram, pulsations have not
generally been considered as a contributing cause for the Be phenomenon. Papers presented by Baade and by Bolton, in which probable
β-Cephei type pulsations in Be stars were reported, force us to
consider this possibility seriously. As Prof. Kippenhahn noted in a
comment, in a rapidly rotating star, there is a wide range of atmospheric conditions from pole to equator, and it is likely that in some
regions in some Be stars, the instabilities associated with stellar
pulsations may be encountered.

The evolutionary status of the stars was not widely discussed.
The review by Slettebak, and reports by Mermilliod and Endal stressed
that the Be stars are on or very near the main sequence, and the
latter author discussed the effects of rotation on stellar structure,
but clearly this is an area that needs further work.

4. CIRCUMSTELLAR ENVELOPES

The presence and nature of circumstellar envelopes is, of course,
the principal characterisitic of Be stars and by far the greatest
number of reports at the Symposium had to do with the observational
properties of the envelopes. Here again great caution must be exer-
cised, because of the unknown geometry of the circumstellar material,
and because of the difficulty in separating effects due to the envelope from those due to the star.

Confusion with stellar emission has the greatest impact on
observations of the flux distributions of circumstellar envelopes,
because it can be exceedingly difficult to ascertain the intrinsic
continuum flux distribution of the underlying star. Things are
complicated even more by the problem of determining the interstellar
extinction for a star that may have intrinsic reddening due to Paschen
continuum emission or an infrared excess.

Despite these difficulties, some clear results were shown.
Several authors, including Mendoza and Houziaux in their review
papers, referred to the effects of flux redistribution by the circumstellar envelopes, which tend to produce ultraviolet deficiencies (an
effect illustrated statistically by Zorec et al.) and infrared excesses through electron scattering processes. The Paschen and Balmer
continua and discontinuities were discussed by Divan and co-workers,
whose BCD spectrophotometric index was shown to be a good means of
distinguishing between the stellar and the circumstellar Balmer
jumps; and by Chkhikvadze, who discussed the different properties of
the Paschen continuum in Be and shell stars.

The most obvious manifestation of the presence of circumstellar
envelopes is to be found in spectroscopic data, which reveal the
emission and absorption lines that qualify stars as members in the Be

or shell-star class. The majority of papers in the Symposium dealt
with spectroscopic observations of the envelopes, either of emission
lines, shell absorption features, or both. Andrillat and Fehrenbach
presented illustrations of the fine H_α emission-line profiles obtained
electronically at Haute Province, and showed the importance of using
high spectral resolution, revealing the presence in several cases of
heretofore undetected narrow absorption components. Barker presented
results from his own extensive program of Be-star spectroscopy, and
Andrillat et al. showed rather conclusively that the Oe stars are
similar spectroscopically to the Be stars.

In addition to these survey projects, a large number of analyses
of individual stars were presented, too numerous to be mentioned here
in any detail. A note of caution was raised, particularly in comments
by Doazan, about the danger of analyzing or comparing data obtained
at different times.

One of the major themes of the Symposium was the question of
non-spherical symmetry of the circumstellar envelopes. Several
papers were presented, some based on samplings of numbers of stars,
and others based on detailed analyses of individual objects, all
supporting or at least consistent with the classical conclusions that
line and continuum emission from the envelope are both enhanced in
stars seen at low polar inclinations to the line of sight, while
shell absorption is enhanced in stars seen nearly equator-on. Coyne,
in his review of polarization, and Poeckert, in his subsequent summation of models for Be stars, showed how polarization measurements
also mandate an equatorial concentration of the circumstellar
material.

New forms of evidence were presented, from all parts of the
electromagnetic spectrum, for dense, relatively cool disks around Be
stars. The arguments for this include infrared data, shown by Persi
et al. to be consistant with an expanding disk; by Rappaport, who in
his review of X-ray emission from binary companions of Be stars
showed that the X-ray fluxes in some cases are best explained by
assuming the presence of a thick equatorial disk with low expansion
velocity; and by Poeckert, who illustrated very successful models in
which Be stars have disks sufficiently dense and cool to contain
neutral gas. Further evidence supporting this picture emerged from
the contrasting mass-loss rates reported from infrared data by Persi
et al. and from ultraviolet line profile analysis by Snow, which can
only be reconciled if the observed regions are entirely distinct from
each other, one being a high-density, low-velocity region, and the
other being a low-density, very hot, rapidly-expanding zone.

Metz, in contrast, argued that polarization could arise by scattering the non-spherically symmetric flux distribution from a rapidly-
rotating star through a spherically symmetric shell; and Doazan,

Stalio, and Thomas presented arguments in support of a spherically
symmetric geometry to account for all the observed phenomena.

5. WINDS, CORONAE, AND MASS LOSS

At the last Symposium, six years before this one, the first
direct evidence for the general existence of mass loss from Be stars
was just becoming available, (even though the idea, and lots of
indirect evidence for it, has a much longer history). At this meeting
mass loss was treated as an established phenomenon, and the discussions centered more on its nature, its relationship to the stationary
material in the circumstellar envelopes, and the rates of mass loss.

Marlborough reviewed the ultraviolet spectroscopic evidence for
winds, Snow showed that nearly all Be stars in his survey of nearly
20 have ultraviolet wind indicators, and several authors, notably
Peters, Paterson-Beeckmans, and Hirata, added reports on winds in
individual stars.

The important question of whether non-thermal heating is present
was discussed in the review of Marlborough and in contributions by
Doazan et al., by Henrichs, and by Peters, all of whom showed that
many Be star winds are characterized by highly-ionized species, such
as CIV and NV, which cannot be produced in radiative equilibrium with
the stellar photospheres. Unfortunately, no information was available
on the presence or absence of soft X-ray emission from Be stars, a
universal indicator of coronal gas in the winds from O stars and B
supergiants.

The mass-loss rates were summarized by Snow, who added values
derived from nearly 20 stars observed in the ultraviolet. His
results, the survey of lower limits on mass-loss rates by Franco et
al., and the values reported for individual stars by Hirata and by
de Freitas Pacheco, all were in strong contrast with the rates
inferred from infrared excess emission by Persi et al., a discrepancy
that probably can be explained by concluding that the ultraviolet
absorption and infrared emission form in different locations.

6. TIME VARIABILITY

Awareness of time variability in Be stars no doubt reached an
all-time record level during this Symposium. Most authors of observational papers were quick to point out either the presence or the
possibility of variations, and if they did not, almost invariably
someone in the audience rectified the omission.

A great amount of effort has gone into establishing the frequency
of occurrence and the timescale of variability. This is difficult

business, especially in cases where historical data from the litera
ture are used. Because of uncertainties in the data and the incom-
pleteness of the time coverage, it is often possible to find variations ·or periods of variation that may not actually be well-founded.

Photometric variability was summarized in a statistical sense in
the review of Mendoza and the papers by Jerzykiewicz, Ponomareva,
Feinstein, and others. Studies of individual stars, either photometrically or spectroscopically, were reported by numerous authors.
It is especially useful to combine, where possible, photometric and
spectroscopic data in attempting to interpret variations, and this
was done by Hubert-Delplace, Ballereau, and Doazan and co-workers, as
well as by Daminelli Neto and de Freitas Pacheco.

The question of timescales, and especially of periodicity, is an
important one. Changes over characteristic times of decades, often
in the form of extended emission-line episodes, were well documented,
and in many cases these appear to represent some sort of cyclical
behavior. True periodicity has been found in a number of Be stars,
many of which are binaries. Harmanec reviewed the subject of binary
phenomena in Be stars, Peters added results on a number of systems,
and Bossi, Koubsky, and Galkina each reported on binary behavior in
individual cases. The best-established examples are those that are
well-observed over several orbital cycles by a consistent and uniform
technique, but Pastori et al. were able to find evidence for long-
period orbital motions in a few cases, using historical literature
data.

A fascinating new interpretation of periodicity was offered by
Baade and by Bolton, each of whom reported on stars whose periodic
behavior can best be explained in terms of non-radial pulsations,
with the surface wave travelling around the star in the direction
opposite of rotation. These reports opened the possibility of an
entirely new class of theories regarding the origin of the Be
phenomenon.

In very many cases, however, the observed variations undoubtedly
reflect changes in the circumstellar envelope that are quite distinct
from orbital or pulsational behavior. Often spectroscopic changes
are accompanied by photometric variations, and the relationship of
the two was discussed by numerous authors, including Dachs and Hirata,
who in separate papers commented on the general relationship;
Hubert-Delplace, who with different co-authors has studied variation
episodes in a few stars; and by Hirata, who has also analyzed in
detail the changes in the circumstellar envelopes of a few individual
stars. Variations in polarization have also been found, as summarized
by Coyne, and Poeckert in his review showed that these may be precursors of the onset of new shell episodes, and hence ·should be watched
for more closely.

It has become clear that the rates of mass loss also vary, and
Doazan et al. and Henrichs reported on short-term fluctuations in
high-velocity winds. Stellar wind variations can also be inferred
from observations of changes in infrared excess fluxes, as reported
by Persi et al. and from variations in X-ray emission from mass-ac-
creting secondary stars, as noted in the review of Rappaport.
Guarnieri et al. reported one case where an optical change in a Be
star was followed in 8 days by an X-ray flare-up in its compact
secondary companion; perhaps the time delay represents the wind flow
time from the Be star to the orbital radius of the secondary.

7. MODELS

The question of interpreting all the assorted data on Be stars,
to develop a physical picture, is obviously very complex. The problems are enhanced by the unknown inclination angles and by the time
variability, which together make it very difficult to isolate general
properties (if any exist!) of the Be phenomenon.

There are two distinct questions one can ask: (1) what physical
picture can account for the observed properties of Be stars; and (2)
what is the underlying cause of the Be phenomenon? The reviews of
Coyne and Poeckert both illustrated a general class of interpretations
of polarization data, in which a disk structure is invoked to provide
the asymmetry needed to explaining the observed polarization.
Poeckert further described a physical picture in which the disk may
contain the neutral gas responsible for both Balmer emission and
shell absorption lines, although the location of the highly-ionized
material that produces the ultraviolet spectroscopic wind indicators
remains unknown.

The presence of distinct velocity components in the highly-
ionized species, with the simultaneous existence of low-velocity
Balmer emission and shell lines, was interpreted by Doazan, Stalio,
and Thomas as being due to a spherically symmetric envelope with
acceleration in a coronal zone near the star and a deceleration zone
well away from the star, where a high-density "balloon" builds up
until it disperses, accounting for the well-observed shell episodes
in some Be stars. Barker pointed out the potential influence of
magnetic fields on mass outflow from rotating stars, illustrating
that fields much too small to detect can have important effects,
which depend strongly on the wind velocity law.

On the question of the underlying cause of the Be phenomenon,
Harmanec in his review described three alternatives: rotation,
expansion (stellar wind), and mass exchange in binary systems; and
the papers of Baade and of Bolton suggested a fourth: pulsational
instability. None of these has been developed into a comprehensive
model, although it is clear that at least some Be stars are binaries,

with mass exchange in some cases. Individual systems were described
by Harmanec, Peters, Molaro et al., Selvelli et al., and Plavec et
al. On the other hand, the case for non-radial pulsations in some
objects seems well-established, it is clear that most or all Be stars
have winds, and it was demonstrated by Endal that rapid rotation may
lead naturally to the formation of a circumstellar envelope. Perhaps
elements of all four hypotheses are correct under different circumstances, or perhaps other, as yet unidentified, factors are responsible for Be stars.

8. RELATED OBJECTS

     Several classes of objects not included in the standard defini-
tion of Be stars were discussed. These included the peculiar Be
stars (designated B[e], with forbidden FeII emission lines) described
by Swings; the so-called Herbig Ae and Be stars, pre-main sequence
emission-line objects discussed by Finkenzellar and by Baschek et
al.; the famous binary β Lyrae, whose properties were described by
Mendoza and by Bahyl; and extreme B supergiant emission-line stars,
such as the clone of P Cygni discussed by Stahl et al. and the
variable supergiant described by Wolf et al.

     Other objects not classified as Be stars may be more closely
related than those just mentioned. The Oe and A shell stars have
already been discussed, as probably extensions of the Be phenomenon
to hotter and cooler spectral types. It was stressed by Henrichs and
by Snow that the Be stars must also bear some relationship to O stars
in general, because of the similarities of their stellar winds. The
Be star winds seem to represent an extrapolation of the winds of O
stars to objects of lower luminosity.

9. SUMMARY OF PROBLEMS

     A number of particularly important general problems emerged from
the discussions at this Symposium. Underlying all of them is the
basic question of the cause of the Be phenomenon, a question still
entirely without a satisfactory answer. The list of possible mechanisms grew to four during the Symposium, with the addition of pulsational instabilities to the older ideas of rotation, expansion, and
binary mass exchange, and there is no completely-developed theory
showing how any of them can account for all the properties of the
stars.

     The question of the evolutionary status of the stars is inti-
mately related to the origin of the Be characteristics, and probably
has not received the attention it deserves.

Time variability is an unavoidable complication, that probably should be viewed more as a fundamental property, one that is closely linked to the basis of the phenomenon. If the cause of the variations were understood, quite possibly so would the cause of the Be phenomenon itself.

A more specific problem is presented by the geometry of the circumstellar environment, and this became a general theme of the Symposium. Be stars show a bewildering assortment of properties, requiring an assortment of circumstellar regimes, and the difficulties of fitting them all together into one coherent picture are immense. Somehow we must reconcile the simultaneous presence of Balmer emission, stationary shell absorption, polarization, infrared excess emission, and high-velocity winds with high ionization.

Related to the problem of geometry is the difficulty in ascertaining the inclination angles and true rotational velocities of the stars. The development of a reliable technique for doing this would be enormously helpful in the development of a proper geometrical description of the circumstellar material, and would also be important in elucidating the role of rotation in general. A glimmer of hope has been offered by analyses of equatorial darkening such as those presented at this Symposium, although it appears that further refinement is needed.

9.   WHERE DO WE GO FROM HERE?

It is appropriate to complete this summary with an outline of important research that should be pursued in the ongoing effort to understand the Be stars.

On the observational side, two general strategies were discussed, and are being implemented. One is to carry out extensive observing campaigns, both photometric (as described by Harmanec) and spectroscopic (discussed by Barker). Such campaigns provide archival data, allowing future observers to trace the historical behavior of individual objects; they provide sufficient time coverage to allow searches for long-term periodic behavior (<u>if</u> the data are of uniform quality); and they can help ensure that new episodes of activity will be noticed at an early date, so that more intensive observations can be carried out.

The second general approach is to select individual objects for detailed scrutiny, in hopes of developing physical insight into the geometry and the time variability of the Be phenomenon. Such studies are especially important if good time coverage is obtained, particularly if many parts of the electromagnetic spectrum are observed, preferably over the same time intervals. From such observing programs, it may be possible to unravel the physical basis for the time

variations, and hence gather information on the fundamental cause.
One particularly extensive program, involving a number of observers
and a range of instruments (including IUE and several observatories
on the ground) has been described by Doazan et al., who are inten-
sively studying the star 59 Cygni. This object has been varying in
spectacular fashion, on rather short timescales, and there is hope
that significant new understanding will emerge from the study.

More specific observational questions should be mentioned.
While some data probably exist on the presence or absence of soft X-
ray emission from Be stars, the information was not forthcoming
during the week of the Symposium. Such data could help establish the
existence of coronal gas in the circumstellar envelopes of Be stars,
something suspected from the high ionization that is observed in the
winds and from the apparent similiarity of the Be star winds to those
of the O stars.

The question of magnetic fields in Be stars is entirely an open
one, and very difficult to pursue observationally, but the attempts
must continue. Magnetic fields may play an essential role in governing the circumstellar environment. No model of the Be stars can
really be complete until the presence and strength of magnetic fields
can be included.

On the theoretical side, several areas of needed work can be
indentified. The entire question of non-radial pulsations in Be
stars is new, and needs to be pursued; it was clear that more detailed
calculations of stellar structure with rotation are needed: improvements are needed in the theory of equatorial darkening in rotating
stars; and the entire subject of stellar winds requires further
analysis, since the origin of these winds, which might have a lot to
do with the cause of the Be phenomena, is itself unknown.

Perhaps some of the fundamental questions will ultimately be
answered by a direct probe of the circumstellar environment around a
Be star. Discussions of an unmanned mission to a nearby star are
already under way; it may be possible to conceive of such a journey
to a Be star in the future. Maybe in the year 2066, when astronomers
gather on the occasion of the two-hundredth anniversary of the discovery of Be stars for IAU Symposium 523, as suggested in Slettebak's
review, this probe will be on its way. Perhaps in 2166, when the
three-hundredth anniversary occurs, the data will be in, and at last
it will not be necessary to conclude that more observations are
needed.